TRADE-OFF ANALYTICS

WILEY SERIES IN SYSTEMS ENGINEERING AND MANAGEMENT

Andrew P. Sage, Editor

ANDREW P. SAGE and JAMES D. PALMER
Software Systems Engineering

WILLIAM B. ROUSE
Design for Success: A Human-Centered Approach to Designing Successful Products and Systems

LEONARD ADELMAN
Evaluating Decision Support and Expert System Technology

ANDREW P. SAGE
Decision Support Systems Engineering

YEFIM FASSER and DONALD BRETINER
Process Improvement in the Electronics Industry, Second Edition

WILLIAM B. ROUSE
Strategies for Innovation

ANDREW P. SAGE
Systems Engineering

HORST TEMPELMEIER and HEINRICH KUHN
Flexible Manufacturing Systems: Decision Support for Design and Operation

WILLIAM B. ROUSE
Catalysts for Change: Concepts and Principles for Enabling Innovation

UPING FANG, KEITH W. HIPEL, and D. MARC KILGOUR
Interactive Decision Making: The Graph Model for Conflict Resolution

DAVID A. SCHUM
Evidential Foundations of Probabilistic Reasoning

JENS RASMUSSEN, ANNELISE MARK PEJTERSEN, and LEONARD P. GOODSTEIN
Cognitive Systems Engineering

ANDREW P. SAGE
Systems Management for Information Technology and Software Engineering

ALPHONSE CHAPANIS
Human Factors in Systems Engineering

YACOV Y. HAIMES
Risk Modeling, Assessment, and Management, Third Edition

DENNIS M. SUEDE
The Engineering Design of Systems: Models and Methods, Second Edition

ANDREW P. SAGE and JAMES E. ARMSTRONG, Jr.
Introduction to Systems Engineering

WILLIAM B. ROUSE
Essential Challenges of Strategic Management

YEFIM FASSER and DONALD BRETTNER
Management for Quality in High-Technology Enterprises

THOMAS B. SHERIDAN
Humans and Automation: System Design and Research Issues

ALEXANDER KOSSIAKOFF and WILLIAM N. SWEET
Systems Engineering Principles and Practice

HAROLD R. BOOHER
Handbook of Human Systems Integration

JEFFREY T. POLLOCK and RALPH HODGSON
Adaptive Information: Improving Business Through Semantic Interoperability, Grid Computing, and Enterprise Integration

ALAN L. PORTER and SCOTT W. CUNNINGHAM
Tech Mining: Exploiting New Technologies for Competitive Advantage

REX BROWN
Rational Choice and Judgment: Decision Analysis for the Decider

WILLIAM B. ROUSE and KENNETH R. BOFF (editors)
Organizational Simulation

HOWARD EISNER
Managing Complex Systems: Thinking Outside the Box

STEVE BELL
Lean Enterprise Systems: Using IT for Continuous Improvement

J. JERRY KAUFMAN and ROY WOODHEAD
Stimulating Innovation in Products and Services: With Function Analysis and Mapping

WILLIAM B. ROUSE
Enterprise Tranformation: Understanding and Enabling Fundamental Change

JOHN E. GIBSON, WILLIAM T. SCHERER, and WILLAM F. GIBSON
How to Do Systems Analysis

WILLIAM F. CHRISTOPHER
Holistic Management: Managing What Matters for Company Success

WILLIAM W. ROUSE
People and Organizations: Explorations of Human-Centered Design

MO JAMSHIDI
System of Systems Engineering: Innovations for the Twenty-First Century

ANDREW P. SAGE and WILLIAM B. ROUSE
Handbook of Systems Engineering and Management, Second Edition

JOHN R. CLYMER
Simulation-Based Engineering of Complex Systems, Second Edition

KRAG BROTBY
Information Security Governance: A Practical Development and Implementation Approach

JULIAN TALBOT and MILES JAKEMAN
Security Risk Management Body of Knowledge

SCOTT JACKSON
Architecting Resilient Systems: Accident Avoidance and Survival and Recovery from Disruptions

JAMES A. GEORGE and JAMES A. RODGER
Smart Data: Enterprise Performance Optimization Strategy

YORAM KOREN
The Global Manufacturing Revolution: Product-Process-Business Integration and Reconfigurable Systems

AVNER ENGEL
Verification, Validation, and Testing of Engineered Systems

WILLIAM B. ROUSE (editor)
The Economics of Human Systems Integration: Valuation of Investments in People's Training and Education, Safety and Health, and Work Productivity

ALEXANDER KOSSIAKOFF, WILLIAM N. SWEET, SAM SEYMOUR, and STEVEN M. BIEMER
Systems Engineering Principles and Practice, Second Edition

GREGORY S. PARNELL, PATRICK J. DRISCOLL, and DALE L. HENDERSON (editors)
Decision Making in Systems Engineering and Management, Second Edition

ANDREW P. SAGE and WILLIAM W. ROUSE
Economic Systems Analysis and Assessment: Intensive Systems, Organizations, and Enterprises

BOHDAN W. OPPENHEIM
Lean for Systems Engineering with Lean Enablers for Systems Engineering

LEV M. KLYATIS
Accelerated Reliability and Durability Testing Technology

BJOERN BARTELS, ULRICH ERMEL, MICHAEL PECHT, and PETER SANDBORN
Strategies to the Prediction, Mitigation, and Management of Product Obsolescence

LEVANT YILMAS and TUNCER OREN
Agent-Directed Simulation and Systems Engineering

ELSAYED A. ELSAYED
Reliability Engineering, Second Edition

BEHNAM MALAKOOTI
Operations and Production Systems with Multipme Objectives

MENG-LI SHIU, JUI-CHIN JIANG, and MAO-HSIUNG TU
Quality Strategy for Systems Engineering and Management

ANDREAS OPELT, BORIS GLOGER, WOLFGANG PFARL, and RALF MITTERMAYR
Agile Contracts: Creating and Managing Successful Projects with Scrum

KINJI MORI
Concept-Oriented Research and Development in Information Technology

KAILASH C. KAPUR and MICHAEL PECHT
Reliability Engineering

MICHAEL TORTORELLA
Reliability, Maintainability, and Supportability: Best Practices for Systems Engineers

DENNIS M. BUEDE and WILLIAM D. MILLER
The Engineering Design of Systems: Models and Methods, Third Edition

JOHN E. GIBSON, WILLIAM T. SCHERER, WILLIAM F. GIBSON, and MICHAEL C. SMITH
How to Do Systems Analysis: Primer and Casebook

GREGORY S. PARNELL, Editor
Trade-off Analytics: Creating and Exploring the System Tradespace

TRADE-OFF ANALYTICS

Creating and Exploring the System Tradespace

Edited by

GREGORY S. PARNELL

Copyright © 2017 by John Wiley & Sons, Inc. All rights reserved

Published by John Wiley & Sons, Inc., Hoboken, New Jersey
Published simultaneously in Canada

No part of this publication may be reproduced, stored in a retrieval system, or transmitted in any form or by any means, electronic, mechanical, photocopying, recording, scanning, or otherwise, except as permitted under Section 107 or 108 of the 1976 United States Copyright Act, without either the prior written permission of the Publisher, or authorization through payment of the appropriate per-copy fee to the Copyright Clearance Center, Inc., 222 Rosewood Drive, Danvers, MA 01923, (978) 750-8400, fax (978) 750-4470, or on the web at www.copyright.com. Requests to the Publisher for permission should be addressed to the Permissions Department, John Wiley & Sons, Inc., 111 River Street, Hoboken, NJ 07030, (201) 748-6011, fax (201) 748-6008, or online at http://www.wiley.com/go/permission.

Limit of Liability/Disclaimer of Warranty: While the publisher and author have used their best efforts in preparing this book, they make no representations or warranties with respect to the accuracy or completeness of the contents of this book and specifically disclaim any implied warranties of merchantability or fitness for a particular purpose. No warranty may be created or extended by sales representatives or written sales materials. The advice and strategies contained herein may not be suitable for your situation. You should consult with a professional where appropriate. Neither the publisher nor author shall be liable for any loss of profit or any other commercial damages, including but not limited to special, incidental, consequential, or other damages.

For general information on our other products and services or for technical support, please contact our Customer Care Department within the United States at (800) 762-2974, outside the United States at (317) 572-3993 or fax (317) 572-4002.

Wiley also publishes its books in a variety of electronic formats. Some content that appears in print may not be available in electronic formats. For more information about Wiley products, visit our web site at www.wiley.com.

Library of Congress Cataloging-in-Publication Data:
Names: Parnell, Gregory S., editor.
Title: Trade-off analytics : creating and exploring the system tradespace /
 [edited by] Gregory S. Parnell.
Description: Hoboken, New Jersey : John Wiley & Sons Inc., [2017] | Includes
 bibliographical references and index.
Identifiers: LCCN 2016023582| ISBN 9781119237532 (cloth) | ISBN 9781119238300
 (epub) | ISBN 9781119237556 (Adobe PDF)
Subjects: LCSH: Systems engineering–Decision making. | Multiple criteria decision making.
Classification: LCC TA168 .T73 2017 | DDC 620.0068/4–dc23 LC record available at
https://lccn.loc.gov/2016023582

Typeset in 10/12pt TimesLTStd by SPi Global, Chennai, India

Printed in the United States of America

10 9 8 7 6 5 4 3 2 1

CONTENTS

List of Contributors	xix
About the Authors	xxi
Foreword	xxxi
Preface	xxxiii
Acknowledgments	xli
About the Companion Website	xlv

1 Introduction to Trade-off Analysis   1

Gregory S. Parnell, Matthew Cilli, Azad M. Madni, and Garry Roedler

    1.1 Introduction, 2
    1.2 Trade-off Analyses Throughout the Life Cycle, 3
    1.3 Trade-off Analysis to Identify System Value, 3
    1.4 Trade-off Analysis to Identify System Uncertainties and Risks, 6
    1.5 Trade-off Analyses can Integrate Value and Risk Analysis, 6
    1.6 Trade-off Analysis in the Systems Engineering Decision Management Process, 8
    1.7 Trade-off Analysis Mistakes of Omission and Commission, 9
        1.7.1 Mistakes of Omission, 12
        1.7.2 Mistakes of Commission, 15
        1.7.3 Impacts of the Trade-Off Analysis Mistakes, 18

1.8   Overview of the Book, 20
      1.8.1   Illustrative Examples and Techniques Used in the Book, 24
1.9   Key Terms, 24
1.10 Exercises, 25
      References, 26

2   A Conceptual Framework and Mathematical Foundation for Trade-Off Analysis   29
Gregory S. Parnell, Azad M. Madni, and Robert F. Bordley

2.1   Introduction, 29
2.2   Trade-Off Analysis Terms, 30
2.3   Influence Diagram of the Tradespace, 31
      2.3.1   Stakeholder Needs, System Functions, and Requirements, 33
      2.3.2   Objectives, 33
      2.3.3   System Alternatives, 34
      2.3.4   Uncertainty, 36
      2.3.5   Preferences and Evaluation of Alternatives, 37
      2.3.6   Resource Analysis, 44
      2.3.7   An Integrated Trade-Off Analyses, 44
2.4   Tradespace Exploration, 46
2.5   Summary, 46
2.6   Key Words, 47
2.7   Exercises, 48
      References, 48

3   Quantifying Uncertainty   51
Robert F. Bordley

3.1   Sources of Uncertainty in Systems Engineering, 51
3.2   The Rules of Probability and Human Intuition, 52
3.3   Probability Distributions, 56
      3.3.1   Calculating Probabilities from Experiments, 56
      3.3.2   Calculating Complex Probabilities from Simpler Probabilities, 58
      3.3.3   Calculating Probabilities Using Parametric Distributions, 59
      3.3.4   Applications of Parametric Probability Distributions, 62
3.4   Estimating Probabilities, 66
      3.4.1   Using Historical Data, 66
      3.4.2   Using Human Judgment, 68
      3.4.3   Biases in Judgment, 70
3.5   Modeling Using Probability, 72
      3.5.1   Bayes Nets, 72
      3.5.2   Monte Carlo Simulation, 75
      3.5.3   Monte Carlo Simulation with Dependent Uncertainties, 76

3.5.4 Monte Carlo Simulation with Partial Information on Output Values, 77
3.5.5 Variations on Monte Carlo Simulation, 78
3.5.6 Sensitivity Analysis, 78
3.6 Summary, 81
3.7 Key Terms, 81
3.8 Exercises, 83
References, 86

4 ANALYZING RESOURCES 91

Edward A. Pohl, Simon R. Goerger, and Kirk Michealson

4.1 Introduction, 91
4.2 Resources, 92
    4.2.1 People, 92
    4.2.2 Facilities, 95
    4.2.3 Costs, 95
    4.2.4 Resource Space, 99
4.3 Cost Analysis, 99
    4.3.1 Cost Estimation, 102
    4.3.2 Cost Estimation Techniques, 108
    4.3.3 Learning Curves, 120
    4.3.4 Net Present Value, 125
    4.3.5 Monte Carlo Simulation, 130
    4.3.6 Sensitivity Analysis, 134
4.4 Affordability Analysis, 135
    4.4.1 Background, 136
    4.4.2 The Basics of Affordability Analysis Are Not Difficult, 137
    4.4.3 DoD Comparison of Cost Analysis and Affordability Analysis, 138
    4.4.4 Affordability Analysis Definitions, 139
    4.4.5 "Big A" Affordability Analysis Process Guide, 141
4.5 Key Terms, 147
4.6 Excercises, 149
References, 152

5 Understanding Decision Management 155

Matthew Cilli and Gregory S. Parnell

5.1 Introduction, 155
5.2 Decision Process Context, 156
5.3 Decision Process Activities, 157
    5.3.1 Frame Decision, 159
    5.3.2 Develop Objectives and Measures, 163
    5.3.3 Generate Creative Alternatives, 171

		5.3.4	Assess Alternatives via Deterministic Analysis, 180
		5.3.5	Synthesize Results, 183
		5.3.6	Develop Multidimensional Value Model, 187
		5.3.7	Identify Uncertainty and Conduct Probabilistic Analysis, 190
		5.3.8	Assess Impact of Uncertainty, 192
		5.3.9	Improve Alternatives, 196
		5.3.10	Communicating Trade-Offs, 197
		5.3.11	Present Recommendation and Implementation Plan, 197

- 5.4 Summary, 199
- 5.5 Key Terms, 199
- 5.6 Exercises, 200
  References, 201

6 Identifying Opportunities — 203

Donna H. Rhodes and Simon R. Goerger

- 6.1 Introduction, 203
- 6.2 Knowledge, 205
  - 6.2.1 Domain Knowledge, 205
  - 6.2.2 Technical Knowledge, 205
  - 6.2.3 Business Knowledge, 205
  - 6.2.4 Expert Knowledge, 206
  - 6.2.5 Stakeholder Knowledge, 206
- 6.3 Decision Traps, 207
- 6.4 Techniques, 210
  - 6.4.1 Interviews, 210
  - 6.4.2 Focus Groups, 213
  - 6.4.3 Surveys, 215
- 6.5 Tools, 219
  - 6.5.1 Concept Map, 219
  - 6.5.2 System Boundary, 220
  - 6.5.3 Decision Hierarchy, 220
  - 6.5.4 Issues List, 221
  - 6.5.5 Vision Statement, 221
  - 6.5.6 Influence Diagram, 222
  - 6.5.7 Selecting Appropriate Tools and Techniques, 223
- 6.6 Illustrative Examples, 223
  - 6.6.1 Commercial, 223
  - 6.6.2 Defense, 226
- 6.7 Key Terms, 228
- 6.8 Exercises, 230
  References, 230

7 Identifying Objectives and Value Measures — 233
Gregory S. Parnell and William D. Miller

- 7.1 Introduction, 233
- 7.2 Value-Focused Thinking, 234
  - 7.2.1 Four Major VFT Ideas, 235
  - 7.2.2 Benefits of VFT, 235
- 7.3 Shareholder and Stakeholder Value, 236
  - 7.3.1 Private Company Example, 237
  - 7.3.2 Government Agency Example, 237
- 7.4 Challenges in Identifying Objectives, 238
- 7.5 Identifying the Decision Objectives, 239
  - 7.5.1 Questions to Help Identify Decision Objectives, 239
  - 7.5.2 How to Get Answers to the Questions, 240
- 7.6 The Financial or Cost Objective, 241
  - 7.6.1 Financial Objectives for Private Companies, 241
  - 7.6.2 Cost Objective for Public Organizations, 242
- 7.7 Developing Value Measures, 243
- 7.8 Structuring Multiple Objectives, 243
  - 7.8.1 Value Hierarchies, 244
  - 7.8.2 Techniques for Developing Value Hierarchies, 245
  - 7.8.3 Value Hierarchy Best Practices, 247
  - 7.8.4 Cautions about Cost and Risk Objectives, 248
- 7.9 Illustrative Examples, 248
  - 7.9.1 Military Illustrative Example, 248
  - 7.9.2 Homeland Security Illustrative Example, 250
- 7.10 Summary, 250
- 7.11 Key Terms, 252
- 7.12 Exercises, 253
  - References, 255

8 DEVELOPING AND EVALUATING ALTERNATIVES — 257
C. Robert Kenley, Clifford Whitcomb, and Gregory S. Parnell

- 8.1 Introduction, 257
- 8.2 Overview of Decision-making, Creativity, and Teams, 258
  - 8.2.1 Approaches to Decision-Making, 258
  - 8.2.2 Cognitive Methods for Creating Alternatives, 260
  - 8.2.3 Key Concepts for Building and Operating Teams, 260
- 8.3 Alternative Development Techniques, 263
  - 8.3.1 Structured Creativity Methods, 263
  - 8.3.2 Morphological Box, 266

- 8.3.3 Pugh Method for Alternative Generation, 270
- 8.3.4 TRIZ for Alternative Development, 271
- 8.4 Assessment of Alternative Development Techniques, 275
- 8.5 Alternative Evaluation Techniques, 276
  - 8.5.1 Decision-Theory-Based Approaches, 276
  - 8.5.2 Pugh Method for Alternative Evaluation, 276
  - 8.5.3 Axiomatic Approach to Design (AAD), 277
  - 8.5.4 TRIZ for Alternative Evaluation, 280
  - 8.5.5 Design of Experiments (DOE), 280
  - 8.5.6 Taguchi Approach, 282
  - 8.5.7 Quality Function Deployment (QFD), 283
  - 8.5.8 Analytic Hierarchy Process AHP, 287
- 8.6 Assessment of Alternative Evaluation Techniques, 290
- 8.7 Key Terms, 290
- 8.8 Exercises, 290
  References, 293

9 An Integrated Model for Trade-Off Analysis 297

Alexander D. MacCalman, Gregory S. Parnell, and Sam Savage

- 9.1 Introduction, 297
- 9.2 Conceptual Design Example, 298
- 9.3 Integrated Approach Influence Diagram, 300
  - 9.3.1 Decision Nodes, 300
  - 9.3.2 Uncertainty Nodes, 303
  - 9.3.3 Constant Node, 310
  - 9.3.4 Value Nodes, 314
- 9.4 Other Types of Trade-Off Analysis, 322
- 9.5 Simulation Tools, 322
  - 9.5.1 Monte Carlo Simulation Proprietary Add-Ins, 324
  - 9.5.2 The Discipline of Probability Management, 324
  - 9.5.3 SIPmath^{TM} Tool in Native Excel, 324
  - 9.5.4 Model Building Steps, 325
- 9.6 Summary, 329
- 9.7 Key Terms, 330
- 9.8 Exercises, 331
  References, 335

10 EXPLORING CONCEPT TRADE-OFFS 337

Azad M. Madni and Adam M. Ross

- 10.1 Introduction, 337
  - 10.1.1 Key Concepts, Concept Trade-Offs, and Concept Exploration, 341
- 10.2 Defining the Concept Space and System Concept of Operations, 345
- 10.3 Exploring the Concept Space, 346

        10.3.1   Storytelling-Enabled Tradespace Exploration, 346
        10.3.2   Decisions and Outcomes, 347
        10.3.3   Contingent Decision-Making, 347
  10.4 Trade-off Analysis Frameworks, 348
  10.5 Tradespace and System Design Life Cycle, 349
  10.6 From Point Trade-offs to Tradespace Exploration, 351
  10.7 Value-based Multiattribute Tradespace Analysis, 351
        10.7.1   Tradespace Exploration and Sensitivity Analysis, 353
        10.7.2   Tradespace Exploration and Uncertainty, 354
        10.7.3   Tradespace Exploration with Spiral Development, 356
        10.7.4   Tradespace Exploration in Relation to Optimization and Decision Theory, 356
  10.8 Illustrative Example, 359
        10.8.1   Step 1: Determine Key Decision-Makers, 359
        10.8.2   Step 2: Scope and Bound the Mission, 360
        10.8.3   Step 3: Elicit Attributes and Utilities (Preference Capture), 360
        10.8.4   Step 4: Define Design Vector Elements (Concept Generation), 362
        10.8.5   Step 5: Develop Model(s) (Evaluation), 362
        10.8.6   Step 6: Generate the Tradespace (Computation), 364
        10.8.7   Step 7: Explore the Tradespace (Analysis and Synthesis), 365
  10.9 Conclusions, 369
  10.10 Key Terms, 371
  10.11 Exercises, 372
        References, 372

11 Architecture Evaluation Framework   377

James N. Martin

  11.1 Introduction, 377
        11.1.1   Architecture in the Decision Space, 378
        11.1.2   Architecture Evaluation, 379
        11.1.3   Architecture Views and Viewpoints, 380
        11.1.4   Stakeholders, 382
        11.1.5   Stakeholder Concerns, 382
        11.1.6   Architecture versus Design, 383
        11.1.7   On the Uses of Architecture, 384
        11.1.8   Standardizing on an Architecture Evaluation Strategy, 384
  11.2 Key Considerations in Evaluating Architectures, 385
        11.2.1   Plan-Driven Evaluation Effort, 386
        11.2.2   Objectives-Driven Evaluation, 387
        11.2.3   Assessment versus Analysis, 387
  11.3 Architecture Evaluation Elements, 389
        11.3.1   Architecture Evaluation Approach, 389
        11.3.2   Architecture Evaluation Objectives, 390
        11.3.3   Evaluation Approach Examples, 391

11.3.4 Value Assessment Methods, 391
11.3.5 Value Assessment Criteria, 393
11.3.6 Architecture Analysis Methods, 394
11.4 Steps in an Architecture Evaluation Process, 396
11.5 Example Evaluation Taxonomy, 398
11.5.1 Business Impact Factors, 398
11.5.2 Mission Impact Factors, 398
11.5.3 Architecture Attributes, 399
11.6 Summary, 400
11.7 Key Terms, 400
11.8 Exercises, 402
References, 402

12 Exploring the Design Space   405

Clifford Whitcomb and Paul Beery

12.1 Introduction, 405
12.2 Example 1: Liftboat, 406
12.2.1 Liftboat Fractional Factorial Design of Experiments, 406
12.2.2 Liftboat Design Trade-Off Space, 409
12.2.3 Liftboat Uncertainty Analysis, 411
12.2.4 Liftboat Example Summary, 411
12.3 Example 2: Cruise Ship Design, 411
12.3.1 Cruise Ship Taguchi Design of Experiments, 411
12.3.2 Cruise Ship Design Trade-Off Space, 412
12.3.3 Cruise Ship Example Summary, 416
12.4 Example 3: NATO Naval Surface Combatant Ship, 417
12.4.1 NATO Surface Combatant Ship Stakeholder Need, 418
12.4.2 NATO Surface Combatant Ship Box–Behnken Design of Experiments, 420
12.4.3 NATO Surface Combatant Ship Cost-Effectiveness Trade-Off, 421
12.4.4 NATO Surface Combatant Ship Design Tradespace, 421
12.4.5 NATO Surface Combatant Ship Design Trade-Off, 422
12.4.6 NATO Surface Combatant Ship Trade-Off Summary, 430
12.5 Key Terms, 431
12.6 Exercises, 433
References, 435

13 SUSTAINMENT RELATED MODELS AND TRADE STUDIES   437

John E. MacCarthy and Andres Vargas

13.1 Introduction, 437
13.2 Availability Modeling and Trade Studies, 439
13.2.1 FMDS Background, 439
13.2.2 FMDS Availability Trade Studies, 449

13.2.3 Section Synopsis, 453
13.3 Sustainment Life Cycle Cost Modeling and Trade Studies[14], 454
    13.3.1 The Total System Life Cycle Model, 454
    13.3.2 The O&S Cost Model, 456
    13.3.3 Life Cycle Cost Trade Study, 459
13.4 Optimization in Availability Trade Studies, 464
    13.4.1 Setting Up the Optimization Problem, 464
    13.4.2 Instantiating the Optimization Model, 465
    13.4.3 Discussion of the Optimization Model Results, 468
    13.4.4 Deterministic Sensitivity Analysis, 469
13.5 Monte Carlo Modeling, 471
    13.5.1 Input Probability Distributions for the Monte Carlo Model, 471
    13.5.2 Monte Carlo Simulation Results, 472
    13.5.3 Stochastic Sensitivity Analysis, 473
13.6 Chapter Summary, 475
13.7 Key Terms, 476
13.8 Exercises, 478
References, 482

14 Performing Programmatic Trade-Off Analyses   483
Gina Guillaume-Joseph and John E. MacCarthy

14.1 Introduction, 483
14.2 System Acceptance Decisions and Trade Studies, 485
    14.2.1 Acceptance Decision Framework, 486
    14.2.2 Calculating the Confidence That a System Is "Good", 491
    14.2.3 Acceptance Test Design and Trade Studies, 493
    14.2.4 A "Delay, Fix, and Test" Cost Model, 499
    14.2.5 The Integrated Decision Model, 504
    14.2.6 Conclusions, 511
14.3 Product Cancelation Decision Trade Study, 512
    14.3.1 Introduction, 512
    14.3.2 Significance, 513
    14.3.3 Defining Failure, 514
    14.3.4 Developing the Predictive Model, 519
    14.3.5 Research Results, 522
    14.3.6 Model Implementation In Industry, 528
    14.3.7 Predictive Model Deployment in Industry, 530
    14.3.8 When the Decision Has Been Made to Cancel the System, 536
    14.3.9 Conclusion, 537
14.4 Product Retirement Decision Trade Study, 538
    14.4.1 Introduction, 538
    14.4.2 Legacy HR Systems, 539
    14.4.3 The US NAVY Retirement and Decommission Program for Nuclear-Powered Vessels, 544

14.4.4 Decision Analysis for Decommissioning Offshore Oil and Gas Platforms in California, 551
14.4.5 System Retirement and Decommissioning Strategy, 559
14.4.6 Conclusion, 561
14.5 Key Terms, 562
14.6 Exercises, 564
References, 566

15 SUMMARY AND FUTURE TRENDS 571

Gregory S. Parnell and Simon R. Goerger

15.1 Introduction, 571
15.2 Major Trade-Off Analysis Themes, 572
    15.2.1 Use Standard Systems Engineering Terminology, 572
    15.2.2 Avoid the Mistakes of Omission and Commission, 572
    15.2.3 Use a Decision Management Framework, 572
    15.2.4 Use Decision Analysis as the Mathematical Foundation, 573
    15.2.5 Explicitly Define the Decision Opportunity, 573
    15.2.6 Identify and Structure Decision Objectives and Measures, 574
    15.2.7 Identify Creative, Doable Alternatives, 574
    15.2.8 Use the Most Appropriate Modeling and Simulation Technique for the Life Cycle Stage, 575
    15.2.9 Include Resource Analysis in the Trade-Off Analysis, 575
    15.2.10 Explicitly Consider Uncertainty, 575
    15.2.11 Identify the Cost, Value, Schedule, and Risk Drivers, 575
    15.2.12 Provide an Integrated Framework for Cost, Value, and Risk Analyses, 576
15.3 Future of Trade-Off Analysis, 576
    15.3.1 Education and Training of Systems Engineers, 577
    15.3.2 Systems Engineering Methodologies and Tools, 577
    15.3.3 Emergent Tradespace Factors, 580
15.4 Summary, 581
References, 581

Index   **583**

LIST OF CONTRIBUTORS

Paul Beery, Systems Engineering Department, Naval Postgraduate School, Monterey, CA, USA

Robert F. Bordley, Systems Engineering and Design, University of Michigan, Ann Arbor, MI, USA; Booz Allen Hamilton, Troy, MI, USA

Matthew Cilli, U.S. Army Armament Research Development and Engineering Center (ARDEC), Systems Analysis Division, Picatinny, NJ, USA

Simon R. Goerger, Institute for Systems Engineering Research, Information Technology Laboratory (ITL), U.S. Army Engineer Research and Development Center (ERDC), Vicksburg, MS, USA

Gina Guillaume-Joseph, MITRE Corporation, McLean, VA, USA

Alexander D. MacCalman, Department of Systems Engineering, United States Military Academy, West Point, NY, USA

John E. MacCarthy, Systems Engineering Education Program, Institute for Systems Research, University of Maryland, College Park, MD, USA

Azad M. Madni, Department of Astronautical Engineering, Systems Architecting and Engineering and Astronautical Engineering, Viterbi School of Engineering, University of Southern California, Los Angeles, CA, USA

James N. Martin, The Aerospace Corporation, El Segundo, CA, USA

Kirk Michealson, Tackle Solutions, LLC, Chesapeake, VA, USA

William D. Miller, Innovative Decisions, Inc., Vienna, VA, USA

Gregory S. Parnell, Department of Industrial Engineering, University of Arkansas, Fayetteville, AR, USA

Edward A. Pohl, Department of Industrial Engineering, University of Arkansas, Fayetteville, AR, USA

Donna H. Rhodes, Sociotechnical Systems Research Center, Massachusetts Institute of Technology, Cambridge, MA, USA

C. Robert Kenley, School of Industrial Engineering, Purdue University, West Lafayette, IN, USA

Garry Roedler, Corporate Engineering LM Fellow, Engineering Outreach Program Manager, Lockheed Martin Corporation King of Prussia, PA

Adam M. Ross, Systems Engineering Advancement Research Initiative (SEAri), Massachusetts Institute of Technology (MIT), Cambridge, MA, USA

Sam Savage, School of Engineering, Stanford University, Stanford, CA, USA

Andres Vargas, Department of Industrial Engineering, University of Arkansas, Fayetteville, AR, USA

Clifford Whitcomb, Systems Engineering Department, Naval Postgraduate School in Monterey, Monterey, CA, USA

ABOUT THE AUTHORS

Gregory S. Parnell is Director, M.S. in Operations Management and Research Professor in the Department of Industrial Engineering at the University of Arkansas. He teaches systems engineering, decision analysis, operations management, and IE design courses. He coedited Decision Making for Systems Engineering and Management, Wiley Series in Systems Engineering, 2nd Ed, Wiley & Sons Inc., 2011, and cowrote the Wiley & Sons Handbook of Decision Analysis, 2013. Dr Parnell has taught at West Point, the United Stated Air Force Academy, Virginia Commonwealth University, and the Air Force Institute of Technology. He is a fellow of the International Committee on Systems Engineering (INCOSE), the Institute for Operations Research & Management Science, Military Operations Research Society, the Society for Decision Professionals, and the Lean Systems Society. During his Air Force career, he served in a variety of R&D positions and operations research positions including at the Pentagon where he led two analysis divisions supporting senior Air Force leadership. He is a retired Colonel in the US Air Force. Dr Parnell received a B.S. in Aerospace Engineering from the University of Buffalo, an M.E. in Industrial & Systems Engineering from the University of Florida, an M.S. in Systems Management from the University of Southern California, and a Ph.D. in Engineering-Economic Systems from Stanford University.

Robert F. Bordley is an adjunct professor of decision analysis and systems engineering at the University of Michigan and a full-time consultant for Booz Allen Hamilton. Bob was formerly technical Fellow at General Motors and a Program Director at the National Science Foundation. His Ph.D., M.S., and MBA in Operations Research are from the University of California, Berkeley with an M.S. in Systems Science, B.S. in Physics, and B.A. in Public Policy from Michigan State University. He is an INCOSE-certified expert systems engineering professional (ESEP), an

INFORMS-certified analytic professional (CAP), a professional statistician (PStat), and a certified Project Management Professional (PMP). Bob is a Fellow of the Institute for Operations Research and Management Sciences, a Fellow of the American Statistical Association, and a Fellow of the Society of Decision Professionals. Bob also received the 2004 Best Decision Analysis Publication Award. At the National Science Foundation, he served as Program Director for Decision, Risk and Management Sciences. As Technical Fellow at General Motors, he received GM's Chairman Award, President's Council Award, Research Award of Excellence, GM's Engineering Award of Excellence, and UAW-GM Quality Award. Bob led the mission analysis group in Project Trilby, which helped launch GM's vehicle systems engineering effort as well as its R&D portfolio management group. Bob was also a Technical Director on GM's corporate strategy staff and served as internal consultant to GM's marketing, product planning, and quality engineering staffs. At Booz Allen Hamilton, Bob supports requirements management and concept selection for the Army.

Matthew Cilli received his Ph.D. in Systems Engineering from Stevens Institute of Technology in Hoboken, NJ, and leads an analytics group at the US Army's Armament Research Development and Engineering Center (ARDEC) in Picatinny, NJ. His research interests are focused on improving strategic decision-making through the integrated application of holistic thinking and analytics. Prior to his current position, Dr Cilli accumulated over 20 years of experience developing proposals, securing resources, and leading effective technology development programs for the US Army. Dr Cilli graduated from Villanova University, Villanova, PA, with a Bachelor of Electrical Engineering and a Minor in Mathematics in May 1989. He is also a graduate of the Polytechnic University, Brooklyn, NY, with a Master of Science – Electrical Engineering received in January 1992 and in May 1998, graduated from the University of Pennsylvania, Wharton Business School, Philadelphia, PA, with a Masters of Technology Management.

Simon R. Goerger is the ERDC Director of the Institute for Systems Engineering Research (ISER) at the Information Technology Laboratory (ITL) of the Engineer Research and Development Center (ERDC) in Vicksburg, MS. He has been an Operations Research Analyst with the US Army Corps of Engineers since 2012. Prior to working for the Corps of Engineers, he was a Colonel in the US Army serving as the Director of the Department of Defense Readiness Reporting System (DRRS) Implementation Office (DIO). Simultaneously, he served as Senior Defense Readiness Analyst in the Office of the Undersecretary of Defense (Personnel and Readiness). Simon has served as an Assistant Professor and the Director of the Operations Research Center of Excellence in the Department of Systems Engineering at the United States Military Academy, West Point, NY, before deploying to serve as the Joint Multinational Networks Division Chief, Coalition Forces Land Combatant Command/US Army Central Command, Kuwait. He received his Bachelor of Science from the United States Military Academy, his Master of Science (M.S.) in National Security Strategy from the National War College, and his M.S. in Computer Science and Doctorate of Philosophy in Modeling and Simulations from the Naval Postgraduate School. He is a board member for the Military Operations Research Society. His research interests include decision analysis, systems modeling, tradespace analysis, and combat modeling and simulations.

Dr Gina Guillaume-Joseph is an Information Systems Engineer at The MITRE Corporation in McLean, Virginia. In her current role, she acts as a trusted advisor to senior leadership in Federal Agencies by partnering with them to design enhancements to their work systems. Dr Guillaume-Joseph's work has led to improvements that allow the systems and processes to operate more efficiently and effectively in fulfillment of specific functions. Her various roles have included project manager, software developer, test engineer, and quality assurance engineer within the private, government consulting, nonprofit, and telecommunications arenas. Dr Guillaume-Joseph is President of the INCOSE Washington Metro Area (WMA) Chapter. Dr Guillaume-Joseph has a strong record of success based on direct personal contributions. She leads and develops teams that are adaptive, flexible, and highly responsive in the exceptionally dynamic environment of Government support. Her accomplishments and successes are based on strong program performance, leadership discipline, a commitment to developing relevant, innovative and adaptive solutions, and a vigilant focus on best value solutions for her clients. Dr Guillaume-Joseph has advanced knowledge of software development lifecycle activities, such as agile, waterfall, iterative, incremental, and associated processes including planning, requirements management, design and development, testing, and deployment. Her strong communication skills make her adept at conveying specialized technical information to nontechnical audiences. Dr Guillaume-Joseph received her B.A. in Computer Science from Boston College and M.S. in Information Technology Systems from the University of Maryland. She obtained her Ph.D. in Systems Engineering from George Washington University with a topic focused on Predicting Software Project Failure Outcomes using Predictive Analytics and Modeling.

C. Robert Kenley is an Associate Professor of Engineering Practice in Purdue's School of Industrial Engineering in West Lafayette, IN. He teaches courses in systems engineering at Purdue and has over 30 years of experience in industry, academia, and government as a practitioner, consultant, and researcher in systems engineering. He has published papers on systems requirements, technology readiness assessment and forecasting, Bayes nets, applied meteorology, the impacts of nuclear power plants on employment, agent-based simulation, and model-based systems engineering. Professor Kenley holds a Bachelor of Science in Management from Massachusetts Institute of Technology (MIT), a Master of Science in Statistics from Purdue University, and a Doctor of Philosophy in Engineering-Economic Systems from Stanford University.

Azad M. Madni is a Professor of Astronautical Engineering and the Technical Director of the multidisciplinary Systems Architecting and Engineering (SAE) Program at the University of Southern California's Viterbi School of Engineering. He is also a Professor of USC's Keck School of Medicine and Rossier School of Education. Dr Madni is the founder and Chairman of Intelligent Systems Technology, Inc., a high-tech company specializing in modeling, simulation, and gaming technologies for education and training. His research has been sponsored by several prominent government agencies including DARPA, DHS S&T, MDA, DTRA, ONR, AFOSR, AFRL, ARI, RDECOM, NIST, DOE, and NASA, as well as major aerospace companies including Boeing, Northrop Grumman, and Raytheon. He is the Co-Editor-in-Chief of Engineered Resilient Systems: Challenges and Opportunities in the 21st Century, Procedia Computer Science, 2014. His recent awards include the 2011 INCOSE Pioneer Award and the 2014 Lifetime Achievement Award from INCOSE-LA. He is a Fellow of AAAS, AIAA, IEEE, INCOSE, SDPS, and IETE. He is the Strategic Advisor of the INCOSE Systems Engineering Journal. He received his B.S., M.S., and Ph.D. degrees from the University of California, Los Angeles. He is also a graduate of AEA/Stanford Executive Institute for senior executives.

Alexander D. MacCalman is an Army Special Forces Officer in the Operations Research System Analyst Functional Area and has a Masters in Operations Research and a Ph.D. in Modeling and Simulations from the Naval Postgraduate School. He served in various assignments within the Special Operations and Army Analytical communities. He is currently an Assistant Professor in the Department of Systems Engineering at the United States Military Academy and works as the Systems Engineering Program Director. His research interests are in simulation experiments and how they can inform decision analysis and trade decisions.

John MacCarthy is currently serving as the Director of the Systems Engineering Education Program at the University of Maryland's Institute for Systems Research (College Park). Prior to taking this position, he completed a 28-year career as a systems engineer that included serving as a research staff member at the Institute for Defense Analyses, a senior technology and policy advisor for an senior government executive, as well as a variety of systems engineering leadership positions within Northrop Grumman and TRW (e.g., Senior Systems Engineer/Manager, Lead Systems Engineer, Manager of Proposal Operations, Deputy Director of the Center for Advanced Technology, and others). He has extensive experience in applying the full range of systems engineering processes to diverse domains that included very large defense systems and system of systems, a national nuclear waste disposal system, and a number of smaller state and local government systems. During his last 8 years in the industry, Dr MacCarthy taught a variety of graduate-level systems engineering courses as an Adjunct Professor at the University of Maryland, Baltimore County. He began his career as an Assistant Professor of physics at Muhlenberg College. He holds a Ph.D. in Physics from the University of Notre Dame, an M.S. in Systems Engineering from George Mason University, and a B.A. in Physics from Carleton College. His professional experience and interests include systems engineering; systems analysis, modeling and simulation; communications and sensor networks; sustainment engineering; life cycle cost analysis; the acquisition process; and science and engineering education.

James N. Martin is a Principal Engineer with The Aerospace Corporation. He teaches courses for The Aerospace Institute on architecting and systems engineering. Dr Martin is an enterprise architect and systems engineer developing solutions for information systems and space systems. He previously worked for Raytheon Systems Company as a lead systems engineer and architect on airborne and satellite communications networks. He has also worked at AT&T Bell Labs on wireless telecommunications products and underwater fiber optic transmission products. His book, Systems Engineering Guidebook, was published by CRC Press in 1996. He is an INCOSE Fellow and was leader of the Standards Technical Committee. Dr Martin is the founder and current leader of INCOSE's Systems Science Working Group. He received from INCOSE the Founders Award for his long and distinguished achievements in the field. Dr Martin was a key author on the BKCASE project in development of the SE Body of Knowledge (SEBOK). His main SEBOK contribution was the articles on Enterprise Systems Engineering. Dr Martin led the working group responsible for developing ANSI/EIA 632, a US national standard that defines the processes for engineering a system. He is the INCOSE representative to ISO for international standards on architecture, one of which is dealing with architecture evaluation, the topic of the chapter he wrote for this book. Dr Martin received his Ph.D. from George Mason University in Enterprise Architecture as well as a BS from Texas A&M University and an M.S. from Stanford University.

Kirk Michealson is the President of Tackle Solutions, LLC, a consulting firm for operations research analysis, project management, and training. He is an Operations Research Analyst, Fellow of the Military Operations Research Society (MORS), Lean Six Sigma Black Belt, retired Naval Officer, and an Adjunct Professor for the University of Arkansas' M.S. in Operations Management program teaching Decision Support Tools, Analytics, and Decision Models. He has degrees in Operations Research, graduating with a B.S. from the United States Naval Academy and an M.S from the Naval Postgraduate School. As a MORS Fellow, he leads the Affordability Analysis Community of Practice developing an affordability analysis process for government and industry and received the Clayton J. Thomas Award for lifetime achievement as an Operations Research Practitioner. Kirk was formerly a technical Fellow for Operations Research Analysis at Lockheed Martin where he was responsible for designing an Operations Analysis (OA) Practitioner's Success Profile and Competency Model determining the necessary skills and expertise to be a successful OA at Lockheed Martin and developing the corporate-wide experimentation process as the corporation's experimentation lead. Kirk is a retired Commander in the US Navy, and during his naval career, he was a surface warfare officer serving on ships and in various operations research positions supporting senior Navy and Department of Defense leadership.

William D. Miller is Executive Principal Analyst at Innovative Decisions, Inc. and adjunct professor at Stevens Institute of Technology in the School of Systems and Enterprises. He teaches courses in systems engineering fundamentals, system architecture and design, and systems integration. He coauthored with Dennis M. Buede *The Engineering Design of Systems: Models and Methods*, 3rd Ed, Wiley Series in Systems Engineering, Wiley & Sons Inc., 2016. He contributed content for *Systems Engineering: Coping with Complexity*, Prentice Hall Europe 1998, by Richard Stevens, Peter Brook, Ken Jackson, and Stuart Arnold. Mr Miller has over 40 years of professional experience as an engineer, manager, and consultant in the conceptualization and engineering of communications and information technologies, products, and services. His systems engineering work has encompassed resource allocation, R&D priorities, strategic planning, trade studies, requirements definition, system modeling, system design, system acquisition, system development, system integration, and system test. He previously worked at Bell Labs and AT&T in systems integration of telecommunications systems and services as well as specific developments of wireless and fiber optic technologies. He is the editor-in-chief of INCOSE *INSGHT* systems engineering practitioner's magazine, former INCOSE Technical Director, and was elected INCOSE secretary for several terms. He is a Life Member of the IEEE. Mr Miller received B.S. and M.S. degrees in electrical engineering from the Pennsylvania State University.

Edward A. Pohl is a Professor and holder of the 21st Professorship in the Department of Industrial Engineering at the University of Arkansas. Prior to his appointment as Department Head, Ed served as the Director of the Operations Management Program, the largest graduate program at the University of Arkansas. Ed currently serves as Director of the Center for Innovation in Healthcare Logistics (CIHL), Codirector for the Emerging Institute for Advanced Data and Analytics at the University of Arkansas, and the Director of Distance Education in the College of Engineering. He has participated and led several reliability, risk, and supply chain related research efforts at the University of Arkansas. Before coming to Arkansas, Ed spent 20 years in the United States Air Force, where he served in a variety of engineering, analysis, and academic positions during his career. Previous assignments include the Deputy Director of the Operations Research Center at the United States Military Academy, Operations Analyst in the Office of the Secretary of Defense where he performed independent cost schedule, performance, and risk assessments on Major DoD acquisition programs, and as a munitions logistics manager at the Air Force Operational Test Center. Ed received his Ph.D. in Systems and Industrial Engineering from the University of Arizona. He holds an M.S. in Systems Engineering from the Air Force Institute of Technology, and an M.S. in Reliability Engineering from the University of Arizona, an M.S. in Engineering Management from the University of Dayton, and a B.S. in Electrical Engineering from Boston University. His primary research interests are in reliability, engineering optimization, supply chain risk analysis, decision-making, and quality. Ed is a Fellow of IIE, a Senior Member of IEEE and ASQ, and a member of ASEM and INCOSE.

Donna H. Rhodes is a principal research scientist and senior lecturer at the Massachusetts Institute of Technology and director of the Systems Engineering Advancement Research Initiative (SEAri), a research group focused on advancing theories, methods, and practice of systems engineering applied to complex sociotechnical systems. She conducts research and consults on innovative approaches and methods for architecting complex systems and enterprises, designing for uncertain futures, and creating anticipatory capacity in enterprises. During her prior industry career, Dr Rhodes held senior management positions at IBM, Lockheed Martin, and Lucent. She has been very much involved in the evolution of the systems engineering field, including corporate education and development of several university graduate programs. She has served on numerous industry, academic, and government advisory boards focused on advancement of systems practice and education, as well as on study panels for issues of national and international importance. Dr Rhodes is a

Past President and Fellow of the International Council on Systems Engineering (INCOSE). She is a recipient of the INCOSE Founders Award and associate editor of the INCOSE journal Systems Engineering. She received her Ph.D. in Systems Science from the T.J. Watson School of Engineering at Binghamton University.

Garry Roedler is a Fellow and the Engineering Outreach Program Manager for Lockheed Martin and the President-elect for the International Council on Systems Engineering (INCOSE). His systems engineering (SE) experience spans the full life cycle and includes technical leadership roles in both programs and systems engineering business functions. Garry holds degrees in mathematics education and mechanical engineering from Temple University and the ESEP certification from INCOSE. Garry is an INCOSE Fellow, author of numerous publications and presentations, and the recipient of many awards, including the INCOSE Founders Award, Best SE Journal Article, IEEE Golden Core, Lockheed Martin Technical Leadership Award, and Lockheed Martin NOVA Award. His other leadership roles across many technical organizations include Past Chair of the INCOSE Corporate Advisory Board; member of the INCOSE Board of Directors; steering group member for the National Defense Industrial Association Systems Engineering Division; working group chair for the IEEE Joint Working Group for DoD Systems Engineering Standardization; editor of ISO/IEC/IEEE 15288, Systems Life Cycle Processes, and several other standards; and key editor roles in the development of the Systems Engineering Body of Knowledge (SEBoK) and the INCOSE Systems Engineering Handbook. This unique set of roles has enabled Garry to influence the technical coevolution and consistency of these key SE resources.

Adam M. Ross is a senior innovator at The Perduco Group and cofounder and former lead research scientist for the MIT SEAri, a research group focused on advancing the theories, methods, and effective practice of systems engineering applied to complex sociotechnical systems through collaborative research with industry and government. Dr Ross has published over 90 papers in the areas of space systems design, systems engineering, and tradespace exploration. He has received numerous paper awards, including the Systems Engineering 2008 Outstanding Journal Paper of the Year. He has led over 15 years of research and development of novel systems engineering methods, frameworks, and techniques for evaluating and valuing system tradespaces and the "ilities" across alternative futures during early phase design. He uses a transdisciplinary approach, leveraging techniques from engineering design, operations research, behavioral economics, and interactive data visualization. He serves on technical committees with both AIAA and IEEE and is recognized as a leading expert in system tradespace exploration and change-related "ilities." He consults for government agencies, applying analytic techniques for decision support and optimization for acquisition planning. Application domains have included civil transportation, defense and civil aerospace, and commercial and defense maritime systems. Dr Ross holds a dual bachelor's degree in Physics and Astrophysics from Harvard University, two master's degrees in Aeronautics and Astronautics Engineering and Technology & Policy from MIT, as well as a doctoral degree in Engineering Systems from MIT.

Sam Savage is the author of "The Flaw of Averages: Why We Underestimate Risk in the Face of Uncertainty" (John Wiley & Sons, 2009, 2012), in which he defines the discipline of probability management. He is also the inventor of the SIP, a data structure that allows simulations to communicate with each other. Dr Savage has been a Consulting Professor in the School of Engineering at Stanford University since 1990 and taught at the Graduate School of Business at the University of Chicago since 1974. He is also a Fellow of the Judge Business School at Cambridge University. In 2012, he incorporated ProbabilityManagement.org, as a 501(c)(3) nonprofit that has been cited in the MIT Sloan Management Review for "improving communication of uncertainty." He serves as its Executive Director and is joined on the board by Harry Markowitz, Nobel Laureate in Economics. Sponsoring organizations include Chevron, General Electric, Lockheed Martin, and Wells Fargo Bank. Dr Savage holds a Ph.D. in the area of computational complexity from Yale University.

Clifford Whitcomb is a Professor in the Systems Engineering Department at the Naval Postgraduate School in Monterey, CA. Dr Whitcomb's research interests include design thinking, model-based systems engineering, naval construction and engineering, and leadership, communication, and interpersonal skills development for engineers. He is the coauthor of "Effective Interpersonal and Team Communication Skills for Engineers" published by John Wiley and Sons and has published several other textbook chapters. He is a principal investigator for the US Navy Office of Naval Research, Office of the Joint Staff, Office of the Secretary of the Navy, and the Veteran's Health Administration. He is an INCOSE Fellow, has served on the INCOSE Board of Directors, and was a Lean Six Sigma Master Black Belt for Northrop Grumman Ship Systems. Dr Whitcomb was previously the Northrop Grumman Ship Systems Endowed Chair in Shipbuilding and Engineering in the department of Naval Architecture and Marine Engineering at the University of New Orleans, a senior lecturer in the System Design and Management (SDM) program at MIT, as well as an Associate Professor in the Ocean Engineering Department, at MIT. Dr Whitcomb is also a retired naval officer, having served 23 years as a submarine warfare officer and Engineering Duty Officer. He earned his B.S. in Engineering from the University of Washington, Seattle, WA, in 1984, an Engineer degree in Naval Engineering and S.M. in Electrical Engineering and Computer Science from MIT in 1992, and Ph.D. in Mechanical Engineering from the University of Maryland, College Park, MD, in 1998.

Paul T. Beery is a Faculty Associate for Research in the Systems Engineering Department at the Naval Postgraduate School. He has a Masters in Systems Engineering Analysis from Naval Postgraduate School and is working toward his doctoral degree in Systems Engineering at the Naval Postgraduate School. His research interests are the design and analysis of complex systems, simulation analysis, and the applications of advances in design of experiments.

Andres Vargas is a research assistant and graduate student in the Department of Industrial Engineering at the University of Arkansas. His research interests are focused on decision analysis, network optimization, and supply chain transportation strategies. Andres has worked for The Coca-Cola Company in his home country Bolivia, where he served as an occupational safety supervisor and quality control analyst. He has also worked with Sam's Club, where he was involved in a project initiative to redesign and improve self-checkout station operations using queueing theory concepts and simulation. Andres received his B.S. in Industrial Engineering from the University of Arkansas. As an undergraduate researcher, he worked for the Center for Excellence in Logistics and Distribution as a web application developer.

FOREWORD

Though many may not recognize it as such, we live in the systems age. Those engaged with systems engineering understand the challenges and opportunities of today and tomorrow are truly systems challenges. On the grand scale, we must address clean energy, clean water, food, resource allocation, health care, and more. None would argue that these are systems engineering challenges, but they are systems challenges indeed. On a somewhat smaller scale, we see unprecedented opportunities fueled by the ever-increasing rate of technology infusion and the opportunity to connect existing systems in new and novel ways. Our stakeholders demand more from us, and technology allows us to deliver: end-to-end connected transportation enabled by autonomous vehicles; new efficiencies in energy generation, storage, and distribution unlocked by new sensor technologies and insights from big data; innovations in personal health care through wearables and other technologies.

Scientists, architects, specialists, and engineers of all types collaborate in a wide range of complicated and often complex situations to address issues and deliver against stakeholder demands with upgraded and innovative systems. As Randy Pausch reminded us in *The Last Lecture*, "engineering is not about perfect solutions, it's about doing the best you can with limited resources." In this quest to serve our customers and stakeholders, the fundamental purpose of systems engineering is neither process, enabler, specification, nor any other tool or artifact. Systems engineering is charged with delivering the required value in an effective and efficient manner, making the best possible use of the resources available. In doing so, making the inevitable trade-offs should not be treated as a necessary evil. Informed trade-offs based on appropriate analysis properly performed are a critical enabler delivering the required value efficiently and effectively. They are a key tool in our systems engineering toolbox – one that we must embrace and improve.

Systems engineering and all those who practice it are connectors, and connecting diverse approaches across multiple perspectives requires these trade-offs, both large and small. In connecting processes and analytics to properly understand the true problem and architect the right solution, what processes do we select and what level of fidelity do we pursue to balance investment made with value delivered? In assessing both the problem space and the possible solution space, how do we prioritize the "right" blend of desires, constraints, approaches, technologies, and specialties necessary to solve the problem within the bounds of capability, schedule, and budget? Looking to the evolution of the environment and our solution, what resilience and adaptability must we account for and what range of sensitivity can we accept?

As we continue to embrace and expand systems practices across diverse communities, our problems move from the complicated to the complex. We engage an even broader range of subject matter experts with their particular perspectives, tools, and techniques. We have greater technical, economic, and social considerations as we address both bounded problems with defined requirements and fuzzy problems characterized by market behaviors and stakeholder concerns. As the scope continues to expand, these questions become more challenging. Properly performing trade-off analysis from problem definition through ultimate solution delivery becomes even more critical.

In this text, Parnell et al. bring together in a systems engineering context the fundamental foundations, the processes and principles, the techniques and examples necessary to help us perform better trade-off analyses. They recognize the broad scope including cost, value, schedule, and risk drivers and provide tools to deal with the inherent uncertainty within which systems engineers operate. Put simply, Parnell and his colleagues provide a complete and cohesive treatment of this critical enabler for systems engineering, moving us from sometimes disjoint, ad hoc approaches to an informed, disciplined approach to explicitly define our decision opportunities and alternatives in the journey to making better decisions.

As we connect teams and technologies to better meet stakeholder needs in an ever-evolving environment, it is not a question of whether or not we perform trade-off analysis. It is a question of how well we do so: whether we make errors of omission and commission, whether we are implicit or explicit, whether we rely on unsound approaches or informed practices. All those who practice systems engineering serve as the "guardians of why," balancing multiple options and considerations as we collaborate with others to match the right solution to the real problem. Through informed trade-off analysis, we better leverage the talents, techniques, insights, and perspectives of those around us, ultimately driving better decisions and enabling the delivery of systems to meet the diverse and complex challenges of today and tomorrow.

David Long
President, Vitech Corporation
INCOSE President (2014 & 2015)

PREFACE

NEED FOR MORE EFFECTIVE TRADE STUDIES

Today's complex systems are multidisciplinary systems involving challenging missions, advanced technologies, significant uncertainties, and multiple stakeholders with conflicting objectives. Decision-making is central to generating creative alternatives, creating value, managing risks, and meeting affordability goals. Systems engineering trade-offs are needed throughout the system life cycle to inform these system decisions. In the absence of a formal framework, trade-off studies are sometimes performed in an ad hoc manner. Also, some systems engineers may not have an in-depth understanding of trade-off analysis techniques. As a result, some use unsound techniques.

This project began with a need identified by a professional society. The International Council on Systems Engineering (INCOSE) (www.incose.org) has nearly 10,000 members and about 95 members of its Corporate Advisory Board. The INCOSE Corporate Advisory Board documented the need for more effective trade studies. They believed there was a lack of best practices information that crossed the life cycle and aligned with ISO standard (ISO/IEC/IEEE 15288, 2015), the Systems Engineering Handbook (INCOSE, 2015), and the Systems Engineering Body of Knowledge (SEBok, Systems Engineering Body of Knowledge (SEBoK), 2015).

This textbook presents a Decision Management process based on decision theory and cost analysis best practices and is aligned with ISO/IEC 15288, the Systems Engineering Handbook, and the Systems Engineering Body of Knowledge. We introduce key concepts and demonstrate these trade-off analysis concepts in the different life cycle stages using illustrative examples from defense and commercial domains.

AUDIENCE

The audience for this book are graduate students (systems engineering, industrial engineering, engineering management, other engineering disciplines); professional systems engineers, operations analysts, project managers, and engineering managers; and undergraduate students (systems engineering, industrial engineering, engineering management, other engineering disciplines). We assume that the reader has had an introduction to systems engineering, an undergraduate knowledge in probability and statistics, a course in systems modeling, and a course in engineering economy and/or life cycle cost. However, Chapter 3 reviews probability and Chapter 4 presents important resource analysis techniques required for cost analysis and affordability analysis.

THEMES

We had several major themes that provided the foundation for this book.
1. **Use standard SE terminology.** We have attempted to use terminology from the ISO standard (ISO/IEC/IEEE 15288, 2015), the Systems Engineering Handbook (INCOSE, 2015), and the Systems Engineering Body of Knowledge (SEBok, Systems Engineering Body of Knowledge (SEBoK) wiki page, 2015).
2. **Avoid trade-off analysis mistakes of omission and commission.** The mistakes of omission are errors made by not doing the right things. The mistakes of commission are errors made by doing the right things the wrong way.
3. **Use a decision management process.** Systems decisions are made throughout the life cycle. Many of these systems decisions are difficult decisions that include multiple competing objectives, numerous stakeholders, substantial uncertainty, significant consequences, and high accountability. These decisions can benefit from a structured decision management process.
4. **Use decision analysis as the mathematical foundation for trade-off analyses.** A credible trade-off analysis should be based on a sound mathematical foundation. Ad hoc methods and unsound mathematics provide a base of sand for a trade-off study and, therefore, a base of sand for the decision-makers. Since trade-off studies involve complex alternatives, multiple objectives, and major uncertainties, we believe that decision analysis is the operations research technique that provides this sound mathematical foundation for trade-off analyses.
5. **Explicitly define the decision opportunity.** Every trade-off study begins with an implicit understanding of the problem or opportunity. The initial problem is never the final problem. It is important to clearly define the decision problem as a broader decision opportunity.
6. **Identify and structure decision objectives and measures.** Once the opportunity is explicitly identified, the next step is to identify and structure the decision objectives of the decision-makers and stakeholders. The decision

opportunity and stakeholder values determine the objectives. Measures that align with the objectives are required to perform the trade-off analysis.
7. **Develop creative, doable alternatives.** The key to trade-off analysis is developing good alternatives that span the tradespace. The generation of the tradespace is a critical trade-off analysis task that requires participation of the entire trade-off analysis team and support from decision-makers, stakeholders, and subject matter experts.
8. **Include resource analysis in the trade-off analysis.** Organizations do not have unlimited resources. Therefore, affordability analysis is almost always a critical part of the trade-off analysis.
9. **Explicitly consider uncertainty.** Systems development, deployment, operation, and retirement involve many uncertainties. The systems life cycle may be years to decades. The major uncertainties include technology performance, integration with other systems, markets/missions, environments, and the actions of competitors/ adversaries.
10. **Identify the cost, value, schedule, and risk drivers**. The purpose of a trade-off analysis is to provide insights for system decision-making. Decision-makers need to understand the cost, value, schedule, and risk drivers of the system.
11. **Provide an integrated for cost, value, and risk analysis.** Unfortunately, most of the current systems engineering practice develops and performs separate cost, value, and risk analyses. We recommend an integrated framework for cost, value, and risk analysis.

BOOK ORGANIZATION

The book is organized into three sections and a summary (Figure 1). The first section discusses the trade-off analysis foundations. Chapter 1 provides an introduction to trade-off analysis and includes common mistakes of commission and omission made in trade studies. Chapter 2 provides a conceptual framework for trade-off analysis and presents key decision theory concepts required for a sound mathematical foundation. As mentioned earlier, Chapter 3 reviews probability and Chapter 4 presents resource analysis techniques and affordability analysis.

The Decision Management process is presented in the second section of the book. Chapter 5 introduces the INCOSE Decision Management process and provides a detailed illustrative example of the process. Chapter 6 provides the principles and techniques for identifying the decision opportunity that the trade-off analysis supports. Chapter 7 provides principles and techniques for identifying objectives and value measures that assess how well the alternatives meet the objectives. These measures are the foundation for assessing the trade-offs. Chapter 8 reviews and evaluates the techniques for generating and evaluating alternatives. Many of these techniques are illustrated in the third section of the book. Chapter 9 illustrates a model for trade-off analysis that integrates value and cost analysis.

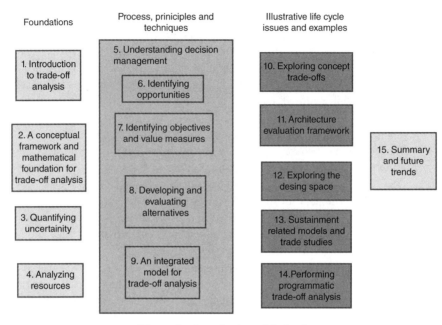

Figure 1 Organization of the book

The third section provides trade-off analysis issues and illustrative examples in the life cycle stages. The scope and information available for trade-off analysis are different in each life cycle stage. Chapter 10 presents trade-off analysis methods to explore the trade-offs in the early life cycle stages when many system concepts and architectures need to be evaluated to determine the most affordable concept for further development. Chapter 11 presents processes and techniques for evaluating system architectures. Chapter 12 presents illustrative examples for system design trade-off analysis. Chapter 13 presents an illustrative sustainment model with deterministic and probabilistic analysis. Chapter 14 provides several illustrative examples of programmatic trade-offs that focus on system acceptance and termination.

Chapter 15 summarizes the major themes of the book and identifies some potential trends that may impact trade-off analyses in the future.

COURSE OUTLINES USING THE TEXTBOOK

In this section, we offer some possible course outlines that could be developed using this textbook. Of course, the content presented in the course should be selected based on the academic/professional education program objectives and the course objectives.

PREFACE   xxxvii

We present a notional set of course objectives and offer some potential course outlines.

NOTIONAL COURSE OBJECTIVES

1. Understand the role of trade-off analyses to support system decisions in each stage in the system life cycle.
2. Identify and avoid the mistakes of omission and commission in trade-off analysis.
3. Understand and use decision analysis as the mathematical foundation for trade-off analysis.
4. Understand the sources of uncertainty in the system life cycle and be able to identify, assess, and model uncertainty using probability.
5. Understand the advantages and disadvantages of common systems engineering approaches used to generate and evaluate system alternatives.
6. Identify and structure stakeholder objectives and develop single objective and multiobjective decision analysis models to evaluate alternatives.
7. Identify and define a system decision opportunity that requires a trade-off analysis.
8. Understand the advantages and disadvantages of tradespace exploration techniques for trade-off analysis of concepts, architectures, designs, operations, and retirement.
9. Understand the need for an integrated decision model that incorporates design features, value, cost, and risk.
10. Perform a trade-off analysis using the INCOSE Decision Management Process using both deterministic and probabilistic techniques.
11. Communicate the insights of a trade study and the important trade-offs to senior stakeholders and decision-makers.

ILLUSTRATIVE ACADEMIC COURSE OUTLINES

In addition to the course objectives, the coverage of course topics will depend on the role of the course in the curriculum (required or elective), the prerequisites, the location of the course (early or late in program), and the type of course (lecture, project, or combined). The textbook could be used to prepare for a capstone design course. The textbook presents more material that can probably be covered in a one semester course. I would recommend covering all of Chapters 1, 2, 5–7. The instructor would select the sections to read for other chapters. Depending on the academic curriculum, Chapters 3 and 4 could be reviewed or covered in more detail.

Week	Systems Analysis Project Course	System Design Project Course	Systems Analysis Lecture Course
Pre-reqs	Undergrad probability and statistics	Undergrad probability and statistics	None
1	Introduction (Chapter 1)	Introduction (Chapter 1)	Introduction (Chapter 1)
2	Framework and Mathematical Foundations (Chapter 2)	Framework and Mathematical Foundations (Chapter 2)	Framework and Mathematical Foundations (Chapter 2)
3	Uncertainty (Chapter 3)	Uncertainty (Chapter 3)	Uncertainty (Chapter 3)
4	Resource Analysis (Chapter 4)	Resource Analysis (Chapter 4)	Uncertainty (Chapter 3)
5	Decision Management Process I (Chapter 5)	Decision Management Process I (Chapter 5)	Resource Analysis (Chapter 4)
6	Decision Management Process II (Chapter 5)	Opportunity Definition (Chapter 6)	Resource Analysis (Chapter 4)
7	Opportunity Definition (Chapter 6)	Objectives and Measures (Chapter 7)	Decision Management Process II (Chapter 5)
8	Objectives and Measures (Chapter 7)	Class Project – Opportunity Presentations	Opportunity Definition (Chapter 6)
9	Class Project – Opportunity Presentations	Generation and Evaluation of Alternatives (Chapter 8)	Objectives and Measures (Chapter 7)
10	Generation and Evaluation of Alternatives (Chapter 8)	Integrated Value, Cost, and Risk Analysis (Chapter 9)	Generation and Evaluation of Alternatives (Chapter 8)
11	Integrated Value, Cost, and Risk Analysis (Chapter 9)	Concept Evaluation (Chapter 10)	Integrated Value, Cost, and Risk Analysis (Chapter 9)
12	Concept and Architecture Evaluation (Chapters 10 & 11)	Architecture Evaluation (Chapter 11)	Concept Evaluation (Chapter 10)
13	Design Evaluation (Chapter 12)	Design Evaluation (Chapter 12)	Architecture Evaluation (Chapter 11)
14	Sustainment Trade-Offs (Chapter 13)	Design Evaluation (Chapter 12)	Design Evaluation (Chapter 12)
15	Programmatic Trade-Offs (Chapter 14)	Sustainment Trade-Offs (Chapter 13) Programmatic Trade-Offs (Chapter 14)	Sustainment Trade-Offs (Chapter 13)
16	Class Project – Trade-off Analysis Presentations (Chapter 15)	Class Project – Trade-off Analysis Presentations (Chapter 15)	Programmatic Trade-Offs (Chapter 14) Summary (Chapter 15)

ILLUSTRATIVE PROFESSIONAL SHORT COURSE OUTLINE

The textbook can also be used as a textbook/reference for professional short courses. The topics presented in the course would depend on the needs of the organization and the students' academic and professional backgrounds. The course could be taught as a seminar to present new material or as a project course with student's applying the material they learn in the course on a notional trade-off analysis or trade-off analyses they are working or will work in the future. The following outline is for a 1-week project course with trade-off analysis modeling using notional data (provided to students or developed by students).

	Monday	Tuesday	Wednesday	Thursday	Friday
Morning	Introduction (Chapter 1) Framework and Mathematical foundations (Chapter 2)	Decision Management Process I (Chapter 5)	Resource Analysis (Chapter 4)	Concept and Architecture Evaluation (Chapters 10 & 11)	Sustainment Trade-Offs (Chapter 13)
	Opportunity Definition (Chapter 6)	Decision Management Process II (Chapter 5)	Uncertainty (Chapter 3) Monte Carlo Simulation	Design Evaluation (Chapter 12)	Programmatic Trade-Offs (Chapter 14)
Afternoon	Objectives and Measures (Chapter 7)	Generation and Evaluation of Alternatives (Chapter 8)	Integrated Value, Cost, and Risk Analysis (Chapter 9)	Class Project – Development of Notational Life Cycle Cost Model	Class Project – Monte Carlo Simulation of Value and Cost Models
	Class Project – Opportunity Presentation	Class Project – Generation of Alternatives	Class Project – Development of Notional tradespace Exploration Model	Class Project – Integration of Cost and Value Model	Class Project – Trade-Off Analysis Presentations (Chapter 15)

Gregory S. Parnell, PhD, INCOSE Fellow
Editor
University of Arkansas
Fayetteville, AR
September 2016

REFERENCE

ISO/IEC/IEEE 15288 (2015). *Systems and Software Engineering – System Life Cycle Processes*, International Organization for Standardization (ISO)/International Electrotechnical Commission (IEC)/Institute of Electrical and Electronics Engineers (IEEE), Geneva, Switzerland.

SEBoK (2015). Systems Engineering Body of Knowledge (SEBoK) wiki page. SEBoK: http://www.sebokwiki.org (accessed 06 June 2016).

ACKNOWLEDGMENTS

The development and writing of this book was an intense 1-year team effort by 18 contributors. As the editor, I would like to acknowledge some special contributions.

INTERNATIONAL COUNCIL ON SYSTEMS ENGINEERING (INCOSE) CORPORATE ADVISORY BOARD (CAB)

The INCOSE CAB identified the need for more effective trade studies as one of their top five needs. **Garry Roedler** (one of the contributors) was CAB Chair when the need was identified. **Max Berthold,** the next CAB Chair, has continued the support. This book is one of the INCOSE activities to help meet this need. The INCOSE CAB has reviewed our plan for the book, and three INCOSE CAB representatives have become chapter authors, including **Azad Madni**, **Cliff Whitcomb**, and myself.

INCOSE TECHNICAL DIRECTORS

The Technical Director is a voting member of the INCOSE Board of Directors and has functional responsibility for Technical Operations. As the Technical Director at the time, **Bill Miller** (later one of our contributors) assigned this need to the INCOSE Decision Analysis Working Group. **Paul Schreinemakers**, the next Technical Director, has continued to support our effort.

INCOSE DECISION ANALYSIS WORKING GROUP

The INCOSE Decision Analysis Working Group, led by **Frank Salvatore**, has played a central role. Frank has supported our entire effort including the Decision

Management input to the Systems Engineering Handbook, the Decision Management input to the Systems Engineering Body of Knowledge (SEBoK), and using working group meetings to recruit chapter authors, provide feedback on the book plan, and recruit chapter reviewers. **Matt Cilli** (chapter author) developed the Decision Management framework for the SE Handbook and SEBoK and provided more useful information on the framework in Chapter 5. **Clifford Marini** spearheaded all the computer code development efforts needed to generate many of the visualizations used in the examples of Chapter 5.

CHAPTER AUTHORS

\looseness=-1{}The chapter authors are systems engineering thought leaders and fully employed with senior academic, corporate, or government positions (see About the Authors). The role of the chapter authors was much more than writing their chapters. The authors provided recommendations on the content of the book, and each author has peer-reviewed one or two chapters, revised their chapters based on my reviews, and the reviews of two internal reviewers (authors of other chapters) and two external reviewers. Several authors contributed to more than one chapter. Based on when I received emails, most of their work on the book was done at night and on the weekends.

CHAPTER REVIEWERS

Our goal was to have every chapter reviewed by an academic and a practitioner who had not been participating in the book project. We asked for and received very helpful reviews. Many times, these reviews caused the authors to add material or reorganize the chapter to be clearer. The following individuals provided valuable chapter reviews that improved the quality of the book: Tyson Browning, Roger Burk, Chuck Ebeling, Tony Farina, Paul Garvey, Melissa Garber, Daniel McCarthy, Kim Needy, George Rebovich, Frank Salvatore, James "Jed" Richards, Valarie Sitterle, Gerardo Siva, Michael Vinarcik, and Adam Whitlock.

DEPARTMENT OF INDUSTRIAL ENGINEERING AT THE UNIVERSITY OF ARKANSAS

I would like to acknowledge the support of my department head, **Ed Pohl** (also a chapter author), who supported my participation in this project and provided a research assistant to help with the effort.

RESEARCH ASSISTANT

This book would not have been possible in 1 year without the dedicated help of **Andres Vargas**, my research assistant for the past year. Andres performed an excellent literature search that provided references for all chapter authors. In addition, he performed many of the tasks required to align each chapter to the final format in the book. He is also preparing the solutions manual and the Excel files that will be provided on the Wiley book site.

FINAL NOTE

Any remaining errors of omission or commission are of course my responsibility.

Gregory S. Parnell, PhD, INCOSE Fellow
Editor
University of Arkansas
September 2016

ABOUT THE COMPANION WEBSITE

This book is accompanied by a companion website:
www.wiley.com/go/Parnell/Trade-off_Analytics

The website includes:

- An instructor website with a solutions manual
- A student website with Excel files

1

INTRODUCTION TO TRADE-OFF ANALYSIS

GREGORY S. PARNELL
Department of Industrial Engineering, University of Arkansas, Fayetteville, AR, USA

MATTHEW CILLI
U.S. Army Armament Research Development and Engineering Center (ARDEC), Picatinny, NJ, USA

AZAD M. MADNI
Department of Astronautical Engineering, Systems Architecting and Engineering and Astronautical Engineering, Viterbi School of Engineering, University of Southern California, Los Angeles, CA, USA

GARRY ROEDLER
Corporate Engineering LM Fellow, Engineering Outreach Program Manager, Lockheed Martin Corporation King of Prussia, PA

> The complexity of man-made systems has increased to an unprecedented level. This has led to new opportunities, but also to increased challenges for the organizations that create and utilize systems.
>
> (ISO/IEC/IEEE 15288, 2015)

Trade-off Analytics: Creating and Exploring the System Tradespace, First Edition. Edited by Gregory S. Parnell.
© 2017 John Wiley & Sons, Inc. Published 2017 by John Wiley & Sons, Inc.
Companion website: www.wiley.com/go/Parnell/Trade-off_Analytics

1.1 INTRODUCTION

This book is about trade-off analyses in the life cycle of a system. It is written from the perspective of engineers, systems engineers, and other decision-makers involved in the life cycle of a system. In this book, we present the best practices for performing systems engineering trade-off analyses in a step-by-step, structured manner. Our intent is to make it an easy-to-understand and useful reference for students, practitioners, and researchers.

Systems are developed to create value for stakeholders by providing desired capabilities. Stakeholders include investors, government agencies, customers/acquirers, end users/operators, system developers/integrators, trainers, and system maintainers, among others. Decisions are ubiquitous across the system life cycle. System decision-makers (DMs) are those individuals who make important decisions pertaining to the technical and management compromises that shape the concept definition, system definition, system realization, deployment and use, and product and service life management (including maintenance, enhancement, and disposal).

When there are multiple stakeholders, there are often competing objectives and requirements. To achieve a certain attainment level on one objective, a sacrifice or trade-off may be required in the attainment level of other objectives. Similarly, complex system designs may offer multiple alternatives to achieve the system's objectives, and this, too, requires analysis to achieve the best balance among the trade-offs. The process that leads to a reasoned compromise in these situations is commonly referred to as a "trade-off analysis" or a "trade study."

This book project began with a request by the International Council on Systems Engineering (INCOSE) (INCOSE Home Page, 2015) Corporate Advisory Board (CAB) to the INCOSE Decision Analysis Working Group. The CAB identified the lack of effective trade-off analysis methods as a key concern and requested help in documenting best practices. This book project was also motivated by the need to formalize systems engineering trade-off analysis to help make it an integral part of the systems engineering life cycle. It provides essential elaboration of the decision management process in ISO/IEC/IEEE 15288, Systems and Software Engineering – System Life Cycle Processes, the INCOSE Systems Engineering Handbook Version 4, and the Systems Engineering Body of Knowledge (SEBoK, 2015).

Decision-makers (DM), especially program managers and systems engineers, stand to benefit from a collaborative decision management process that engages all stakeholders (SH) who have a say in system design decisions. In particular, systems engineers can exploit trade-off studies to help define the problem/opportunity, characterize the solution space, identify sources of value, identify and evaluate alternatives, identify risks, acquire insights, and provide recommendations to system SHs and other DMs.

This book focuses on engineering trade-off analysis techniques for both systems and systems of systems (Madni and Sievers, 2014a,b; Ordoukhanian and

Madni, 2015). We recommend that trade-off studies be consistent with SE standards (ISO/IEC/IEEE 15288:2015), based on a formal lexicon, have a sound mathematical foundation, and provide credible and timely data to DMs and other SHs. We provide such a lexicon and a formal foundation (Chapter 2) based on decision analysis for effective and efficient trade-off studies. Our approach supplements decision analysis, a central part of decision-based design (Hazelrigg, 1998), with Value-Focused Thinking (Keeney, 1992) within a model-based engineering framework (Madni & Sievers, 2015).

1.2 TRADE-OFF ANALYSES THROUGHOUT THE LIFE CYCLE

New system development entails a number of interrelated decisions. Table 1.1 provides a partial list of decisions opportunities to improve the system value that are commonly encountered throughout a system's life cycle. Many of these decisions stand to benefit from a holistic perspective that combines the systems engineering discipline with a composite decision model that aggregates the data produced by engineering, performance, and cost models and translates them into terms that are relevant and meaningful to the various stakeholders, especially DMs. This holistic perspective is especially valuable in gate (go/no-go funding) decisions to ensure that affordable alternatives are available for the next life cycle stage.

1.3 TRADE-OFF ANALYSIS TO IDENTIFY SYSTEM VALUE

Systems provide value through the capabilities they provide or the products and services they enable (Madni, 2012). Decision analysis is an operations research technique that provides models to define value and a sound data-driven, objective, defensible, mathematical foundation for trade-off analyses. The graphic shown in Figure 1.1 helps visualize the importance of opportunity definition (Chapter 6) to value creation. For example, Chevron uses the "Eagle's Beak ", as shown in this figure, to convey the importance of project definition and project execution. The five phases shown in Figure 1.1 constitute the project life cycle used by Chevron (Lavingia, 2014). The process leads to value identification and value realization. At Chevron, decision analysis plays an important role in the three phases of value identification: identify opportunity; generate and select alternatives; and develop the preferred opportunity. The Chevron process employs stages and gates similar to those found in most system life cycles. Each phase consists of activities that produce information; clearly defined deliverables; and an explicit decision to proceed, exit, or recycle. Chevron employs project management in all five phases of the Chevron Project Development and Execution Process (Decision-Making in an Uncertain World: A Chevron Case Study, 2014). Similarly, for the system life cycle, value

Table 1.1 Partial List of Decision Opportunities throughout the Life Cycle

Life Cycle Stage	Decision Opportunity
Exploratory research	Assess technology opportunity/initial business case • Of all the potential system concepts or capabilities that could incorporate the emerging technology of interest, do any offer a potentially compelling and achievable market opportunity? • Of those that do, which should be pursued, when, and in what order?
Concept	Inform, generate, and refine a capability • What requirements should be included? What are the desired parameters? • What really needs to be accomplished and what is able to be traded away to achieve it within anticipated cost and schedule constraints? • How should requirements be expressed such that they are focused yet flexible? • How can the set of requirements be demonstrated to be sufficiently compelling while at the same time achievable within anticipated cost and schedule constraints? • Which concepts are affordable? Create solution class alternatives and select preferred alternative • After considering the system-level consequences of the sum of solution class alternatives across the full set of stakeholder values (to include cost and schedule), which solution class alternative should be pursued? • Is the solution class still affordable?
Development	Select/define system elements • After considering the system-level consequences of the sum of system element design choices across the full set of stakeholder values (to include cost and schedule), which system element alternatives should be pursued? (Repeated for each recursive level of the system structure.) Select/design verification and validation methods • Is prototyping warranted? • What verification and validation methods should be performed (test, demonstration, analysis/simulation, inspection)? • What are the verification and validation plans?

(*continued overleaf*)

Table 1.1 (*Continued*)

Life Cycle Stage	Decision Opportunity
Production	Craft production plans • What is the target production rate? • To what extent will low-rate initial production be utilized? • What is the ramp-up plan? • What production process will be used? • Who will produce the system? • Where will the system be produced? • Is the system still affordable?
Operation, support	Generate maintenance approach • What is the maintenance strategy? • What is the logistics concept? • What is the preventive-maintenance plan? • What is the corrective-maintenance plan? • What is the spare-parts plan? • Is the system still affordable?
Retirement	Retirement plan • When is it time to retire the system? • How will disposal of materials be accomplished?

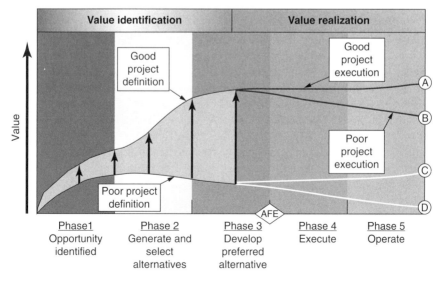

Figure 1.1 Eagle's beak chart

identification occurs during the concept definition and system definition phases, and decision analysis plays the same important role.

Figure 1.1 highlights five important points. First, the problem or opportunity definition (see Chapter 6) is an important first step in value identification. Second, the generation of good alternatives (see Chapter 8) is critical to identifying higher value. Third, the development, evaluation, and selection of preferred alternatives can significantly increase value. Fourth, good project execution is required to realize potential value. Fifth, project execution is performed in the face of uncertainties (see Chapter 3). In this book, we focus on the value of using trade-off studies to help in the identification of both value and risk, as the timely identification of risk can help implementers mitigate potential barriers to value realization.

1.4 TRADE-OFF ANALYSIS TO IDENTIFY SYSTEM UNCERTAINTIES AND RISKS

System risks can affect performance, schedule, and cost. Building on several frameworks, Table 1.2 provides a list of the sources of systems risk (Parnell, 2009). The first column in Table 1.2 lists the potential source of risk. The second column lists the major questions defining the risk. The third column lists some of the major potential uncertainties for this risk source. The major questions and the uncertainties are meant to be illustrative and not all inclusive. Many of these risks create uncertainties, which should be considered in trade-off analyses. Chapter 3 provides techniques for using probability to model these uncertainties in trade-off analyses. Later chapters explicitly consider these uncertainties in illustrative trade-off analyses.

1.5 TRADE-OFF ANALYSES CAN INTEGRATE VALUE AND RISK ANALYSIS

Program managers for the development of a new system must consider performance, cost, and schedule, as they are all interrelated. We know that performance problems can cause cost increases and schedule delays. Similarly, schedule changes can increase costs. Finally, cost estimate increases can result in reduced performance targets or schedule delays to make the system more affordable. Trade-off analysis, cost analysis, and risk analysis are frequently separate analyses performed by different analysts. Cost analysts typically perform a cost-risk analysis using Monte Carlo simulation. Many trade-off studies ignore uncertainty and risk.

A major theme of this book is that trade-off analyses should be used to identify both system value and system risks and that the analysis needs to be performed in a more integrated manner. In Chapter 9, we discuss and provide examples of how system value, system costs, and system risks can be integrated by identifying the system features that impact value, cost, and risk.

Table 1.2 Sources of Systems Risk

Sources of Risk	Major Questions	Potential Uncertainties
Business	Will political, economic, labor, social, technological, environmental, legal, or other factors adversely affect the business environment?	Changes in political viewpoint (e.g., elections) Economic disruptions (e.g., recession) Global disruptions (e.g., supply chain) Changes to law Disruptive technologies Adverse publicity
Market	Will there be a market if the product or service works?	Consumer demand Threats from competitors (quality and price) and adversaries (e.g., hackers and terrorists) Continuing stakeholder support
Performance (technical)	Will the product or service meet the required/desired performance?	Defining future requirements in dynamic environments Understanding technical baseline Technology maturity to meet performance. Adequate modeling, simulation, test, and evaluation capabilities to predict and evaluate performance Impact to performance from external factors (e.g., interoperating systems) Availability of enabling systems needed to support use
Schedule	Can the system that provides the product or service be delivered on time?	Concurrency in development Impact of uncertain events on schedule Time and budget to resolve technical and cost risks
Development and production cost	Can the system be delivered within the budget? Will the cost be affordable?	Changes in concept definition (mission or needs) Technology maturity Stability of the system definition Hardware and software development processes Industrial/supply chain capabilities Production/facilities capabilities Manufacturing processes
Management	Does the organization have the people, processes, and culture to manage a major system?	Organization culture SE and management experience and expertise Mature baselining (technical, cost, schedule) processes Reliable cost-estimating processes

(*continued overleaf*)

Table 1.2 (*Continued*)

Sources of Risk	Major Questions	Potential Uncertainties
Operations and support cost	Can the owner afford to operate and support the system?	Increasing operations and support (e.g., resource or environmental) costs Trades of performance versus ease/cost of operations and support Adaptability of the design Changes in maintenance or logistics strategy/needs
Sustainability	Will the system provide sustainable future value?	Availability of future resources and impact on the natural environment

1.6 TRADE-OFF ANALYSIS IN THE SYSTEMS ENGINEERING DECISION MANAGEMENT PROCESS

Successful systems engineering requires sound decision making. Many systems engineering decisions are difficult because they include multiple competing objectives, numerous stakeholders, substantial uncertainty, significant consequences, and high accountability. In these cases, sound decision making requires a formal decision management process. The purpose of the decision management process, as defined by ISO/IEC/IEEE 15288:2015, is " … to provide a structured, analytical framework for objectively identifying, characterizing and evaluating a set of alternatives for a decision at any point in the life cycle and select the most beneficial course of action." The process presented in this book aligns with the structure and principles of the decision management process of ISO/IEC/IEEE 15288, the INCOSE Systems Engineering Handbook v4.0 (INCOSE, 2015), the Systems Engineering Body of Knowledge (SEBoK), and an INCOSE proceedings paper that elaborated this process (Cilli & Parnell, 2014). This process was designed to use best practices and to avoid the trade-off analysis mistakes discussed in the next section.

The INCOSE decision management process, introduced in Figure 1.2, is presented in more detail in Chapter 5. The purpose of the process is to "provide a structured, analytical framework for objectively identifying, characterizing, and evaluating a set of alternatives for a decision at any point in the life cycle and select the most beneficial course of action." The white text within the outer green ring identifies elements of a systems engineering process while the 10 blue arrows represent the 10 steps of the decision management process. Interactions between the systems engineering process and the decision management process are represented by the small, dotted green (outer ring to inner ring) or blue arrows (inner ring to outer ring). (*The reader is referred to the online version of this book for color indication.*)

TRADE-OFF ANALYSIS MISTAKES OF OMISSION AND COMMISSION

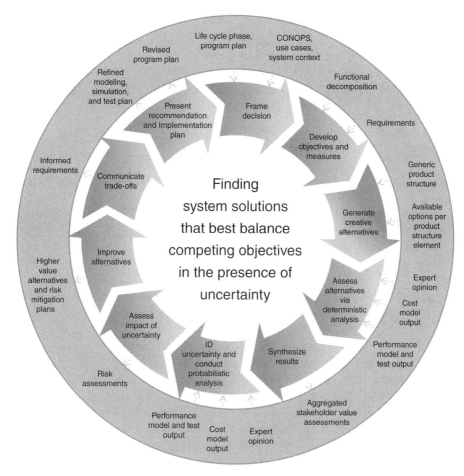

Figure 1.2 INCOSE decision management process

The steps in the decision management process are briefly described in Table 1.3, with references to the primary chapters that provide additional details about each step. Chapter 5 describes and illustrates the INCOSE decision management process.

1.7 TRADE-OFF ANALYSIS MISTAKES OF OMISSION AND COMMISSION

Using the INCOSE decision management process, we identify and discuss the most common trade-off study mistakes of omission and commission (Parnell et al., 2014).

Table 1.3 Decision Management Process

Process Step	Description	Primary Chapters
Frame decision	Describe the decision problem or opportunity that is the focus of the trade-off analysis in a particular system life cycle stage	Chapter 6
Develop objectives and measures	Use mission and stakeholder analysis and the system artifacts in the life cycle stage (e.g., function, requirements) to define the objectives and value measures for each objective alternative needed to satisfy	Chapter 7
Generate creative alternatives	Use a divergent–convergent process to develop creative, feasible alternatives	Chapter 8
Assess alternatives via deterministic analysis	Use a value model to perform deterministic analysis for trade-off analyses	Chapters 9–14
Synthesize results	Provide an assessment of the value of each alternative and the cost versus value to identify the dominated alternatives	Chapters 9–14
Identify uncertainty and conduct probabilistic analysis	Identify the major scenarios and system features that are uncertain and conduct probability analysis	Chapters 9, 12–14
Assess impact of uncertainty	Assess the impact of the uncertainties on value and cost	Chapters 9, 12–14
Improve alternatives	Improve the alternatives by increasing their system value and/or reducing their associated system risk	Chapters 9–14
Communicate trade-offs	Communicate the trade-off analysis results to decision-makers and other stakeholders	Chapters 9–14
Present recommendations and implementation plan	Provide decision recommendations and an implementation plan to describe the next steps to implement the decision	Chapter 5

Mistakes of omission are errors made by not doing the right things, while mistakes of commission are errors made by doing the right things the wrong way. For each step in the decision process, Table 1.4 provides a list of trade-off mistakes, the type of mistake (omission or commission), and the potential impacts.

Table 1.4 Trade-Off Mistakes

Step	Mistakes	Omission/ Commission	Impacts
Overall process	Not having a decision management process	Omission	No trade-off studies or variable trade-off study quality of those conducted Poor decisions; potential selection of a poor design Increased cost and schedule; inadequate performance Loss of SE credibility
Frame decision	Not obtaining access to key DM and SH	Omission	No trade-off studies or trade-off studies on the wrong issues
	Decision frame not defined	Omission	Incorrect selection criteria Loss of trade-off study and SE credibility
Develop objectives and measures	Objectives and/or measures not credible	Commission	Loss of trade-off study and SE credibility Potential selection of a poor design
Generate creative alternatives	Decision space not defined	Omission	Potential selection of poor design
	Doing an advocacy study	Commission	Potential increased cost and schedule Loss of trade-off study and SE credibility
Assess alternatives via deterministic analysis	Using non-normalized value functions	Commission	Potential selection of poor designs
	Not using swing weights	Commission	Loss of trade-off study and SE credibility
	No sensitivity analysis	Omission	
Synthesize results	Lack of a sound mathematical foundation	Omission	Potential selection of poor designs Loss of trade-off study and SE credibility

(continued overleaf)

Table 1.4 (*Continued*)

Step	Mistakes	Omission/ Commission	Impacts
Identify uncertainty and conduct probabilistic analysis	Not identifying uncertainties	Omission	Loss of trade-off study and SE credibility
	Improper assessment of uncertainty	Commission	Potential selection of poor designs
Assess impact of uncertainty	Not integrating with system/program risk assessments	Omission	Potential selection of poor designs Loss of SE credibility
Improve alternatives	Not improving alternatives	Omission	Loss of trade-off study and SE credibility Potential selection of poor designs
Communicate trade-offs	Results not timely or understood	Commission	Recommendations not implemented Loss of SE credibility
Present recommendations and implementation plan	Recommendations not implemented	Commission	Loss of trade-off study and SE credibility
Overall process	Not using trade-off study models on subsequent trade-off studies	Omission	Loss of trade-off study and SE credibility

1.7.1 Mistakes of Omission

There are 10 common mistakes of omission.

1.7.1.1 Not Having a Decision Management Process One of the most fundamental trade-off analysis mistakes is not having a decision management process that provides a foundation for all studies. The decision management process should have the acceptance and participation of the decision-makers and other stakeholders. To achieve stakeholder acceptance, the process should be tailorable to the needs of each specific trade-off analysis. Having a sound decision management process can save time while allowing for organizational learning and development of best practices. The INCOSE decision management process, shown in Figure 1.2, is an example of this kind of process. Without such a process, engineers in an organization are essentially free to use their own, invariably unsound process, and unsound processes can have a long lifetime! Since systems engineers are the ones who frequently perform trade-off analysis for critical system decisions, a natural home for the decision management process is the systems engineering organization.

1.7.1.2   Not Obtaining Access to Key DM, SH, and Subject Matter Experts (SMEs)
Framing any system decision can be a challenge, especially without the right stakeholders involved. Therefore, it is critically important to have access to key decision makers, stakeholders, and SMEs to ensure that the opportunity is adequately defined and the important objectives have been identified. Challenges include gaining access to leaders and senior decision makers despite their busy schedules, including stakeholders who are critical to the system or its impact on them, and assuring access to SMEs in all steps of the trade-off study. To achieve this end, experiential opportunities that allow all stakeholders to readily understand the context and situation without having to understand SE notations are an imperative (Madni, 2016).

1.7.1.3   Decision Frame Not Defined   The first step in the decision management process is to identify and describe the decision opportunity in the context of the problem space. In decision analysis, we call this framing the decision. Experience has taught us that the initial problem is never the final problem (Madni, 2013; Madni et al., 1985). The frame describes how we look at the problem. A good decision frame begins with thorough research and mission/stakeholder analysis (Parnell et al., 2011). A decision hierarchy (Parnell et al., 2013), which lists the past decisions, the current decisions, and the subsequent decisions, can also be useful. A short paragraph, written in clear terms that define the problem, can be quite helpful to decision-makers, other stakeholders, and study participants.

1.7.1.4   Lack of a Sound Mathematical Foundation   To be credible and have defensible results, a trade-off study should be based on a sound mathematical foundation comprising both deterministic and probabilistic analyses. Several operations research and engineering analysis techniques (e.g., optimization, simulation, decision analysis) are potentially appropriate for trade-off studies. If all the objectives can be converted into dollars, then a net present value model would serve as a sound foundation. If not, then the mathematics of multiple objective decision analysis (MODA) offers a sound foundation for trade-off studies. Chapter 2 discusses this further.

1.7.1.5   Undefined Decision Space   Some trade-off studies list alternatives that are not explicitly connected to the decision space. In many studies, alternatives are listed as bullets on a PowerPoint chart. In these cases, there is no explicit understanding of the decision space. The best techniques to help develop good alternatives are those that explicitly define the decision space (see Chapter 8). One best practice technique is called Zwicky's Morphological Box or Alternative Generation Table

(Parnell et al., 2011). In decision analysis, the technique is called the Strategy Table (Parnell et al., 2013), and it seeks to design alternatives that span the decision space. When the decision space is explicitly defined, it becomes possible to explore the decision space, identify more decision options, and come up with a better set of alternatives (Madni, 2012; Madni et al., 1985). The impact of not defining the decision space is the loss of the opportunity to create better alternatives to achieve the desired system value and/or reduce risk.

1.7.1.6 Absence of Sensitivity Analysis Any deterministic trade-off study has to make multiple assumptions about parameters in the model(s). The parameters typically include shapes of the value curves, swing weights, scores on the performance measures, and other variables that are used to calculate the scores. There may be some uncertainty about what numerical value each parameter should have. The best practice is to perform sensitivity analysis to determine if the best alternative changes when the parameter settings are varied across a reasonable range. Based on the sensitivity analysis, additional effort should be devoted to understanding and modeling the most sensitive variables (Madni, 2015).

1.7.1.7 Not Identifying Uncertainty and Performing Probabilistic Analysis Deterministic trade-off studies ignore uncertainties. Since uncertainty and risk are inherent in the life cycle of new systems, this omission is problematic. When decision analysis is used, it is easy to identify key uncertainties in deterministic models using deterministic sensitivity analysis, assess the uncertainties, and perform probabilistic analysis using Monte Carlo simulation, decision trees, influence diagrams, or probability management decisions (Parnell et al., 2011; Parnell et al., 2013). The impact of not modeling uncertainty is that we forgo the opportunity to understand the sources of risk early in the system life cycle when it is invariably easier to avoid, mitigate, or manage risks.

1.7.1.8 Not Improving Alternatives Several trade-off studies assess only proposed alternatives and never consider improving them. With several bad alternatives, even a "correctly performed" trade-off study can do no better than identify a bad alternative! Keeney calls the focus on existing options Alternative-Focused Thinking and advocates using Value-Focused Thinking to define our values, create decision opportunities, use our values to create better alternatives, and improve the proposed alternatives (Keeney, 1992). The decision analysis model provides useful data for Value-Focused Thinking, since it defines the ideal alternative and the gaps between the best alternative and the ideal alternative.

1.7.1.9 Failure to Integrate Trade-Off Study Uncertainty Analysis with System/Program Risk Assessments Uncertainty analysis performed in trade-off studies should be integrated with the system/program risk assessment process.

Unfortunately, many times trade-off studies do a good job of analyzing uncertainty, but the results are not integrated into the system risk management process. On many programs, risk analysis is performed using a simple risk matrix with likelihood on the rows (columns) and consequences on the columns (rows). In this case, the risks being analyzed may or may not be linked to trade-off studies. An alternative approach is to use the trade-off analysis value and cost models to perform risk assessment. This approach may result in better assessment of the likelihood and consequences (the loss in potential value) of the risk. In addition, the results of the risk analysis can be used to identify the need for additional trade-off studies to mitigate or manage risk.

1.7.1.10 Failure to Use Trade-Off Models on Subsequent Studies On some programs, each trade-off study is unique and there is no traceability between the results in one life cycle stage and the subsequent stages. This means the systems engineering organization might have been using very different value trade-offs for the same system without knowing it. A great deal of effort can go into developing trade-off study value models in early life cycle stages. The best practice is to use information from previous trade-off study value models (if available) and improve and tailor the model for subsequent studies. Using improved models can make the analysis results more accurate as well as more credible to decision-makers, stakeholders, and SMEs.

1.7.2 Mistakes of Commission

In addition to the 10 mistakes of omission, there are 6 common mistakes of commission.

1.7.2.1 Performing an Advocacy Study Trade-off studies work best when a creative set of alternatives that span the decision space are developed (Madni, 2013). It is worth noting that the final decision will be only as good as the alternatives that are considered. Some project managers and systems engineers inappropriately convert a trade-off study to a biased advocacy study (Parnell et al., 2013). They advocate the alternative they recommend and use the study to highlight the weaknesses of other alternatives. Advocacy studies put a significant burden on the decision-makers and stakeholders to identify and ask the hard questions to make sure that the other potential alternatives do not provide higher value/lower risk than the advocated alternative. Decision-makers and stakeholders should insist on a clear definition of the opportunity and on a set of creative, feasible alternatives that cover the full range of possibilities to create value, including verified and validated data and selection criteria that are free of bias.

1.7.2.2 Objectives and/or Measures Not Credible Trade-off studies require the development of a complete set of system objectives and measures. To meet the mathematical requirements of MODA, a nonoverlapping set of direct objectives is needed. In systems engineering, a great deal of effort is spent on identifying and analyzing system functions. The list of system functions can provide a good foundation for the development of objectives and value measures by constructing a functional value hierarchy (Buede & Miller, 2009; Parnell et al., 2011). The functional hierarchy has functions at the top level(s), then the objectives for each function, and value measures for each objective.

1.7.2.3 Using Measure Scores Instead of Normalized Value Functions Trade-off studies require the ability to compare performance on one measure with performance on other measures. If we have converted every measure level into a common currency, for example, dollars, we can use dollars as the metric. If decision-makers are unwilling to use dollars, we can use MODA to quantify the value as a function of the capability versus the cost. MODA uses the value functions to enable this trade-off analysis. The value functions (sometimes called scoring functions) convert a value measure score into a normalized measure of value on a common scale. The most common scales are 0–1, 0–10, and 0–100. Value functions assess returns to scale on the range of the value measure score. Value functions usually are of four types: linear, diminishing returns, increasing returns, and S-curve (increasing, then linear, then decreasing returns). The value function will be increasing (for a maximize objective) or decreasing (for a minimize function). The value functions allow us to compare apples and oranges. These functions must at least be on an interval scale (Keeney, 1992). Zero value on an interval scale means the minimum acceptable value and does not mean the lack of value. If a ratio scale is used, zero value would mean no value. The best practice is to obtain the shape of the curve and the rationale for the curve shape before you assess points on the curve. This will provide very useful information when a decision-maker or stakeholder challenges the value judgments of one or more alternatives. See Chapter 2 for additional information.

1.7.2.4 Use of Importance Weights Instead of Swing Weights A critical mistake in trade-off studies is using importance weights instead of swing weights. MODA quantitatively assesses the trade-offs between conflicting objectives by evaluating the alternative's contribution to the value measures (a score converted to value by single-dimensional value functions) and the importance of each value measure across the range of variation of the value measure (the swing weight). Every MODA book identifies this as a major problem. For example, "some experimentation with different ranges will quickly show that it is possible to change the rankings of the alternatives by changing the range that is used for each evaluation measure. This does not seem reasonable. The solution is to use swing weights" (Kirkwood, 1997). Swing weights play a key role in the additive value model presented in Chapter 2. The swing weights depend on the measure scales' importance and range. The word "swing" refers to varying the range of the value measure from its minimum acceptable level to its ideal level. If we hold constant all other measure ranges and reduce the range of one of the measure scales, the measure's relative swing weight decreases, and the swing weight assigned to the others increases since the weights have to add to 1.0. The following story explains the need for swing weights (Parnell et al., 2013).

Using Swing Weights – Greg's Car-Buying Example

Recently, Greg and his wife decided to consider buying a car. Greg wanted to buy an SUV with awesome off-road capability. His wife preferred a minivan on the grounds that it would provide a convenient transport for their children and grandchildren. Once they agreed to buy a minivan, they talked about the choice criteria. The criteria they selected were cost, safety, performance, and comfort. Before they could assign swing weights, they had to define the range of the value measure scores for each criterion. The swing weight that they assigned to each measure depended on the importance (an intuitive assessment) of the "swing in range" of the measure (a factual assessment). Let us now see how Greg and his wife did this the right way.

Let us begin with safety. Suppose they measure safety using a 5-star scale and assign a value of 0 to a safety score of 1 star and a value of 100 to 5 stars. The variation in this measure, from 1 to 5 stars, represents a significant difference in the likelihood of personal injury in an accident (a factual judgment). Given this variation in safety, they would say that a high safety score is very important to them because their family is expected to be in the vehicle frequently (intuitive importance assessment). Therefore, they would assign a high weight to safety, since the measure has high importance given the significant "swing" (1 to 5 stars means bottom 20% to top 20%).

Suppose they think about it some more and decide to eliminate from consideration 1- and 2-star vehicles (the bottom 40% by safety rating). Clearly, their intuitive assessment of the importance of safety has not changed, but the range of the measure has been reduced from 1–5 stars to 3–5 stars. So, if they now assign a value of 0–3 stars and keep a value of 100 for 5 stars, they would then assign less weight to safety than before since they are guaranteed to buy at least a 3-star vehicle. Finally, suppose they think some more and decide to consider only vehicles with 5-star safety ratings. Their importance assessment has not changed, but now there is no variation in safety rating because they have made the 5-star safety rating a screening criterion. Therefore, they would assign a swing weight of 0 to safety since there is no longer any "swing" in safety in the decision. In conclusion, they always assess weights based on the swings in the measure range.

Not using swing weights can have significant consequences on the credibility of the trade-off study.

1.7.2.5 Improper Assessment of Uncertainty Uncertainty assessment requires an understanding of heuristics and cognitive biases that humans exhibit when dealing with uncertain information (Kirkwood, 1997). For example, humans anchor on irrelevant data and do a poor job of assessing the range of uncertainty. Therefore, an assessment should never begin by asking an individual for the mean of the distribution, since this will anchor the individual on the mean. Once anchored on the mean, humans seldom identify the bounds that capture the true range of uncertainty. The best practice is to start with the extremes of the distribution and work toward the middle to avoid anchoring. Another useful best practice is to use an uncertainty assessment form that captures the key information on the assessment (Parnell et al., 2013).

1.7.2.6 Results Not Timely or Not Understood Performing a system trade-off study, developing insights, and communicating key insights to decision-makers and stakeholders is a challenging and important task that is usually performed under significant time pressure. Late studies have no impact. Complex technical charts may be difficult to grasp for individuals who do not use them all the time or who are not wired to think that way (Madni and Sievers, 2016). In addition, some engineers tend to provide detailed, technical information that decision-makers and stakeholders neither need nor want. The analyst should take actions to understand what level of information the decision-makers want to support their decisions and then identify and communicate key insights as clearly and concisely as possible. So perhaps the most challenging task for the analyst is identifying the important insights and determining how to convey them to decision-makers and stakeholders. The ability to identify insights and display quantitative information is a soft skill that those conducting trade-off studies need to develop. One of the best sources of advice for excellence in the presentation of quantitative data is the work of Edward Tufte (Tufte, 1983).

1.7.3 Impacts of the Trade-Off Analysis Mistakes

Earlier, we discussed mistakes of omission and commission as distinct errors. Any one of these errors can have significant consequences in system design and, ultimately, on the program and the system (Madni, 2010, 2011). Unfortunately, it is quite common to find multiple mistakes made in trade-off studies with some errors leading to, or cascading with, other errors. These cascading errors can lead to adverse impacts for the trade-off study team, decision-makers, stakeholders, and the system or program. In addition, repeating these trade-off mistakes can undermine the credibility of the SE organization/enterprise (Madni et al., 2005). In Figure 1.3, we show the dependencies among these errors. A dotted arrow represents correlation. A solid arrow represents high correlation. These dependencies show how the mistakes can cascade to the impacts described as follows.

1.7.3.1 Potential Selection of Poor Designs The ultimate impact of mistakes made during the conduct of a system design trade-off study is the selection of poor designs. This includes missed opportunities to increase value and reduce risk. The selection

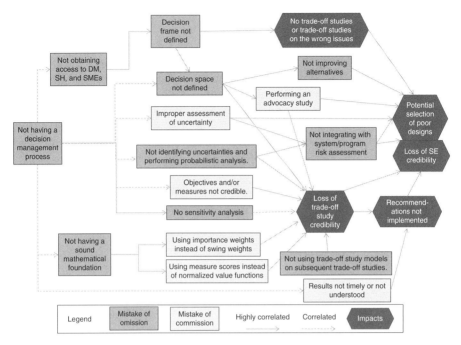

Figure 1.3 Relationships among trade-off study mistakes and impacts

of a poor design may then lead to significant impacts to cost, schedule, and technical performance. The primary sources of this shortcoming are poor decision frame, alternatives that fail to span the decision space, not taking the time to improve alternatives, not integrating trade-off studies with system/program risk assessments, failure to implement recommendations, and results not produced in timely fashion or imperfectly understood. The most obvious, but often neglected, cause of poor designs is poor alternatives. There are two primary causes of poor alternatives. The first is beginning with a limited set of alternatives that does not span the decision space and then not applying techniques to expand the alternative set (Madni et al., 1985; Madni, 2013). The second cause is not systematically trying to improve and expand the alternatives during the trade-off study. The identification and improvement of a good set of alternatives are sometimes impeded by the biases of the people involved. For example, there are often preconceptions of adequate solutions that limit the thought process from looking at the full decision space. The alternative generation table and decision analysis models provide excellent information for Value-Focused Thinking to improve the alternatives (Parnell et al., 2011).

1.7.3.2 Loss of Systems Engineering (SE) Credibility The long-term organizational impact of trade-off study mistakes is the loss of SE credibility in the program, organization, or enterprise, and ultimately with the customer or acquisition organization. Nearly any of the trade-off study errors can result in a loss of SE credibility, since they impact the quality of the analysis and the resulting decision. Once the SE credibility is lost, it takes a significant effort to restore it.

1.7.3.3 No Trade-Off Studies Conducted or Trade-Off Studies Conducted on Wrong Issues The impact of a poor problem definition (decision frame) can lead to no trade-off studies being conducted or trade-off studies being conducted on the wrong issues. Defining the problem(s) is difficult and time-consuming; however, it is critical. A great solution to the wrong problem may not be even a feasible solution to the real problem. If the error is not detected, it can lead to the selection of a poor design. The causes of poor problem definition are usually the lack of a clear problem statement and a poor framing of the decision. The typical cause for a poor decision frame is not having access to (or not listening to) decision-makers and stakeholders. While the decision frame can be improved late in the study, it will likely have schedule and cost impacts.

1.7.3.4 Loss of Trade-off Study Credibility The loss of trade-off study credibility can usually be traced to multiple causes. The first is not having a sound mathematical foundation for the analysis. This typically results in not using normalized value functions and/or swing weights. The second is lack of confidence on the part of decision makers and/or stakeholders in the objectives and measures, which they might perceive as incomplete or not credible. Not showing the linkage of objectives to system functions can contribute to this problem. The third is the lack of good alternatives that span the decision space. A symptom of this problem is an advocacy trade-off analysis that advocates the team's preferred alternative and denigrates the others. The fourth is not identifying uncertainties and performing risk analysis. Trade-off studies can be improved or even reconducted late in the study but most likely not without adverse schedule and cost impacts.

1.7.3.5 Not Implementing Recommendations Not having the trade-off study recommendations implemented can be a disheartening outcome. The reasons may lie with analysts, decision-makers, and/or program managers. The analyst may not have worked the right problem, may not have developed a credible model, may not have sensitivity/uncertainty/risk analysis to provide confidence in the decision, may not have communicated well, may not have provided an implementation plan, or may have delivered the study too late. The decision-maker may not have provided the analyst enough guidance to work on the right problem, may not have bothered to understand the rationale for the recommendation, or may have already made up their own mind about the decision before or during the study. The program manager may have failed to secure the requisite funds to implement the recommendations (Neches and Madni, 2013).

1.8 OVERVIEW OF THE BOOK

Figure 1.4 provides a graphical overview of the organization of this book. The first four chapters provide the foundations for understanding trade-off analyses. This

OVERVIEW OF THE BOOK 21

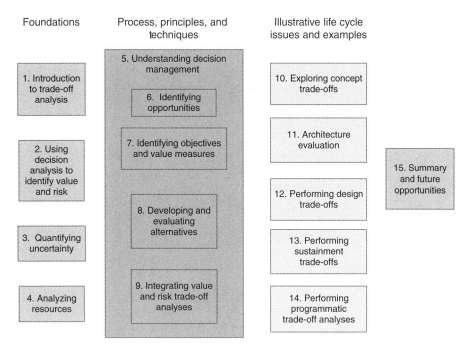

Figure 1.4 Outline of the book

chapter has described the need for trade-off analysis and identified errors of omission and commission to avoid. Chapter 2 provides the mathematical foundations for trade-off analyses; Chapter 3 provides a review of the probability theory, cognitive biases, and probability assessment necessary to understand and model uncertainty; and Chapter 4 provides techniques for conducting resource analyses that are essential to assess system affordability.

Chapter 5 presents the INCOSE decision management process for conducting a trade-off analysis, and provides a detailed illustrative example using some of the techniques illustrated in this book. Chapter 6 presents techniques for defining the opportunity space that frames the trade-off analysis. Chapter 7 describes how to identify and structure objectives and, then, how to develop value measures that quantify the attainment of these objectives. Chapter 8 provides commonly used techniques to generate and evaluate alternatives and compares the ability of these techniques to generate and evaluate alternatives. Chapter 9 provides an example of how we can perform an integrated model of value that can be used to perform value and risk trade-off analysis. The next five chapters discuss the life cycle stage context and provide examples of techniques that can be used to perform trade-off analyses in that stage. Since different types of information are available in the different life cycle stages, different trade-off analysis techniques are more appropriate. Chapter 10 discusses trade-off analysis that is conducted in the conceptual design phase. Chapter 11 discusses architecture trade-off analysis. Chapter 12 discusses design trade-off

Table 1.5 Illustrative Examples

Chapter	Title	Illustrative Examples	Qualitative Techniques	Deterministic Techniques	Probabilistic Techniques
5	Understanding decision management	Unmanned aeronautical vehicle (UAV)	Functional value hierarchy	Multiattribute value and life cycle cost	Monte Carlo simulation (@risk)
6	Identifying opportunities	Drone performing community-based forest monitoring	Decision hierarchy		
		UAV to support dismounted soldiers on patrol	Decision hierarchy		
7	Identifying objectives and value measures	Army squad enhancement problem	Functional value hierarchy		
		Global nuclear detection architecture	Objectives value hierarchy		
9	An integrated model for trade-off analyses	Army squad enhancement problem	Influence diagram	Multiattribute value and life cycle cost	Agent-based simulation Monte Carlo simulation (probability management)
10	Exploring concept trade-offs	Maritime security system	Objectives hierarchy	Optimization	Agent-based simulation Utility

#		Case		
11	Architecture evaluation framework		Objectives hierarchy	
12	Exploring the design space	Liftboat	Design parameter trade-offs	Design of experiments (DOE) (JMP)
		Commercial ship	Design parameter trade-offs	DOE (JMP)
		NATO ship	Multiattribute value	Monte Carlo (Crystal Ball), DOE (JMP)
13	Performing programmatic trade-off analysis	Forest monitoring drone system	Influence diagram	Monte Carlo simulation (@risk)
14	Performing programmatic trade-off analyses	System acceptance model System cancellation	Operating characteristic curve and cost modeling	Statistics, hypothesis testing, decision trees
		Software failure predictive model		Logistics regression model(R)
		HR systems cancellation	NPV, return on investment, and break-even analysis	
		Decommissioning offshore oil and gas platforms in California	Multiattribute value	

analysis. Chapter 13 discusses sustainment trade-off analyses. Chapter 14 discusses programmatic trade-off analyses (acceptance and retirement). Chapter 15 provides a summary and discusses future developments that may impact trade-off analyses.

1.8.1 Illustrative Examples and Techniques Used in the Book

This book uses illustrative examples to demonstrate how trade-off analysis techniques can be used to define the opportunity, identify value, identify uncertainties, evaluate alternatives, and provide insights about the best alternatives. There are many qualitative and quantitative (deterministic and probabilistic) techniques that have been proposed for trade-off analyses. The chapter authors reference many of these techniques in their chapters. However, for their illustrative examples, the authors have selected qualitative techniques that have proven useful and quantitative techniques that use sound mathematics and have been used effectively to provide actionable insights to decision-makers and stakeholders. Table 1.5 lists the illustrative examples we have selected to illustrate how to perform trade-off analyses.

1.9 KEY TERMS

Decision: A choice among alternatives that results in an allocation of resources.

Decision Management Process: "a structured, analytical framework for objectively identifying, characterizing, and evaluating a set of alternatives for a decision at any point in the life cycle and selecting the most beneficial course of action." The decision management process uses the systems analysis process to perform the assessments (ISO/IEC/IEEE 15288, 2015).

Life Cycle Model: "An abstract functional model that represents the conceptualization of the need for the system, its realization, utilization, evolution, and disposal" (ISO/IEC/IEEE 15288, 2015).

Mistake of Omission: Mistakes of omission are errors made by not doing the right things.

Mistake of Commission: Mistakes of commission are errors made by doing the right things the wrong way.

Stakeholder: "An individual or organization having a right, share, claim or interest or in its possession of characteristics that meet their needs or expectations" (ISO/IEC/IEEE 15288, 2015).

System: "Systems are manmade, created and utilized to provide products or services in defined environments for the benefits of users and other stakeholders" (ISO/IEC/IEEE 15288, 2015).

Risk: "The effect of uncertainty on objectives" (ISO/IEC/IEEE 15288, 2015).

Uncertainty: Imperfect knowledge of the outcome of some future variable.

Trade-off: "Decision making actions that select from various requirements and alternative solutions on the basis of net benefit to the stakeholders" (ISO/IEC/IEEE 15288, 2015).

Trade-off Study: An engineering term for an analysis that provides insights to support system decision-making in a decision management process.

Value: The benefits provided by a product or service to the stakeholders (customers, consumers, operators, etc.).

Value Identification: The determination of the potential value of a new capability.

Value Realization: The delivery of the value of a new capability.

1.10 EXERCISES

1.1. Trade-off analysis
   (a) Provide a definition.
   (b) Why are trade-offs needed?
   (c) Who should participate in a trade-off analysis?
   (d) Why is trade-off analysis different in different life cycle stages?
   (e) Should trade-off analysis be performed to support life cycle gate decisions?
   (f) Who should make the trade-off decisions?

1.2. Decision management
   (a) What is a decision management process?
   (b) Why does an organization need a decision management process?
   (c) What are the benefits of having a decision management process?
   (d) What is the impact of not having a decision management process?

1.3. Identify a system that is currently operational and answer the following questions:
   (a) Identify the system stakeholders.
   (b) Define value for the system.
   (c) Describe the products and services of the system that provide value.
   (d) Identify system risks.
   (e) What is the anticipated life of the system?
   (f) How difficult will it be to retire the system?

1.4. Identify a system early in its life cycle and answer the following questions:
   (a) Identify the system stakeholders.
   (b) Define value for the system.
   (c) Describe the products and services of the system that provide value.
   (d) Identify potential system risks.
   (e) What is the anticipated life of the system?
   (f) List one major decision in each future life cycle stage that could require a trade-off analysis.

1.5. Identify a system in development and answer the following questions:
  (a) Describe the difference between value identification and value realization.
  (b) What products and services will provide value?
  (c) How are risks identified?
  (d) What risks does the system have?
  (e) What is the anticipated life of the system?
  (f) List one major decision in each future life cycle stage that could require a trade-off analysis.

1.6. Trade-off analysis mistakes
  (a) Why are trade-off analysis mistakes made?
  (b) Describe the difference between a mistake of omission and a mistake of commission.
  (c) What is the reason that trade-off analysis mistakes cascade?

1.7. Review a trade-off analysis paper in the proceedings of an engineering conference.
  (a) How was value defined?
  (b) What life cycle stage was considered?
  (c) Were uncertainty and risk considered?
  (d) Identify any mistakes of commission or omission in the paper.
  (e) What trade-off analysis insights were provided?
  (f) Was the trade-off analysis convincing?
  (g) What decision was recommended or made?

1.8. Review a published trade-off analysis paper in a refereed engineering journal.
  (a) How was value defined?
  (b) What life cycle stage was considered?
  (c) Were uncertainty and risk considered?
  (d) Identify any mistakes of commission or omission in the paper.
  (e) What trade-off analysis insights were provided?
  (f) What decision was recommended or made?
  (g) Was the trade-off analysis convincing?

REFERENCES

Buede, D.M. and Miller, W.D. (2009) *The Engineering Design of Systems: Models and Methods*, 2ndVol. Wiley Series in Systems Engineering edn, Wiley & Sons, Hoboken.

Cilli, M. and Parnell, G. (2014). Systems Engineering Tradeoff Study Process Framework. *International Council on Systems Engineering (INCOSE) International Symposium*. June 30–July 3. Las Vegas, NV.

REFERENCES

Decision-making in an uncertain world: A Chevron Case Study. (2014). From Business Case Studies: http://businesscasestudies.co.uk/chevron/decision-making-in-an-uncertain-world/#axzz3NDPqSBNF (accessed 28 December 2014).

Hazelrigg, G.A. (1998) A framework for decision-based engineering design. *Journal of Mechanical Design*, **120** (4), 653–658.

INCOSE (2015) *INCOSE systems engineering handbook v. 4.0*. INCOSE, SE Handbook Working Group, INCOSE.

INCOSE Home Page. (2015). http://www.incose.org/: http://www.incose.org/ (accessed 06 June 2015).

ISO/IEC/IEEE (2015) *Systems and Software Engineering - System Life Cycle Processes*, International Organization for Standardization (ISO)/International Electrotechnical Commission (IEC)/Institute of Electrical and Electronics Engineers (IEEE), Geneva, Switzerland.

Keeney, R.L. (1992) *Value-Focused Thinking: A Path to Creative Decision Making*, Harvard University Press, Cambridge, MA.

Keeney, R. and Raiffa, H. (1976) *Decision with Multiple Objectives: Preference and Value Tradeoffs*, Wiley & Sons, New York.

Kirkwood, C. (1997) *Strategic Decision Making with Multiobjective Decision Analysis with Spreadsheets*, Duxbury Press, Belmont, CA.

Lavingia, N. J. (2014). Business Success through Excellence in Project Management. Critical Facilities Roundtable: http://www.cfroundtable.org/ldc/040706/excellence.pdf (accessed 28 December 2014).

Madni, A.M. (2010) Integrating humans with software and systems: technical challenges and a research agenda. *Systems Engineering*, **13** (3), 232–245.

Madni, A.M. "Integrating humans with and within software and systems: challenges and opportunities," (Invited Paper) *CrossTalk, The Journal of Defense Software Engineering*, May/June 2011, "People Solutions."

Madni, A. (2012) Adaptable platform-based engineering: key enablers and outlook for the future. *Systems Engineering*, **15** (1), 95–107.

Madni, A.M. (2013) Generating novel options during systems architecting: psychological principles, systems thinking, and computer-based aiding. *Systems Engineering*, **16** (4), 1–9.

Madni, A.M. (2015) Expanding stakeholder participation in upfront system engineering through storytelling in virtual worlds. *Systems Engineering*, **18** (1), 16–27.

Madni, A.M., Brenner, M.A., Costea, I., MacGregor, D., and Meshkinpour, F. "Option Generation: Problems, Principles, and Computer-Based Aiding," *Proceedings of 1985 IEEE International Conference on Systems, Man, and Cybernetics*, Tucson, Arizona, November, 1985, pp 757–760.

Madni, A.M., Sage, A., and Madni, C.C. "Infusion of Cognitive Engineering into Systems Engineering Processes and Practices," *Proceedings of the 2005 IEEE International Conference on Systems, Man, and Cybernetics*, October 10–12, 2005, Hawaii.

Madni, A.M. and Sievers, M. (2014a) Systems integration: key perspectives, experiences, and challenges. *Systems Engineering*, **17.1**, 37–51.

Madni, A.M. and Sievers, M. (2014b) System of systems integration: key considerations and challenges. *Systems Engineering*, **17.3**, 330–347.

Madni, A. M. and Sievers, M. N. (2015). Model based systems engineering: motivation, current status and needed advances. *Systems Engineering*, Accepted for publication, 2016.

Madni, A.M. and Sievers, M. (2016) Model based systems engineering: motivation, current status and needed advances. *Systems Engineering*, Accepted for publication, 2016.

Neches, R. and Madni, A.M. (2013) Towards affordably adaptable and effective systems. *Systems Engineering*, **16** (2), 224–234.

Ordoukhanian, E. and Madni, A. M. (2015) System Trade-offs in Multi-UAV Networks. *AIAA SPACE 2015 Conference and Exposition*.

Parnell, G. (2009). Evaluation of risks in complex problems. in *Making Essential Choices with Scant Information: Front-end Decision-Making in Major Projects* (eds T. Williams, K. Sunnevåg, and K. Samset), Palgrave MacMillan, Basingstoke, UK, pp. 230–256.

Parnell, G.S., Bresnick, T.A., Tani, S.N., and Johnson, E.R. (2013) *Handbook of Decision Analysis*, Wiley & Sons.

Parnell, G., Cilli, M., and Buede, D. (2014). Tradeoff Study Cascading Mistakes of Omission and Commission. *International Symposium*. June 30–July 3. Las Vegas, NV: INCOSE.

Parnell, G.S., Driscoll, P.J., and Henderson, D.L. (eds) (2011) *Decision Making for Systems Engineering and Management*, 2nd edn, Wiley & Sons.

SEBoK. (2015). Systems Engineering Body of Knowledge (SEBoK) wiki page. SEBoK: http://www.sebokwiki.org (accessed 06 June 2016).

Tufte, E.R. (1983) *The Visual Display of Quantitative Information*, Graphics Press, Cheshire, CT.

2

A CONCEPTUAL FRAMEWORK AND MATHEMATICAL FOUNDATION FOR TRADE-OFF ANALYSIS

GREGORY S. PARNELL
Department of Industrial Engineering, University of Arkansas, Fayetteville, AR, USA

AZAD M. MADNI
Systems Architecting and Engineering and Astronautical Engineering, Viterbi School of Engineering, University of Southern California, Los Angeles, CA, USA

ROBERT F. BORDLEY
Systems Engineering and Design, University of Michigan, Ann Arbor, MI, USA; Booz Allen Hamilton, Troy, MI, USA

> Truly successful decision making relies on a balance between deliberate and instinctive thinking.
>
> (Malcolm Gladwell)

2.1 INTRODUCTION

Systems are developed to create value for stakeholders that include customers, end users/operators, system developers/integrators, and investors. Decisions are ubiquitous in the system life cycle beginning from systems definition to the systems concept development to design to delivery of products and services to retirement of

---

Trade-off Analytics: Creating and Exploring the System Tradespace, First Edition. Edited by Gregory S. Parnell.
© 2017 John Wiley & Sons, Inc. Published 2017 by John Wiley & Sons, Inc.
Companion website: www.wiley.com/go/Parnell/Trade-off_Analytics

the system (see Chapter 1). System decision-makers (DMs) are those individuals who need to make important decisions pertaining to system definition, concept, architecture, design, test, implementation, operation, maintenance, improvement, and disposal.

Enterprise decision-makers, program managers, and systems engineers stand to benefit from a collaborative decision-making process that engages all stakeholders (SHs) who have a say in system trade-off analysis (e.g., customers, operators/users, system architects and engineers, subject matter experts (SMEs)). In particular, systems engineers can exploit trade-off studies to identify the decision opportunity, identify sources of value, create alternatives, evaluate alternatives, identify risks, and provide insights and recommendations to system DMs and other SHs.

The focus of this chapter is on the use of decision analysis for systems engineering trade-off analysis techniques for both systems and system of systems. We propose that trade-off studies be based on a formal lexicon, a sound mathematical foundation, supported by credible data from DMs, SHs, and SMEs. We provide such a lexicon and a formal foundation based on decision analysis for effective and efficient trade-off studies. Our approach supplements decision analysis, a central part of decision-based design (Hazelrigg, 1998), with Value-Focused Thinking (Keeney, 1992) within a model-based engineering rubric (Madni & Sievers, 2016).

The proposed framework also accommodates trade-offs that vary with context (Madni & Freedy, 1981) (Madni et al., 1982). At the outset, we also note that System of Systems (SoS) trade-offs are complicated by the fact that different systems have different governances and systems can often enter and exit the SoS constellation based on mission requirements (Madni & Sievers, 2013). This means that context is continually changing and, therefore, trade-offs have to be revisited when the context changes (Madni & Freedy, 1981). These changes in context have a fundamental impact on system concept and architectural trade-offs.

2.2 TRADE-OFF ANALYSIS TERMS

In this section, we provide a brief definition and some examples of the key terms that we will use in this book. Table 2.1 provides a list of the key terms and examples from oil and gas, aerospace, Air Force, and Army domains. A decision opportunity is a potential activity that we consider expending resources to achieve organizational objectives. A concept focuses on the system need, purpose, functions, and operations prior to defining the architecture and the design. An architecture comprises the fundamental concepts or properties of a system in its environment embodied in its elements, relationships, and in the principles of its design and evolution. A system design defines the architecture, system elements, interfaces, and other characteristics of the system (ISO/IEC/IEEE, 2015). The objectives identify the goals of the stakeholders for the system. The value measures identify how the organization can measure the potential achievement of the objectives at the time of the decision to pursue the decision opportunity.

Table 2.1 Key Terms with Examples

Terms	Oil and Gas	Aerospace	Air Force	Army
Decision opportunity	Develop energy resources	Design and manufacture of a commercial aircraft	Provide air superiority to achieve national security objectives	Provide intelligence support to ground forces
Concept	Deepwater project in the Gulf of Mexico	Long-range, mid-size jetliner	Manned stealth fighter	Portable unmanned aircraft system
Architecture	Topside and semisubmersible technology	Wide-body, twin-engine jet airliner	Supersonic speed, agility, and integrated sensors	Hand-launched and controlled unmanned aircraft system
Design	Jack/St. Malo Project Design	The Boeing 787 Dreamliner	Lockheed F-35	RQ-11A/B Raven
Objectives (illustrative)	Maximize shareholder value, maximize safety, and minimize impact on environment	Maximize shareholder value, increase fuel efficiency, and increase humidity	Maximize air superiority, improve affordability, replace several aging airframes	Maximize loiter time, maximize range, and maximize probability of threat detection
Value measures (illustrative)	Profit, reliability, number of hazards, and probability of spill	Profit, fuel efficiency, and humidity	Radar cross section, probability of kill against specific future threat, and life cycle cost	Loiter time, range, and probability of threat detection

2.3 INFLUENCE DIAGRAM OF THE TRADESPACE

Defining the decision opportunity is the most important step in any trade-off analysis and determines the boundary for the analysis. The systems engineering decision process, similarly to every decision process, starts with an understanding of the problem or opportunity.[1] This is referred to as the following:

(a) Framing and Tailoring the Decision in the Decision Management Process in the Systems Engineering Body of Knowledge (Systems Engineering Body of Knowledge)

---

[1] In this paper, we use the word opportunity since it is broader than problem. Some system designs begin with a problem, and some begin with an opportunity. Also, sometimes we are able to convert a problem to an opportunity.

(b) Problem Definition in the Systems Decision Process (Parnell et al., 2011)
(c) Decision Framing (Parnell et al., 2013) in the decision analysis literature.

Chapter 6 describes the best practices for defining the decision opportunity.

A common term in trade-off analysis is the tradespace. The tradespace is a multidimensional space that defines the context for the decision, defines bounds to the region of interest, and enables Pareto optimal solutions for complex, multiple stakeholder decisions. The tradespace can be described in many dimensions including the requirements, decisions, system performance measures, value, or utility.

Influence diagrams, a decision analysis technique (Buede, 1994; Buede, 2000), have been used for trade-off analysis. These diagrams are used to show the important decisions, uncertainties, and values (Parnell et al., 2013). We use an influence diagram to provide a conceptual framework and an integrated model for trade-off analysis. Figure 2.1 provides an influence diagram that identifies the tradespace concepts discussed in this chapter along with their primary relationships. This influence diagram is used in Chapter 9 to provide an illustrative example of a trade-off analysis. In specific studies, some of these nodes may not be used and others may be added. In early stages of the life cycle, more nodes will be decisions and uncertainties. In later stages, some of the nodes will be known and considered to be constants.

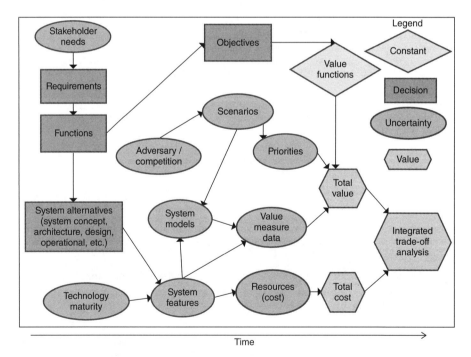

Figure 2.1 Overview of integrated trade-off value, cost, and risk analysis

In the next few sections, we briefly describe each node in the influence diagram and present the use of decision analysis as a mathematical foundation for trade-off analysis. First, we discuss the stakeholder needs, functions, and requirements. Second, we discuss the stakeholder objectives for the system in the decision opportunity. Third, we describe the decision alternatives to meet the objectives. We emphasize the concepts, architectures, designs, and other system alternatives. Fourth, we discuss the major uncertainties inherent in the scenarios, including adversary/competitor behavior. Fifth, we discuss the evaluation of alternatives using value measures, value functions, and utility functions (not shown in Figure 2.1). Sixth, we discuss the important resource issues for trade-off analysis. Seventh, we discuss our proposed framework for integrated trade-off analyses, that is, trade-off analyses that can be used to identify value, cost, and risks as opposed to separate and disconnected analyses.

2.3.1 Stakeholder Needs, System Functions, and Requirements

For systems decision-making, the mission analysis includes stakeholder needs analysis, functional analysis, and requirements analysis to identify the stakeholder needs, functions that the system must perform to create value, the requirements that must be met, and the interfaces with other systems. The results of this analysis available in the life cycle stage are important inputs to any trade-off analysis. Of course, in trade-off analysis, the affordability of any need, function, or requirement may be part of the trade-off analysis.

2.3.2 Objectives

As noted earlier, the fundamental purpose of a system is to create value for its stakeholders (Madni et al., 1985; Madni, 2015). One of the key principles of Value-Focused Thinking is to begin with objectives (Madni, 2013) before developing alternatives (e.g., system concepts, architectures, system designs) (Keeney, 1992). Beginning with alternatives runs the risk of missing actual, important objectives from key stakeholders (Madni et al., 1985). It is important to define the value in terms that are understandable by all DMs, SHs, and SMEs. Decision analysts typically use a value hierarchy (also called a value tree or an objectives hierarchy) to structure the objectives and to identify value measures to assess how well alternatives might achieve our objectives (Keeney, 1992). The value hierarchy has at least three levels: the decision purpose; the objectives that define value; and the value measures for each objective to assess potential value.

We use the term objectives to include goals, criteria, and other similar terms used in the literature. An objective is specified by a verb followed by an object (noun). Common objectives include "maximize profit," "maximize performance," "minimize time," and "minimize cost."

The identification of objectives is challenging for a new system that is expected to be in service for many years and survive the actions of competition/adversaries.

Again, the systems engineering (Parnell et al., 2011; Madni, 2015) and decision analysis literature (Keeney, 1992; Parnell et al., 2013) offer useful techniques for engaging DMs, SHs, and SMEs using research, interviews, focus groups, and surveys to identify potential objectives and then structure them in value hierarchies and functional value hierarchies. These are described in Chapter 7.

It is important to remember that the objectives are developed based on the decision opportunity, NOT the decision alternatives. The objectives need to be independent of decision alternatives. This allows evaluation of very different alternatives using the same model. Also, the objectives should be fundamental objectives (i.e., what we care about) and not means objectives (i.e., how we achieve a fundamental objective). A fundamental objective for a small Unmanned Air Vehicle might be to "maximize detection range." Detection range is important because a soldier would want advance warning of threats. A means objective would be to "maximize use of composite materials in the airframe." The use of composite materials is a means to reduce weight, which in turn can increase the range.

For systems engineering applications, we have found the functional value hierarchy to be a useful technique for systems of systems and complex systems that perform multiple functions (Parnell et al., 2011). The functional value hierarchy has four levels: the system purpose; the necessary and sufficient functions and their interactions required to meet the system purpose; the objectives for each function that define value; and the value measures for each objective to assess the potential value.

2.3.3 System Alternatives

Developing alternatives that span the decision space (sometimes called tradespace generation) is one of the most important, and sometimes overlooked, tasks in systems engineering trade-off analyses. The decision tradespace is much more than the baseline or the baseline and two ad hoc alternatives. The decision space needs to be as large as possible to offer the most potential to create value. Since extraneous system constraints and requirements reduce the decision tradespace, it is important to identify and remove them to the extent possible.

Systems engineering and decision analysis provide useful techniques for defining a large decision tradespace and identifying creative, feasible initial alternatives for analysis that span the tradespace (Madni et al., 1985; Madni, 2013). One important technique has been called Zwicky's morphological box, the alternative generation table (Parnell et al., 2011), and, in decision analysis, the strategy table (Parnell et al., 2013). Typically, the choices within each column range from the very conservative (mild) to the very innovative (wild). A decision alternative is defined by making different choices from each of the columns of the table. Certain combinations of choices are infeasible. In other cases, making a choice in one column necessitates the making of a certain choice in another column. It is critical to remember that these are only the initial alternatives. An important purpose of the trade-off analysis is to identify higher value, more affordable, and lower risk alternatives after we learn from the development of our model and our analysis. In that sense, the selection and evaluation of the initial alternatives is primarily an information-gathering exercise intended to discover which aspects of the different alternatives create the most value, which features drive the cost, and which features introduce the most risk. This makes it important to select alternatives that allow the project team to explore as much of the tradespace as possible. Chapter 8 provides a review of the techniques for alternative generation.

In this chapter, we define decision alternatives that depend on the system life cycle. For example, early in the life cycle, we need to decide which concept to pursue. Concept tradespace includes all the possible system concepts that could reasonably be considered to perform the system functions in the time horizon of the study. The next step is to consider alternative architectures. Later in the life cycle, the architecture is selected and alternative design decisions considered. We may select an existing system, modify an existing system, or design a new system. However, in the early stages of the life cycle, the majority of important decisions are made that determine a large percentage of the system capabilities and cost. In general, we can use the variable y to identify a decision alternative at any stage in the life cycle. Chapters 10–14 provide the unique features and illustrative example of trade-off analysis by life cycle stage.

It is important to note that a trade-off analysis might include multiple spaces. For example, the trade-off analysis might include exploring multiple architectures for each concept. See Chapter 8 for an illustrative example.

2.3.3.1 Concept Tradespace
The system concept significantly impacts the potential system value, schedule, risks, and life cycle costs. For example, a small unmanned aircraft system can be developed quicker and at a significantly lower cost than a manned fighter aircraft. While it will have less risk, it will also have fewer capabilities. Consequently, it is imperative to explore system concepts in early trade-off analyses.

We use the index $c = 1, \ldots, C$ to indicate candidate system concepts. After exploring various system concepts, one system concept, c, is selected.

y_c is the cth alternative of system concept c.

2.3.3.2 Architecture Tradespace
Architecture is the fundamental organization of a system embodied in its components, their interdependencies and relationships to each other, and to the environment, and the principles that guide its design and evolution (ISO/IEC 15288, 2015). The best practice is to consider alternative architectures.

We use the index $a = 1, \ldots, A$ to indicate candidate system architectures. Once architecture evaluation is completed, a particular architecture, a, for system concept c is selected.

y_{ca} is the ath alternative architecture of system concept c.

2.3.3.3 Design Tradespace
We use the index $d = 1, \ldots, D$ to indicate candidate system designs. Once the system design evaluation is completed, there is one system design for architecture a of system concept c. During the design phase, the selection

of system concept c and architecture a significantly reduces the size of the decision tradespace.

y_{cad} is the dth design for architecture a of system concept c.

2.3.3.4 Other Decision Alternatives in Later Life Cycle Stages While the system concept, system architecture, and system design trade-off analyses are arguably the most critical, systems engineers conduct trade-off analyses throughout the system life cycle. Examples might include the following: testing, manufacturing, deployment, operations, maintenance, and system retirement alternatives. The trade-off studies assume concept c, architecture a, and design d. In general, the variable y can be used to identify a decision alternative at any stage in the life cycle. We remove the indices c, a, and d and use the following instead:

$y_j, j = 1, \ldots, m$ is the alternative index, where m is the number of alternatives.

2.3.4 Uncertainty

Surprisingly, many trade-off studies do not consider uncertainty or consider uncertainty in only one objective, for example, cost. But considering uncertainty is especially important when systems development includes the consideration of new technologies whose cost, schedule, and value are poorly understood. In addition, system developers will typically be uncertain of all the circumstances under which users might employ the system.

There are many sources of uncertainty in systems design and operation that should be considered in trade-off analyses. The uncertainties include independent uncertainties (e.g., what scenario will the system be used in, how well will a technology work, what will be the cost of a new component) and dependent uncertainties (e.g., system value, cost, and schedule) that depend on the independent uncertainties. In general, there is greater independent and dependent variable uncertainty in the earlier stages in the system life cycle. As the life cycle progresses, the uncertainties are reduced (Keisler & Bordley, 2014). Some of the major uncertainties include the following:

(a) Technology maturity. One of the major challenges for system development is technology maturity. These uncertainties can directly impact value, cost, and schedule.
(b) Requirements. In the early life cycle stages, the requirements are preliminary or not yet defined. Requirements can have a major impact on design features, cost, and schedule.
(c) Design features. Many system concepts and architectures have design feature requirements, for example, two engines, and goals, for example, UAV weight. Depending on the requirements imposed on the system and design decisions, there is uncertainty about the final design features.
(d) Adversaries. In security applications, we should consider the potential actions of our adversaries over time.

(e) Competition. In commercial applications, we should consider the potential actions of our competition.
(f) Scenarios. Scenarios define when, where, how, and how long a system will be used and the capabilities of the adversary/competition. When and where the system is used determines the physical environment of system operation. Many systems are designed for one or more scenarios. However, since system lifetimes are very long, systems are modified to be used for other scenarios and changing adversary/competition capabilities.
(g) System performance. The system performance is uncertain. It depends on the technology maturity, design features, adversary/competition, and scenarios.
(h) Value. Value uncertainty is a function of the problem uncertainties and their interactions. Models can be used to propagate the uncertainty in the independent variables to the value model.
(i) Schedule. It is well known that system development also offers significant schedule uncertainty.
(j) Cost. It is also well known that system development offers significant cost uncertainty. The best practice for cost analysis is to develop a life cycle cost analysis and to perform Monte Carlo simulation by assigning probability distributions to the independent variables.

2.3.5 Preferences and Evaluation of Alternatives

Once the tradespace has been generated, we need to explore the tradespace. Using decision analysis, we use our preferences to evaluate alternatives and guide the exploration. In performing an evaluation of alternatives, the preferences of the DMs and SHs are used implicitly or explicitly. The mere identification of value measures is a statement of preferences. For example, if miles per gallon is used as a value measure of a vehicle, maximize fuel efficiency (a natural measure) could be the objective, or a proxy measure for minimizing the impact on the environment could be used. Many times value measures are developed in an ad hoc fashion. Instead, the value measures should be derived from the objective identification and structuring described in Section 2.2.2. If SHs identify a value measure not linked to the objectives, it may well be that there is a missing objective.

A key aspect of preferences and evaluation of alternatives is mathematical scales. There are three types of mathematical scales that can be employed in alternative evaluation. The first is an ordinal scale. In an ordinal scale, we define a preference ordering without defining the strength of preference. Consider, for example, a five-point ordinal scale, 1, 2, 3, 4, and 5 in increasing preference. All that can be said is that $5 > 4 > 3 > 2 > 1$. Actually, the use of numbers in an ordinal scale is deceptive and should be avoided. It is preferable to use a, b, c, d, and e. Thus, the ordinal preferences could be $e > d > c > b > a$. The second is an interval scale. An interval scale provides the strength of preference but does not define an absolute zero. Common examples of an interval scale are the temperature scales of Celsius and Fahrenheit. The third scale is a ratio scale. A ratio scale provides both strength of preference and an absolute zero. The ratio scale for temperature is Kelvin. The type of scale defines the mathematical operations that can be performed. With an interval scale, a positive affine transformation can be used (e.g., we can transform an interval scale from 2–12 to 0–100 by subtracting 2 and multiplying by 10). With a ratio scale, only multiplication by any positive transformation is possible.

Since value measures require an interval or ratio scale (see next section), it is useful to note that an ordinal ranking can be converted to an interval or a ratio scale by performing a value assessment or other techniques. For example, Barron & Barrett (1996) suggest using the rank-ordered centroid technique to construct a ratio scale using only the information in an ordinal scale. With n alternatives, rank-order centroid assigns the following:

1. The lowest ranked (or nth highest ranked) alternative, a score proportional to $(1/n)$
2. The $(n-1)$ highest ranked alternative, a score proportional to $(1/n) + 1/(n-1)$
3. ...
4. The top-ranked alternative, a score proportional to $(1/n) + 1/(n-1) + \ldots + 1$.

Scores are then normalized to sum to 1. While there are other methods for estimating ratio scales from ordinal scales, empirical evidence suggests that this method is more consistent with the ratio scores that individuals would actually assign to alternatives (if they were asked).

2.3.5.1 Value Measures There are several terms used in the literature as synonyms for measures. Some common terms include value measures, performance measures, attributes, figures of merit, effectiveness measures, and metrics. This chapter uses value measures to include all measures used to assess the objectives of interest to the DMs and SHs. Value measures assess how well an alternative achieves an objective that we care about. If all potential outcomes can be converted into dollars, we can use dollars as our measure of value. If not, we can use multiple value measures. Example value measures are profit, range, probability of detection, reliability, safety, sustainability, maintainability, cost, schedule, and safety. An alternative's scores on a value measure for an objective may be deterministic or uncertain.

We use the following notation for value measures:

i = index of system value measure, $1, \ldots, n$
x_{ji} = alternative j's score in the ith value measure
$\mathbf{x}_j$ = vector of the scores of each alternative j on all n value measures.

Systems engineering trade-off studies can be conducted using value measures. However, these tend to be two or three value measures at a time. To make decisions based on value measures requires the decision-makers to make judgments about preferences for a level of one value measure compared to the levels of the other value measures. This assumes that the DMs and SHs understand the nuances of the scales being used and the ranges of each value measure. For example, comparing an ordinal scale with an interval and ratio scale would be cognitively challenging. In addition, changing the range of a value measure can make significant visual differences in the trade-off charts.

To increase the number of value measures being considered and to make the preference judgments more explicit, decision analyst and systems engineers can employ the value model for trade-off analyses.

2.3.5.2 Value Models Value can be single-dimensional (e.g., dollars) or multidimensional using many value measures derived from the objectives. For single-dimensional values such as dollars, it is possible to calculate the net present value using a discount rate to normalize dollars to a standard year.

The majority of the trade-off analysis literature use multiple objectives – sometimes called multiattribute value or multiattribute utility. Multiobjective decision analysis (MODA) defines value as the normalized returns to scale on a value measure (Kirkwood, 1997).

For multidimensional value, two kinds of value dimensions need to be distinguished:

1. Screening dimensions: Screening is used to eliminate alternatives. Screening reduces the decision tradespace and needs to be used very carefully. The screening criteria (requirements) can be applied simultaneously or sequentially. If the screening criteria are applied simultaneously, the alternatives that do not meet the criteria are eliminated. One sequential screening approach would be to rank screening criteria from highest to lowest importance. In deciding between alternatives, only those alternatives that are good enough under the most important screening criteria need to be considered. For example, when buying a car, the sequential screening criteria are off-road vehicles and 25 miles per gallon and in that order. Thus, the steps are to first find the off-road vehicles and then find the one that had greater than or equal to 25 mpg. (If alternatives provide roughly the same value of the most important value measure, then we make the choice based on the next most important screening dimension.)
2. "Tradable" dimensions: These are dimensions that act in trade-off fashion. Once all screening dimensions are satisfied, attention shifts to other dimensions to balance improvement on one against improvements on the others. For example, a little performance may be sacrificed to achieve increased affordability, and a little affordability can be sacrificed to achieve an increase in performance. If this is the case, then affordability and performance are tradable dimensions. It is important to realize that the tradable dimensions can also have screening criteria. For example, there may be a minimum level of performance that is required.

For tradable dimensions, the most common multidimensional value model is the additive value model (Keeney & Raiffa, 1976). Many other value model forms exist (Kirkwood, 1997), but they are most complex and require more time for DMs and

SHs to assess preference and are more difficult to interpret. The additive value model requires mutual preferential independence (Kirkwood, 1997). In simple terms, mutual preferential independence means that our preferences for one value measure do not depend on the scores on other value measures. From important theoretical findings, we know that only $n-1$ questions need to be asked to ensure that this requirement is met. In practice, this requirement is met by constructing the value measures to be collectively exhaustive and mutually exclusive. For example, both reliability and availability would not be used because reliability is included in the calculation of availability. In addition, preferential dependence can be avoided by combining two preferential dependent value measures into one new value measure. (It is important to note that preferential independence has nothing to do with probabilistic independence, which is discussed in Chapter 3.)

The additive value model uses Equations 2.1 and 2.2. A single-dimensional value function, $v_i(x_i)$, can be developed for each value measure. Each single-dimensional value function can be normalized and then weighted by each contribution, w_i. The total value is the weighted sum of the single-dimensional value.

$$v(\mathbf{x}_j) = \sum_{i=1}^{n} w_i v_i(x_{ij}) \qquad (2.1)$$

$$\sum_{i=1}^{n} w_i = 1 \qquad (2.2)$$

where

i = index of the value measures, 1, ..., n
j = index of the alternatives, 1, ..., m
$v(\mathbf{x}_j)$ = the multidimensional value of value measure i
x_{ji} = alternative's score in the ith value measure
$v_i(x_{ji})$ = normalized single-dimensional value of the score of x_{ji}
w_i = normalized swing weight of ith value measure

To obtain a precise mathematical meaning for $v(x_j)$, appropriate scales must be used for the single-dimensional value functions and the swing weights. The additive value model places requirements on the two scales. Each single-dimensional value function must use an interval or a ratio scale. In practice, interval scales are usually used for the practical reason that if a ratio scale did exist, it would usually result in a much larger scale than required for the trade-off study. This would result in unused value tradespace. It is important to note that swing weights must use a ratio scale, and zero weight means no value.

It is also important to note that neither the value functions v_i nor the weights w_i depend on the alternatives. Therefore, a value model is useful when comparing significantly different system concepts, architectures, designs, and alternatives in any life cycle stage.

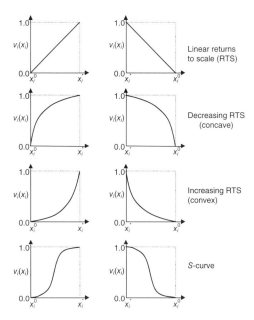

Figure 2.2 Single-dimensional value functions

The six most common single-dimensional value functions are shown in Figure 2.2. The four curve shapes are linear returns to scale, decreasing returns to scale (concave), increasing returns to scale (convex), and the S-curve. In all of the functions, x_i^0 is the minimum acceptable score on the value measure and x_i^* is the ideal. The minimum and maximum scores should be determined based on the decision opportunity and the time horizon of the system development.

Since interval scales are used for the value functions and ratio scales for the swing weights, a precise mathematical measure can be defined and associated with a multidimensional alternative value. Value has a very clear meaning relative to trade-off analyses. If the value range is 0–10, a hypothetical alternative with the minimum acceptable score on each measure has a value of 0 and a hypothetical alternative with the ideal score on each measure has a value of 10. The additive value model defines the n-dimensional tradespace. Therefore, a value of 7.5 means that the alternative achieves 75% of the weighted multidimensional value between a hypothetical minimum acceptable alternative and the ideal alternative. This interpretation is a very useful way to communicate value and explain the trade-offs to DMs and SHs. Other approaches to multiattribute value analysis have been developed using relative scoring instead of value functions (Buede & Choisser, 1992).

One of the most common and critical mistakes in trade-off studies is using importance weights instead of swing weights (Parnell et al., 2014). MODA quantitatively assesses the trade-offs between conflicting objectives by evaluating the alternative's contribution to the value measures (a score converted to value by single-dimensional value functions) and the importance of each value measure across the range of variation of the value measure (the swing weight). Every decision analysis book identifies this as a major problem. For example, "some experimentation with different ranges will quickly show that it is possible to change the rankings of the alternatives by changing the range that is used for each evaluation measure. This does not seem reasonable. The solution is to use swing weights" (Kirkwood, 1997). Swing weights play a key role in the additive value model in Equation 2.1. The swing weights depend on the measure scales' importance and range. The word "swing" refers to varying the range of the value measure from its minimum acceptable level to its ideal level. If all other measure ranges are held constant and the range of one of the measure scales is reduced, then the measure's relative swing weight decreases, and the swing weight assigned to the others increases since the weights have to sum to 1.0.

Not using swing weights can have dramatic consequences on the credibility of the trade-off study. Parnell et al. (2014) were asked to consult on an Analysis of Alternatives performed on an Army ground vehicle. The analysts had used importance weights and assessed the weights top-down. There were five criteria at the top of the hierarchy, each was assessed an importance weight of about 0.20. Mission and reliability were two of the top-level objectives. One of the four mission's objective measures was vehicle operator safety. Using top-down importance weights, the analysts assessed that passenger safety had a weight of 0.04 and reliability a weight of 0.20. The weights and the results of the study were dismissed by senior reviewers since the primary purpose of the new vehicle was to improve the vehicle operator safety (large difference between the current capability and the desired capability) and the reliability was a secondary factor (small difference between the current capability and the desired capability). When the analysis was redone using the swing weight matrix (Parnell et al., 2011), passenger safety was assessed the highest weight (about 0.25) and reliability a weight of 0.05. With the new swing weights, the client was able to successfully complete their trade-off study. The use of swing weights made a significant change in the credibility of the study and in the alternative rankings.

The swing weight matrix technique is one of many good "direct elicitation" approaches for eliciting ratio scale weights. There are also indirect elicitation techniques focused on inferring weights from choices individuals make in hypothetical settings. These techniques, called conjoint analysis, have been widely used in marketing and involve the following:

1. Presenting an individual with a hypothetical choices
2. Observing the individual's actual choice
3. Defining a new set of hypothetical choices – designed to gather as much relevant information as possible based on what the past choices have revealed about the individual
4. Repeat steps 2 and 3 until sufficient information is gathered about the individual's weights.

An advantage of indirect methods over direct methods is that they ask experts simpler questions and can sometimes be fielded as surveys. The disadvantage is that they require many more questions, which can fatigue respondents (and lead to inaccurate responses.)

These are only two of many techniques being used in decision analysis. Hybrids of direct and indirect elicitation are also possible. Proposed assessment techniques must be evaluated based on their logical defensibility as well as whether empirical evidence shows that they faithfully reflect the priorities of the DMs. In addition, the method has to be easy enough so that clients continue to use it and transparent enough to convince them that the results are credible.

2.3.5.3 Utility Models Utility can be used in place of value in Figure 2.1. Utility considers the decision-maker's preferences for the value of the outcome and the probability of the outcome. As decision-based design (Hazelrigg, 1998) emphasizes, consistent decision-making (in the sense of satisfying the von Neumann–Morgenstern axioms (von Neuman & Morganstern, 1953)) requires that the selected decision have the highest utility score. The value defined in the previous section is a special case of utility – a risk-neutral decision-maker. Alternatively, a decision-maker using axiomatic design would specify utility to be the probability of satisfying all functional requirements. This is an example of a target-oriented utility (Castagnoli and LiCalzi, 1996; Bordley & Li Calzi, 2000).

Standard safety system methods for assessing failure risk can be used to decompose the probability of meeting functional requirements into the probability of satisfying detailed requirements. As Bordley and Kirkwood (2004) showed, the resulting decomposed probability will be consistent with the classic multi-attribute utility (Keeney and Raiffa, 1976) in decision theory with the attributes being the dimensions described by the detailed requirements. But satisfying requirements, instead of being the end goal, is often a means of satisfying customer objectives – with there being some uncertainty about whether meeting requirements satisfies stakeholder objectives. To extend target-oriented utility to this case, utility could be defined as the probability of meeting stakeholder objectives.

But this is only straightforward as long as there is an unambiguous definition of what it means to satisfy stakeholder objectives. When there are varying degrees to which stakeholders can be satisfied (when they are not either satisfied or dissatisfied), developing such an unambiguous definition may be difficult. In such cases, utility can still be treated as simply a measure of returns to scale and risk preference (Kirkwood, 1997).

The three types of risk preference are risk-neutral, risk-averse, and risk-seeking. Value is appropriate for a risk-neutral decision-maker. Utility functions have been defined for single and multiple objectives (Keeney & Raiffa, 1976). For example, for a tradespace exploration using utility and uncertainty, see (Ross & Hastings, 2005). The trade-off analysis literature shows a very limited use of utility functions (using the decision analysis definition of utility). There are several reasons that can explain this finding.

1. Many systems engineers and decision-makers believe that a risk-neutral preference is most appropriate for large organizations making a large number of decisions.

2. The cumulative probability distribution of value can be used to assess deterministic and stochastic dominance (Clemen & Reilly, 2001).
3. Some systems engineers may not know the difference between value and utility.
4. Many decision-makers and stakeholders involved in the system decision may have very different risk preferences. For very good reasons, the project manager may be risk-adverse, but the director of R&D may be risk-neutral or even risk-seeking for some projects.
5. Assessing risk preference is challenging and time-consuming. This is especially true when we have several single-dimensional utility functions and multiple DMs and SHs with conflicting risk preferences.

Based on the first two reasons, we use value instead of utility in Figure 2.1. We will define risk as the probability of obtaining a low value, and we will use the cumulative probability distribution of value to assess deterministic and stochastic dominance. See Chapters 5, 9, and 13 for examples.

2.3.6 Resource Analysis

Affordability is a major concern for most systems. Systems engineering trade-off analyses need to consider finite resources. Minimizing resources required for a system is usually an objective. The primary resource is financial (cost). However, other resources including manpower and facilities must be included. The best practice for cost analysis is to develop a life cycle cost analysis model that converts the operations concept and the features of the system under consideration to life cycle cost.

It is useful to note that mathematically cost could be one of the measures in the additive value model. However, many decision-makers prefer to visually see the value for the cost. In DoD, the policy of Cost as an Independent Variable (CAIV) and affordability analysis requires the separation of cost from value. The chart with value plotted on the vertical dimension and cost along the horizontal dimension becomes a particularly useful way to think about the value added of alternatives and is an essential tool for affordability analysis.

2.3.7 An Integrated Trade-Off Analyses

Many trade-off studies do not provide an integrated assessment of value, cost, and risk (Parnell et al., 2014). Figure 2.1 illustrates how an influence diagram is used to model the relationships between decisions, uncertainties, and value and provide an integrated trade-off analysis model. The influence diagram was selected as the preferred representation because it offers a probabilistic decision analysis model that identifies the major variables and their dependencies.

We can now provide a more precise discussion of Figure 2.1. Several important features of this influence diagram model are as follows:

1. The decisions include the functions, objectives, requirements, and system decision alternatives. Adding the functions, objectives, and requirements as

decisions is very important, especially in the concept, architecting, and design decisions. In later life cycle stages, the functions, objectives, and requirements may be constants.
2. There are several major uncertainties: stakeholder needs, scenarios, adversary actions, competition actions, technology maturity, system features, system models, value measure data, priorities, and resources. In general, the priorities depend on the scenario. The scenarios will depend on the objectives and the adversary actions.
3. The influence diagram assumes a multiple objective model using the additive value model. The value can be calculated based on decision, the value measure data, the value functions, and the priorities (swing weights) using the additive value model. If all objectives can be monetized, we would delete the priorities (swing weight) node and add a constant node for the discount rate. The single value would be Net Present Value. Similarly, if we want to model the risk preference other than risk neutral, we could replace value by utility.
4. If the stakeholders are willing to agree on a set of priorities, the priorities would be constant. Another approach is to have multiple priority sets for different stakeholder groups and/or different scenarios. This approach allows the identification of alternatives preferred by different stakeholders and alternatives preferred for different scenarios. The value measure scores would depend on the scenario, the competitor's actions, and the outputs of system models and simulations. The design features are uncertain until the later life cycle stages. The resources (cost) will depend on the system features. If the system value is higher, fewer systems may be needed.
5. The value functions are modeled as constant for the decision opportunity, objectives, and requirements. As the decision opportunity, objectives and requirements change, and the value functions change. See, for example, Ross et al. (2015).
6. The primary tradespace is between value and cost. The analysis can be deterministic or probabilistic. Risk can be value or cost risk. The decision tradespace can be explored by optimization, experimental design, or designing creative decisions that span the decision space.
7. In probabilistic analysis, probability distributions can be assigned to the uncertain independent variables. The uncertainty is then propagated through system models and the value model by Monte Carlo simulation. The decision space can be explored by optimization, simulation optimization, experimental design, or designing creative decisions that span the space. After the probabilistic integrated value and cost analyses are performed, the value and risk drivers can be analytically identified using tornado diagrams (Clemen & Reilly, 2001).

A framework to implement integrated trade-off analyses has been added to the Systems Engineering Body of Knowledge (Decision Management). For an illustration of this integrated framework, see Chapters 5 and 9, and (MacCalman & Parnell, 2016).

2.4 TRADESPACE EXPLORATION

In this chapter, we advocate the use of value (or utility) to evaluate each alternative in the decision space. However, we do not advocate the use of decision analysis as the only technique to explore the tradespace. For complex systems and system of systems, value measure scores may come from operational data, test data, simulations, models, or expert opinion. In addition, tradespace exploration requires the tailored use of other mathematical techniques in conjunction with decision analysis including optimization, simulation optimization, and simulation with design of experiments.

Optimization (Bordley & Pollock, 2009, 2012) is an analytical technique to explore the decision tradespace. This technique can be very effective and efficient. However, it requires the model be able to mathematically express the objective, the essential relationships, and the constraints. Optimization provides the one best alternative unless there are alternative optimal solutions. Additional analysis is required to provide information about nearly optimal alternatives that may be of interest to the decision-makers.

Another important tradespace exploration technique is to perform an experimental design using models (Bordley & Bier, 2009) and simulations (Parnell et al., 2011). The statistical method of design of experiments is a widely used technique in all areas of science to help understand complex behavior. These experiments allow systems engineers to efficiently explore design spaces to identify features that have the most impact on the desired system behavior.

Chapter 8 presents and compares several techniques for defining and exploring the tradespace. Chapter 9–14 provides illustrative examples of the techniques our authors have found to be the most useful in each stage of the life cycle.

2.5 SUMMARY

This chapter has four themes. The first is a standard terminology to discuss trade-off analyses. The second is the use of decision analysis as the mathematical foundation for trade-off analysis. The third is the use of integrated trade-off analysis to identify value, cost, and risk based on a probabilistic decision analysis with a value and cost model developed for trade-off analyses. The fourth is the use of tradespace exploration techniques combined with decision analysis.

Systems are designed to last decades, and system development involves many uncertainties. However, many trade-off analyses are deterministic and do not consider the risk and opportunities provided by uncertainty. The integrated framework for value and risk analysis is an important step to value risks and opportunities and to provide a sound foundation for risk analysis. The framework will need to be tailored to the opportunity, system, life cycle stage, and decision-makers.

It is well known that trade-off analyses are very important in systems development but are not consistently well done. We believe that a major contributor is the lack of a standard lexicon and a sound mathematical foundation. The proposed lexicon offers the opportunity to provide a terminology that is mathematically sound and has been proven effective in many important systems engineering trade-off analyses.

Decision analysis has a successful track record for providing insights to decision-amakers and stakeholders in public and private organizations. Decision analysis is the mathematical foundation for the INCOSE SE Handbook Decision Management section and the Systems Engineering Body of Knowledge Decision Management Section.

The purpose of a system (or a system of systems) is to provide value in the form of capabilities and services to clients, operators, and owners. Decision-making is central to value identification and value realization. Trade-off analyses, a fundamental systems engineering technique, inform system decisions throughout the system life cycle. Systems engineering is a trade-off rich field. Trade-off analysis at present is performed in somewhat ad hoc fashion. This chapter presents an integrated trade-off analysis approach grounded in decision sciences. The same analysis can identify the value and risk drivers. The approach employs a clear lexicon and a sound mathematical foundation for trade-off analyses. The integrated model for trade-off analyses presented in this chapter considers the decisions, the uncertainties, and the values of decision-makers. Both systems and system-of-systems issues are addressed in this chapter.

2.6 KEY WORDS

Architecture: The fundamental concepts or properties of a system in its environment embodied in its elements, relationships, and in the principles of its design and evolution.

Concept: A concept focuses on the system need, purpose, functions, and operations prior to defining the architecture and the design.

Design: A design defines the architecture, system elements, interfaces, and other characteristics of the system (ISO/IEC/IEEE, 2015).

Decision: A choice among alternatives that results in an allocation of resources.

Decision Opportunity: Potential activity that we consider expending resources on to achieve organizational objectives.

Influence Diagram: An influence diagram is a compact graphical representation of a decision opportunity that identifies the decisions, uncertainties, and value and shows their relationships.

Objectives: The goals of the stakeholders for the system. Objectives are defined by a verb and an object.

Risk: A potential and undesirable deviation of an outcome from some anticipated value.

Tradespace: A multidimensional space that defines the context for the decision, defines bounds to the region of interest, and enables Pareto optimal solutions for complex, multiple stakeholder decisions.

Uncertainty: Imperfect knowledge of the value of some variable.

Value Function: Returns to scale on a value measure.

Value Measure: Identifies how the organization can measure the potential achievement of the objectives at the time of the decision.

Utility Function: Returns to scale and risk preference on a value measure.

2.7 EXERCISES

2.1. Consider Table 2.1 and a problem domain of interest to you. Develop a similar table for the following:
   (a) Two existing systems
   (b) One system in development
   (c) One future system.

2.2. This book uses decision analysis as the mathematical foundation for quantifying the tradespace and performing a trade-off analysis. (Alternative approaches for evaluating alternatives are presented in Chapter 8.)
   (a) Identify the advantages of this approach.
   (b) Identify the disadvantages of this approach.

2.3. This chapter advocates an integrated approach to value, cost, and risk analysis. Compare this approach to separate value, cost, and risk analyses.
   (a) Identify the advantages of the integrated approach.
   (b) Identify the disadvantages of integrated approach.

2.4. Find a trade-analysis article in a refereed journal, for example, Systems Engineering. Use the influence diagram in Figure 2.1 as a framework to assess the trade-off analysis.
   (a) What decisions were considered in the article?
   (b) What uncertainties were considered in the article?
   (c) What type of resource analysis was performed?
   (d) What preference function was used to evaluate the trade-offs?
   (e) Was the risk of the alternatives explicitly considered?
   (f) Was the trade-off analysis an integrated analysis?

2.5. Consider a trade-off analysis you have performed in the past or will be performing in the future.
   (a) Draw an influence diagram for the analysis.
   (b) Compare your influence diagram with Figure 2.1. What are the similarities and differences?

REFERENCES

Barron, F.H. and Barrett, B.E. (1996) Decision quality using ranked attribute weights. *Management Science*, **42**, 1515–1523.

REFERENCES

Bordley, R. and Bier, V. (2009) Updating beliefs about variables given new information on how those variables relate. *European Journal Of Operational Research*, **193** (1), 184–194.

Bordley, R. and Li Calzi, M. (2000) Decision analysis using targets instead of utility functions. *Decisions in Economics and Finance*, **23**, 53–74.

Bordley, R.F. and Kirkwood, C.W. (2004) Multiattribute preference analysis with performance targets. *Operations Research*, **52** (6), 823–835.

Bordley, R. and Pollock, S. (2009) A decision analytic approach to reliability-based design optimization. *Operations Research*, **57** (5), 1262–1270.

Bordley, R. and Pollock, S. (2012) Assigning resources and targets to an organization's activities. *European Journal of Operational Research*, **220** (3), 752–761.

Buede, D.M. (1994). Engineering design using decision analysis. *1994 IEEE International Conference. 2*, (pp. 1868–1873), IEEE, San Antonio.

Buede, D.M. (2000) *The Engineering Design of Systems: Models and Methods*, John Wiley & Sons, 2016.

Buede, D.M. and Choisser, R.W. (1992) Providing an analytic structure for key system design choices. *Journal of Multi-criteria Decision Analysis*, **1**, 17–27.

Castagnoli, E. and LiCalzi, M. (1996) Expected utility without utility. *Games and Economic Behavior*, **41** (3), 281–301.

Clemen, R.T. and Reilly, T. (2001) *Making Hard Decisions with Decision Tools*, Duxbury Press, Belmont, CA.

Decision Management (n.d.). Systems Engineering Body of Knowledge: http://www.sebokwiki.org/wiki/Decision_Management (accessed 15 November 2014).

Hazelrigg, G.A. (1998) A framework for decision-based engineering design. *Journal of Mechanical Design*, **120** (4), 653–658.

ISO/IEC (2015) *Systems and Software Engineering — System Life Cycle Processes*, International Organization for Standardization (ISO)/International Electrotechnical Commission (IEC), Geneva, Switzerland.

ISO/IEC/IEEE (2015) *Systems and Software Engineering — System Life Cycle Processes*, International Organization for Standardization (ISO)/International Electrotechnical Commission (IEC)/Institute of Electrical and Electronics Engineers (IEEE), Geneva, Switzerland.

Keeney, R.L. (1992) *Value-Focused Thinking: A Path to Creative Decisionmaking*, Harvard University Press, Cambridge, MA.

Keeney, R.L. and Raiffa, H. (1976) *Decisions with Multiple Objectives Preferences and Value Tradeoffs*, Wiley & Sons, New York, NY.

Keisler, J.M. and Bordley, R.F. (2014) Project management decisions with uncertain targets. *Decision Analysis*, **12** (1), 15–28.

Kirkwood, C. (1997) *Strategic Decision Making: Multiobjective Decision Analysis with Spreadsheets*, Duxbury Press, Belmont, CA.

MacCalman, A. and Parnell, G. (2016). Multiobjective Decision Analysis with Probability Management for Systems Engineering Trade-off Analysis. *Hawaii International Conference on System Science - 49*.

Madni, A.M. (2013) Generating novel options during systems architecting: psychological principles, systems thinking, and computer-based aiding. *Systems Engineering*, **16** (4), 1–9.

Madni, A.M. (2015) Expanding stakeholder participation in upfront system engineering through storytelling in virtual worlds. *Systems Engineering*, **18** (1), 16–27.

Madni, A. M., Brenner, M. A., Mac Gregor, D., and Meshkinpour, F. (1985). Option Generation: Problems, Principles, and Computer-Based Aiding. *Proceedings of 1985 IEEE International Conference on Systems, Man, and Cybernetics*, (pp. 757–760). Tucson, AZ.

Madni, A. M. and Freedy, A. (1981). Decision Aids for Airborne Intercept Operations in Advanced Aircrafts. *Proceedings of the International Conference on Cybernetics and Society* (pp. 224–234). Sponsored by IEEE Systems, Man, and Cybernetics Society.

Madni, A.M., Samet, M.G., and Freedy, A. (1982) A trainable on-line model of the human operator in information acquisition tasks. *IEEE Transactions of Systems, Man, and Cybernetics*, **12** (4), 504–511.

Madni, A.M. and Sievers, M. (2013) System of systems integration: key considerations and challenges. *Systems Engineering*, **17** (2), 330–347.

Madni, A.M. and Sievers, M.N. (2016). Model based systems engineering: motivation, current status and needed advances. *Systems Engineering,* accepted for publication, 2016.

Parnell, G.S., Bresnick, T.A., Tani, S.N., and Johnson, E.R. (2013) *Handbook of Decision Analysis*, Wiley & Sons.

Parnell, G., Cilli, M., and Buede, D. (2014). Tradeoff Study Cascading Mistakes of Omission and Commission. *International Council on Systems Engineering (INCOSE) International Symposium.* Las Vegas: INCOSE.

Parnell, G.S., Driscoll, P.J., and Henderson, D.L. (eds) (2011) *Decision Making for Systems Engineering and Management*, 2nd edn, Wiley & Sons.

Ross, A. M. and Hastings, D. E. (2005). 11.4. 3 The Tradespace Exploration Paradigm. *INCOSE International Symposium*, *15*.

Ross, D. H., Rhodes, D. H., and Fitzgerald, M. E. (2015). Interactive Value Model Trading for Resilient Systems Decisions. *13th Conference on Systems Engineering Research.* Hoboken, NJ.

von Neuman, J. and Morganstern, O. (1953) *Theory of Games and Economic Behavior*, Princeton University Press, Princeton, NJ.

3

QUANTIFYING UNCERTAINTY

ROBERT F. BORDLEY

Integrated Systems & Design, University of Michigan, Ann Arbor, MI, USA; Defense & Intelligence Group, Booz Allen Hamilton, Troy, MI, USA

> **Laplace:** *The theory of probability is at bottom nothing but commonsense reduced to calculus.*

3.1 SOURCES OF UNCERTAINTY IN SYSTEMS ENGINEERING

Best practice in systems engineering identifies all relevant factors outside the control of the design team. Some of these factors, for example, certain external interfaces, are completely known to the team. But "*many phenomena that are treated as deterministic or certain are, in fact, uncertain,*" (*OMB guidance, circular 94, 2015*). The inappropriate treatment of uncertain quantities as deterministic, also called the flaw of averages (Savage et al., 2012), can lead to design decisions that could prove very costly if the actual value of this quantity is very different from the average. So the team must acknowledge their uncertainty about both

1. what is feasible and affordable, for example, technological, integration, interface, and supply uncertainties and
2. what is desirable, for example, uncertainties about stakeholder preferences, adversary reactions, and environmental changes impacting stakeholders and adversaries.

Trade-off Analytics: Creating and Exploring the System Tradespace, First Edition. Edited by Gregory S. Parnell.
© 2017 John Wiley & Sons, Inc. Published 2017 by John Wiley & Sons, Inc.
Companion website: www.wiley.com/go/Parnell/Trade-off_Analytics

There is added intrateam uncertainty due to different specialists within the design team not knowing how they will be impacted by the choices and discoveries of other specialists.

In addition to acknowledging these uncertainties, the team must continuously update their understanding of these uncertainties as they learn more information. Information about uncertainties can be learned from risk management activities that address both

1. uncertainty about outcomes and
2. the severity of the consequences associated with outcomes.

Since these two dimensions addressed by risk management are typically but not always (Bordley & Hazen, 1991) unrelated, this section only focuses on outcome uncertainty.

3.2 THE RULES OF PROBABILITY AND HUMAN INTUITION

Probability theory was created (Bernstein, 1998) by mathematicians (Galileo, Pascal, Cardano) working as consultants to affluent gamblers. The uncertainties gamblers address in gambling houses as illustrated in Figure 3.1 are generated from analog mechanisms such as spinning roulette wheels, rolling dice, dealing out playing cards, and, in modern times, digital mechanisms such as random number generation software operating on embedded microprocessors. While the gambler may not understand these mechanisms, the owners of the gambling house (or their consultants) do understand these mechanisms. These uncertainties have several convenient properties:

1. All of the possible outcomes of the gambling device (e.g., all of the possible outcomes of spinning a roulette wheel) are known.
2. There are a set of elementary outcomes (e.g., the slots in the roulette wheel or a number on the roll of a single die) that are equally likely.
3. The outcome of one trial of a gambling device (one roll of a die) is independent of the outcome of a second trial of a gambling device.

Figure 3.1 Uncertainty in gambling

Uncertainties that are analogous to the uncertainties arising from gambling devices (such as rolling a dice) are often called aleatory uncertainties from the Latin word for dice.

Because of these properties, the aleatory probability of a gambler winning a bet equals the fraction of elementary outcomes consistent with the gambler winning the bet. If a gamble is played repeatedly, then this fraction eventually reflects the long-run frequency with which different outcomes will occur. This makes this fraction useful in developing long-run strategies for both gamblers and gambling houses.

All the equally likely outcomes of the gambling device can be arranged as points on a square. Each of these points represents a different "state" of the world. Suppose that some of these outcomes, if they occur, would be acceptable to the systems engineer's stakeholders. Draw a circle labeled A around the acceptable stakeholder outcomes (and let A^c denote the set of remaining unacceptable outcomes).

Suppose an engineering test was conducted with certain outcomes passing the test and others not passing the test. While satisfying stakeholders is the goal of the systems engineer, passing the test is often the goal of a supplier to the systems engineer. Draw a circle labeled E about the outcomes passing the engineering test (with E^c denoting the outcomes outside the circle that fail the test). Then the problem can be represented with the classic Venn diagram shown in Figure 3.2.

If E and A include exactly the same points, then an outcome that failed the engineering test would also be unacceptable to users (and vice versa). In reality, there will be some outcomes, $E\&A^c$, that pass the engineering tests but are unacceptable to stakeholders as well as outcomes, $E^c\&A$, that fail the engineering tests but are acceptable to stakeholders.

Note that $P(E\&A)$ can never exceed $P(E)$. While this is obvious from the Venn diagram, psychological experiments establish that human intuition often violates this principle. For example, in one such experiment, individuals were asked to estimate the probability of a very liberal individual either being a bank teller or being a feminist bank teller, that is, being both a feminist and a bank teller. Individuals consistently judged being a feminist bank teller as more likely than being a bank teller (Tversky & Kahneman, 1974). So in this example, human intuition was in error.

Since all outcomes are treated as equally likely, the proportion of outcomes passing the engineering test will equal the probability, $P(E)$, that the engineering test is passed.

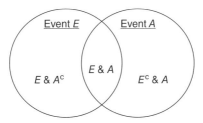

Figure 3.2 Venn diagram

Since the sum of the proportion of outcomes passing the test and the proportion of outcomes not passing the test is 1, $P(E)+P(E^c)=1$. Also note that summing

- the proportion of outcomes passing the test that are acceptable to stakeholders
- the proportion of outcomes passing the test that are not acceptable to stakeholders

gives the proportion of outcomes passing the test. As a result,

$$P(E\&A) + P(E\&A^c) = P(E) \qquad (3.1)$$

This seems obvious from the Venn diagram in Figure 3.2. But psychological evidence shows that human intuition often violates this principle. In another psychological experiment,

- a group of individuals estimated the probability of death from natural causes
- a second group estimated the probability of death from heart disease, the probability of death from cancer, and the probability of death from a natural cause other than heart disease or cancer.

Summing up the probabilities from this second group consistently gives a higher probability of death than probability assessed by the first group (Tversky & Koehler, 1994). So once again, human intuition was found to be in error.

Define $P(A|E) = P(E\&A)/P(E))$ and $P(A^c|E) = P(E\&A^c)/P(E)$. Then

$$P(A|E) + P(A^c|E) = 1 \qquad (3.2)$$

Of those outcomes passing the engineering test, the quantity $P(A|E)$ corresponds to the proportion that is also acceptable to stakeholders. If $P(A|E) = P(A|E^c)$, then

$$P(A) = P(A\&E) + P(A\&E^c) = P(A|E)P(E) + P(A|E^c)P(E^c) = P(A|E) \qquad (3.3)$$

In this case, events A and E are independent where independence is defined as follows.

Independence: Events A and E are called independent if $P(A) = P(A|E)$, that is, if knowing that the occurrence of outcome in E does not alter the probability of occurrence of an outcome in A.

When events A and E are independent, both $P(A|E) = P(A)$ and $P(E|A)=P(E)$. But events are typically not independent. Specifically,

$$P(A|E) = \frac{P(A\&E)}{P(E)} = \left[\frac{P(A\&E)}{P(A)}\right]\left[\frac{P(A)}{P(E)}\right] = P(E|A)\left[\frac{P(A)}{P(E)}\right] \qquad (3.4)$$

Since $P(E) = P(E|A)P(A) + P(E|A^c)P(A^c)$, substituting for $P(E)$ gives an expression known as Bayes' rule:

Bayes' rule:
$$P(A|E) = \frac{P(E|A)P(A)}{P(E|A)P(A) + P(E|A^c)P(A^c)}$$

If an unacceptable outcome always fails the engineering test (and $P(E|A^c)=0$), then $P(A|E) = 1$.

When $P(E|A^c)$ is nonzero, then one can define the likelihood ratio $L(A;E) = P(E|A)/P(E|A^c)$ as a measure of how much the probability of passing the test differs between acceptable and unacceptable outcomes. Then Bayes' rule can be rewritten as

$$P(A|E) = \frac{L(A;E)P(A)}{L(A;E)P(A) + P(A^c)} \tag{3.5}$$

When $L(A;E) = 1$ (and the test does not discriminate between acceptable and unacceptable outcomes), $P(A|E) = P(A)$, that is, the test result is independent of whether an outcome is acceptable. But as $L(A;E)$ increases (with $P(A)$ positive), $P(A|E)$ increases.

According to equation 3.4, $P(A|E)$ will only equal $P(E|A)$ when $P(A)=P(E)$. So even if the test was very discriminating with $P(E|A)=1$ and $P(E|A^c)=0.05$ (and $L(A;E) = 20$), no conclusion could be drawn about $P(A|E)$ until $P(A)$ was specified. But experimentation shows that people typically treat $P(A|E)$ as equaling $P(E|A)$ without considering the value of $P(A)$ (Villejoubert & Mandel, 2002). So human intuition would interpret a positive test result as indicating that event A is likely to be true. This has led to very serious errors both in medicine (where patients were diagnosed as having cancer based on a test result) and in legal cases (where defendants were judged guilty based on circumstantial evidence).

To show why $P(A|E)$ does not always equal $P(E|A)$, suppose that $P(A) = 1\%$, that is, the probability of a randomly chosen outcome satisfying the stakeholders is quite small. If there are 2000 possible outcomes, then

- 1% of 2000 or 20 of these outcomes will be acceptable
  - If the test is administered to all 2000 outcomes, then all 20 of the acceptable outcomes pass the test.
- 99% of 2000 or 1980 outcomes will be unacceptable
  - If the test is administered to all 2000 outcomes, then 5% of the 1980 unacceptable outcomes (or 99 unacceptable outcomes) pass the test.

So $20 + 99 = 119$ outcomes would pass the test if it were administered to all 2000 outcomes. Since only 20 of these 199 outcomes are acceptable, the probability of an outcome being acceptable given it passed the test, or $P(A|E)$, is approximately $20/119 = 17\%$. Thus, an outcome passing the test is still unlikely to meet stakeholder expectations. This example shows that both $L(A;E)$ and $P(A)$ must be specified to determine $P(E|A)$. For historical reasons, the probability $P(A)$ is commonly called the prior probability while $P(A|E)$ is called the posterior probability.

These rules of probability can be derived from Kolmogorov's three axioms (Billingsley, 2012):

(A1) Suppose $E_1, \ldots, E_n$ are disjoint events, that is, at most, one of these events can occur. Then,

$$P(E_1, \ldots, E_n) = P(E_1) + \ldots + P(E_n) \tag{3.6}$$

(A2) If U is an event that includes all possible outcomes, then $P(U) = 1$.
(A3) For any event E, $P(E) \geq 0$.

For example, equation 3.1 can be deduced from (A1) by defining $E_1 = E\&A$ and $E_2 = E\&A^c$. While these axioms simply formalize commonsense, this section highlighted three cases where informal commonsense contradicted the rules implied by these axioms. Because human intuition about probability is often misleading, careful, and consistent calculation of probabilities is critical in properly accounting for uncertainty.

3.3 PROBABILITY DISTRIBUTIONS

3.3.1 Calculating Probabilities from Experiments

When a gambler makes a wager (e.g., places a bet on odd numbers appearing on the roulette wheel), the gambler is running an experiment where the outcomes of the experiment are payoffs, that is, how much money the gambler wins or loses. By repeatedly placing a wager and comparing the results with repeated use of different wagers, the gambler generates statistics useful in evaluating the payoffs from different gambling strategies. All the results of interest are easily measured and all the possible outcomes of the experiment (called the experimental sample space) are known in advance.

But while the gambler can make the same wager over and over again, there are many realistic settings where it is not possible to repeat the same decision under the same circumstances. Furthermore, even if it were possible to repeat the same decision, it might not be possible to measure the resulting outcomes of this repeated decision. So the conditions that hold in a gambling house do not necessarily hold in realistic settings. Fortunately, new concepts and measurement devices had been invented, which allowed these conditions to be approximately satisfied in settings outside the gambling house. These inventions allowed probability to be applied across a wide variety of scientific and engineering contexts. As Laplace wrote, "*it is remarkable that a science which began with the consideration of games of chance should have become the most important object of human knowledge.*"

Scientific experiments typically involve the comparison of experimental treatments with control treatments on the dimension of interest. Making these comparisons is considerably easier when the dimension is an ordinal scale where objects can be ranked from low to high. There are many ways to create an ordinal scale. The Mohs scratch test scale (Hodge & McKay, 1934) ranks the hardness of rocks based on which rock can scratch which other rock. Thus, diamond ranks higher than quartz (because diamond can scratch quartz) while quartz ranks higher than gypsum because quartz can scratch gypsum.

For illustrative purposes, suppose the experimental group consists of belted drivers and the control group consists of unbelted drivers. Let the ordinal dimension of interest be the injuries sustained in collisions. Then all possible outcomes of both treatments can be ranked. Suppose that in this ranking, minor injuries outrank moderate injuries and moderate injuries outrank major injuries. This immediately gives a ranking of any treatment guaranteed to lead to only one outcome. To rank other treatments, define the state of an accident by the kind of vehicles involved in the collision, vehicle speeds, and other factors. Suppose the following property holds.

State-Wise Dominance: For each possible state, the outcome of the experiment treatment never ranked lower (and, in at least one case, ranked higher) than the outcome of the reference treatment.

When this property holds, the experimental treatment **outranks** the control treatment.

But in the seatbelt example, state-wise dominance does not hold because there are some states, for example, a burning or submerged vehicle, in which belted drivers may have greater injuries than unbelted drivers. To introduce a more generally applicable notion of dominance, define $P_0(X)$ as the probability of the reference treatment leading to outcome x and $P_1(X)$ as the probability of the experimental treatment leading to outcome x. Also define the cumulative probability as

Cumulative Probability: The cumulative probability, $F_1(x)$, is the probability of the experimental treatment leading to an outcome ranked no higher than x. This probability is defined over all possible outcomes x in the sample space.

We similarly define $F_0(x)$ as the probability of the reference treatment scoring no better than x. Suppose the following statistics were collected.

Table 3.1 includes injuries from cars burning, being submerged, and other accidents.

Table 3.1 Distribution of Injuries From a Major Accident

Treatment	Minor Injuries (%)	Moderate Injuries (%)	Severe Injuries (%)
Belted	30	20	10
Unbelted	10	10	60

Since being unbelted leads to fewer minor or moderate injuries than being belted, being belted does not state-wise dominate being unbelted. But note the following:

1. For severe injuries, $F_1(\text{severe}) = 10\% < F_0(\text{severe}) = 60\%$.
2. For moderate to severe injuries, $F_1(\text{moderate}) = 30\% < F_0(\text{moderate}) = 70\%$.
3. For minor to severe injuries, $F_1(\text{minor}) = 60\% < F_0(\text{minor}) = 80\%$.

So the probability of having a less severe injury (i.e., an outcome ranked higher) is greater for belted drivers. We can formalize this property as follows.

> **Stochastic Dominance:** For any reference outcome t, the probability of the experimental treatment leading to an outcome outranking outcome t is at least as high as the probability of the control treatment leading to an outcome outranking outcome t (i.e., $F_1(t) < F_0(t)$ for all t).

If the experimental treatment were state-wise dominant with respect to the control treatment, then it would be stochastically dominant. But while stochastic dominance will in general allow more treatments to be ranked than state-wise dominance, it still does not allow all possible treatments to be ranked.

To rank a treatment in the absence of stochastic dominance, the American Psychological Association mandated (Wilkinson & Task Force on Statistical Inference, 1999) the use of the effect size measure (which is different from the widely known measure of statistical significance). The simplest and most general formulation of the effect size measure (McGraw & Wong, 1992; Grissom & Kim, 2011) is the probability of the experimental treatment leading to an outcome outranking the outcome of the control treatment.

To describe the effect size measure, define X_1 (and X_0) as a random variable leading to outcome x with probability $P_1(x)$ (and $P_0(x)$) for all possible outcomes x. The experiment treatment's effect size is the probability that X_1 outranks X_0. If there are m experimental treatments numbered $k = 1, \ldots, m$, then treatment k's effect size is defined as the probability that X_k outranks X_0 – where each experimental treatment k is compared with the control treatment. The experimental treatments can be compared with one another based on their effect size. If $P_k(x)$ is the kth treatment's probability of leading to outcome x, then the kth treatment's effect size is $\sum_x F_0(x) P_k(x)$, that is, the expected probability of outranking the control treatment on the dimension of interest.

3.3.2 Calculating Complex Probabilities from Simpler Probabilities

Applying these concepts requires that $F(x)$ be specified. Sometimes the probability for a complex event can be estimated from known probabilities for simpler events. For example, suppose the probability of a single component failing in some time period is prespecified as equaling r. Suppose there are two components in a system, each with this same probability of failure. Also, suppose that the components are independent. Then there are several possible outcomes:

1. Both components could fail. This occurs with probability r^2.
2. Only one component remains functional. This will occur if either

(a) the first component fails while the second remains functional. This will happen with probability $r(1-r)$ or
(b) the second component fails while the first remains functional. This will happen with probability $(1-r)r$.

So the overall probability of only one component being functional is $2r(1-r)$.

3. Both components could remain functional. This occurs with probability $(1-r)^2$.

This specifies the probability of two, one, and zero failures in the system.

In the more general case of a system with N components, calculating the probability of x components failing (and $(N-x)$ remaining functional) requires listing the number of distinct outcomes consistent with exactly x components failing. This includes such outcomes as follows:

- The first x components fail while the last $N-x$ components remain functional.
- The first $(N-x)$ components remain functional while the last x components fail.

Let $C(N,x)$ be the number of possible distinct outcomes (or permutations) where exactly x out of N components fail. The probability of each of these outcomes occurring is $r^x(1-r)^{N-x}$. Hence, the total probability of any of these outcomes occurring, $P(x)$, is described by the following.

The Binomial Distribution: $P(x) = C(N,x) r^x (1-r)^{N-x}$

If the system only fails when at least x out of N components fail, the probability of the system not failing, that is, having $x-1$ failures or fewer, will be $F(x-1)$.

Using the binomial distribution is difficult when N is very large (e.g., greater than 100). But instead of treating r as a parameter whose value is fixed, suppose the expected number of failures, $a = rN$, is treated as fixed. Then for large N and a small (e.g., $a < 20$), the binomial distribution can be approximated by the simpler Poisson distribution. Because the Poisson distribution allows N to be continuous, the Poisson distribution can be used to estimate the number of failures occurring over some time interval N.

The probability of zero failures in time interval N in the Poisson is $e^{-a} = e^{-rN}$. So the probability that the time until the first failure exceeds N is e^{-rN}. Thus, the Poisson distribution for discrete outcomes also implies a distribution (referred to as the exponential distribution) over the continuous variable N. This distribution will be a building block for some of the distributions considered in the next section.

3.3.3 Calculating Probabilities Using Parametric Distributions

When x is large (or continuous), estimating $F(x)$ can be difficult. As a result, it is convenient to define $F(x)$ so that it is completely specified by a small number of parameters. While there are many possible functions, statistical practice has identified a small number of basic distribution functions $F(x)$ that provide the flexibility needed to address a wide variety of practical problems. In complex problems, the actual function $F(x)$ may have to be written as a weighted average (or mixture) of these basic distribution functions.

These basic distribution functions are completely specified by a location parameter, M, and a scaling parameter, s. There are three kinds of basic distribution functions:

1. Unbounded. In this case, the value of the uncertain variable x has neither a lower bound nor an upper bound. The location parameter equals the expected value (or mean) of x while the scaling parameter equals the standard deviation of x.
2. Bounded from below. In this case, the value of x has a lower bound but no upper bound. The scaling parameter will still equal the standard deviation, but the location parameter now equals the lower bound. (If x has an upper bound but no lower bound, redefining F to be a function of $-x$ allows the uncertainty to be treated as if it were bounded from below.)
3. Bounded from above and below. In this case, the value of x has both a lower and an upper bound. The location parameter will equal the lower bound, but the scaling parameter now equals the difference between the upper and lower bounds.

Once M and s are identified, define the standardized value of x by $r = (x - M)/s$ and the standardized distribution by $F^*(r) = F(x)$.

In the unbounded case, $F^*(r)$ is assumed to be the normal distribution (Figure 3.3) whose values are available in look-up tables:

According to the central limit theorem, the normal distribution describes the average of many independent and identically distributed errors (when the mean and variance of these errors are finite.) In applications where the normal distribution understates the degree to which extreme outcomes or black swans (Taleb, 2007) occur, the t or Cauchy distribution may be used instead.

In the case where there is a lower bound on x but no upper bound, $F(r)$ is often assumed to follow the Weibull distribution (Figure 3.4).

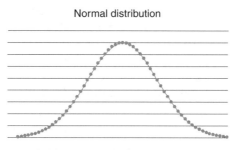

Figure 3.3 Normal distribution

PROBABILITY DISTRIBUTIONS   61

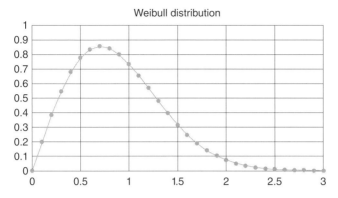

Figure 3.4 Weibull distribution

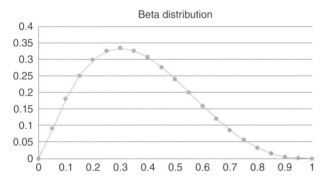

Figure 3.5 Beta distribution

In some cases, the Gamma, Fréchet, or lognormal distributions are used instead of the Weibull distribution. The shape of these distributions is similar to the shape of the Weibull distribution.

If there are both a lower bound and an upper bound on x, it is common to assume that $F^*(r)$ is the beta distribution (Figure 3.5):

The beta distribution requires the specification of two parameters (in addition to the lower and upper bounds.) In project management applications where only a single parameter, that is, the best estimate of x, is specified, the Beta Pert distribution is used. The Beta Pert distribution corresponds to a Beta distribution where the Beta distribution parameters are set to yield an expression for the variance used in the PERT scheduling tool. As an alternative to either the Beta distribution or the Beta Pert distribution, some practitioners use the more easily explained triangular distribution (Figure 3.6):

The analytic expressions for these distributions are given in Table 3.2.

These and other distributions are compactly described in the classic textbook of Johnson et al., (1995).

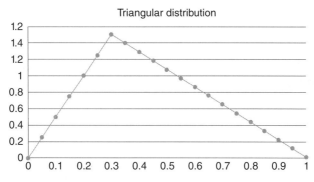

Figure 3.6 Triangular distribution

3.3.4 Applications of Parametric Probability Distributions

Application of the Gaussian Distribution: System validation is concerned with ensuring that stakeholders are truly satisfied with the system when they use it. Suppose the systems engineer develops a design that is used to create several products that are sold to different customers. To satisfy these customers, the systems engineer must consider two kinds of variability:

(a) Variability in the quality of the different units of product created from the design.

(b) Variability in the level of quality expected by different customers. For example, customers driving on cold snowy roads place very different demands on their car than customers driving in a dry hot desert.

The customer will only be satisfied when the quality of the product purchased by the customer meets or exceeds that customer's expectations (Bordley, 2001). If x is the difference between product quality and customer expectations, the customer will be satisfied when $x > 0$. If Q is the mean quality customers expect and q is the mean quality the system provides, then the mean of X is $M = q - Q$. Let V be the variance of the quality customers expect, and let v be the variance in the quality the system provides. If product quality is uncorrelated with customer expectations, the variance of x will be $v + V$ with s being the square root of that variance. Since the customer is satisfied when $x > 0$ and since $r = (x - M)/s$, the customer will be satisfied when $r > -M/s$. In order to maximize the number of satisfied customers, the engineer will maximize (M/s).

Improving mean quality q will always improve (M/s). But note the following:

1. If $M > 0$, reducing product variability will improve (M/s) and thus customer satisfaction.

2. If $M < 0$, reducing product variability reduces (M/s) and thus customer satisfaction.

Hence, contrary to intuition, reducing variability can reduce customer satisfaction if $M = q - Q < 0$, that is, if the product's mean quality is less than the customer's mean expectations.

Table 3.2 Distributions and Their Parameters and Functions

Name	Variable	First Parameter	Second Parameter	$P(x)$	Mean	Application
Binomial	x	a: Probability of success	N: # successes & failures	$N!/[x!(N-x)!]\, a^x(1-a)^{N-x}$ $C(x,N-x) = N!/[x!(N-x)!]$	Na	Discrete # successes in discrete time
Poisson	x	a: Mean successes		$\exp(-a)\, a^x/x!$	a	Discrete # of successes in continuous time
Standardized normal	$r = (x-M)/s$			$(1/2\pi)^{1/2} \exp(-r^2/2)$	0	Continuous unbounded variable
Standardized Weibull	$r = (x-M)/s$	a: Shape parameter		$a\, r^{a-1} \exp(-r^a)$	Gamma function at $(1+1/a)$	Continuous variable with only a lower bound
Standardized beta	$r = (x-M)/s$	a: Success parameter	b: Failure parameter	$r^{a-1}(1-r)^{b-1} C(a-1, b-1)$ $C(a,b) = (a+b)!/[a!b!]$	$a/(a+b)$	Continuous variable with upper and lower bound
Standard triangular	$r = (x-M)/s$	a: Best estimate		$2r/a$ for $r < a$, $2(1-r)/(1-a)$ for $r > a$	$(1+a)/3$	Continuous variable with upper and lower bound

Application of the Weibull Distribution: A core problem in systems engineer is decomposing a system requirement into subsystem requirements. For example, suppose the systems engineer must assign targets t and t^* to two engineering teams. The systems engineer is uncertain about the following:

1. Whether or not each engineering team can achieve these targets. If the target is more than the team can achieve, then the project fails. But if the target is less than what the team can achieve, the team will only deliver the target level of performance.
2. Whether or not the combined output, $t+t^*$, will satisfy stakeholder needs. If the output fails to satisfy stakeholder needs, the project fails.

Let x and x^* represent the uncertain amount each team can deliver, and let X represent what the stakeholder will expect when the system is finished. Then to maximize the probability of setting targets, which are both achievable and acceptable to the stakeholder, t and t^* should be chosen to maximize the joint probability

$$P(x > t, x \ast > t\ast, t+t \ast > X) = P(x > t)P(x^* > t^*)P(t+t^* > X) \quad (3.7)$$

which is the product of three probabilities.

To solve this problem (Bordley & Pollock, 2012), let s and s^* be the standard deviations describing the uncertain performance of the first and the second team, respectively. Let M and M^* be the lower bounds on what the first and the second team, respectively, can achieve.

Define $r = [(x-M)/s]$ to be the standardized performance of the first team, and define $r^* = [(x^* - M)/s^*]$ to be the standardized performance of the second team. Then the objective function can be rewritten as

$$P(x > t, x \ast > t\ast, t+t \ast > X) = P\left(r > \frac{t-M}{s}\right) P\left(r^* > \frac{t^*-M^*}{s^*}\right)$$
$$P(t+t^* > X) \quad (3.8)$$

Suppose the systems engineer can also specify an upper bound, U, on how demanding customer expectations might be. The quantity $(U-X)$ describes the uncertain amount by which actual stakeholder expectations will be less stringent than U. The systems engineer describes the likelihood of $(U-X)$ being less stringent than the most extreme expectations by specifying a scaling factor S. The scaling factor may partially reflect the systems engineer's confidence in how well customer expectations have been managed. Define $R = [(U-X)/S]$. Then the objective function can be further rewritten as

$$P(x > t, x \ast > t\ast, t+t \ast > X) = P\left(r > \frac{t-M}{s}\right) P\left(r^* > \frac{t^*-M^*}{s^*}\right)$$
$$P\left(R < \frac{[U-t-t^*]}{S}\right) \quad (3.9)$$

If r, r^*, and R are described by Weibull distributions, calculus can be used to solve for the values of t and t^* maximizing this probability. In this solution, both t and t^* will be proportional to $U-M-M^*$, which corresponds to the maximum gap between customer expectations and the worst-case product performances. As a result, the solution assigns each team responsibility for making up some fraction of the maximum gap. Because there is some possibility that customer expectations will not be as stringent as U, the teams will not be required to make up all of the gap.

Application of the Beta Distribution: Define the reliability of a supplier to the project as the supplier's probability, r, of producing a part meeting design specifications. Suppose the reliability of this supplier is unknown but the historical reliability of other suppliers is known. Suppose the fraction of those other suppliers having a reliability of r is proportional to a beta distribution $P(r)$ with parameters a and b. If we have no reason to believe that the project's supplier is systematically better or worse than these other suppliers, it may be reasonable to assume that this supplier is comparable to a supplier randomly drawn from this population. Given this assumption, the probability of the supplier having a reliability of r can be described by a beta distribution with parameters a and b. The supplier is asked to produce a small batch of n parts. The parts are tested, and k of these products are found to meet specifications while the other $(n-k)$ do not. The probability of observing this test result, if the reliability of the supplier were r, is given by a binomial probability $P(k|r,n)$. Given this information, the original probability $P(r)$ can be updated using Bayes' rule to

$$P(r|k,n) = \frac{P(k|r,n)P(r)}{P(k)} \quad (3.10)$$

Substituting for the binomial and beta distribution implies that $P(r|k,n)$ is proportional to

$$r^k(1-r)^{n-k}r^{a-1}(1-r)^{b-1} = r^{k+a-1}(1-r)^{n-k+b-1} \quad (3.11)$$

which is a beta distribution with parameters $a^* = a+k$ and $b^* = b+n-k$. So the new information has not changed the fact that r is described by a beta distribution. It has only changed the value of the parameters. This parameter change adjusts the probability of one of the supplier's part meeting specification from $E[r] = a/(a+b)$ to $E[r|k,n] = a^*/(a^*+b^*)$. Defining $w = n/(a+b+n)$ implies

$$E[(r|k,n] = \frac{a^*}{a^*+b^*} = \frac{(a+k)}{(a+b+n)} = \frac{(a+fn)}{(a+b+n)} = wf + (1-w)E[r] \quad (3.12)$$

As a result, the revised probability of the supplier's part meeting specifications, $E[r|k,n]$, is a weighted average of the frequency, f, estimated from experiment and the probability of meeting specifications, $E[r]$, based on the population of factories. See Raiffa and Schlaifer (1961) for a classic discussion of this and other updating formulas.

3.4 ESTIMATING PROBABILITIES

3.4.1 Using Historical Data

Note that
$$E[(r|k,n] = wf + (1-w)E[r] = f + (1-w)[E[r]-f] \tag{3.13}$$

As n increases, w increases toward 1, and the value of $E[r|k,n]$, which initially equals $E[r]$, increasingly approximates the empirical frequency. So $E[r|k,n]$ is an adjustment of the empirical frequency, f, to a value closer to some reference value $E[r]$. If $E[r|k,n]$ had equaled f, it would be an unbiased estimate – since it does not systematically deviate from the empirical data. Instead, it is an estimator that is based on "shrinking" the distance between the unbiased estimator and some reference value. Such a biased estimator is called a shrinkage estimator (Sclove, 1968). While bias is typically considered undesirable, the mean squared error of an estimate is the sum of the square of the bias in an estimator and variance. Hence, a shrinkage estimator could outperform a biased estimator if it had a lower variance, which more than offset the square of its bias. Such shrinkage estimators have been designed and validated in many applications, for example, marketing (Ni, 2007), where sample sizes are limited.

The shrinkage estimator, $E[r|k,n]$, had updated parameter values $a^* = a + k$ and $b^* = b + n - k$. Because k and $n - k$ represent actual observations of success and failure, it is common to refer to the parameters a and b as being pseudo-observations of success and failure with a^* and b^* being the sum of actual and pseudo-observations.

OMB guidance writes that, "*in analyzing uncertain data, objective estimates of probabilities should be used whenever possible.*" Consistent with this guidance, empirical Bayes methods (Casella, 1985) estimate the prior probability (and, in this case, a and b) from the empirical frequencies in some larger population of systems similar to the system of interest. If the system of interest were a new vehicle program, then the parameters a and b of the prior probability might be estimated by calculating the mean and standard deviation of failure rates for past vehicles made by that manufacturer. If there were N observations in this database of similar systems, then $E[r|k,n]$ combines the n observations from the experiment with the N historical observations of similar systems. Since the N observations are used to generate $(a+b)$ pseudo-observations, the average historical observation is given a weight of $w = (a+b)/N$ relative to each of the experimental observations.

Sometimes there are several choices for the historical population, for example,

1. a historical population of vehicles made by the same manufacturer;
2. a larger historical population of all vehicles made by any American manufacturer; and
3. a still larger historical population of vehicles made by any manufacturer.

Suppose we treat these populations as subpopulations of one large population. If we were to randomly pick a system – comparable to the system of interest – from this large population, it would be important to use stratified sampling in order to increase the probability of sampling from more similar subpopulations and decrease the probability of sampling from less similar subpopulations. If a weight is assigned to each subpopulation reflecting the similarity between that subpopulation and the system of interest, this stratified sampling would effectively weight observations from similar subpopulations more than observations from less similar subpopulation – consistent with the concept of similarity-weighted frequencies (Billot et al., 2005). This use of information from all the subpopulations, instead of just the most similar subpopulation, is consistent with the general finding (Poole & Raftery, 2000) that combining multiple different forecasts of an event yields more accurate predictions than using the best of any one of the forecasts (Armstrong, 2001).

Thus, the prior probability can be used to incorporate historical data on both very similar and less similar systems. The importance of incorporating such information was highlighted by the catastrophic failure of the Space Shuttle Challenger. Recognizing that the Space Shuttle was a uniquely sophisticated system, NASA had estimated its failure rate as extremely low using a sophisticated version of the fault tree methods. But when the failure occurred, this estimated failure rate seemed unrealistic. As the Appendix (Feynman, 1996) to the Congressional commission space shuttle catastrophe report noted, the historical frequency with which failures occurred in general satellite launches would have provided a considerably more realistic estimate of failure rate – even though the shuttle is very different from a satellite. These findings suggest that even though some systems engineering projects (e.g., the space shuttle, the Boston Big Dig, the first electric vehicle, first catamaran combat ship, complex software packages) are truly one-of-a-kind systems, historical data on less similar systems can still be helpful. In fact, Lovallo & Kaheman (2003) argued that forecasts of a project's completion time based on the historical completion time frequency of somewhat similar projects were considerably more realistic than model-based forecasts, which considered only important details about the project and ignored the historical information about the past, less similar projects.

While this section focused on updating beta probabilities, these same convenient updating rules can be generalized (Andersen, 1970) to apply to the larger "exponential" family of distributions. The exponential family of distributions includes most (but not all) of the probability distributions used in engineering, for example, the normal, gamma distribution, chi-squared, beta, Dirichlet, categorical, Poisson, Wishart and inverse Wishart distributions, the Weibull (when the shape parameter is fixed), and the multinomial distribution (when the number of trials is fixed). In these more general distributions, k is replaced by a vector of *sufficient statistics, that is, statistics* that describe all the information in the n observations relevant to the model parameters.

The generalizability of this updating rule across so many different probability distributions has motivated the definition of the Bayesian average as the sum of actual and pseudo values divided by the number of actual and pseudo-observations.

3.4.2 Using Human Judgment

The previous section discussed procedures for assessing prior probabilities from historical information. But suppose there is no historical information. To determine how to specify the prior probability, consider an analyst using prior probabilities to create a probability model that will be presented to the systems engineer. The systems engineer will then use this model to make various decisions. Consider the following two scenarios:

1. The results of this probability model were never shown to the engineers. As a result, the engineers relied on their experiences and intuitions to make decisions. Since these are the decisions the engineers would make prior to receiving new information, we refer to these decisions as the "default decisions."
2. The probability model was shown to the engineers. But the engineers felt that none of the information used by the probability model was credible. Furthermore, sensitivity analysis showed that the recommendations of the model were very sensitive to the inputs provided to that model. Consistent with the Garbage-In/Garbage-Out (GIGO) Principle, the engineers rejected the model. The engineers then made the default decisions (Figure 3.7).

In many applications, the default decisions involve continuing with current plans and not making any changes. In addition to being consistent with the Hippocratic dictum (Bordley, 2009) of "doing no harm," avoiding change may also be the most legally defensible course of action. In these applications, different systems engineers will agree on the default decisions. But in other applications, the default decisions may vary depending upon which systems engineer is making the decisions. In other words, the default decision depends upon the experiences and intuitions of the decision-maker. This potential subjectivity in the default decision highlights the importance of having *experienced* systems engineer with good intuition.

If the model is not credible, then these default decisions will be appropriate – as long as they were made rationally. If a rational decision is a decision satisfying the Bayesian axioms of rationality (Savage, 1954), then the default decisions will be consistent with the decisions supported by some hypothetical probability model where the prior probability in this model is equal to what is often called the personal

Figure 3.7 Default decision

probability. This personal probability, such as the default decision, can vary from engineer to engineer.

Since the default decisions are the appropriate decisions in this case, the analyst can ensure that the model leads to the appropriate decision by setting the prior probability equal to this personal probability. The model will then reproduce what the engineer would do if there were no credible data. If the model does have credible information, then the posterior probability implied by the model will update that personal probability based on that credible information.

For example, suppose an engineer will do nothing unless the probability of an innovation succeeding exceeds 95%. Then the engineer's personal probability of the innovation succeeding is less than 95%. But in this case, as in most cases, knowing the default decisions typically does not give the analyst sufficient information to specify the personal probability. As Abbas (2006) suggests, probabilities might be chosen to maximize entropy subject to those probabilities being consistent with various decisions. But even if the analyst uses maximum entropy arguments, the analyst will typically need more information on what decisions the engineer would make in other related decision situations. To gather this information, the analyst may present the engineer with hypothetical choice experiments and observe the engineer's response.

To illustrate these hypothetical choice experiments, consider a simpler situation where the analyst wants to assess the prior probability of some event occurring. Then in a typical experiment, the analyst asks a subject matter expert (SME) to choose between

- winning money by betting on the event of interest occurring. (If we are trying to assess a cumulative probability distribution $F(x)$, then the event of interest might be observing an uncertain value exceeding some threshold x) and
- winning money by betting on a casino gamble where the chance of payoff is known to be some stated probability P.

In this simple situation, the SME will always have a clear preference for either betting on the event or betting on the casino gamble unless P is set equal to the probability of the event occurring. Hence, by running this experiment with enough values of P, we can determine the personal probability. To minimize the number of experiments needed, the analyst typically begins with $P = 0.5$, observe the SME's decision and

- if the SME had decided to bet on the event, define the next experiment using a larger value of P (e.g., $P = 0.75$)
- if the SME had decided to be on the casino gamble, define the next experiment using a smaller value of P (e.g., $P = 0.25$).

These stated preference experiments are typically (Wardman, 1988) regarded as less reliable than revealed preference experiments in which individuals actually make a prior choice and receive a payoff (vs stating what they would have chosen); however, revealed preferences are not available in situations when there no analogues from the past. When there is substantial data, the probability will closely approximate historical frequencies (Gelman et al., 2013) and the prior probabilities will become irrelevant (as long as the default probabilities do not equal 0 or 1). Decades of psychological research has focused on improved ways of estimating personal probabilities from such experiments (Lichtenstein et al., 1982).

3.4.3 Biases in Judgment

In a gambling house, the probabilities are calibrated, that is, the asymptotic frequency with which an event occurs will equal the probability it is assigned. But there is no physical mechanism ensuring that this will be the case with personal probabilities. In fact, there is substantial empirical evidence indicating that personal probabilities are systematically biased. For example, probabilities are often based on what people can remember and human memory is systematically biased toward remembering more vivid events at the expense of less memorable events (Lichtenstein et al., 1978). Since individuals will want their personal probabilities to be as calibrated as possible, they will revise their personal probabilities if they are given evidence that they are poorly calibrated.

This willingness to revise a probability has important practical implications for systems safety engineering and, in particular, on the amount of system safety achievable with redundant components. Consider a system that is designed to fail only when 10 different components simultaneously fail. Let $A_1, \ldots, A_{10}$ denote the events of failure of components 1, ..., 10, respectively. Suppose these systems are causally independent, that is, the factors that would cause one system to fail are completely unrelated to the factors that would cause another system to fail. Then it is common to assume that these components are statistically independent so that the chance of the system failing is the product of $P(A_1), \ldots, P(A_{10})$. For simplicity, suppose that each component has an identical probability $r = (1/2)$ of failing. Then the chance of failure is roughly 1 in 1000. This calculation is valid for aleatory probabilities where r represents a long-run frequency.

But suppose these probabilities are personal probabilities. Then the expert will be uncertain about the actual long-run frequency corresponding to the probability r. Suppose this uncertainty about r is described, as before, with a beta distribution. In this beta distribution, let the parameters $a = b = 1$ so that $P(r)$ is a constant, that is, all values of r are equally likely. This is called a uniform distribution and is sometimes interpreted as representing ignorance. The mean value of r is $E[r] = 1/(1+1)$. As a result, $P(A_1), \ldots, P(A_{10}) = (1/2)$ as before.

Now suppose the first component fails. This failure gives the expert information about the realism of their personal probabilities. Using the rule for updating the beta distribution, the parameters of the beta distribution describing the expert's beliefs change to $a = 2, b = 1$. This implies that the probability of failure (the mean value of r) is now $2/(2+1) = (2/3)$. In other words, $P(A_2|A_1) = (2/3)$, which is unequal to $P(A_2) = (1/2)$. As a result, A_1 and A_2 are not statistically independent (Harrison, 1977). Hence, events that are causally independent need not be statistically independent *if their probabilities are assessed by a common measurement device* (in this case, the expert), which has some unknown level of bias.

Now suppose that the first two components fail. Then the parameters of the beta distribution change to $a = 3$, $b = 1$. As a result, $P(A_3|A_1 A_2) = (3/4)$. The same argument can be used to show that $P(A_k|A_1, \ldots, A_{k-1}) = (k/(k+1))$. Hence, the joint probability of all 10 components failing is

$$P(A_1)P(A_2|A_1)P(A_3|A_1A_2) \ldots P(A_{10}|A_1A_2 \ldots A_{10}) = 1/11 \qquad (3.14)$$

The probability of a system with 10 redundant components failing – in the presence of uncertain measurement bias – is now approximately 10%. This is two orders of magnitude greater than the estimated chance of failure when all probabilities were calibrated.

This phenomenon would only be a serious problem if individuals were well-calibrated. Unfortunately, the empirical evidence establishes that individuals are, in fact, poorly calibrated. This is hardly surprising. As the celebrated psychologist Amos Tversky noted, having an individual assess the probability of an event is at least as hard as having a navigator estimate the distance to the shore in a dense fog. Experienced navigators are better at estimating than inexperienced individuals. The same is also true for probability assessment. Weather forecasters are very well-calibrated, that is, roughly 20% of the events to which they assign probability $P = 0.20$ are eventually found to occur (and 80% do not occur), 40% of the events to which they assign probability $P = 0.4$ occur, and so on. This stimulated research on training individuals to become better probability assessors. This research has resulted in formal probability assessment training and certification programs (Hubbard, 2010).

One of the first steps in becoming a calibrated probability assessor is to understand the biases, which can distort probability assessments. There are many different kinds of bias, for example,

1. Motivational Bias. The most qualified expert on whether a research project will be successful is often the people working on the project. But these individuals know that assigning pessimistic probabilities to their project succeeding could lead to its cancellation – which jeopardizes the funding of themselves or their colleagues. So the most qualified expert will also have the greatest difficulty being objective. Instead of asking the expert to assess a probability, the expert should be asked to provide specific details on what they have accomplished on the project and what remains to be accomplished in order for the project to be a success. This can allow a comparison of progress of the project with other projects whose probability of success might be known. This approach is implemented in NASA's technology readiness level system (El-Khoury & Kenley, 2014).

2. Anchoring and adjustment bias. Suppose we are defining the lower, upper, and most likely values as part of assessing probabilities for continuous events. It has been found that individuals – if first asked to specify a most likely value – will

then estimate the lower and upper bounds by adjusting that most likely value. This typically leads to lower and upper bounds that are unrealistically close to the best estimate. To avoid this bias, the lower and upper bounds should be estimated before providing a best estimate.

3. Overconfidence in making assessments. To address this bias when individuals have estimated a lower bound, ask individuals to think about scenarios in which the true value might be lower than this lower bound. This often motivates them to further downwardly adjust their estimate to obtain a more realistic lower bound. Similarly, when individuals have made a tentative estimate of an upper bound, they should be asked to think of scenarios where values could be higher than this upper bound. Only when individuals are satisfied with their lower and upper bounds, should they be asked to specify a most likely value.

These biases – as well as other previously discussed biases – have been documented in more than half a century of psychological work, which was recognized by Daniel Kahneman, who received the Nobel Prize in 2002 (Flyvbjerg, 2006).

The Kolmogorov axioms (Billingsley, 2012) were enunciated to describe the aleatory uncertainties generated by randomization mechanisms (such as gambling devices) whose properties were well understood. But there are also epistemic uncertainties reflecting the lack of knowledge about the mechanisms (Kiurehgian & Ditlevsen, 2009) themselves. These biases in intuition occur regardless of whether the uncertainties are aleatory or epistemic. (There are also biases that only occur with epistemic probabilities (Ellsberg, 1961).)

This motivates using probability to remedy the biases associated with both aleatory and epistemic uncertainties. In fact, the Bayesian axioms of rationality require the use of probability for modeling all uncertainties. However, fuzzy set theory (Zimmermann, 2001) and Dempster–Shafer theory (Shafer, 1976) have been used as alternatives to probability theory for reasoning on epistemic uncertainties. Since most scientists and engineers are familiar with probability theory, this book uses probability theory to describe all uncertainties. This also allows the use of the Bayesian axioms of rationality as a unifying foundation for all the techniques discussed in this chapter.

3.5 MODELING USING PROBABILITY

3.5.1 Bayes Nets

The available information is often not captured in a manner that is adequate to make an informed decision. Fortunately, the rules of probability can be useful in converting the available information into the information that is adequate for making decisions. Converting this information requires us to exploit the interdependencies between the events about which we have information and the events on which we need to have

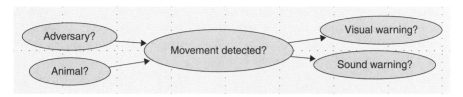

Figure 3.8 Bayes net example

information. Because of the sheer number of events and the complexity of the relationships involved, graphical representations can be quite useful.

Bayes Nets: Bayes nets (Pearl, 2000) are a graphical representation of the probabilistic interdependencies between events.

Bayes nets (Ben-Gai, 2007) are extremely useful in converting whatever information happens to be available into the information needed to make informed engineering decisions. For example, consider a warning system for a military base (Figure 3.8). There is a measurement device that is sensitive to the movements of adversaries and animals. Let $P(B) = 20\%$ be the probability of a bear being in the vicinity, let $P(A) = 5\%$ be the probability of an adversary being in the vicinity, and assume that these two events are probabilistically independent. Let $P(D|B) = 50\%$ be the probability of the measurement device detecting motion given there is an animal, let $P(D|A) = 90\%$ being the probability given an adversary, let $P(D|B\&A) = 95\%$ when both are present, and let $P(D|B^*\&A^*) = 0$ when neither are present. If movement is detected, there is a visual warning (VW) system providing a warning with probability $P(VW|D) = 80\%$ and a sound system providing a warning with a probability $P(SW|D) = 90\%$. Suppose there is a probability $P(VW|D^*) = P(SW|D^*)$ of both systems given warnings when no motion is detected. We can represent this problem graphically:

The absence of an arc connecting Adversary and Animal in the Bayes Net is a representation of the assumption that the two events are independent, that is, the probability of both occurring is given by

$$P(A\&B) = P(A)P(B) = 20\%(5\%) = 1\% \quad (3.15)$$

The probability that the device detects motion using the law of total probability is

$$P(D|A\&B)P(A\&B) + P(D|A\&B^*)P(A\&B^*) + P(D|A^*\&B)P(A^*\&B) \quad (3.16)$$
$$+ P(D|A^*\&B^*)P(A^*\&B^*) = 95\%(20\%)(5\%) + 90\%{}^*80\%(5\%)$$
$$+ 50\%{}^*95\%{}^*20\% + 10\%(80\%)(95\%) = 21.65\%$$

Note that while arcs connect the two events, sound warning (SW) and VW, to the detection event (D), no arcs connect SW or VW to either A and B. So once the outcome of D is known, the occurrence of either a VW or a SW will be independent of whether or not there was an animal or adversary. As a result,

$$P(SW|D, A, B) = P(SW|D, A, B^*) = P(SW|D, A^*, B) \quad (3.17)$$
$$= P(SW|D, A^*, B^*) = P(SW|D) = 90\%$$
$$P(VW|D, A, B) = P(VW|D, A, B^*) = P(VW|D, A^*, B) \quad (3.18)$$
$$= P(VW|D, A^*, B^*) = P(VW|D) = 80\%$$

Also, note that there is no arc connecting VW to SW. As a result, the events VW and SW are assumed to be conditionally independent – given movement is detected (D) and also given it is undetected (D^*), that is,

$$P(VW\&SW|D, A, B) = P(VW|D, A, B)P(SW|D, A, B) = P(VW|D)P(SW|D) \quad (3.19)$$

$$= (80\%)(90\%) = 72\%$$
$$P(VW\&SW|D^*, A, B) = P(VW|D^*, A, B)P(SW|D^*, A, B) = P(VW|D^*)P(SW|D^*) \quad (3.20)$$

$$= (10\%)(10\%) = 1\%$$

To show that VW and SW, while being conditionally independent, are not unconditionally independent, we compute the probability of occurrence of both the VW and SW as

$$P(VW\&SW) = P(VW\&SW|D)P(D) + P(VW\&SW|D^*)P(D^*) \quad (3.21)$$
$$= 72\%(21.65\%) + 1\%(78.35\%) = 16.35\%$$

We now show that $P(VW \& SW) > P(VW)*P(SW)$ by computing

$$P(VW)^*P(SW) = [P(VW|D)P(D) + P(VW|D^*)P(D^*)] \quad (3.22)$$
$$* [P(SW|D)P(D)(SW|D^*)P(D^*)]$$
$$= [80\%(21.65\%) + 10\%(78.35\%)]$$
$$* [90\%(21.65\%) + 10\%(78.35\%)] = 6.87\%$$

VW and SW are not independent because both warnings depend upon whether or not the detector was triggered.

This graphical representation allows for the computation of the joint probability of the events: SW, VW, D, B, and A. If the joint probability is known, then all the probabilities associated with any combination of these events can also be calculated.

For example, it is possible to calculate the probability of an adversary, but not an animal being present, given there was a VW. Since these probabilities are based on the technologies used in designing the detectors and warning systems, they may change if these technologies were replaced with newer technologies. As a result, the Bayes net could be used to specify how different technologies change the probability of soldiers being warned given an adversary is present as well as the probability of a false alarm (the probability of soldiers being warned when an adversary is not presented). When rectangular nodes corresponding to the choice of technologies (and other decisions) are explicitly added to the Bayes net, the Bayes net becomes an influence diagram (Shachter, 1986).

3.5.2 Monte Carlo Simulation

In deterministic modeling, a mathematical or physical model is constructed, which calculates the variables of interest from input variables with known values in a certain setting. The model is calibrated by testing to ensure that it gives the correct answers for settings where both the input and output values are known. If the input values are uncertain but we know the probability distribution of the input values, then the parameters of the distribution of output variables (e.g., the expected value and the variance) can sometimes be calculated. But if there are many input variables or the functional form of the input–output model is not restricted to relatively simple, closed forms such as linear and quadratic equations, then the exact calculation of expected values and variances is very difficult. (This is the curse of dimensionality and nonlinearity.) In fact, if the model is complex enough, calculating output parameters from a small number of input parameters can be very difficult.

To address this computational problem, we return once again to the gambling house context, which was the motivation behind probability theory. Instead of developing a model relating input parameters to output parameters, this procedure only uses the original deterministic model relating input values to output values.

Suppose we are trying to estimate the maximum velocity of a boat through the water. This will typically depend on the length, width (beam), depth (draft), and weight of the boat as well as the temperature, turbulence, and many other aspects of the water. Since the equations required to calculate the resistance of the water to the boat are notoriously complicated – and usually require test data – closed-form solutions are generally not feasible. To estimate velocity with the simplest version of the Monte Carlo process, assume that our input random variables X_k are independent and have a known probability distribution $F(X_k)$. In this case, the Monte Carlo process

1. uses random number generators to generate n independent and uniformly distributed random variables $(U_1, \ldots, U_n)$;
2. assigns the ith input variable value x_1 for which $F(x_i) = u_i$ for $i = 1, \ldots, n$;
3. uses the model to calculate an output value y_1 from these input values;
4. stores these values for the ith iteration in a table and repeats step 1 until the desired number of iterations has been implemented.

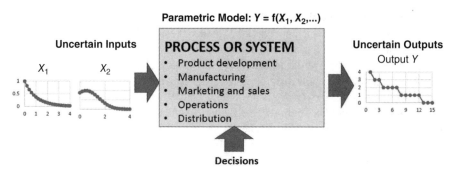

Figure 3.9 Monte Carlo simulation

By the central limit theorem, the mean of the frequency distribution of the output values in the table will approximate the normal distribution with a variance inversely proportional to the number of parameters. As a result, the sample mean and variance will then give us the desired parameters of the approximate output distribution of mean of the outputs (see Figure 3.9).

In realistic settings, it is important to consider four refinements of this approach.

3.5.3 Monte Carlo Simulation with Dependent Uncertainties

To motivate the first refinement, note that the length, width, draft, and weight of a boat were assumed independent. But the weight of the boat will be less than the product of the length, width, draft, and mass density. As a result, the weight is not independent of the length, width, and draft. This shows that it is necessary to allow for dependency between the input variables $X_1, \ldots, X_n$. We now use the following Monte Carlo procedure to generate random variables with the appropriate multivariate distribution.

The Monte Carlo methods generated independent and random values of $X_1, \ldots, X_n$ by first generating uniformly and independently distributed values of $U_1, \ldots, U_n$. To generate values of $X_1, \ldots, X_n$, which are dependent, we must generate dependent values of $U_1, \ldots, U_n$ from some multivariate distribution, which describes uniform random variables that are dependent. We denote this multivariate distribution by $C(U_1, \ldots, U_n)$ where $C(U_1, \ldots, U_n) = U_1 \ldots U_n$ if the variables are independent and equals $\min(U_1, \ldots, U_n)$ if the variables are perfectly positively dependent. Copula theory (Nelsen, 2007) establishes that C will be that function for which $F(X_1, \ldots, X_n) = C(F(X_1), \ldots, F(X_n))$.

Hence, to use Monte Carlo methods when $X_1, \ldots, X_n$ are dependent, we first specify the multivariate distribution $F(X_1, \ldots, X_n)$ and the univariate distributions $F(X_1), \ldots, F(X_n)$, then infer the function C, and then generate the uniform random variables $U_1, \ldots, U_n$ from the distribution $C(U_1, \ldots, U_n)$. We then use the ith uniform random variable generated from distribution C to assign a value to x_i by setting $F(x_i) = u_i$.

3.5.4 Monte Carlo Simulation with Partial Information on Output Values

To motivate the second refinement, suppose the model indicated that our sailing boat would achieve an average maximum speed of 700 knots (which is more than 10 times the world's record.) This unusually high-speed prediction will naturally make us skeptical of either the model or some of the input values used in the model (Kadane, 1992). If we felt the model was accurate and if there were no uncertainty in our input values, then we would simply accept the model's prediction as valid and ignore our initial skepticism. If we were absolutely confident in our information about maximum boat speeds, we would use that information about boat speeds to adjust (or calibrate) our prior input values. But in most cases, we will be uncertain about both inputs and outputs.

This will require a procedure for simultaneously adjusting both input values and output values. Let X denote the uncertain input variables and Y the output variables with $P(X,Y)$ being a prior probability, which is not based on the model but is based on all other judgmental and historical data. Suppose we consider the model without any information on the values of input variables. Then the model only gives us new information about previously unrecognized relationships between the input and output values (Bordley & Bier, 2009). To represent this information, let $Y(x)$ be a random variable describing the possibly uncertain value of y, which the model generates given the input value x.

First, suppose we describe the information from the model by the output of a single simulation run. In this run, random values x_1 for each input variable are generated based on our prior beliefs $P(X)$. The model is then used to infer the output value y_1 associated with those input values. In the absence of the model, the probability of observing this output value y_1 – given the knowledge of X and Y and the input value x_1 – would be

$$P(y_1|x_1, XY)$$

As Bordley (2014) proved, this probability is proportional to the likelihood of the model being accurate when the true input and output values are known to be X and Y.

Now suppose we describe the model with m simulation runs. Then $(x_1, \ldots, x_m)$ represents the m inputs of each of those runs while $(y_1, \ldots, y_m)$ are the output values which the model associates with these input values. The likelihood of the model would be proportional to

$$P(y_1|x_1, XY) \ldots P(y_m|z_m, XY)$$

Multiplying the likelihood by $P(XY)$ gives the updated probability for XY given the model. Integration over X then gives the updated probability of Y. When all uncertainties are Gaussian, $P(y_k|x_k, XY)$ can be written in terms of the probability of deviations between the simulated inputs and true input values leading to some level of deviation between the simulation output and the true output value. In this Gaussian case, the expected value of the output variable will be a weighted average of two forecasts:

1. A forecast based on only the model and input information.
2. A forecast based on all other (prior) information about the output value.

3.5.5 Variations on Monte Carlo Simulation

Our last two refinements involve more substantive changes to Monte Carlo analysis. To discuss the third refinement, note that each iteration in the standard Monte Carlo process generates a new sequence of input values by randomly drawing from the same input distribution. But sometimes, this input distribution, $P(x)$, is difficult to calculate. Fortunately, there is an alternative approach that allows us to perform this sampling while generating observations using more tractable expressions for the probability distribution.

1. Let $k = 1$. Generate the first value of $x = x_1$ based on some convenient density $q(x)$.
2. Define $k = k + 1$. Generate the value of x_k for the kth iteration using the more easily calculated conditional probability $p(x_k | x_{k-1})$.
3. Repeat step 2.

Because the value generated in the $(k+1)$st iteration only depends on the value generated in the kth iteration, this iterative process forms a Markov chain. For Markov chains, the probability with which x_k is generated on the kth iteration will approximate some stationary distribution $P(x)$ for large k – as long as $q(x)$ and $P(x_k | x_{k-1})$ assign nonzero probabilities to all combinations of observations. This Monte Carlo Markov chain approach (Brooks et al., 2011) has been widely used in the recent decades.

For the fourth refinement, note that the Monte Carlo approach randomly selects input values, uses the model to calculate an output value, and then reruns the model for different selections of input values. An alternate approach is to model each single input value as a vector with the components of that vector being randomly selected values for that input. The model could then be applied once to the vector for each input value in order to generate a vector of output values. While a variation of this approach was implicitly implemented in the Analytica Bayes Net software (Morgan & Henrion, 1998), Savage developed a technique for enabling Excel to perform these operations. Unlike the Monte Carlo Markov chain, this technique, called probability management, is a recent new development, which is discussed in more detail in Section 9.5.

3.5.6 Sensitivity Analysis

Logical and physical models are used to model the performance at each higher level as a function of the performance at each lower level. However, we cannot calculate

MODELING USING PROBABILITY

the overall performance without specifying the values for the inputs. While the Monte Carlo method was designed to avoid us having to specify exact values for inputs, we still need to specify a probability distribution for these input values. Since collecting information takes time and resources, a preliminary analysis is performed with very limited information (e.g., with information on only the lower and upper bounds of the input variables.)

Suppose we are interested in determining whether we are at risk of violating our mass target for a vehicle. Vehicle mass is driven by the mass of various subsystems. Suppose there were only three vehicle subsystems, that is, door modules, engine, and body. Suppose we only set a subsystem's mass to its lower or upper bound. This leads to $2^3 = 8$ different combinations of vehicles (Table 3.3).

In a full factorial design of experiments (Barrentine, 1999) with two levels, a separate simulation run would be used to calculate vehicle mass for each of these possible combinations. Because there were only three input variables, there were only eight vehicle mass estimates. In the actual application, there were 13 input variables and 8192 (2^{13}) different mass estimates. A separate simulation run was used to calculate vehicle mass for each of these possible combinations.

We then pick one input variable, for example, seats.

1. In half of the runs (or 4096 cases), seat mass was set to its lower bound. We calculate the average vehicle mass for those classes. This gives us the average vehicle mass given that the seat mass is set to its lower bound.
2. In the other half of the runs (or 4096 cases), the seat mass was set to its upper bound. We similarly calculate the average vehicle mass for those cases. This gives us the average vehicle mass given that the seat mass is set to its upper bound.

The difference between the higher and lower of these two numbers gives the variation in vehicle mass associated with variations in seat mass.

We repeat this procedure for each of the input variables and obtain the average vehicle masses when each variable's mass is set to the lower bound and when each variable's mass is set to the upper bound. As Figure 3.10 shows, we then plot the upper

Table 3.3 Vehicle Mass Combinations

Run	Door	Engine	Body
1	Lower Bound	Lower Bound	Lower Bound
2	Upper Bound	Lower Bound	Lower Bound
3	Lower Bound	Upper Bound	Lower Bound
4	Upper Bound	Upper Bound	Lower Bound
5	Lower Bound	Lower Bound	Upper Bound
6	Upper Bound	Lower Bound	Upper Bound
7	Lower Bound	Upper Bound	Upper Bound
8	Upper Bound	Upper Bound	Upper Bound

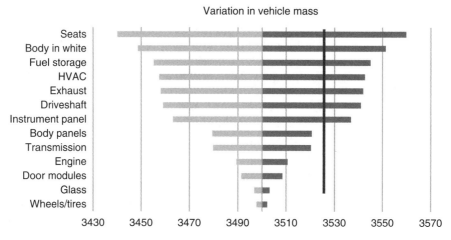

Figure 3.10 Tornado diagram for sensitivity analysis

and lower vehicle masses associated with each variable – with the highest variation variables plotted at the top of the chart. This chart highlights the variables whose variation has the greatest impact on vehicle mass. The data point at the junction of the light gray bar and the dark gray bar for each variable is the estimate of the expected value using all 8192 (2^{13}) estimates.

The chart includes a red line for a mass target, which identifies those variables that could potentially lead to violations in that target. But in some cases, the red line – instead of representing the target – might represent the mass associated with an alternative design. In this case, the seat mass might – if it were known to equal its worst-case estimate – lead us to consider switching to the alternative design. However varying the map of the body panels will not cause us to switch to the alternative design. In fact, the bottom six variables – even if they were set to their upper value – would not change our decision while the top seven variables might. Thus, the sensitivity analysis has identified those variables which could – if we learned more about their actual values – potentially change the decision suggested by a deterministic analysis. This allows attention to be focused on better understanding and quantifying the uncertainty in these two factors – using the methods discussed in this section.

The sensitivity analysis may also be used for a second purpose. Once the critical uncertainties are identified, they can be used to construct a small number of intuitively meaningful scenarios with each scenario reflecting the outcome of the uncertainties assuming various extreme values. These scenarios are communicated to the stakeholders. Use cases can then be constructed, which allow stakeholders to describe how they want the system to behave in the presence of these scenarios.

As noted previously, calculation of these estimated vehicle masses when there are 13 variables requires 8192 observations. When the separate experiment required to generate each observation is time-consuming and costly, performing this analysis may not be cost-effective. Fractional factorial designs (Kulahci & Bisgaard, 2005) provide an ingenious way of conducting the sensitivity analysis with fewer observations.

3.6 SUMMARY

Uncertainty is implicit, although frequently ignored, in many aspects of systems engineering. In reliability-based design optimization (Bordley & Pollock, 2009), the project manager often minimizes two kinds of risks:

1. Project Risk: The risk of not satisfying stakeholder expectations.
2. Safety Risk: The risk of injury to individuals or damage to property.

While originally developed to describe the uncertainties encountered in gambling, probability is generally applicable to all uncertainties. Because human intuition is easily misled in thinking about uncertainties, systematic use of applied probability is essential. Probabilities should be calculated using

- models to relate the event of interest, Y, to other events, X;
- experimental or field information on past occurrences of Y or X or past occurrences of events partially similar to X or Y;
- judgmental probabilities – collected with psychologically validated elicitation protocols.

Monte Carlo and related techniques allow for the integration of analytical modeling with probability assessments. Sensitivity analysis can then be used to determine which uncertainties have the biggest impact on the outputs of the Monte Carlo model.

3.7 KEY TERMS

Anchoring and Adjustment Bias: Psychological tendency to underestimate the range on a variable by anchoring on a single number and then making upward and lower adjustments in that number.

Bayes Nets: Graphical model specifying each variable using a node with the probabilistic relationship between different variables represented by arcs between different nodes.

Beta Distributions: Widely used probability distribution for a continuous variable with lower and upper bounds.

Binomial Distributions: Probability distribution for number of successes in a discrete number of trials.

Cognitive Biases: Psychologically innate tendencies to deviate from how a rational person would make a decision or describe their beliefs and values about a situation.

Cumulative Distribution: Probability of a variable being less than some input value x.

Design of Experiments: A systematic method for specifying the input values to a series of experiments, which extracts the maximum amount of information on quantities of interest.

Decomposition Bias: A systematic tendency for the sum of the psychologically assessed probability of different events exceeding the psychologically assessed probability of any one of those events occurring.

Bayes Rule: A rule for calculating the conditional probability of an event given new information from

- the probability of that event and alternative events prior to getting that information
- the likelihood of that information if each of these possible events were true.

Deterministic Dominance: The value of a random variable is never less (and in some cases exceeds) than that of another random variable.

Effect Size: A measure of how much the outcome of one treatment differs from the outcome of another.

Eliciting Probabilities from Experts: A scientific procedure for describing an expert's beliefs about an uncertainty based on the expert's observed responses to different questions or problems.

Ignoring Base Rate Bias: Assess the probability of an event without considering the probability with which identical (or nearly identical) events have occurred in the past.

Laws of Probability: An extension of the rules of logic for reasoning about uncertain quantities.

Monte Carlo Simulation: Technique for calculating the probability distribution of some possibly probabilistic function by repeatedly sampling the input values from probability distributions.

Motivational Bias: A tendency to assess probabilities so as to favor one's self-interest.

Normal Distribution: A distribution that – when the number of observations is large – closely approximates the sum of the values assigned to those observations. This distribution is also used to describe the distribution of quantities with no lower or upper bounds.

Optimism Bias: Tendency to overestimate the probability of desirable events occurring.

Probability: A measure of a rational individual's willingness to gamble on an event's occurrence.

Probability Assessment: A probability estimated based on an individual's judgment.

Redundancy: Deliberately creating multiple components to perform a function that could be achieved by one function. Typically used to increase the probability of that function being performed given the occurrence of unanticipated events.

System Safety: The probability of a system not leading to unanticipated injury or loss of life.

Uncertainty: Imperfect knowledge of the value of some variable.

Representativeness Bias: Estimation of the probability of an observation being drawn from some population based on the degree to which the observation is similar to other units in that population (Kahneman & Tversky, 1972).

Risk: A potential and undesirable deviation of an outcome from some anticipated value.

Sensitivity Analysis: A procedure for determining how much a variable of interest changes as a function of changes in the values of other variables.

Stochastic Dominance: A relationship between two uncertain quantities where the first uncertain quantity never has a smaller probability (and in at least one case, has a better probability) of achieving an outcome ranked higher than the second quantity.

Triangular Distributions: A popular distribution in which the probability of achieving a higher score increases linearly in that score up to some peak value and then decreases linearly.

Updating Probabilities Based on Data: A process for revising a probability given new information (data) specified by the rules of probability.

Weibull Distributions: Popular distribution for describing an uncertainty with a lower bound.

3.8 EXERCISES

3.1. Suppose the probability of a typical project being successful was 40%. Suppose that of those projects which have eventually been successful, 80% of them have had a strong sponsor. In contrast, of those projects which have not eventually been successful, only 40% of them had a strong support.

   (a) If your project has a strong sponsor, what is the probability of it being successful?

   (b) What is the probability of the typical project having a strong sponsor?

   (c) What is the probability your project fails if it does not have a strong sponsor?

(d) Calculate the odds of success (the ratio of the probability of success to the probability of failure) both for a typical project and for a project with a strong sponsor. How much does a strong sponsor change the odds of success? How does this compare with the likelihood ratio?

3.2. Suppose your system uses five parts. There is a 5% probability of each part being defective. Part failures are independent.

(a) If your system fails if any part fails, what is the probability of the system failure?

(b) If your system fails if any two parts fail, what is the probability of the system failure?

(c) Suppose the probability of a part failing doubles to 10%. If your system fails when any two parts fail, what is the probability of the system failure?

(d) How do these three probabilities of system failure compare?

3.3. Your system will be used for 10 years. The probability of failing at any point in time does not change. The expected number of system failures in the course of 10 years is expected to be 2.

(a) What is the probability of the system having one failure in its lifetime? What is the probability of the system having five failures in its lifetime?

(b) If there are two failures in 10 years, the rate of failure is 1/5. What is the probability of the system not having any failures in 2 years?

(c) What is the expected number of failures in 2 years?

3.4. The weight of a component could, once it is designed, range anywhere from 10 pounds to 20 pounds. Your best estimate of the component weight is 12 pounds.

(a) If you assume that the weight follows a triangular distribution, what is the probability of the component weight being less than 15 pounds?

(b) How about if you use assume that the weight follows a Beta Pert distribution?

3.5. You are assigning mass targets t, t^*, and t^{**} to three engineering teams. Your estimates of the maximum amount of mass the teams ought to require are 10 pounds, 20 pounds, and 30 pounds. Your best estimate of how much mass they will need is 5 pounds, 10 pounds, and 20 pounds. In the worst case, your customer might require that the system only weighs 30 pounds. However, your best estimate is that the customer will require that the system weighs 40 pounds.

(a) How should you assign targets t, t^*, and t^{**} to the three engineering teams?

(b) How does your answer change if your best estimate of the customer's weight requirement is 50 pounds?

(c) How does your answer change if your best estimate of what the first team can deliver changes from 5 pounds to 8 pounds?

EXERCISES

3.6. Your process is capable of delivering quality with a mean of 20 and a standard deviation of 5. Your customer expectations for quality have a mean of 15 and a standard deviation of 10.

(a) What is the proportion of customers whose expectations you will meet?

(b) Suppose you focused on reducing variability and the standard deviation of quality is reduced from 5 to 2. How does your answer change?

(c) Suppose you focused on improving capability and the mean quality increased from 20 to 25. How does your answer change?

(d) Suppose your key competitor produced a vehicle with a quality of 25. Suppose this shifts the mean quality expected by customers from 25. How does your answer change? Would you focus on reducing variability in your process or on improving quality?

3.7. Suppose you are uncertain about the actual reliability of your system. You believe it can be described by a beta distribution with parameters $a = 5$ and $b = 3$.

(a) What is the expected reliability of your system?

(b) What is the probability of your system producing three good products in a row?

(c) If your system does produce three products in a row, what would be your revised assessment of the expected reliability of the system?

3.8. Bayes Net: Draw a Bayes Net describing the factors influencing a company's profit.

(a) In this model, profit is the product of price and sales minus fixed costs

(b) Sales is the minimum of actual capacity and product demand.

(c) Fixed costs is the product of intended capacity and cost per unit of capacity.

(d) Actual capacity is the product of intended capacity and availability.

(e) Availability is 1 minus the percentage downtime.

(f) Product demand is the product of total demand and product market share.

(g) Total demand is 10% of Gross National Product.

(h) Product market share is product appeal divided by the sum of the product's appeal and the rival product's appeal.

(i) Your product's appeal is product quality minus price.

3.9. Updating probability, given a model

(a) Suppose product sales, S, are normally distributed with a mean of 1000 and a standard deviation of 100. Suppose the price, P, of gasoline is normally distributed with a mean of $3 and a standard deviation of 20 cents. Construct the prior probability density over sales and price.

(b) Suppose a model specifies that $S = 1000\,P + e$, where e is an error term that is normally distributed with mean 0 and a standard deviation of 10. So if $S = s$ and $P = p$, the error in the model is $e = s - 1000\,p$. Using your prior probability density, what is the probability density for having $P = p$, $S = s$, and a model error of $= s - 1000p$? Rewrite this probability in the form of a normal density defined over the two variables S and P.

(c) What is the correlation between S and P, given the model?

(d) What is the mean value of S and P, given the model?

(e) How would your answer change if the standard deviation on price changes from 20 cents to 1 cent?

(f) How would your answer change if the standard deviation on sales changed from 100 to 10?

(g) How would your answer change if the standard deviation of model error changed from 10 to 50?

3.10. An assembled leaf spring was passed through a furnace. You have two variables:

(a) Variable A (or temperature), which could be high (+) or low (−)

(b) Variable B (or time in furnace), which could be long (+) or short (−).

Define an interaction variable AB, which is + when A is + and B is + or when A is − and B is −. Otherwise, AB is −. You run several experiments and obtain the following results.

Treatment		Interaction	Test Result
A	B	AB	Score
−	−	+	32
+	−	−	35
−	+	−	38
+	+	+	31

What is the average test result when A is at its − level? How about at its + level? Answer the same questions for B and AB. Perform sensitivity analysis and display the results on a tornado chart.

REFERENCES

Abbas, A. (2006) Maximum entropy utility. *Operations Research*, **54**, 277–290.

Andersen, E.B. (1970) Sufficiency and exponential families for discrete sample spaces. *Journal of the American Statistical Association*, **65**, 1248–1255.

Armstrong, S. (2001) *Principles of Forecasting*, Kluwer Academic Publishing.

Barrentine, L. (1999) *Introduction to Design of Experiments*, ASQ Quality Press, Wisconsin.

Ben-Gai, I. (2007) Bayesian networks, in *Encyclopedia of Statistics in Quality and Reliability* (ed. F.F. Ruggeri), John Wiley & Sons, New York.

REFERENCES

Bernstein, P. (1998) *Against the Gods: The Remarkable Story of Risk*, John Wiley & Sons.

Billingsley, P. (2012) *Probability and Measure*, John Wiley & Sons, New Jersey.

Billot, A., Gilboa, I., and Schmeidler, D. (2005) Probabilities as similarity-weighted frequencies. *Econometrica*, **73**, 1125–1136.

Bordley, R. (2001) Integrating gap analysis and utility theory in service research. *Journal of Service Research*, **3**, 300–309.

Bordley, R. (2009) The hippocratic oath, effect size and utility. *Journal of Medical Decision Making*, **29**, 377–379.

Bordley, R. (2014) Reference class forecasting: resolving a challenge to statistical modeling. *American Statistician*, **68**, 221–229.

Bordley, R. and Bier, V.M. (2009) Updating beliefs about variables given new information on how those variables relate. *European Journal of Operations Research*, **193**, 184–194.

Bordley, R. and Hazen, G.B. (1991) SSB and weighted linear utility as expected utility with suspicion. *Management Science*, **37**, 396–408.

Bordley, R. and Pollock, S.M. (2009) A decision analytic approach to reliability based design optimization. *Operations Research*, **57**, 1262–1270.

Bordley, R. and Pollock, S.M. (2012) Assigning resources and targets to an organization's activities. *European Journal of Operations Research*, **220**, 752–761.

Brooks, S., Gelman, A., Jones, G.L., and Meng, X.-L. (2011) *Handbook of Markov Chain Monte Carlo*, Chapman and Hall, Florida.

Casella, G. (1985) An introduction to empirical Bayes data analysis. *The American Statistician*, **39**, 83–87.

El-Khoury, B. and Kenley, R. (2014) TRL-based cost and schedule models. *Journal of Cost Analysis and Parametrics*, **7**, 160–179.

Ellsberg, D. (1961) Risk, ambiguity and the savage axioms. *Quarterly Journal of Economics*, **75**, 643–669.

Feynman, R. P. (1996). Personal Observations on the Reliability of the Shuttle. *Commission Report on Space Shuttle Accident: Subcommittee on Science Technology and Space*.

Flyvbjerg, B. (2006) From nobel prize to project management: getting risks right. *Project Management*, **7**, 5–15.

Gelman, A., Carlin, J.B., Stern, H.S. *et al.* (2013) *Bayesian Data Analysis*, CRC Press, New York.

Grissom, R.J. and Kim, J.J. (2011) *Effect Sizes for Research: A Broad Practical Perspective*, Lawrence Erlbaum, Matwah, New Jersey.

Harrison, J.M. (1977) Independence and calibration in decision analysis. *Management Science*, **24**, 320–328.

Hodge, H.C. and McKay, J.H. (1934) The microhardness of materials comprising the mohs scale. *American Mineralogist*, **61**, 161–168.

Hubbard, D.W. (2010) *How to Measure Anything: The Value of Intangibles in Business*, John Wiley & Sons, New York.

Johnson, N.L., Kotz, S., and Balakrishnan, N. (1995) *Continuous Univariate Distributions*, John Wiley & Sons, New York.

Kadane, J.B. (1992) Healthy skepticism as an expected utility explanation of the phenomenon of Allais and Ellsberg. *Theory and Decision*, **77**, 57–64.

Kahneman, D. and Tversky, A. (1972) Subjective probability: a judgement of representativeness. *Cognitive Psychology*, **185**, 430–454.

Kiurehgian, A.D. and Ditlevsen, O. (2009) Aleatory or epistemic: does it matter. *Structural Safety*, **31**, 105–112.

Kulahci, M. and Bisgaard, S. (2005) The use of plackett-burman designs to construct split-plot designs. *Technometrics*, **17**, 495–501.

Lichtenstein, S., Fischhoff, B., and Phillips, L.D. (1982) Calibration of probabilites: the state of the art, in *Judgement under Uncertainty: Heuristics and Biases* (ed. P.S.D. Kahneman), Cambridge University, Cambridge, pp. 306–334.

Lichtenstein, S., Slovic, P., Fischhoff, B. *et al.* (1978) Judged frequency of lethal events. *Journal of Experimental Psychology: Human Learning and Learning*, **4**, 551–578.

Lovallo, D. and Kahneman, D. (2003) Delusions of success: how optimism undermines executives' decisions. *Harvard Business Review*, **81**, 56–63.

McGraw, K.O. and Wong, S.P. (1992) A common language effect size measure. *Psychological Bulletin*, **111**, 361–368.

Morgan, G. and Henrion, M. (1998) Analytica: a software tool for uncertainty analysis and model communication, in *Uncertainty: A Guide to Dealing with Uncertainty in Quantitative Risk and Policy Analysis* (ed. G.A. Morgan), Cambridge University Press, New York.

Nelsen, R.B. (2007) *An Introduction to Copulas*, Springer Series in Statistics.

Ni, J. (2007). Extensions of hierachical bayesian shrinkage estimation with applications to a marketing science problem. Los Angeles: PhD Dissertation, University of Southern California.

Pearl, J. (2000) *Causality: Models, Reasoning*, Cambridge University Press, Cambridge.

Poole, D. and Raftery, A.E. (2000) Inference for deterministic simulation models: the bayesian melding approach. *Journal of the American Statistical Association*, **95**, 1244–1255.

Raiffa, H. and Schlaifer, R. (1961) *Applied Statistical Decision Theory*, Harvard University Press, Cambridge.

Savage, L.J. (1954) *The Foundation of Statistics*, John Wiley & Sons.

Savage, S.L., Danziger, J., and Markowitz, H.M. (2012) *The Flaw of Averages: Why We Underestimate Risk in the Face of Uncertainty*, John Wiley & Sons.

Sclove, S.L. (1968) Improved estimators for coefficients in linear regression. *Journal of the American Statistical Association*, **65**, 596–606.

Shachter, R.D. (1986) Evaluating influence diagrams. *Operations Research*, **34**, 871–882.

Shafer, G. (1976) *A Mathematical Theory of Evidence*, Princeton University Press, Princeton.

Taleb, N.N. (2007) *The Black Swan: The Impact of the Highly Improbable*, Random House, New York.

Tversky, A. and Kahneman, D. (1974) Judgement under uncertainty: heuristics and biases. *Science*, **185**, 1124–1131.

Tversky, A. and Koehler, D.J. (1994) Support theory. *Psychological Review*, **101**, 547–567.

Villejoubert, G.A. and Mandel, D.R. (2002) The inverse fallacy: an account of deviations from bayes rule and the additivity principle. *Memory and Cognition*, **30**, 171–178.

Wardman, M. (1988) A comparison of revealed preference and stated preference models of travel behavior. *Journal of Transport Economics and Policy*, **22**, 71–91.

Wilkinson, L.A. (1999) Statistical methods in psychology journals: guidelines and explanations. *American Psychologist*, **54**, 598–604.

Zimmermann, H. (2001) *Fuzzy Set Theory and its Applications*, Kluwer Academic Press, Massachusetts.

4

ANALYZING RESOURCES

EDWARD A. POHL
Department of Industrial Engineering, University of Arkansas, Fayetteville, AR, USA

SIMON R. GOERGER
Institute for Systems Engineering Research, Information Technology Laboratory (ITL), U.S. Army Engineer Research and Development Center (ERDC), Vicksburg, MS, USA

KIRK MICHEALSON
Tackle Solutions, LLC, Chesapeake, VA, USA

> Decision-making ... is the irrevocable commitment of resources today for results tomorrow.
> (George K. Chacko (Chacko, 1990, p. 5))

> For which of you, intending to build a tower, does not first sit down and estimate the cost, to see whether he has enough to complete it? – Unknown

4.1 INTRODUCTION

A fundamental fact of decision-making includes the commitment of resources. Resources are essential assets committed to perform a trade-off analysis and to execute the subsequent decision. They come in many forms and include the following: money, facilities, time, people, and cognitive effort. Resources required for possible resolution of an issue are included in the resource space (Figure 4.1). To better understand the cost of each alternative, it is necessary to identify the required

---

Trade-off Analytics: Creating and Exploring the System Tradespace, First Edition. Edited by Gregory S. Parnell.
© 2017 John Wiley & Sons, Inc. Published 2017 by John Wiley & Sons, Inc.
Companion website: www.wiley.com/go/Parnell/Trade-off_Analytics

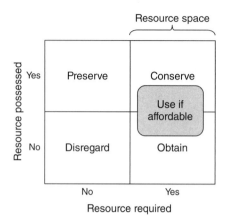

Figure 4.1 Resource space

set(s) of resources, define the resource space, and determine which resources will be committed for each alternative.

This chapter discusses the resource categories that comprise a resource space, techniques to determine the cost of resources for proposed alternatives, and means for assessing the affordability of the alternatives. Using these techniques in a logical and repeatable manner helps to understand the resource impacts of trade-off analysis alternatives.

4.2 RESOURCES

A resource is an asset accessible for use in producing the benefits of a decision. In identifying resources, it is useful to have a framework to ensure that you more fully capture the type and quantity of resources. The type and amount of resources available to support a decision are defined as the trade-off analysis resource space. This section discusses three components of the resource space: people, facilities, and costs. Figure 4.2 illustrates the three components. People and facilities can also be considered as "cost." In this section, they are broken down as separate resources to facilitate their description.

4.2.1 People

People and the skills they possess are essential to effectively implement a decision. They provide the means to leverage assets and accomplish the vision and goals of the decision. Therefore, it is crucial to identify what skills are on hand, the skills each solution requires to be successfully implemented, and what skills will need to be obtained via external organizations. Personnel skills can be binned into generalizable capabilities and further broken down into subspecialties. General skill bins are hard skills and soft skills. Soft skills are interpersonal and tend be more intrinsic than

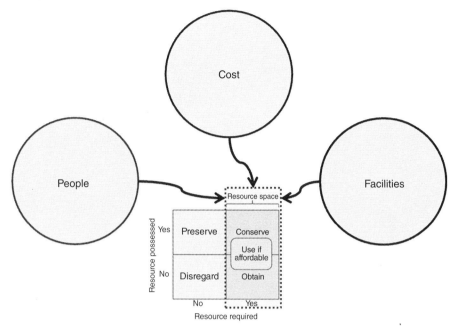

Figure 4.2 The three components of the resource space

hard skills, which are more learned knowledge and quantifiable abilities. Both are required for success in a job. Figure 4.3 provides an example of hard and soft skills for people resources.

The soft skills bin consists of numerous interpersonal skills. Table 4.1 lists some examples of these skills.

Table 4.2 is an example list of hard skills one should consider when determining the team capabilities and skills required to execute a decision. These skills can be physical or analytical in nature, but require education or training to attain or maintain.

The skills people possess allow them to perform various roles in the execution of a decision such as executive, management, customer service, communication, operation, maintenance, and logistics. These roles often require a combination of soft and hard skills. Based on changes that occur during the execution of a decision, managers often require the use of soft skills such as active listening, critical thinking, flexibility, problem solving, and conflict resolution as well hard skills such as technical knowledge of the area of interest to identify issues and solutions that will help execute a decision. Table 4.3 is an example list of hard and soft skills that a manager may need to facilitate the execution of a decision. Executive, management, customer service, and communication roles tend to include more soft skills while those personnel performing the roles of operations, maintenance, and logistics tend to have more hard skills.

Table 4.4 is an example set of functions for the roles present in an organization. Based on these roles and the personnel in these roles, an inventory of hard and soft skills can be conducted.

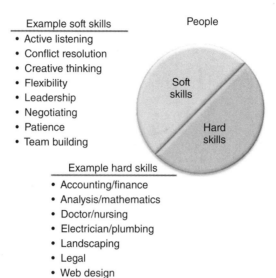

Figure 4.3 Example of skills for people resources

Table 4.1   Example Soft Skills

Active listening	Collaboration	Conflict management
Conflict resolution	Consulting	Counseling
Creative thinking	Customer service	Diplomacy
Flexibility	Instructing	Interviewing
Leadership	Mediating	Mentoring
Negotiating	Networking	Nonverbal communication
Patience	Persuasion	Problem solving
Team building	Teamwork	Verbal communication

Table 4.2   Example Hard Skills

Accounting	Analysis	Computer programming
Construction	Doctor/nursing	Electrician
Finance	Flying	Heavy equipment operator
Landscaping	Law	Machining
Mathematics	Plumbing	Typing
Web design	Welding	Writing

Table 4.3 Example Set of Hard and Soft Skills for Management

(S) Adaptability	(H) Administrative	(H) Analytical ability
(S) Assertiveness	(H) Budget management	(H) Business management
(S) Collaboration	(S) Conflict management	(S) Conflict resolution
(S) Coordination	(S) Critical thinking	(S) Decision-making
(S) Delegation	(S) Empowerment	(H) Financial management
(S) Flexibility	(S) Focus	(S) Goal setting
(S) Innovation	(S) Interpersonal	(S) Leadership
(H) Legal	(S) Listening	(S) Nonverbal communication
(S) Obstacle removal	(S) Organizing	(H) Planning
(S) Problem-solving	(H) Process management	(H) Product management
(S) Professionalism	(H) Project management	(H) Scheduling
(S) Staffing	(S) Team building	(S) Team manager
(S) Team player	(H) Technical knowledge	(H) Time management
(S) Verbal communication	(S) Vision	(S) Writing

(H) – Hard skills; (S) – Soft skills.

Inventorying the skills and quantity of the skills possessed by an organization is half the question. For each course of action, it is imperative to assess the skills and quantify requirements. This information will be used to help assess what additional people skills will be required to execute each option.

4.2.2 Facilities

Facility (or capital asset) resources are durable assets used in the production of products and/or services. Examples include tools, vehicles, roads, ships, plans, waterways, airports, machines, office space, communications equipment and infrastructure, power grid, and factories. These can be binned as infrastructure or equipment. Figure 4.4 is an illustration of the types of facility resources.

Table 4.5 provides an example list of these facility categories by bin. As with personnel, management must ascertain the types, numbers, and capacity of each facility asset they control.

4.2.3 Costs

Most people think of currency when they hear the term cost (e.g. Dollars, Euros, Pounds, Renminbi, Rubles, and Bitcoin). However, cost refers to any term used to represent resources of an organization. These include people (labor), facilities, and hours. Cost is an essential factor of a trade-off analysis as it can help to place resources, products, and services into a unifying quantitative measure that is more easily understood by analysts and decision makers. (Parnell et al., 2011, pp. 143–144)

There are several types of costs whether you are enhancing an existing system or developing a new one. The types of costs and their magnitude differ based on the type of system and the life cycle phase of the system. The Department of Defense defines five phases for its Acquisition Life Cycle and subsequent cost modeling. These

Table 4.4 Example Set of Roles and Functions for People Resources

Role	Example Functions
Executive	• Analyze competitive threats and opportunities
	• Analyze and assess strengths and weaknesses of the organization
	• Determine how to position the organization to effectively compete in market
	• Develop the strategic plan for the organization and provide vision
	• Ensure that effective internal controls and management information systems are in place
	• Develop and interface with external partners
Management	• Develop detailed means to implement the strategic plan and accomplish the organization's vision
	• Set objectives and determine the course of action for achieving these objectives
	• Educate, train, and equip subordinates to execute the course of action
	• Lead the workforce and supervise subordinates
	• Provide operational guidance and conflict resolution as required
	• Inspect for safety and quality control
Budget	• Track and manage the flow of cash, raw material, and products
	o Execute payment to suppliers in a timely manner
	o Manage receipt of payments to the organization
	• Track work effort and pay for employees
Customer service	• Interact with clients, suppliers, and sponsors to identify needs, resolve low-level issues, and inform leadership if additional action is required
	• Assist with product sales
Communication	• Ensure that instructor and processes exist to communicate internally and externally to the organization in a timely and effective manner
	• Develop and execute an internal communications plan to communicate the vision and goals of the organization
	• Develop and execute an external communications plan to communicate the vision and goals of the organization to customers and clients
Operations	• Execute the course of action in order to achieve the vision of the organization
	• Inform management of task completion or issues requiring their involvement in a timely manner
Maintenance	• Perform timely preventive and unplanned maintenance on equipment to ensure that it remains safely operational by other members of the team
	• Perform timely preventive and unplanned maintenance on facilities to ensure that they remain safely operational by other members of the team
Logistics	• Track and manage supplies to ensure that required raw material and supplies are on hand in a timely manner
	• Track and manage products to ensure that they are delivered in a timely manner

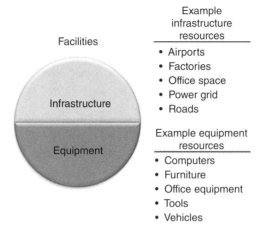

Figure 4.4 Example of types of facility resources

Table 4.5 Facility Examples

Infrastructure	Equipment
Airports	Cars
Communications infrastructure	Communications equipment
Factories	Computers
Office space	Furniture
Power grid	Machines
Power plants	Office equipment
Roads	Planes
Schools	Robots
Stores	Ships
Warehouses	Tools
Waterways	Trucks

five phases are as follows: (i) Material Solution Analysis (MSA), (ii) Technology Maturation and Risk Reduction (TMRR), (iii) Engineering and Manufacturing Development (EMD), (iv) Production and Deployment (P&D), and (v) Operations and Support (O&S). Each phase is preceded by a milestone or decision point. During the phases of the Acquisition Life Cycle, a system goes through research, development, test, and evaluation (RDT&E); production; fielding or deployment; sustainment; and disposal (Acquisition Life Cycle, 2015). When considering the entire life cycle of a system, we need to consider five cost classifications to identify the sources and effects of these sources on a system's life cycle: development, construction, acquire, O&S, and system retirement.[1]

---

[1] The Project Management Institute subdivides program management into five process groups: initiating, planning, executing, monitoring and control, and close (Project Management Institute, 2013, p. 8).

Stewart et al. divided cost into four separate classes, which occur across the phases of a program or system life cycle: (i) acquisition, (ii) fixed and variable, (iii) recurring and nonrecurring, and (iv) direct and indirect (Stewart et al., 1995). These are not four elements of the same analysis, but instead four separate ways to classify costs. The remainder of this section defines these classes of cost. Section 4.3 discusses the use of these classes for calculating and using these costs in assessing the resource space.

4.2.3.1 Acquisition (Operate, Sustain, and Dispose) Costs These are the total costs associated with the concept, design, development, production, or deployment of system or process (e.g., buildings, bridges, communications systems, vehicles, etc.). It does not include the cost to operate, sustain, or dispose of a product or process.

4.2.3.2 Fixed and Variable Costs An organization incurs fixed or sunk cost, no matter the phase of the system life cycle a product is in or the quantity of products produced. These independent costs of the program may include the cost of maintaining a research team, long-term rental cost for facilities, depreciation of equipment value, taxes (local, state, and federal), insurance for permanent assets, and site security. Variable costs vary based on the number and types produced or operated. These costs can easily be associated with each unit produced. Examples of variable costs include direct labor, material, and energy for the production of a product.

4.2.3.3 Recurring and Nonrecurring Costs Similarly to variable costs, recurring costs are associated with each unit produced or each time the process is executed. Unlike variable costs, recurring costs may not vary with the number or type of products produced. For example, the annual property taxes are recurring and fixed costs. Nonrecurring costs are those incurred only once in the life cycle or expected to be incurred only once in a life cycle. An example of nonrecurring costs include the resources required for initial design and testing as these tasks occur only once for each product.

4.2.3.4 Direct and Indirect Costs Direct costs are associated with a specific system, product, process, or service. These costs are often subdivided into direct labor, direct material, or direct expense costs. These cost subdivisions are similar to examples of variable costs. Labor costs associated with a specific product are considered direct labor cost, while fixed labor costs such as plant security and janitorial services are often considered indirect costs. Thus, indirect costs are costs that cannot easily be assigned to a specific product or process. Overhead costs such as executive leadership, human resources, accounting, annual training, and grounds maintenance are traditionally categorized as indirect costs. An example of an indirect, variable, recurring cost would be the energy used for the guard shack and lights on the overflow yard/warehouse used to stage excess products produced for the holiday surge in demand. Cost estimates routinely do a better job of identifying direct costs, but often fall short with identifying accurate indirect costs. To obtain more accurate indirect cost estimates, a life cycle cost (LCC) technique called activity-based costing may be used. This technique subdivides indirect costs by

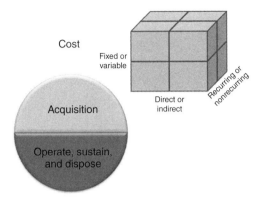

Figure 4.5 Cost resource by class

functional activities executed during the system life cycle (Canada et al., 2005). The costs associated with each activity are further defined by defined cost drivers to help identify which costs could be appreciated against a specific product or service.

Based on these classes of cost, any single cost could be categorized into several classes (Figure 4.5). For example, management cost for a multiyear product line could be classified as acqusiton, fixed, reccuring, indirect costs as it is performed across all phases of the life cycle as part of the organization's structure and processes. Management costs could also be variable depending on how many shifts are needed to produce a product. The management costs could be nonrecurring and direct if a part-time manager was hired to temporaraily run the night shift.

4.2.4 Resource Space

Once resources have been identified, one must determine if they reside within the nexus of resources for the organization. This nexus is the preliminary resource space and consists of the list of the resources, the required quantities, the duration and time of their use, and the cost of their use. The products from this effort are used for resource analyses. Figure 4.6 is an example resource framework to facilitate the identification and cost classification of resources in a preliminary resource space.

4.3 COST ANALYSIS

A system is defined as "an integrated set of elements that accomplishes a defined objective. System elements include products (hardware, software, firmware), processes, people, information, techniques, facilities, services, and other support elements" (International Committee for Systems Engineering (INCOSE), 2015). All systems have a specific life cycle. There are many system life cycle models in the literature. For example, a common system life cycle in the system engineering literature consists of seven stages: (Parnell et al., 2011, pp. 7–9) conceptualization, design,

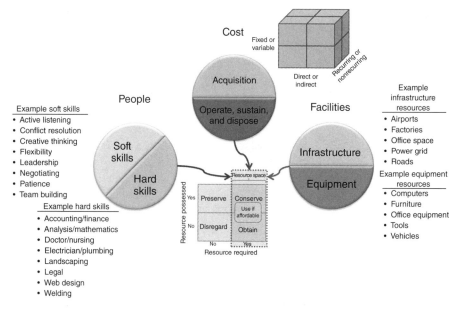

Figure 4.6 Example resource framework

development, production, deployment, operation, and retirement of the system. Throughout each of these stages, various levels of LCCs occur and trade-offs are made that impact the future costs of development, production, support, and disposal costs of the system.

Capturing system LCCs is necessary to make reasonable trade-offs during design, development, production, and operations. A LCC model is used by a systems engineering team in a trade-off analysis to estimate whether new alternatives or proposed system modifications meet a specific set of functional requirements at an affordable total cost over the duration of its anticipated life. When successfully performed, life cycle costing is an effective trade-off analysis tool that can be used throughout the life cycle of a system. The Society of Cost Estimating and Analysis (Glossary, 2015) defines a LCC estimate in the following way.

> Life cycle cost estimate is a cost estimate that covers all of the costs projected for a system's life cycle, and which aids in the selection of a cost-effective total system design, by comparing costs of various trade-offs among design and support factors to determine their impact of total system acquisition and ownership costs.

The concept map for life cycle costing is provided by Parnell et al. (2011, p. 138), as shown in Figure 4.7. This figure provides a pictorial illustration of the key elements that contribute to develop a comprehensive life cycle assessment. The LCC assessment centers around the development of a system's cost estimate, which in conjunction with a project schedule is used to manage a system's design,

COST ANALYSIS

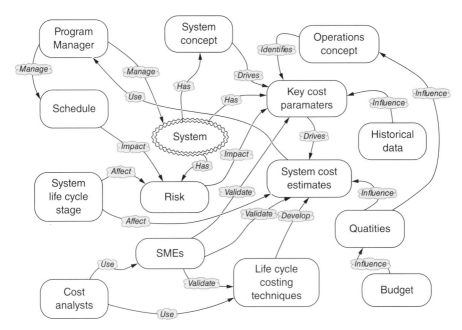

Figure 4.7 Life cycle costing concept map (Source: Parnell et al. 2011. Reproduced with permission of John Wiley & Sons)

development, production, as well as operation, and disposal. System design and operational concepts drive the key cost parameters, which in turn identify the data required for developing a system cost estimate. As part of a systems engineering and trade-off analysis team, cost analysts and engineers rely on historical data, subject matter experts (SME), system schedules, and budget quantities to provide data to use for life cycle costing techniques. In addition, risk plays a key role in LCC estimates. Risk affects the key cost parameters that drive the system cost estimate and is largely driven by the stage of development the system is in as well as the complexity and technology being utilized in the system design.

Cost estimation is a critical activity and key to the perceived success of complex public and private projects. Cost estimates should be developed and refined for all stages of a system life cycle. For example, cost estimates are used to develop a budget for the development of a new system or technology, to prepare a bid on a complex system proposal, to negotiate a purchase price for a system of systems, and to provide a baseline from which to track and manage actual costs and make trade-offs during all stages of development and operations of large-scale complex systems.

Selection of the most appropriate LCC technique depends largely on the quantity and type of data available as well as the perceived system risks. Data can be system specification and/or from historic cost data or models. As each stage of the life cycle progresses, additional information concerning system design and system performance becomes available, and some uncertainty is resolved while new uncertainties may be introduced. Therefore, selecting an appropriate LCC technique depends on the stage of the life cycle that the system is currently in as well as the availability of data and the uncertainty associated with it. Parnell et al. (2011, p. 140) summarize their recommendations of LCC techniques by life cycle stage in Table 4.6 along with appropriate references for each technique.

The AACE International (2015) has established a cost estimation classification system that can be generalized for applying estimate classification principles to system cost estimates in support of trade studies at various stages of a system's life cycle (United States Department of Labor, 2016). Using this classification system, the level and detail associated with the system's definition are the primary characteristics for classifying cost estimates. Other secondary characteristics (Table 4.7) include the use of the estimate, the specific estimating methodology, the expected accuracy range, and the expected effort to prepare the estimate.

AACE International groups the estimates into classes ranging from Class 1 to Class 5. Class 5 estimates are the least precise as they are based on preliminary information and are based on the lowest level of system definition, while Class 1 estimates are very precise because they are based on information from the full system definition as it nears design maturity. Ordinarily, successive estimates are prepared as the level of system definition increases until a final system cost estimate is developed at a specific stage of the system's life cycle.

The "Level of System Definition" column provides ranges of typical completion percentages that systems within each of the five classes will generally fall into, yielding information about the maturity and extent of available input data at each respective stage. The "End Usage" column describes how those cost estimates are typically used at that stage of system definition, that is, Class 4 estimates are generally only used for concept or feasibility analysis while Class 1 estimates might be used to check bidder estimates or make bids or offers. The "Methodology" column defines some of the characteristics of the typical estimating methods used to generate each class of estimate. The "Expected Accuracy Range" column indicates the relative uncertainty associated with the various estimates. Specifically, the column defines the degree to which the final cost outcome for a given system estimate is expected to vary from the estimated cost. The values in this column do not represent percentages as generally given for expected accuracy but instead represent an index value relative to a best range index value of 1. For example, if a given industry expects a Class 1 accuracy range of +15/−10, then a Class 5 estimate with a relative index value of 10 would have an accuracy range of +150/−100%. The final characteristic and column, "Preparation Effort", provides an indication of the level of effort and resources required to prepare the estimate such as cost and time. Similarly to the "Expected Accuracy" column, this is a relative index value.

4.3.1 Cost Estimation

Stewart et al. (1995) defines cost estimation as "the process of predicting or forecasting the cost of a work activity or work output." The Cost Estimator's Reference Manual (Stewart et al., 1995) outlines a 12-step process for developing a LCC estimate and is an excellent reference for developing LCC estimates. The book contains extensive

Table 4.6 LCC Techniques by Life Cycle Stage

LCC Techniques	Life Cycle Stages					
	Concept	Design	Development	Production	Deployment	Operation	Retirement
Expert judgment		Estimate by analogy	Estimate by analogy	Estimate by analogy			
Cost estimating relationships (Stewart, Wyskida, & Johannes, 1995)	Prepare initial cost estimates	Refine cost estimates		Create production estimates			
Activity-based costing (Canada, Sullivan, Kulonda, & White, 2005)				Provides indirect product costs		Use for operational trades	
Learning curves (Ostwald, 1992; Lee, 1997)			Provide development and test unit costs	Provide direct labor production costs			
Breakeven Analysis (Park, 2004)			Use in design trades	Provide production quantities		Use for operational trades	
Uncertainty and risk analysis (Kerzner, 2006)			Affects development cost	Affects direct and indirect product costs	Affects deployment schedules	Affects O&S costs projections	
Replacement analysis (United States Department of Labor, 2016)							Determine retirement date

Source: Parnell et al. 2011. Reproduced with permission of John Wiley & Sons.

Table 4.7 AACE International Cost Estimate Classification Matrix

	Primary Characteristic	Secondary Characteristic			Preparation Effort
	Level of System Definition	End Usage	Methodology	Expected Accuracy Range	
Estimate Class	Expressed as % of Complete Definition	Typical Purpose of Estimate	Typical Estimating Method	Typical ± Range Relative to Best Index of 1[a]	Typical Degree of Effort Relative to Least Cost Index of 1[b]
Class 5	0–2%	Screening or feasibility	Stochastic or judgmental	4–20	1
Class 4	1–15%	Concept study or feasibility	Primarily stochastic	3–12	2–4
Class 3	10–40%	Budget authorization or control	Mixed but primarily stochastic	2–6	3–10
Class 2	30–70%	Control or bid/tender	Primarily deterministic	1–3	5–20
Class 1	50–100%	Check estimate or bid/tender	Deterministic	1	10–100

Source: United States Department of Labor 2007.
[a] If the range index value of "1" represents +10/-5%, then an index value of 10 represents +100/-50%.
[b] If the cost index value of "1" represents 0.005% of project costs, then an index value of 100 represents 0.5%.

discussion and numerous examples of how to develop a detailed LCC estimate. The 12 steps defined in the manual are (Stewart et al., 1995) as follows:

1. Developing the work breakdown structure
2. Scheduling the work elements
3. Retrieving and organizing historical data
4. Developing and using cost estimating relationships (CERs)
5. Developing and using production learning curves
6. Identifying skill categories, skill levels, and labor rates
7. Developing labor hour and material estimates
8. Developing overhead and administrative costs
9. Applying inflation and escalation (cost growth) factors
10. Computing the total estimated costs
11. Analyzing and adjusting the estimate
12. Publishing and presenting the estimate to management/customer.

While all 12 steps are important, the earlier steps are critical since they define the scope of the system, the appropriate historical data to be used in the estimate, and the appropriate cost models used in the estimate. The identification of technology maturity for each cost element is also a critical element of the process that Stewart et al. (1995) does not explicitly call out. This is important since many cost studies for complex systems cite technology immaturity as the major source of cost and schedule overruns leading to significant errors in the original cost estimates (GAO-07-406SP, Defense Acquisitions: Assessment of Selected Weapon Systems, 2007).

The GAO Cost Estimating and Assessment Guide (GAO-09-3SP: GAO Cost Estimating and Assessment Guide, 2009, pp. 9–11) also defines an analogous 12-step cost estimation process. Their process can be broken into three components; initiation and research, assessment, and analysis and presentation. Their process does account for technology immaturity, risk, and uncertainty explicitly. Each of the components and their associated steps as defined in the GAO Cost Estimating and Assessment Guide are provided as follows:

Initiation and Research

1. Define the estimate's purpose.
   (a) Determine the estimate's purpose, scope, and required level of detail.
2. Develop an estimating plan.
   (a) Determine team, develop estimate timeline, and outline cost estimation approach.

Assessment

3. Define the program/project characteristics.
   (a) Identify technical baseline, projector program purpose, and system configurations for system.

(b) Identify relationships to existing systems and technology implications.
(c) Identify quantities for development, test, and production.
(d) Define operations and maintenance plans.
(e) Define support (manpower, training etc.) and security needs.
(f) Development and fielding schedule
4. Determine estimation structure.
   (a) Define the work breakdown structure (WBS) and a WBS dictionary describing each element.
   (b) Identify likely cost and schedule drivers.
   (c) Identify and choose the best estimation approach for each WBS element.
5. Identify ground rules and assumptions for estimate.
   (a) Clearly define what each estimate includes and excludes.
   (b) Identify all estimating assumptions (base year, schedule, life cycle etc.).
6. Obtain data.
   (a) Create data collection plan that identifies and documents all sources of relevant data.
   (b) Collect and normalize data for cost accounting, inflation, learning, and quantity adjustments.
   (c) Assess data reliability and accuracy.
7. Develop point estimates.
   (a) Utilizing the WBS, build the cost model in constant year dollars.
   (b) Time phase the estimate based on program schedule.
   (c) Verify and validate your estimate results with domain experts.
   (d) Update the model as new data becomes available.

Analysis

8. Conduct sensitivity analysis.
   (a) Analyze sensitivity to key assumptions in model and data values.
9. Conduct risk and uncertainty analysis.
   (a) Identify risky elements in the estimate.
   (b) Develop min, max, and most likely values for risky elements.
   (c) Analyze each risk for severity and probability.
   (d) Determine appropriate risk distribution and defend the reason for use.
   (e) Using Monte Carlo simulation develop a confidence interval around point estimate.
   (f) Develop a risk management plan to account for identified risks.

Presentation

10. Document the estimate.
    (a) Document all steps in enough detail so the estimate can be recreated.

(b) Describe in detail each of the estimating methodologies used in the estimate.
  (c) Discuss how the estimate compares to previous estimates.
11. Present estimate to management for approval.
  (a) Summarize the estimating process, provide the LCC estimate, and include discussion of project baseline and risks and uncertainties.
  (b) Compare the estimate to any other estimates available for the system.
12. Update the estimate to reflect actual costs/charges.
  (a) As program/project begins to incur costs, replace estimates with actual costs using an appropriate earned value management system and revise the estimate at completion for the system.
  (b) Track the progress on meeting cost and schedule estimates and perform an analysis after completion to address differences between estimates and actual costs and schedule. Incorporate into a lessons learned database.

Not all steps are used in every estimate. The system life cycle stage impacts the level of detail available for the cost estimate. Many of these steps are by-products of a properly executed systems engineering and management process and, when executed in a proper and timely fashion, can provide valuable insights on the economic impacts of design trades during the various stages of the system life cycle. In the early planning phase of a system's development, an initial work breakdown structure for the system is established. As discussed in Section 4.2, scarce resource allocation is a primary reason why scheduling is a prerequisite to costing the WBS, and escalation is another. The low level activities are then scheduled in order to develop a preliminary schedule.

Once all the activities have been identified and scheduled, the next task is to estimate their costs. The most reliable approach is to estimate the cost of these low-level activities based on past experience. Ideally, one would likely to be able to estimate the costs associated with the various activities using historical cost and schedule for similar activities from similar or the actual supplier. Finding this data and organizing it into a useful format is one of the most difficult and time-consuming steps in the process. Even once the data is found and then organized, the analysts must still ensure that it is complete and accurate. Part of this accuracy check is to make sure that the data is "normalized."

One form of data normalization is to make sure that the proper inflationary/deflationary indices are used on estimates associated with future costs (step 9). Once the data has been normalized, it is then used to develop statistical relationships between physical and performance characteristics of the system elements and their respective costs. Next, steps 4 and 5 are used to establish baseline CERs and then adjust the costs based on the specific quantities purchased. Steps 6–8 are used when a detailed "engineering" level estimate (Class 2 or Class 3 estimate) is being performed on a system. This is an extremely time-consuming task, and these steps are necessary if one wants to build a "bottom-up" estimate by consolidating individual estimates for all of the lower level activities into a total project cost estimate. Similarly to the earlier techniques used in steps 4 and 5, these steps are even more dependent on collecting detailed historical information on the lower level activities and their respective costs.

Finally, steps 11 and 12 are key elements to establishing a sound cost estimate. Step 11 provides the analyst the opportunity to revise and update the estimate as more information becomes available about the system being analyzed. Specifically, this may be an opportunity for the analysts to revise or adjust the estimate as the technology matures (see GAO-15-342SP, Defense Acquisitions: Assessment of Selected Weapon Systems (2015)). Additionally, it provides the analyst the opportunity to assess the risk associated with the estimate. An analyst can account for the data uncertainty quantitatively by performing a Monte Carlo analysis on the key elements of the estimate and then create a distribution for the systems LCC estimate. Step 12: publishing and presenting the estimate is one of the most important steps; it does not matter how good an estimate is if an analyst cannot convince the systems engineering team and project managers that that their estimate is credible.

All assumptions must be clearly stated in a manner that provides insight on the quality of the data sources used. One critical insight associated with a cost estimate is the basic list of ground rules and assumptions associated with that estimate. Specifically, all assumptions, such as data sources, inflation rates, quantities procured, amount of testing, and spares provisioning, should be clearly documented up-front in order to avoid confusion and the appearance of an inaccurate or misleading cost estimate.

Next, we will highlight a few of the key tools and techniques that are necessary to develop a credible cost estimate. Specifically, we will focus on developing and using CERs and learning or cost progress curves. The details associated with developing a comprehensive detailed estimate are extensive and cannot be given justice within a single textbook chapter. Interested readers are referred to Farr (2011), Stewart et al.(1995), and Ostwald (1992).

Once the estimate is developed and approved, it can be used to make design trades, create a bid on a project or proposal, establish the price, develop a budget, or form a baseline from which to track actual costs throughout the project's life cycle. Additionally, it can be used as a tool for cost analysis for future estimates on similar systems and projects.

4.3.2 Cost Estimation Techniques

As illustrated in Table 4.6, selection of an appropriate cost estimation tool or technique is dependent on the specific phase of a system's life cycle and the availability of information related to the system undergoing design, development, or production. To begin with, the cost analyst, working closely with the design engineers and end users (or customers) of the system, must develop a deep understanding of the system's operational concept, maintenance concept, the system's key functions, how the functions are allocated to the physical architecture (hardware, software, or human), the maturity of the technology being utilized in the design, the specific quantities desired, the acquisition strategy, and the system's design life. This information is necessary in

COST ANALYSIS   109

order to develop a credible cost model that can be used to make design, operational, maintenance, and cost trades during the various stages of its life cycle.

In this section, we explore several techniques for developing a LCC estimate. The data elements used by the models to create the estimates can be varied based on design trades during the various stages of the system life cycle. Initially, we begin by discussing the use of expert judgment as a means for establishing initial estimates. Expert judgment is useful for developing initial estimates for comparison of alternatives early in the concept exploration phase. Second, the use of CERs is discussed. CERs are used to estimate the cost of a system, product, or process during design and development. The CERs provide more refined estimates of specific alternatives when selecting between alternatives and are often used to develop the initial cost baseline for a system. Finally, we end with a discussion on the use of learning curves in a production cost estimate. This tool is used to analyze and account for the effect quantity has on cost of an item. This tool is often combined with CERs for the development phase to build a LCC estimate.

4.3.2.1 Estimation by Analogy: Using Expert Judgment to Establish Estimates
Cost analysts are often asked to develop estimates for products and services that are in the very early stages of design, sometimes nothing more than a vague concept. The engineers may have nothing more than a preliminary set of requirements and a rough description of a system concept and a set of anticipated functions. Given this limited information, the cost analyst, the systems engineer, and the engineering design team are often asked to develop a rough order of magnitude LCC estimate for the proposed system in order to obtain approval and preliminary funding to design and build the system. Given the scarcity of information at this stage, cost analysts and design engineers will rely on their own experience and/or the experience of other stakeholders and experts to construct an initial rough order of magnitude cost estimate. The use of expert judgment to construct an initial estimate for a system is not uncommon and is often used for Classes 4 and 5 estimates. This underscores yet another reason why a good deal of time and effort by the cost analyst is dedicated to working with the anticipated user of the system to help define and understand the requirements.

Today more than ever, technological advances often create market opportunities for new systems. When this occurs, the existing system/technology can serve as a reference point from which a baseline cost estimate for a new system may be constructed. If historical cost and engineering data are available for similar systems, then that system may serve as a useful baseline from which modifications can be made based upon the complexity of the advances in technology and the increase in requirements for system performance.

Often, experts will be used to define the increase in complexity by focusing on specific technological elements and/or performance requirements of the new system (e.g., the new television technology is 2.5 times as complex as the current technology). The cost analyst must translate this complexity difference into a cost factor by referencing past experience: for example, "The last time we changed screen technology, it was 3 times as complex and it increased cost by 30% over the earlier generation." A possible first-order estimate may be to take the baseline cost, say $3500, and create a cost factor based on the information elicited from the experts.

- Past cost factor: 3× complexity = 30% increase
- Current estimate: 2.5× complexity may increase cost 25%

These factors are often based on the personal experience of the engineers and cost analysts and available historical data. In this example, the expert is making an assumption that there is a linear relationship between the cost factor and the complexity factor. Assuming this is correct, a baseline estimate for the next-generation television technology might be ($3500 * 1.25 = $4375). This type of estimation is often accomplished at the meta system level as well when the new system/technology has proposed characteristics in common with existing systems. For example, the cost of unmanned aeronautical vehicles (UAVs) could initially be estimated by drawing analogies between missiles and UAVs because UAVs and missiles use similar technologies. Making appropriate adjustments for size, speed, payload, and other performance parameters, one could obtain an initial LCC estimate based on historical missile data.

A major disadvantage associated with estimation by analogy is the significant dependency on the judgment of the expert. The credibility of the estimate is dependent upon the credibility of the individual expert and their experience with the specific technology. Estimation by analogy requires significantly less effort in terms of time and level of effort than the other methods identified in Table 4.6. Thus, it is often used to validate the more detailed estimates that are constructed as the system design matures.

4.3.2.2 Cost Estimating Relationships (CERs) Parametric cost estimates are created by using statistical analysis techniques to estimate the costs of a system or technology. Parametric cost estimation was first introduced in the late 1950s to predict the costs of military systems by the RAND Corporation (Parametric Cost Estimating Handbook, 1995, p. 8). In general, parametric cost estimates are preferred to expert judgment techniques because they are based on historical data. However, if there is insufficient historical data, or the product and its associated technology have changed so dramatically that any existing/available data is not applicable, then constructing a parametric cost estimate may not be possible.

Parametric cost estimation is often used during the early stages of the system life cycle before detailed design information is available. As the system design evolves and matures, a parametric cost estimate can be revised using the evolving detailed design and production information. Because the statistical models are designed to forecast costs into the future, they can be used to estimate operation and support costs as well.

The primary purpose of using a statistical approach is to develop a CER, which is a mathematical relationship between one or more system physical and/or performance parameters and the system cost. For example, the cost of a house is often estimated by forming a relationship between cost and the square footage, location, and number

COST ANALYSIS

Table 4.8 Unmanned Aerial Vehicle (UVA) Work Breakdown Structure (WBS)

Level 1	Level 2	Level 3
UAV system	Air vehicle	Propulsion system
		Sensor payload
		Airframe
		UAV guidance and control
		Integration and assembly

of levels in a house. CERs have been developed to estimate the cost of a satellite as a function of weight, power requirements, payload type, and orbit location.

When constructing a system cost estimation model, one should use the baseline WBS or Cost Element Structure (CES), which includes O&S elements for the system to guide the development of the cost model. This ensures that all the necessary cost elements of the system are appropriately accounted for in the model. As an example, a 3-level WBS for the air vehicle of a UAV system is presented in Table 4.8. This example UAV WBS has been adapted from a missile system WBS that is found in MIL-STD-881C, Department of Defense Standard Practice: Work Breakdown Structures for Defense Materiel Items (MIL-STD-881C, Department of Defense Standard Practice: Work Breakdown Structures for Defense Materiel Items, 2011, p. 155). A WBS for a real UAV system would have many more level 2 components. For example, at WBS level 2, one should also consider the costs of the command and control station, launch components, the systems engineering, program management, system test and evaluation costs, training costs, data costs, support equipment costs, site activation costs, facilities costs, initial spares costs, and operational and support costs as well as system retirement costs. Each of these level 2 elements can be further broken down into level 3 WBS elements as has been done for the air vehicle.

A production-level CER can be developed at any of the three levels of the WBS depending on the technological maturity of the system components, available engineering and cost data, and amount of time available to create the estimate. In general, the further along in the development life cycle, the more engineering data available, the lower the WBS level from which an estimate can be constructed.

In the next section, we outline the process for constructing CERs and provide guidance on how these CERs can be used to develop a system-level estimate. We will utilize our simplified UAV Air Vehicle system as an example.

4.3.2.3 Common Cost Estimating Relationship Forms There are four basic forms for CERs; linear, power, exponential, and logarithmic. Each of these functional forms is discussed briefly as follows. As discussed earlier, a CER is a mathematical function whose parameters are derived using statistical analysis in order to relate a specific cost category to one or more system variables. These system variables must have some logical relationship to the system cost. The data used to estimate the parameters for the CER needs to be relevant to the system and the associated technology being used.

If the data used to estimate the parameters is not relevant, then the CERs will provide poor cost estimates. Similarly to other modeling paradigms, garbage in = garbage out!

4.3.2.3.1 Linear CER with Fixed and Variable Cost A large variety of WBS elements can be modeled by a simple linear relationship, $Y = aX$. Examples include personnel costs, facility costs, and training costs. Personnel costs can be modeled by multiplying labor rates by personnel hours, and facility cost can be modeled by multiplying the cost per square foot by the area of the facility. In some situations, it is necessary to account for a fixed cost in the CER. For example, suppose the cost of the facility also needs to include the cost of the land purchase. Then it would have a fixed cost associated with the land purchase and a variable cost that is dependent on the size of the facility built on the land. The resulting relationship is given by $Y = aX + b$, where b is the fixed cost for the land purchase.

4.3.2.3.2 Power CER with Fixed and Variable Cost Many systems may not have a linear relationship between cost and the selected system parameter. In some situations, an economy of scale effect may occur. For example, in a manufacturing facility, as the manufacturing capacity is increased, there will be a point at which the larger capacity will be less than the linear cost of increasing the manufacturing capacity. Similarly, situations occur where there are diseconomies of scale. For example, as a manufacturing facility produces more products, the costs of transporting the additional products to new markets may increase to the point where it offsets the economies of scale from the increase in production rate. Figure 4.8 illustrates the various shapes that a power CER can take as well as the functional form of the various CERs.

4.3.2.3.3 Exponential CER with Fixed and Variable Cost Another functional form that is sometimes used to create cost estimating relationships is the exponential form. Figure 4.9 illustrates the various shapes an exponential CER can take in modeling a cost relationship.

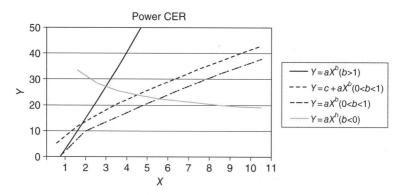

Figure 4.8 Power CERs

COST ANALYSIS

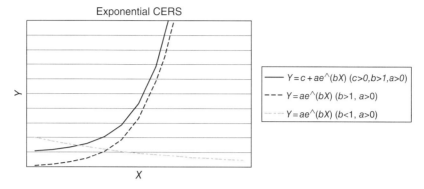

Figure 4.9 Exponential CER

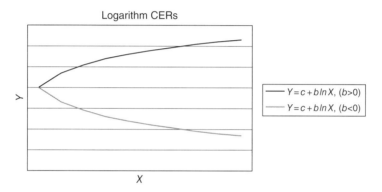

Figure 4.10 Logarithm CER

4.3.2.3.4 Logarithm CERs Finally, another form that may be useful for describing the relationship between cost and a particular independent variable is the logarithmic CER. The shape for a logarithmic CER for a couple of functional forms is provided in Figure 4.10.

4.3.2.4 Constructing a Cost Estimating Relationship As mentioned earlier, in order to construct a CER, we need enough data to adequately fit the appropriate curve. What is adequate is determined by the cost analysts and is a judgmental decision that is largely driven by what data is available. For most of the previous CER model forms, a minimum of three or four data points are sufficient to be able to construct a CER. Unfortunately, a CER constructed from so few points is likely to have a significant amount of error associated with it. Ordinarily, linear regression is used to construct the CER. Linear regression can be used on all of the functional forms by transforming the data in order to establish a linear relationship. Table 4.9, adapted from Stewart et al. (1995), illustrates the relationship between the various CERs and the transformations necessary in order to estimate the parameters for the CER using linear regression.

Once the data has been appropriately transformed, linear regression is used to estimate the parameters for the CERs by fitting a straight line through the set of

Table 4.9 Linear Transformations for CERs

	Linear	Power	Exponential	Logarithmic
Equation form desired	$Y = a + bX$	$Y = ax^b$	$Y = ae^{bX}$	$Y = a + b \ln X$
Linear equation form	$Y = a + bX$	$\ln Y = \ln a + b \ln X$	$\ln Y = \ln a + bX$	$Y = a + b \ln X$
Req'd data transform	X, Y	$\ln X, \ln Y$	$X, \ln Y$	$\ln X, Y$
Regression coef obtained	a, b	$\ln a, b$	$\ln a, b$	a, b
Coef reverse transform req'd	None	EXP(ln a),b	EXP(ln a),b	None
Final coef	a, b	a, b	a, b	a, b

transformed data points. Least squares is used to determine the coefficient values for the parameters a and b of the linear equation. The parameters are determined by using the following formulas:

$$b = \frac{\sum_{i=1}^{n} x_i y_i - \left(\dfrac{\sum_{i=1}^{n} x_i}{n}\right) \sum_{i=1}^{n} y_i}{\sum_{i=1}^{n} x_i^2 - \left(\dfrac{\sum_{i=1}^{n} x_i}{n}\right) \sum_{i=1}^{n} x_i} \qquad (4.1)$$

$$a = \frac{\sum_{i=1}^{n} y_i}{n} - b \left(\frac{\sum_{i=1}^{n} x_i}{n}\right) \qquad (4.2)$$

Most of the time, especially when we have a reasonable size data set, a statistical analysis package such as Excel, Mini-Tab, JMP, or R may be used to perform the regression analysis on the data.

Example: Suppose we have collected the following data on square footage and construction costs for a manufacturing facility in Table 4.10. Establish a CER between square footage and facility cost using the data provided. Analyze the data using a linear model.

We will fit the data to a simple linear model. Figure 4.11 is a regression plot of the data with a regression line fit to the data.

We can estimate the parameters for a line that minimizes the squared error between the line and the actual data points. If we summarize the data, we get the following:

$$\sum_{i=1}^{11} x_i = 804{,}200$$

$$\sum_{i=1}^{11} y_i = 1{,}203{,}750$$

$$\sum_{i=1}^{11} x_i y_i = 1.64699E+11$$

$$\sum_{i=1}^{11} X_i^2 = 1.18469E+11$$

Table 4.10 Square Footage and Facility Costs for Manufacturing Facility Construction

X Square Footage	Y Cost ($)
240,000	450,000
17,500	35,000
24,000	42,000
5,500	14,000
7,000	16,000
57,500	105,750
125,000	225,000
35,000	69,000
89,700	185,000
27,000	62,000
176,000	310,000

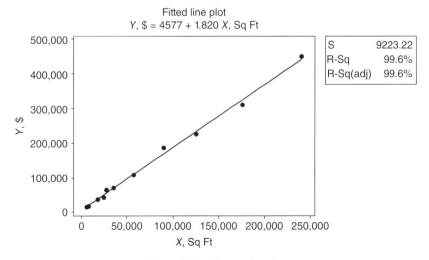

Figure 4.11 Regression plot

Using the summary data, we can calculate the coefficients for the linear relationship.

$$b = \frac{1.64699E+11 - \left(\dfrac{804,200}{11}\right)1,203,750}{1.18469E+11 - \left(\dfrac{804,200}{11}\right)804,200} = 1.8197$$

$$a = \frac{1,203,750}{11} - 1.8197\left(\frac{804,200}{11}\right) = 4577$$

If we enter the same data set into Mini-Tab, we obtain the following output:

Regression analysis: Y, $ versus X, Sq Ft

```
The regression equation is

Y, $ = 4577 + 1.820X, Sq Ft

S = 9223.22  R-Sq = 99.6%  R-Sq(adj) = 99.6%

Analysis of Variance
```

Source	DF	SS	MS	F	P
Regression	1	1.97601E+11	1.97601E+11	2322.86	0.000
Error	9	7.65610E+08	8.50678E+07		
Total	10	1.98366E+11			

```
Coefficients
```

Term	Coef	SE Coef	T-Value	p-Value	VIF
Constant	4577	3918	1.17	0.273	
X, Sq Ft	1.8197	0.0378	48.20	0.000	1.00

Examining the output, we see that the model is significant and that it accounts for approximately 99% of the total variation in the data. We note that the intercept term has a *p*-value of 0.273 and therefore could be eliminated from the model. As part of the analysis, one needs to check the underlying assumptions associated with the basic regression model. The underlying assumption is that the errors are normally distributed, with a mean of zero and a constant variance. If we examine the normal probability plot (Figure 4.12) and the associated residual plot (Figure 4.13), we see that our underlying assumptions seem reasonable. We see that the residual data fall

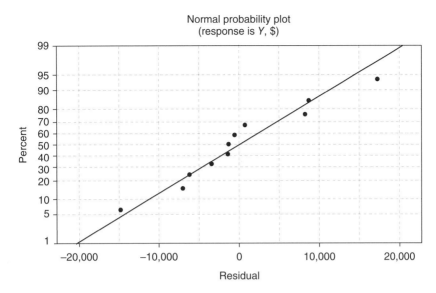

Figure 4.12 Normal probability plot

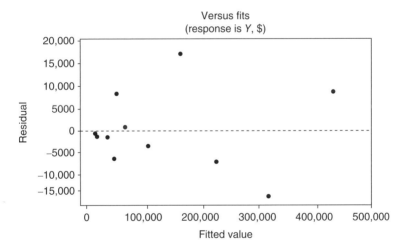

Figure 4.13 Residual plot

along a relatively straight line, passing the "fat pencil" test, and therefore are probably normally distributed. Second, it appears that the variance of the residuals has a mean value of zero, and the variance appears to be relatively constant for this small sample.

4.3.2.5 Cost Estimating Relationship Examples In this section, we provide several examples of hypothetical CERs that could be used to assemble a cost estimate for the air vehicle component of the UAV system described in the WBS given in

Table 4.8. To estimate the unit production cost of the UAV air vehicle component, we sum the first unit costs for the propulsion system, the guidance and control system, the airframe, the sensor payload, and the associated Integration and assembly cost. Suppose the system that we are trying to estimate the first unit production cost has the following engineering characteristics:

Requires 1300 lbs of thrust
Requires a 32 GHz guidance and control computer
Has a 3-in. aperture on the antenna, operating in the narrow band
Airframe weight of 500 lbs
Payload weight of 25 lbs
System uses electro-optics
System checkout requires nine different test procedures

Suppose the following CERs have been developed using data from five different missile programs and one UAV program during the last 10 years.

4.3.2.5.1 Propulsion CER The following CER was constructed using the propulsion costs from four of the five missile programs and the one UAV program. Two of the missile programs were excluded because the technology used in those programs was not relevant for the system currently being estimated. The CER for the propulsion system is given by

$$\text{Mfg \$(FY 2005)} = (\text{Thrust (lbs)})e^{-0.05(\text{Yr}-2000)}$$

$$\text{Mfg \$(FY 2005)} = (1300)e^{-0.05(2015-2000)} = \$614.07 \quad (4.3)$$

The manufacturing cost in dollars for the propulsion system is a function of thrust as well as the age of the motor technology (current year minus 2000).

4.3.2.5.2 Guidance and Control CER The guidance and control CER was constructed using data from the two most recent missile programs and the UAV program. This technology has evolved rapidly, and it is distinct from many of the early systems. Therefore, the cost analysts chose to use the reduced data set to come up with the following CER:

$$\text{Mfg \$(FY 2005)} = 7.43(\text{GHz})^{0.65}(\text{Aperature (in.)})^{0.45} e^{0.7(\text{Band})}$$

$$\text{Mfg \$(FY 2005)} = 7.43(32)^{0.65}(3)^{0.45} e^{0.7(1)} = \$233.37 \quad (4.4)$$

The manufacturing cost in dollars for the guidance and control system is a function of the operating rate of the computer, the diameter of the antenna for the radar, and whether or not the system operates over a wide band (0) or narrow band (1).

COST ANALYSIS

4.3.2.5.3 *Airframe CER* Suppose the following CER was constructed using the airframe cost data from the five missile programs and one UAV program. The CER for the airframe is given by

$$\text{Mfg \$ (FY 2005)} = 3.75(\text{Wt (lbs)})^{0.75}$$
$$\text{Mfg \$ (FY 2005)} = 3.75(500)^{0.75} = \$396.51 \quad (4.5)$$

Thus, the manufacturing costs in dollars for the airframe can be estimated if the analyst knows or has an estimate of the weight of the UAV airframe.

4.3.2.5.4 *Payload Sensor System* The following CER was established using data from two of the previous missile programs and the UAV program. The payload sensing system being estimated is technologically similar to only two of the previous missile development efforts and the UAV program.

$$\text{Mfg \$ (FY 2005)} = 25(\text{payload wt (lbs)})e^{1.25(\text{EO or RF})}$$
$$\text{Mfg \$ (FY 2005)} = 25(25)e^{1.25(1)} = \$2181.46 \quad (4.6)$$

The manufacturing cost for the sensing system in dollars is a function of the weight of the payload and the type of technology used. The term EO/RF is equal to 1 if it uses electro-optic technology and 0 if it uses RF technology.

4.3.2.5.5 *Integration and Assembly* This represents the costs in dollars associated with integrating all of the UAV air vehicle components, testing them as they are integrated, and performing final checkout once the UAV air vehicle has been assembled.

$$\text{Mfg \$ (FY 2005)} = 1.75 \left(\sum \text{Hardware Costs} \right) e^{-(1/\text{\# of system test procedures})}$$
$$\text{Mfg \$ (FY 2005)} = 1.75(614.07 + 233.37 + 396.51 + 2181.46)e^{-(1/9)} = \$5364.11 \quad (4.7)$$

4.3.2.5.6 *Air Vehicle Cost* Using this information, the first unit cost of the UAV air vehicle system is constructed as follows:

$$\text{Mfg \$ (FY 2005)} = 614.07 + 233.37 + 396.51 + 2181.46 + 5364.11 = \$8789.54$$

This cost is in fiscal year 2005 dollars and it must be inflated to current year dollars (2015) using the methods discussed in Section 4.3.4. Once, the cost has been inflated, the initial unit cost can be used to calculate the total cost for a purchase of 500 UAV air vehicles using an appropriate learning curve as discussed in the next section.

4.3.3 Learning Curves

Learning curves are an essential tool for modeling the costs associated with the manufacture of large quantities of complex systems. The "learning" effect was first noticed when analyzing the costs of airplanes in the 1930s (Wright, 1936). Other manufacturing sectors have found similar "learning" effects whereby human performance improves by some constant amount each time the production quantity is doubled (Thuesen & Fabrycky, 1989). For labor-intensive processes, each time the production quantity is doubled, the labor requirements necessary to create a unit decrease by a fixed percentage of their previous value. This percentage is referred to as the *learning rate*.

Typically, each time the production quantity is doubled, a 10–30% cost or labor saving is achieved (Kerzner, 2006). This 10–30% saving equates to a 90–70% learning rate. This learning rate is influenced by a variety of factors, including the amount of preproduction planning, the maturity of the design of the system being manufactured, the level of training of the production workforce, the complexity of the manufacturing process, as well as the length of the production run. Figure 4.14 shows a plot of a 90% learning rate and a 70% learning rate for a task that initially takes 100 h (Parnell et al., 2011). As evidenced by the plot, a 70% learning rate results in significant improvement of unit task times over a 90% curve. Typical learning rates by industry are given as follows (Stewart et al., 1995):

Aerospace – 85%
Repetitive electronics manufacturing – 90–95%
Repetitive machining – 90–95%
Construction operations – 70–90%.

4.3.3.1 Unit Learning Curve Formula The mathematical formula for the learning curve shown in Figure 4.14 is given by

$$T_X = T_1 X^r \tag{4.8}$$

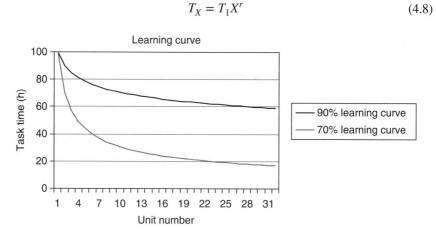

Figure 4.14 Plot of learning curves (Source: Parnell et al. 2011. Reproduced with permission of John Wiley & Sons)

COST ANALYSIS

where

T_X = the cost or time required to build the Xth unit
T_1 = the cost or time required to build the initial unit
X = the number of units to be built
r = negative numerical factor, which is derived from the learning rate and is given by

$$r = \frac{\ln(\text{learning rate})}{\ln(2)} \qquad (4.9)$$

Typical values for r are given in Table 4.11.

The total time required to produce all units in a production run of size N is given as follows:

$$\text{total time} = T_1 \sum_{X=1}^{N} X^r \qquad (4.10)$$

Using the aforementioned equation, with an r-value of -0.152 for a 90% learning rate, we can calculate the unit cost for the first four items. Assuming an initial cost of $100, Table 4.12 provides the unit cost for the first four items as well as the cumulative average cost per unit required to build X units. Figure 4.15 plots the Unit Cost curve and the Cumulative Average cost curve for a 90% learning rate for 32 units.

When using data constructed with a learning curve, the analyst must be careful to note whether they are using cumulative average data or unit cost data. It is easy to derive one from the other, but it is imperative to know what type of data one is working with to calculate the total system cost correctly. Note that the cumulative average curve is above the unit cost curve.

Table 4.11 Factors for Various Learning Rates

Learning Rate (%)	Factor, r
95	−0.074
90	−0.152
80	−0.322
70	−0.515

Table 4.12 Unit Cost and Cumulative Average Cost

Total Units Produced	Cost to Produce X^{th} Unit	Cumulative Cost	Cumulative Average Cost
1	100	100	100
2	90	190	95
3	84.6	274.6	91.53
4	81	355.6	88.9

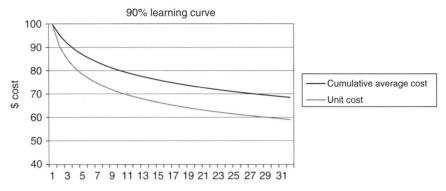

Figure 4.15 Ninety percent learning curve for cumulative average cost and unit cost for 32 units

Example 4.1 Suppose it takes 60 min to assemble the wings for a UAV the first time, and 48 min the second time it is attempted. How long will it take to assemble the eighth unit?

First, the task is assumed to have an 80% learning rate because the cost of the second unit is 80% of the cost of the first. If we double the output again, from 2 to 4 units, then we would expect the fourth unit to be assembled in $(48 \text{ min}) \times (0.8) = 38.4 \text{ min}$. If we double again from four to eight units, the task time to assemble the 8th wing assembly is $(38.4) \times (0.8) = 30.72 \text{ min}$.

Example 4.2 Suppose we wish to identify the assembly time for the 50th unit, assuming a 85% learning rate.

First, we need to define r for an 85% learning rate:

$$r = \frac{\ln(0.85)}{\ln(2)} = -0.235$$

Given r, we can now determine the assembly time for the 50th wing assembly as follows:

$$T_X = T_1 X^r$$
$$T_{50} = (60)(50)^{-0.235}$$
$$T_{50} = 23.93 \text{ min}$$

Example 4.3 It is common for organizations to define a standard of performance based on the Xth production unit. Suppose your organization sets a target assembly time for the 100th unit of 15 h. Suppose that your company has historically

COST ANALYSIS

operated at a 90% learning rate, what is the expected assembly time of the first unit?

$$T_X = T_1 X^r$$
$$T_X X^{-r} = T_1$$
$$T_1 = 15(100)^{-(-0.152)}$$
$$T_1 = 100.69 \, h$$

4.3.3.2 Composite Learning Curves Many new systems are often constructed using a variety of processes, each of which may have their own unique learning rate. A single composite learning rate can be constructed that characterizes the learning rate for the entire system using the rates of the individual processes. Stewart et al. (1995) use an approach that weights each process in proportion to its individual dollar or time value. Using this approach, the composite learning curve is given by

$$r_c = \sum_p \left(\frac{V_p}{T}\right) r_p \qquad (4.11)$$

where

r_c = composite learning rate
r_p = learning rate for process p
V_p = value or time for process p
T = total time or dollars for system.

Example 4.4 Suppose our UAV has a final assembly cost of $75,000 and the final assembly task has a historic learning rate of 80%. Suppose that the UAV engine has an assembly cost of $150,000 and that it has a historic learning rate of 90%. Finally, the guidance section has total cost of $100,000 and a historic learning rate of 95%. Calculate the composite learning rate for the UAV.

$$r_c = \left[\frac{75,000}{325,000}\right](80\%) + \left[\frac{150,000}{325,000}\right](90\%) + \left[\frac{100,000}{325,000}\right](95\%)$$
$$r_c = 89.23\%$$

4.3.3.2.1 Approximate Cumulative Average Formula The formula for calculating the approximate cumulative average cost or cumulative average number of labor hours required to produce X units is given by

$$T_c \approx \frac{T_1}{X(1+r)} \left[(X + 0.5)^{(1+r)} - (0.5)^{(1+r)}\right] \qquad (4.12)$$

This formula is accurate within 5% when the quantity is greater than 10.

Example 4.5 Using the cumulative average formula, compute the cumulative average cost for four units, assuming an initial cost of $100 and a 90% learning rate.

$$T_c \cong \frac{100}{4(1-.152)}\left[(4.5)^{(1-0.152)} - (0.5)^{1-0.152}\right]$$

$$T_C \cong 89.17$$

Note that this value is close to the actual cost found in Table 4.12.

4.3.3.3 Constructing a Learning Curve from Historical Data The previous formulas are all dependent upon having a value for the learning rate. The learning rate can be determined from historical cost and performance data. The basic data requirements for constructing a learning rate include the dates of labor expenditure, or cumulative task hours, and associated completed units.

By taking the natural logarithm of both sides of the learning curve formula, one can construct a linear equation, which can be used to find the learning rate.

$$T_X = T_1 X^r$$
$$\ln(T_X) = \ln(T_1) + r\ln(X) \tag{4.13}$$

The intercept for this linear equation is $\ln(T_1)$ and the slope of the line is given by r. Given r, the learning rate can be found using the following relation:

$$\text{learning rate } \% = 100(2^r) \tag{4.14}$$

This is best illustrated through an example.

Example 4.6 Suppose the data in Table 4.13 is collected from the accounting system for the manufacturing cell for the wing assembly on the floor.

Transforming the data by taking the natural logarithm of the cumulative units and associated cumulative average hours yields Table 4.14.

Table 4.13 Accounting System Data

Week	Cumulative Hours Expended	Cumulative Units Complete	Cum Average Hours
1	50	0	
2	100	1	100/1 = 100
3	190	2	190/2 = 95
4	260	3	260/3 = 86.66
5	320	4	320/4 = 80
6	360	5	360/5 = 72
7	400	6	400/6 = 66.67

COST ANALYSIS

Table 4.14 Natural Logarithm of Cumulative Units Completed and Cumulative Average Hours

Cumulative Units Completed X	$\ln X$	Cumulative Average Hours T_X	$\ln T_X$
1	0	100	4.60517
2	0.693147	95	4.55388
3	1.098612	86.66	4.46199
4	1.386294	80	4.38203
5	1.609437	72	4.27667
6	1.791759	66.67	4.19975

Figure 4.16 is a plot of the transformed data. Performing linear regression on the transformed data yields the following values for the slope and intercept of the linear equation.

Slope	$r = -0.2253$
Intercept	$\ln T_1 = 4.660$
Coefficient of determination	$R^2 = 0.897$

Thus, the learning rate is determined using the following relationship:

$$100 \left(2^{-0.2253}\right) = 85.54\%$$

4.3.4 Net Present Value

To help assess the potential economic benefits of a system to an organization, Net Present Value (NPV) is often used to calculate the present worth from a system or

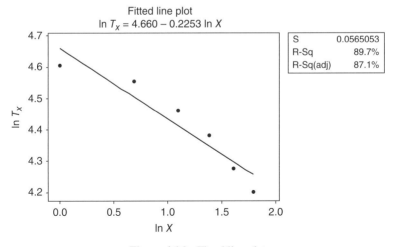

Figure 4.16 Fitted line plot

process based on the summation of cash flows over its lifetime. It differs in other metrics such as Return on Investment (ROI) and its more complex metric Internal Rate of Return (IRR), which are finance metrics and which do not account for risk by including the discount rate. If an NPV for a program is negative, this indicates that the program is not fiscally profitable. If the NPV is positive, it indicates that the program has a good chance of being profitable. Cash flow within the initial 12 months is not discounted for the purpose of calculating NPV. In most cases, the cash flow for the first year is often negative because of initial investments (Khan, 1999). A program's NPV can also be calculated for a prescribed period of time (e.g., 5, 10, or 15 years). This is done when comparing programs with different life expectancies or if attempting to assess when a program becomes profitable.

Calculation of the NPV requires the inclusion of annual inflation and discount/interest rates. Selecting an appropriate discount rate for calculating NPV is an area of continuing research. Example discount rates that may be applicable for calculation of NPV include but are not limited to the following:

- **Hurdle Rate** – often calculated by evaluating existing opportunities in operations expansion, rate of return for investments, and other factors deemed relevant by management
- **Internal Rate of Return (IRR) or Economic Rate of Return (ERR)** – the "annualized effective compounded return rate" or rate of return that makes the NPV of the cash flow for a particular investment or project equal to zero
- **Weighted Average Cost of Capital (WACC)** – the average rate an organization expects to pay to each of its security holders to back the effort

For each year, one must assess the expected value of the year's cash flow (CF) by the inflation and interest rate to normalize the values to a common year. The first step is to calculate the annual cash flow (ACF) or net cash value (NCV) for each year for the program. The ACF is the value of revenues and expenditures of the program for a discrete period of time (e.g., monthly or annually). Revenue can include payments to the program for products or services delivered. Expenses include all cost associated with the effort. A formula for calculating ACF is

$$\text{ACF} = \sum_{i=1}^{N} \text{Rev}_i - \text{Exp}_i \qquad (4.15)$$

where

ACF = Annual Cash Flow (i.e., Net Cash Value)
Rev = Revenue
Exp = Expenditure
i = Expenditure or revenue
N = Total number of expenditure and revenues for the year.

COST ANALYSIS

This can be done by first adjusting for the inflation rate and then adjusting for interest during the calculation of the NPV. Forecasted expenditures and returns are adjusted to the current year's currency value or the annual cash flow after inflation (CFAI) or present value (PV). This is also known as net cash flow. To adjust the expected value of the year's CF by the inflation and interest rate to its current value, the following formula can be used:

$$\text{CFAI} = \text{ACF} * (1 + \text{Inf}_i)^i \tag{4.16}$$

where

CFAI = Cash Flow after Inflation (i.e., Present Value (PV))
ACF = Annual Cash Flow (i.e., Net Cash Value)
Inf = Expected Rate of Inflation for the End of Year (EOY)
i = End of Year (EOY) for the program $(0, \ldots, n)$

Summing the CFAI without accounting for the annual interest rate produces the Discounted Cash Flow (DCF). However, to get a more accurate account for the risk associated with the program, include the interest rate in our summation calculations to generate the NPV. The following equation calculates the NPV by summing the CFAIs for the program after adjusting each for the annual interest rate (Khan, 1999).[2]

$$\text{NPV} = \sum_{i=0}^{N} \frac{\text{CFAI}_i}{(1+\text{IR}_i)^i} \tag{4.17}$$

where

NPV = Net Present Value
CFAI = Cash Flow after Inflation (i.e., Present Value (PV))
IR = Interest Rate for the year (time period)
i = End of Year (EOY) for the program $(0, \ldots, n)$
N = Total number of years (time periods)

Table 4.15 is an example spreadsheet used to show the calculations of the NPV for a program that has an initial investment of $250,000.00 to retool the factory, $52,000.00 annual cost for recurring costs, $86,000.00 in estimated annual sales, and $109,000.00 is expected for recapitalization of facilities and equipment after 7 years. The annual interest is 3%, and the estimated annual inflation rate is 7%.

Many computer-based spreadsheet programs have built-in formulas for ROI, IRR, and NPV. Each spreadsheet calculates NPV using a similar if not exact same methodology; however, they may differ from how you wish to calculate NPV or how you need

---

[2] In some bodies of work, the interest rate is called a discount rate.

Table 4.15 Example Net Present Value Calculations

EOY	Cash Outflows ($)	Cash Inflows ($)	Net Cash Value ($)	Inflation (%)	Interest (%)	Cash Flow after Inflation (CFAI) ($)	Cash Flow Interest ($)	EOY Summation ($)
0	(−) 250,000.00		(−) 250,000.00	3	7	(−) 250,000.00	(−) 250,000.00	(−) 250,000.00
1	(−) 52,000.00	86,000.00	34,000.00	3	7	35,020.00	32,729.69	(−) 217,270.31
2	(−) 52,000.00	86,000.00	34,000.00	3	7	36,071.00	31,504.41	(−) 185,765.90
3	(−) 52,000.00	86,000.00	34,000.00	3	7	37,152.00	30,327.18	(−) 155,438.72
4	(−) 52,000.00	86,000.00	34,000.00	3	7	38,267.00	29,193.89	(−) 126,244.83
5	(−) 52,000.00	86,000.00	34,000.00	3	7	39,416.00	28,103.61	(−) 98,141.22
6	(−) 52,000.00	86,000.00	34,000.00	3	7	40,599.00	27,051.11	(−) 71,090.11
7	(−) 52,000.00	195,000.00[a]	143,000.00	3	7	175,876.00	109,517.99	38,427.88
							NPV at EOY 7	38,427.88

[a] $86,000.00 + $109,000.00 = $195,000.00.

COST ANALYSIS

to reference your raw data. Thus, ensure that you read how each automated spreadsheet calculates NPV. There are also numerous web-based NPV calculators to help you calculate these values and ensure you that understand if you need to precalculate the NCV or ACF for these online calculators.

NPV can be assessed as an aggregate sum of cash flow, individual program resources, or cost categories (e.g., people, facilities, and costs; direct and indirect costs; or science and technology, procurement, and operations and sustainment/maintenance). Figure 4.17 is an example of a tornado chart of the NPV of a 7-year program described in Table 4.15. Tornado charts vary each variable from low to base to high while holding all other variables at their base value. System expenditures and revenue are broken down by direct and indirect cost categories. The chart provides the ability to see where the major cost drivers and savings are for the program. One can see that the largest predicted costs are *Acquisition Costs* for the worst-case (high-case) scenario. The lowest costs are *Indirect Costs* for the low-case (best-case) scenario. In this example, the lower cost case also has the higher *Sales* and *Recapitalization* predictions. The *Base Case* and the *Lower Costs and Higher Sales* forecast a positive NPV for the program after 7 years. The *Inflation* and *Interest* rates used for the calculations are seen in Table 4.16.

As a value based on expected future conditions, uncertainty is involved in the calculation on NPV. This uncertainty should be assessed against each of the annual

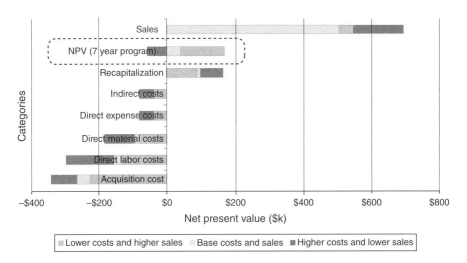

Figure 4.17 Example net present calue (NPV) tornado chart for a 5-year program

Table 4.16 Example Net Present Value Inflation and Interest Rates

Case	Inflation (%)	Interest (%)
Lower costs and higher sales	1.0	9.0
Base costs and sales	3.0	7.0
Higher costs and lower sales	9.0	1.0

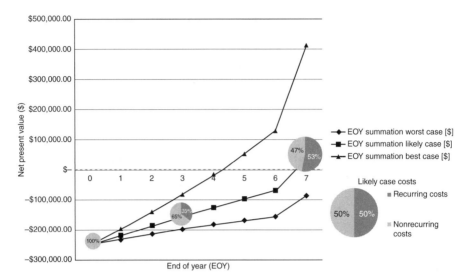

Figure 4.18 Example cumulative net present value hurricane chart

values: expenditures, cash inflow, inflation, and interest. This provides the ability to illustrate the expected best case, worst case, and most likely NPV for a planned investment. Figure 4.18 provides an example of a hurricane chart of the annual cumulative NPV forecast for the program outlined in Table 4.15. The deviation is based on a max and a min range of inflation and interest rates used to calculate the best- and worst-case values with the midline calculation based on the most likely value for each (Table 4.16). The pie charts indicate the percentage of recurring and nonrecurring costs. The size of each pie chart indicates the magnitude of total costs incurred. Although NPV is calculated for the lifetime of the program, the hurricane chart provides an illustration of the annual level of financial risk incurred for the program. The chart indicates that in the worst case, NPV is never positive. In the best case, the NPV is positive starting at EOY 5. The likely case indicates that positive NPV occurs at the end of the program after recapitalization of facilities and equipment.

When comparing the NPV of multiple options or systems, two conditions must be met (Park, 2004):

- Each system assessed must be mutually exclusive of the other (i.e., the choice of one system or alternative must not include the other).
- The lifetime of each system must be the same. If the duration of the lifetime differs between each system, then use the shortest.

4.3.5 Monte Carlo Simulation

Monte Carlo analysis is a useful tool for quantifying the uncertainty in a cost (or NPV) estimate. In Section 3.5.2, Monte Carlo Modeling is introduced and the details

COST ANALYSIS   131

associated with building a Monte Carlo simulation model are summarized in Figure 3.9. For a cost model, the Monte Carlo process rolls up all forms of uncertainty in the cost estimate into a single probability distribution that represents the potential system costs. Given a single distribution for cost, the analyst can characterize the uncertainty and resulting risk associated with the cost estimate and provide management with meaningful insight about the cost uncertainty of the system being studied. Kerzner (2006) provides five steps for conducting a Monte Carlo analysis for models. Kerzner's five steps are as follows:

1. Identify the appropriate WBS level for modeling; as the system definition matures, lower level WBS elements can be modeled in the cost model.
2. Construct an initial estimate for the cost for each of the WBS elements in the model.
3. Identify those WBS elements that contain significant levels of uncertainty. Not all elements will have uncertainty associated with them. For example, if part of your system has off-the-shelf components and you have firm-fixed-price quotes for the material, then there would be no uncertainty with the costs for those elements of the WBS.
4. Quantify the uncertainty for each of the WBS elements with an appropriate probability distribution. Often times, a triangular distribution is used for the first estimate.
5. Aggregate all of the lower level WBS probability distributions into a single WBS level 1 estimate by using a Monte Carlo simulation yielding a cumulative probability distribution for the system cost. This distribution can be used to quantify the cost risk as well as identify the cost drivers in the system estimate.

Kerzner (2006) points out that caution should be taken when using Monte Carlo analysis. As with any model, the results are only as good as the data used to construct the model; "garbage in, yields garbage out" applies to this situation. The specific distribution used to model the uncertainty in WBS elements depends on the information available. As mentioned earlier, many cost analysts default to the use of a triangle probability distribution to express uncertainty. Kerzner (2006) suggests that the probability distribution selected should fit some historical cost data for the WBS element being modeled. When only the upper and lower bounds on the cost for a WBS element are available, a uniform distribution is frequently used in a Monte Carlo simulation to allow all values between the bounds to occur with equal likelihood. The triangular distribution is often adequate for early life cycle estimates where minimal information is available (lower and upper bounds) and an expert is available to estimate the likeliest cost. As the system definition matures, and relevant cost data becomes available, other distributions, such as the Beta distribution, could be considered and the cost estimate updated.

Example: We will continue with our example of estimating the cost for a hypothetical UAV. One key element of the UAV system is estimating the software nonrecurring costs for the system. The following CERs have been developed to estimate the cost of the ground control software for the UAV system based on several previous UAV development efforts, and the mission–embedded flight software relationship is estimated using information from the conceptual design phase of the UAV system. Software development costs are often a function of its complexity and size.

Ground Station Software

$$\text{Person} - \text{Months} = 2.1(\text{EKSLOC})^{1.2}(1.2)^{\text{DoD}} \qquad (4.18)$$

Embedded Flight Software

$$\text{Person} - \text{Months} = 3.2(\text{EKSLOC})^{1.4}(1.5)^{\text{DoD}} \qquad (4.19)$$

The DoD parameter equals 1 if it is a DoD UAV, and 0 otherwise. Since the UAV under consideration is a commercial UAV, the DoD parameter is set equal to 0. EKSLOC is a measure of the size of the software coding effort measured in thousands of source lines of code. During design and development, engineers need to estimate these sizes for their project. Assume that it is early in the design process and the engineers are uncertain about how big the coding effort is. After talking with the design engineers, the cost analyst has chosen to use a triangular distribution to estimate the EKSLOC parameter. The analysts ask the design expert to provide several estimates; an estimate of the most likely number of lines of code, m, which is the mode, and two other estimates, a pessimistic size estimate, b, and an optimistic size estimate, a. The estimates of a and b should be selected such that the expert believes that the actual size of the source lines of code will never be less (greater) than a (b). These become the lower and upper bound estimates for the triangular distribution. Law and Kelton (1991) provide computational formulas for a variety of continuous and discrete distributions. The expected value and variance of the triangular distribution are calculated as follows:

$$\text{Expected value} = \frac{(a + m + b)}{3} \qquad (4.20)$$

$$\text{Variance} = \left(\frac{a^2 + m^2 + b^2 - am - ab - mb}{18} \right) \qquad (4.21)$$

Suppose our expert defines the following values for EKSLOC for each of the software components.

Software Type	Minimum Size Estimate (KSLOC)	Most Likely Size Estimate (KSLOC)	Maximum Size Estimate (KSLOC)
Ground Station Software	10	20	35
Embedded Flight Software	5	15	25

COST ANALYSIS

Using these values and the associated CERs, a Monte Carlo analysis is performed using @Risk (Palisade Corporation, 2014) software package designed for use with Excel. The probability density function for the embedded flight software is shown in Figure 4.19.

The PDF for the estimated labor hours is given in Figure 4.20 for 10,000 simulation runs.

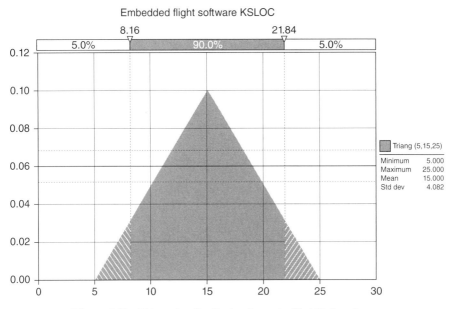

Figure 4.19 Triangular distribution for embedded flight software

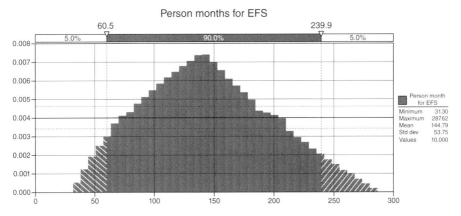

Figure 4.20 PDF for total person-months for embedded flight software development

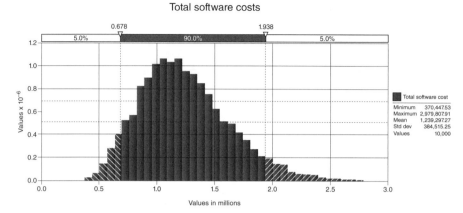

Figure 4.21 PDF for software development cost

Finally, suppose that management is uncertain about the labor cost for software engineers. Management believes the labor cost is distributed as a PERT random variable with a minimum of $25, a maximum of $60, and a most likely value of $35. The PDF and CDF for the total software development cost are given as follows. This estimate assumes that engineers work 36 h in a week on coding and that there are 4 weeks in a month.

The primary observation to make from Figure 4.21 is the spread in possible software development costs due to the uncertainty assumptions imposed on the WBS elements when the Monte Carlo simulation was constructed. For this example, while it is more likely that the actual software development costs will clump around $1.3 million, it is possible for them to be more than two times as much or as little as 500 k because of this uncertainty. In the former case, the project could be threatened; in the latter, the project would continue well within the budget.

Once we have the PDF for the cost of development, we can construct the CDF to calculate the probability that the cost of the software development is less than x. Applications such as @Risk accomplish this task easily. Figure 4.22 contains the CDF for the total software development costs.

Using the CDF, we can make probability statements related to the software development cost. For example, we can state that there is a 50% probability that the software development costs will be less than $1.295 million; similarly, there is a 5% probability that the software development costs will exceed $1.938 million. This information is useful to senior-level management as they assess the costs risks associated with the development program.

4.3.6 Sensitivity Analysis

As discussed in Section 3.5.2, sensitivity analysis can be conducted on the Monte Carlo Simulation model. Specifically, the results can be analyzed to identify those elements that are the most significant cost drivers. In our example, it is relatively

AFFORDABILITY ANALYSIS 135

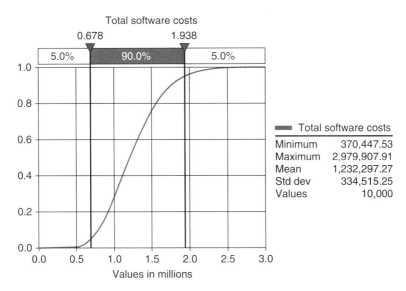

Figure 4.22 CDF for total software development costs

easy as our total cost is only a function of two cost elements and one other factor. But realistic cost estimates may have on the order of 10–50 cost elements/factors and choosing the cost drivers from this set is not so easy. Fortunately, @Risk provides a tool that analyzes the relative contribution of each of the uncertain components to the overall cost and variance for the system estimate. Figure 4.21 shows the sensitivity output for this example.

Examining the chart in Figure 4.23, the uncertainty associated with "Embedded Flight Software Person-Months" (EFKSLOC) is the main contributor to the variability in the total software cost estimate, followed by the uncertainty in labor rate for software engineers. The cost analysis should consider spending more time getting a better estimate for the cost risks associated with the "embedded flight software person-months" since reductions in the uncertainty associated with this WBS element will have the greatest impact on reducing the variability in the total cost estimate seen in Figure 4.21. Better estimates on the labor rate will also substantially reduce the uncertainty in the total system costs.

4.4 AFFORDABILITY ANALYSIS

Building on the resource assessment and cost analysis described earlier, organizations typically desire to understand the impact of cost, performance, risk, and resource allocation in the context of the portfolio of activities and competing priorities. For industry, affordability is the ability to develop a product or provide a service at a profit that balances performance and cost. For government organizations such as DoD, a product or program is affordable if it balances cost, performance, and risk

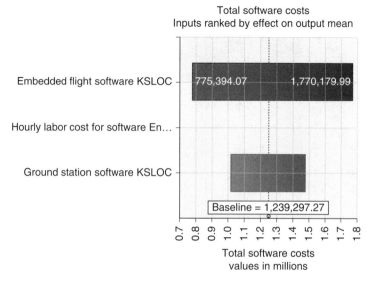

Figure 4.23 Sensitivity chart

while meeting their missions. This section discusses the background of developing affordability analysis, a comparison of cost analysis and affordability analysis, some affordability-related definitions and provides background, definitions, and a framework for conducting affordability analysis.

4.4.1 Background

In 2009, Congress passed the Weapon System Acquisition Reform Act (WSARA) to improve the way DoD contracts and purchases major weapons systems. The law established the Office of Cost Assessment and Program Evaluation (CAPE) and established reforms that were expected to save billions of dollars. As the WSARA formally demanded more fidelity and rigor in acquisition analysis, leaders in DoD asked the Military Operations Research Society (MORS) to engage the Acquisition and Analysis Communities to share and develop a set of best practices that address risk assessment and trade space analysis in support of acquisition. In September 2011, MORS held the workshop "Risk, Trade Space & Analytics in Acquisition," to determine and share a set of best practices for those significant analytical challenges that arise during the acquisition process. One significant conclusion from that workshop was that "affordability analysis" was poorly defined across the community. Leaders in DoD asked MORS for help with definitions and procedures.

In October 2012, MORS conducted an "Affordability Analysis: How Do We Do It?" workshop where the following was learned: (i) no organization was responsible for affordability analysis (OSD-ATL is only responsible for affordability); (ii) every organization defines affordability analysis and conducts it differently; and (iii) there was no process to follow to conduct affordability analysis. In February 2013, the

MORS Affordability Analysis Community of Practice (AA CoP) was formed with members from across government, industry, and academia to continue the research from the October 2012 workshop and develop a "how-to" manual, process, or guide for affordability analysis.

4.4.2 The Basics of Affordability Analysis Are Not Difficult

Let us start with an example to put affordability analysis in perspective. A couple makes $100,000 per year. They see a nice house for $500,000. "Can we afford that?" they ponder. We all know that it depends. It depends on what fraction of their budget they can allocate to the house, the payment terms of the house itself (interest rate, down payment required, etc.), their need for the house, the added value they derive from the house (e.g., utility, change in life style), and the degree to which the individual is willing to give up other budget items so that funds can be allocated to the house.

To summarize, the following are the things you need to know to buy a house:

- Need for house
- Fraction of budget available for house purchase
- Payment terms
- Willingness to give up other spending

While the answer to the couple's question involves several variables, with a few facts, the answers to their questions can generally be figured out with some home-buying analyses. The Department of Defense (DoD) needs a similar capability for affordability analyses, that is, the ease of knowing when something is outside of the fiscally possible. As shown as follows, the things needed to determine affordability-related decisions are related to the things needed to answer whether or not to buy a house.

Following are the things you need to know for affordability decisions:

- Needs and priorities
- Fraction of budget available for need(s)
- Basis of payment (Cost)
- Overall capability implications

As you read the information needed for buying a house and affordability decisions, you can see that affordability analysis is much more than just a straightforward cost analysis. The DoD has the directives to be effective and efficient, but struggles with the dynamic capability (i.e., a force that stimulates change or progress within a system or process) to use affordability as a guide to maximize value within operational, technical, and fiscal constraints. For an affordability analysis to be useful, it must be actionable: it must lead to a well-informed decision or support a specific action, such as a program start or cancellation, or perhaps a new operating concept that provides needed capability at reduced cost.

The remaining paragraphs in this section are DoD-specific. Affordability analysis efforts were started at the request of DoD Leadership and the work completed was conducted by representatives in the DoD, Defense Contractors, and Defense-related academia responding to the request. However, the affordability analysis definitions and framework in this section could be adapted for non-DoD/commercial work similarly to the house buying example.

4.4.3 DoD Comparison of Cost Analysis and Affordability Analysis

From the 2012 MORS Workshop on "Affordability Analysis: How Do We Do It?," consensus of the attendees was that clarity of definition, sufficiency criteria, and regulatory policy were consistently absent from affordability analysis. Affordability was determined not to be a number, but a decision, and may vary depending on the stakeholder or the decision maker (i.e., affordability is in the eye of the beholder). (Michealson, "Big A" Affordability Analysis, May 27, 2015). For example, differences could be:

- Size dimension: DoD, service, mission area, or system
- Measurement dimension: Dollars, lives, or time
- Phase dimension: Requirement, acquisition, or operation

When one conducts cost analysis, the process is normally straightforward; analysts have established guidelines and principles to follow. However, when conducting affordability analysis, approaches vary dramatically. Guidance, processes, and institutional acceptance are needed though tools and methodologies were not considered the binding constraints; without them, there will be varying perspectives on affordability.

Since participants were familiar with current costing techniques, the following definitions were developed during the October 2012 Workshop to provide a foundation for an affordability analysis process:

- Cost Analysis primarily considers the monetary and financial aspects of an acquisition – whether it is acquisition cost, design cost, and/or total ownership cost. Individual cost elements of a system or platform are investigated, estimated, and accumulated to arrive at an overall cost.
- Cost–Benefit Analysis compares cost of numerous alternatives to the benefits they provide to inform the course of action selection.
- Capabilities-Based Assessment is a Joint Capabilities Integration Development System (JCIDS) activity designed to identify and prioritize capability gaps and risks.

It was agreed that none of these are affordability analysis, but they all contribute to affordability analysis. Cost analysis provides the basis for costs used in an affordability analysis. Cost–benefit analysis additionally provides solution advantages, quantifiable and nonquantifiable, which affordability analyses incorporate as value or military worth of the acquisition program. Capability gaps, priorities, and risk output from capabilities-based assessments serve as foundational elements in an affordability analysis. These gaps, priorities, and risks form the basis for evaluating acquisition program costs and benefits against fulfillment of stated capability and reveal how well the acquisition program does or does not satisfy DoD objectives within affordability targets.

As a result of these discussions, in January 2015, the Department of Defense Instruction (DODI) 5000.02 Operation of the Defense Acquisition System, Office of the Secretary of Defense (Acquisition, Technology & Logistics), included an overview on affordability for the first time:

- Affordability analysis and affordability constraints are not synonymous with cost estimation and approaches for reducing costs.
- Cost estimates are generated in a bottom-up or parametric manner and provide a forecast of what a product will cost for budgeting purposes (NOTE from the Naval Center for Cost Analysis that the current wording omits analogy, expert opinion, and extrapolation that are also recognized as cost estimation methods).
- Affordability constraints are determined in a top-down manner.

The updated document discusses cost analysis and the differences between cost analysis and affordability analysis, but it still does not discuss affordability analysis.

4.4.4 Affordability Analysis Definitions

Affordability is an abstract term that most people think they understand but have difficulty defining or explaining. The 2011 MORS Workshop on "Risk, Trade Space & Analytics in Acquisition" revealed a lack of consensus on the definition of affordability and related terms. The MORS AA CoP has developed three key definitions – affordability, affordability analysis, and affordability analysis outcomes. These three terms are critical to understanding why, what, and how tasks are undertaken.

- **Affordability**. The degree to which the resources being allocated to a capability relative to other uses of those resources reflects (i) the importance, urgency, and satisfaction of mission, strategic investment, and organizational needs, and (ii) a prudent balance of performance, cost, and schedule constraints consistent with the time-phased availability (technical, market, and fiscal) of budgeted resources.
- **Affordability Analysis**. A process and assessment that supports resource allocation decision-making. It identifies and quantifies the performance expectations of stakeholders, assigns value to those expectations, and measures the LCC of alternatives relative to both opportunity costs and resourcing actions or plans.
- **Affordability Analysis Outcomes**. Practically, affordability analyses must substantiate resource plans, given a mission scope and a budget scope, while taking advantage of "good buys" and available offsets. Culturally, rewarding

the practice and use of affordability analyses should change the conversations of decisions-makers, enabling them to deliver portfolio outcomes that are more effective and efficient while staying within and informing budget boundaries.

In the first MORS Affordability Analysis Workshop, two interpretations of affordability were also developed.

- Affordability in the "large" means assessing whether a mission, task, function, capability, system of systems, program, or initiative – considering what it is going to cost (or is costing us, i.e., the total costs) – provides sufficient value in the context of all of the other things needed ("**Big A**").
- Affordability in the "small" means being frugal – being cost-efficient in executing a program, from beginning to end and not being extravagant in choosing capabilities and solutions to challenges; getting the most bang for the buck ("**little a**").

As shown in Figure 4.24, the services, contractors, program managers, and others tend to operate in the "little a" realm (i.e., doing things right), while DoD, Congress, and Service leadership usually operate in the "Big A" realm (i.e., doing the right things); the perception is that the majority of "Big A" affordability analysis

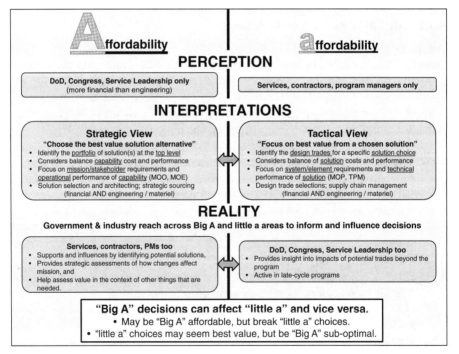

Figure 4.24 "Big A" and "little a" (Source: MORS Affordability Analysis Community of Practice 2015)

AFFORDABILITY ANALYSIS 141

is conducted by the leadership, while "little a" is conducted by program managers. However, that is not quite true. Both leadership and program managers conduct "Big A" affordability analyses, just at different levels in the enterprise, and the leadership is quite active doing "little a" affordability analyses in late-cycle programs. Additionally, program managers support and influence the leadership's "Big A" work, when requested by (i) initially identifying the right solution at the top level to meet the capability, then (ii) throughout the acquisition life cycle when the customer provides changes/new information is learned conducting a strategic assessment to determine how these new changes affect the mission, task, function, capability, system of systems, program, or initiative – that is, considering the analysis of LCC and performance in relation to alternatives to assess value in the context of other things that are needed.

As a result, original "Big A" and "little a" affordability interpretations from the first MORS workshop were updated:

- **Affordability in the Large**:
  - Assessing whether a mission, task, function, capability, system of systems, program, or initiative – considering what it is going to cost (or is costing us, i.e., the total costs) – provides sufficient value in the context of all of the other things needed ("**Big A**").
  - The Strategic View – Identifying the portfolio of solution(s) at the top level, considering LCC and capability/performance in relation to the value provided by other things that are needed.
- **Affordability in the Small**:
  - Being frugal – being cost-efficient in executing a program, from beginning to end and not being extravagant in choosing capabilities and solutions to challenges; getting the most bang for the buck ("**little a**").
  - The Tactical View – Minimizing costs while maximizing capabilities/performance.

With that said, new elements in a portfolio may be "Big A" affordable and break "little a" choices, and conversely, "little a" choices might seem to be the best value but be "Big A" suboptimal. Affordability in the large is a judgment call. That judgment can change over the life of a program for many reasons, some of which may have absolutely nothing to do with the "little a" of a program. The nature of analysis to support the "A's" differs somewhat due to the nature of the associated questions.

4.4.5 "Big A" Affordability Analysis Process Guide

Since (i) the military services, contractors, and program managers have their own processes for conducting "little a" affordability analysis and (ii) documents developed that are coordinated through a professional society (i.e., MORS) cannot be prescriptive, MORS provided considerations for conducting "Big A" affordability analysis with best practices and lessons learned that are supportive/complementary to all organization's "little a" affordability analysis processes.

The MORS "Big A" Affordability Analysis Process Guide (Michealson, "Big A" Affordability Analysis Process Guide, 2015) seeks to be a thinking construct that allows the DoD, at all institutional levels, to have a data-based conversation about affordability and affordability analysis. The principal goal is not to develop a prescriptive, one-size-fits-all "how-to" manual on doing optimal resource allocations, but to: (i) include outcome and constraint quantification, (ii) consider fiscal stewardship, and (iii) demonstrate how to provide high-efficacy decision support. A secondary purpose is to aid decision-makers or decision-supporters, who have data and may not be experts with analytics. The document proposes a simple set of questions that ensures consideration of key facets, which would have a significant impact on the affordability of a system.

Guidelines for high-quality affordability analysis are offered to include in the life cycle process for understanding, but this construct in no way replaces life cycle management. Sufficiency and quality exit criteria are offered in the affordability analysis process to provide rationale to answer some of the questions expected due to scope design, political motivations, stakeholder motivational considerations, and the complexities with data, tools, and analysis.

The goal is not to develop a prescriptive, one-size-fits-all "how-to" document or a manual on doing optimal resource allocations; the overall goal is to develop an affordability analysis process with best practices, lessons learned, considerations, and so on, as well as including ties to the individual military service's new Affordability Policies and DoD Better Buying Power (BBP) initiatives. BBP has been referred to as DoD's mandate to "do more without more." BBP is the implementation of best practices to strengthen DoD's buying power, improve industry productivity, and provide an affordable, value-added military capability to the warfighter. Introduced in 2010, BBP is outlined in a series of three memos from the Under Secretary of Defense for Acquisition, Technology and Logistics (ATL). Affordability is a key tenant of BBP 1.0, 2.0, and 3.0 memos, which further strengthens the need for solid, consistent, repeatable affordability analysis practices. BBP mandates affordability as a requirement and enforces affordability caps.

Affordability analysis is essential to establish requirements and caps. BBP states that affordability constraints are to be based on anticipated future budgets for procurement and support of the program. BBP affordability constraints are the artifact of budget, inventory, and product life cycle analysis within a portfolio context. Affordability constraints force prioritization of requirements and drive performance and cost trades to ensure that unaffordable programs do not enter the acquisition process. BBP 3.0 places emphasis on achieving dominant capabilities through innovation and technical excellence and continues with the core theme stating, "Conduct an analysis to determine whether or not a desired product can be afforded in future budgets – before the program is initiated."

Figure 4.25, the Affordability Analysis Framework, illustrates the question-driven framework that is the basis of the affordability analysis. As shown in the center of the figure, the process is started with the Review Requirements, Needs, and

AFFORDABILITY ANALYSIS

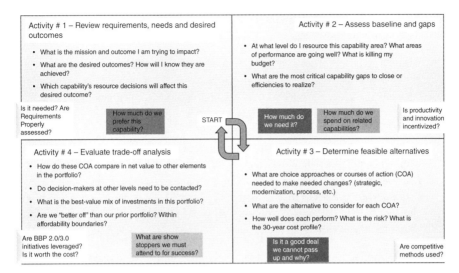

Figure 4.25 Affordability analysis framework (Source: Courtesy of the Military Operations Research Society Affordability Analysis Community of Practice (or MORS AA CoP))

Desired Outcomes Activity and following the arrows working clockwise through the remaining activities: Assess Baseline and Gaps, Determine Feasible Alternatives, and Evaluate Trade-Off Analysis. The bullet questions are subactivities for each activity, and each has several tasks/considerations to help analysts dig for information that assists in drawing conclusions about the affordability of the area or topic in question.

The Review Requirements, Needs, and Desired Outcomes activity sets the stage for the affordability analysis and identifies the resource information readily available by answering the following questions (i.e., the subactivities):

- What is the mission or outcome that will be impacted?
- What are the desired outcomes? How will it be known if they are achieved?
- Which capability's resource decisions will affect the desired outcome?

By answering these questions, this activity (i) generates critical assumptions and shapes the scope of affordability analyses, (ii) identifies analyses needed and the appropriate tradespace to assess, and (iii) affirms that the requirements of the "scope" are properly assessed and capabilities in question are needed. As a result, this activity identifies the resource information readily available and the degree of contention about the AOI.

After an organization has aligned their resources with their goals and targets, they must evaluate their baseline's current resource performance and identify capability gaps. The Assess Baseline and Gaps activity will either identify or validate a mission need and begin the necessary affordability assessments to evaluate alternative resource strategies to meet the emerging needs. Overall, this activity will enable the affordability analysts to understand what is truly needed and incentivize innovation and other high-leverage changes to the baseline by:

- Evaluating baseline performance including baseline effectiveness/performance and cost/affordability of the baseline
- Considering emerging needs that may be driven by many factors to include (i) new or increasing threats, (ii) new responsibilities, roles or missions, (iii) changes in national security objectives, and (iv) aging of resources
- Identifying capability gaps that reflect inability to meet objectives or mission demands and pose risk to accomplishing task and goals
- Considering alternative strategies that not only study new things – platforms/systems/software – that may rectify the gap (problem), but also ways that might change the way business is done in terms of CONOPS, training, doctrine, and so on, to improve the situation.

In the Determine Feasible Alternatives activity, a high-level assessment of affordability of the alternatives proposed helps to gauge which may have more value to the enterprise. As the study of the Affordability options has begun, there is a need to review the alternatives in an analytically rigorous method.

- First, an evaluation the available approaches for desired delta or change (all options) is conducted before addressing any questions of affordability. Are there other possible actions that would result in a change to the bottomline?
- Second, affordability should be included as a key consideration into the design of any solutions. It is easier to assess affordability and affordability options if the environment to study has already been created. The need is to assess alternative capability solutions using proven integrated Capabilities-Based Assessment (CBA) approaches and criteria at a capability level (as opposed to a project level).

The overall goal of this third activity is to determine the feasible solutions to use in the trade-off analysis activity.

The Evaluate Trade-Off Analysis activity focuses on tradespace analysis and a best value evaluation of the affordability assessment in question, to ensure an affordability trade has not been made that produces undesired long-term effects. There are five subactivities associated:

- Prepare for Analysis by checking for completeness of the inputs, normalize data and analysis structures, and evaluate available alternatives.
- Solicit and Determine the Value Structure by analyzing the available information, developing a value framework, and populating the value framework.
- Conduct Trade-Off Analysis by refining candidate portfolios, conducting cost/performance/risk/schedule analysis, conducting the trade-off analysis, and ranking the alternative portfolios.

AFFORDABILITY ANALYSIS 145

- Conduct Sensitivity and Risk Analysis by creating a risk and sensitivity analysis methodology and performing risk and sensitivity analysis.
- Make the Decision by preparing to support the decision-maker, creating a "what-if" approach or tool, and supporting the decision-maker assessment process.

In summary, this last affordability analysis activity analytically "proves" which feasible COA from the previous activity is best for the portfolio area (and affects the portfolio). The data or techniques used should provide a better result as the process matures.

After the specific affordability analysis activity and its associated subactivities and tasks are complete, the exit criteria questions help to assess if each specific activity is actually completed, that is, if we are "doing it right" (the gray boxes in each quadrant) as well as "doing the right thing" (the other white boxes in each quadrant). To determine if the affordability analysts are doing the right things, that is, sufficiency, the MORS AA CoP started with the Government Performance and Results Act Modernization Act (GPRAMA—[16 Dec 13]). GPRAMA requests the following from DoD related to affordability and affordability analysis:

- "Provide information on the analytical underpinnings of DoD reasoning with regard to risk-balancing and other trade-offs involved in aligning available departmental resources with national strategy.
- Explain how DoD has considered the element of budget uncertainty.
- Provide information on decision-making about trade-offs between programs and initiatives, in the context of budget constraints.
- Explain more about how "lessons-learned" analysis has been incorporated into strategic planning, particularly with long-standing problem areas such as financial management and cost-containment, and how it will be used in the future."

As a result, the MORS AA CoP designed their sufficiency exit criteria with the GPRAMA in mind. The affordability analyst can finally get back to the simplicity of the back of the napkin: what are the basic facts that must be known to believe our conclusion? To do this, and support the GPRAMA, five high-level sufficiency criteria support a good affordability analysis for each activity. They are:

- Grounded in a value proposition,
- Addresses the entire life cycle,
- Includes portfolio assessment,
- Is time specific, and
- Contains data-driven analysis.

To ensure that the affordability analyst was doing things right, the MORS Affordability Analysis CoP developed their quality exit criteria:

- Prioritization: The scope and importance of the capability to a relevant value space. **How much do we prefer this capability?**
- Trade demands: The relative need for the capability and supporting commodities to provide it – an assessment of capability gap severity and criticality. **How much do we need it?**
- Dollars per capability: The balance in the portfolio must be graphed – dollars relative to other capabilities in the baseline, where investments lie in life cycle, as well as other discriminating characteristics. **How much do we spend on related capabilities?**
- Cost: An evaluation of net performance/risk/cost bang for the buck; how "buyable" and "best value" is the Course of Action (COA) or option. **Is it a good deal we cannot pass up and why?**
- Behavioral change required to make COA successful: issues about choices, sunk costs, industrial considerations, or cultural risk factors that may not be quantified; yet, they may preclude success of the initiative; a description of this challenge must be included. **What are the showstoppers we must attend to for success?**

These exit criteria are critical – if we are not doing it right, or we are not doing the right thing, then the affordability question is a moot point. If we are doing it right, and it is the right thing, then we need to figure out how to pay for it. What can we give up, given the time duration of the capability in question (it may be less than 30 years) and given the array of uncertainties around the cost and value approximation?

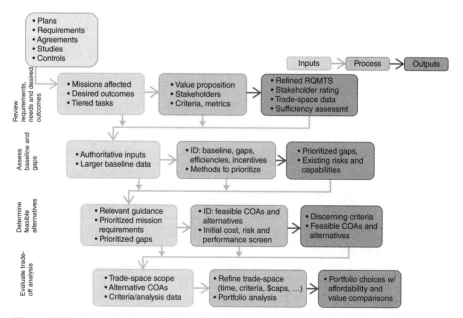

Figure 4.26 Lean six sigma high-level overview product example (Source: Courtesy of the Military Operations Research Society Affordability Analysis Community of Practice (or MORS AA CoP))

Figure 4.26 is a high-level overview product of Lean Six Sigma Value Stream Mapping Activity of the Affordability Analysis Framework described in the question-driven portrayal of Affordability Analysis Framework figure and discussion earlier. This figure shows the activities and artifacts needed to conduct the process and also shows an overview of the "Big A" Affordability Analysis Activities. The "rows" are the four affordability analysis activities and, for each activity, provides an overview of the inputs, process steps (subactivities)/tasks (or considerations), and outputs.

As discussed in the framework and process overview, there are five parts for each affordability analysis activity:

- An overview of the activity (with value stream map or process chart)
- The inputs needed for the activity
- The sub-activities for the activity (the framework questions) with associated tasks and considerations
- The exit criteria to complete the activity (sufficiency & quality)
- The outputs from the activity (may not be used by the next activity, could be used by any of the following activities – or available as references).

4.5 KEY TERMS

Acquisition Cost: A cost class that includes "the total cost to procure, install, and put into operation a system, product, or special piece of infrastructure. These are the costs associated with planning, designing, engineering, testing, manufacturing, and deploying/installing a system or process" (Parnell et al., 2011, p. 144).

Affordability: The degree to which the resources being allocated to a capability relative to other uses of those resources reflect (i) the importance, urgency, and satisfaction of mission, strategic investment, and organizational needs, and (ii) a prudent balance of performance, cost, and schedule constraints consistent with the time-phased availability (technical, market, and fiscal) of budgeted resources.

Affordability Analysis: A process and assessment that supports resource allocation decision-making. It identifies and quantifies the performance expectations of stakeholders, assigns value to those expectations, and measures the LCC of alternatives relative to both opportunity costs and resourcing actions or plans.

Cash flow (CF): The annual summation of revenues and expenditures for a business, project, or financial product.

Cost Analysis: The review and evaluation of the separate cost elements and profit or fee in an offerer's or contractor's proposal to determine a fair and reasonable price or to determine cost realism. Cost analysis includes the application of judgment to determine how well the proposed costs represent what the cost of the contract should be, assuming reasonable economy and efficiency (Cost Analysis, 2014).

Cost: Any term that refers to organizational resource that would be expended for a program/product. Examples include but are not limited to funds, dollars, hours, professional service years, and facility space (Parnell et al., 2011, p. 143).

Cost Estimation or Estimate: An assessment and forecast of the cost of a hardware, software, and/or services (Williams & Barber, 2011).

Cost Objective: One of the numerous possible objectives in a value hierarchy; it is often represented as a mathematical value that the organization seeks to minimize. Treated as a separate value and normalized per unit in a tradespace facilitates its presentation to decision makers (Parnell, 2013, p. 141).

Direct Cost: A cost class that includes costs "that are associated with a specific system, end item, product, process, or service." These costs can be subdivided into "direct labor, material, or expenses" (Parnell et al., 2011, p. 144).

Fixed Cost: A cost class that includes costs "that remain constant independent of the quantity or phase of the life cycle being addressed in the estimate." Examples of fixed costs include "research, lease rentals, depreciation, taxes, insurance, and security" (Parnell et al., 2011, p. 144).

Indirect Cost: A cost class that includes costs "that cannot be associated with a specific product or process." These costs can be clustered into an overhead account and applied to "direct costs" as a burden. Examples include "security, accounting and finance labor, janitorial services, executive management, and training" (Parnell et al., 2011, p. 144).

Learning Curve: A graphical representation of how people and organizations get better at tasks (reduce cost; vertical axis) as experience increases (horizontal axis). (Quantitative Module E: Learning Curves, 2006, p. 772).

Life Cycle Cost: "The total cost of an alternatece across all stages of the altenative (aquisistion) life cycle." It normally does not include sunk costs (Parnell, 2013, p. 134 and 146; Parnell et al., 2011, pp. 137–138).

Monte Carlo: A simulation methodology that is a "realization of a sequence of outcomes by repeated random selection of input scenarios according to experts' probabilities" (Parnell, 2013, p. 146 and 255; Parnell et al., 2011, pp. 171–172; Law & Kelton, 1991, pp. 78–80).

Net Present Value (NPV): A "calculation of the present worth from a sequence of cash flows using a discount rate" (Parnell, 2013, p. 135 and 146; Parnell et al., 2011, p. 401 and 415).

Nonrecurring Cost: A cost class that includes costs "that occur only once in the life cycle of a system". Examples include items that are not anticipated to be experienced more than once in the process such as "the cost associated with design, engineering, and testing." (Parnell et al., 2011, p. 144).

Program (or Project) Evaluation and Review Technique (PERT): A project management tool used to schedule, organize, and coordinate tasks within a project. It is a method to analyze the time required to complete each tasks for a given project. It is used to help calculate the minimum time required to complete a project.

Recurring Space: A cost class that includes costs "that repeats with every unit of product or time a system process occurs." Similarly to "variable costs, they are a function of the quantity of items output" (Parnell et al., 2011, p. 144).

Resource: An asset accessible for use in producing a benefit, the benefit of a decision (e.g., money, facilities, time, people, cognitive effort, etc.).

Resource Space: A collection of resources required for possible resolution of an issue.

Variable Cost: A cost class that includes costs "that increase or decrease as the amount of product or service output from a system increases or decreases." Examples of fixed costs include "direct labor, material, and power" (Parnell et al., 2011, p. 144).

4.6 EXCERCISES

4.1. Identify and describe the difference between Affordability and Cost Analysis.

4.2. List and provide examples of three types of resources.

4.3. List and describe two examples each of soft and hard skills required for those in a customer service roll.

4.4. Name three classes of cost that describe the cost of testing and why these costs fit into these classes.

4.5. List and describe the three terms critical to understanding why, what, and how tasks are undertaken.

4.6. Estimate the cost of a 2500 ft² house in your neighborhood by developing a CER.

4.7. A home appliance company is trying to expand its line of products by incorporating microwaves in their catalog. In order to take the microwaves to the market, the company requires an initial investment of $400,000 to purchase equipment. The annual recurring costs for year 1 are $70,000 and expected to increase by $10,000 every year. Market forecasts predict $110,000 in sales for the first year and increase by $20,000 for each following year. At the end of year 5, $120,000 is expected for recapitalization of equipment. The annual interest rate and estimated annual inflation rate are 5% and 8%, respectively. Consider a 5-year planning horizon. Perform an NPV analysis to this investment decision. Is the project profitable?

4.8. Calculate the projected Net Present Value (NPV) for an effort that you had to invest an initial $100,000 this year. You expect $15,000 for maintenance and supplies expenditures as well as $25,000 of income per year for the next 10 years. At the end of 10 years, you sell your facilities and equipment for $50,000. The estimated annual interest rate is 4% with an inflation rate of 2%. What was your final NPV? In what year does this project return a positive NPV?

4.9. XYZ Automotive takes 2.5 h to assemble a car for the eighth time. As they continue the assembly process, they record the assembly time for the 16th unit as 2.2 h.

(a) Calculate the learning rate percentage.

(b) How much time did XYZ take to assemble the second car?

4.10. AV Technology assembles their first computer for a particular day in 12 minutes. After assembling several units, they determine that their negative numerical factor (r) is –0.1047.

(a) Calculate the learning rate percentage.

(b) How much time will it take to assemble the 35th unit?

4.11. For a production process of 30 units with a learning rate of 85% and a cost of $75 to produce the first unit:

(a) Construct a table displaying the Cost to Produce the Xth Unit, Cumulative Cost, and Cumulative Average Cost for units 1–30.

(b) Using the cumulative average formula, calculate the cumulative average cost for the 25th unit. How much difference is there between the value obtained using the formula and the value obtained in a) for the 25th unit?

4.12. Develop an 85% learning curve for the production of the Q Model Unmanned Aerial Vehicle.

Total Units Produced	Cost to Produce Xth Unit	Cumulative Cost	Cumulative Average Cost
1	100.0	100.0	100.0
2	85.0	185.0	92.5
3	77.3	262.3	87.4
4	72.3	334.5	83.6

4.13. Flying Bikes Company assembles the bicycles they produce in a four-step process: handlebar assembly, pedal assembly, tire assembly, and seat assembly. Handlebar assembly takes 2.5 min at an 88% learning rate. Pedal assembly is done in 3 min at a 95% learning rate. Furthermore, tire assembly requires 4.5 min and has a 90% learning rate. Finally, the seat assembly takes 1 minute, and it is performed at an 80% learning rate. Calculate the composite learning rate for the complete assembly process.

4.14. The results of the painting process for a particular automobile for a 10-week window are shown in the following table:

EXCERCISES

Week	Cumulative Man-Hours Expended	Cumulative Units Completed X
1	285	15
2	585	31
3	860	49
4	1180	71
5	1460	95
6	1760	120
7	2040	147
8	2355	176
9	2640	207
10	2920	240

Find the learning rate percentage for the automobile painting process using the aforementioned data.

4.15. An analyst is trying to determine a CER between road repair costs and miles repaired in the state of Arkansas. He has obtained data regarding previous repairs that have been done in the state in the last 2 years as shown as follows:

Miles Repaired	Repair Costs (M)
4.1	$9.43
6.0	$15.60
1.2	$3.96
3.2	$9.28
5.2	$17.16
7.0	$16.80
9.8	$25.48
5.7	$15.39
3.3	$9.90
7.5	$21.75
8.4	$26.04
5.1	$11.22
6.3	$17.33
9.2	$27.23
2.4	$7.54

(a) Perform a linear regression to obtain the CER.
(b) Construct a normal probability and residual plots. Are the underlying assumptions of a basic regression model met?

REFERENCES

AACE International (2015) From AACE International: http://www.aacei.org/ (accessed 24 Aug 2015).

Acquisition Life Cycle (2015) From ACQuipedia: https://dap.dau.mil/acquipedia/Pages/ArticleDetails.aspx?aid=30c99fbf-d95f-4452-966c-500176b42688#anchorDef (accessed 12 June 2015).

Assistant Secretary of Defense for Acquisition (2014) *Program Management Empowerment and Accountability*. From Office of the Assistant Secretary of Defense for Acquisition: http://www.acq.osd.mil/asda/initiatives/factsheets/program_mgr_empowerment/index.shtml (accessed 12 June 2015).

Canada, J.R., Sullivan, W.G., Kulonda, D.J., and White, J.A. (2005) *Capital Investment Analysis for Engineering and Management*, 3rd edn, Prentice-Hall, Upper Saddle River, NJ.

Chacko, G.K. (1990) *Decision-Making under Uncertainty: An Applied Statistics Approach*, Praeger Publishers, New York.

Cost Analysis (2014) From ACQuipedia: https://dap.dau.mil/acquipedia/Pages/ArticleDetails.aspx?aid=5b784aef-ad2d-4a99-8537-ff09e9d86757 (accessed 15 July 2015).

Farr, J.V. (2011) *Systems Life Cycle Costing: Economic Analysis, Estimation and Management*, 1st edn, CRC Press, New York.

GAO-07-406SP (2007) *GAO-07-406SP, Defense Acquisitions: Assessment of Selected Weapon Systems*, United States Government Accountability Office, Washington, DC.

GAO-09-3SP (2009) *GAO-09-3SP: GAO Cost Estimating and Assessment Guide*, United States Government Accountability Office, Washington, DC.

GAO-15-342SP (2015) *GAO-15-342SP, Defense Acquisitions: Assessment of Selected Weapon Systems*, United States Government Accountability Office, Washington, DC.

Glossary (2015) From Society of Cost Estimating and Analysis: http://www.sceaonline.org/prof_dev/glossary.html (accessed 23 August 2016).

INCOSE (2015) From International Committee for Systems Engineering (INCOSE): www.incose.org (accessed 23 August 2016).

Kerzner, H.R. (2006) *Project Management: A Systems Approach to Planning, Scheduling, and Controlling*, 9th edn, John Wiley & Sons, Hoboken.

Khan, M.Y. (1999) *Theory & Problems in Financial Management*, McGraw Hill Higher Education, Boston.

Law, A.M. and Kelton, W.D. (1991) *Simulation Modelling and Analysis*, 3rd edn, McGraw Hill Higher Education, New York.

Lee, D.A. (1997) *The Cost of Analyst's Compariosn*, Logistics Management Institute, McLean, VA.

Michealson, K. (2015) "Big A XE "**Big A**" " Affordability Analysis: Analytical Considerations for Conducting "Big A" Affordability. Military Operations Research Society (MORS) Affordability Analysis Community of Practice, Alexandria, VA.

MIL-STD-881C, Department of Defense Standard Practice: Work Breakdown Structures for Defense Materiel Items. (2011) Washington, DC.

REFERENCES

Naval Sea Systems Command (1995) *Parametric Cost Estimating Handbook*, 2nd edn, Naval Sea Systems Command, Arlington.

Ostwald, P.F. (1992) *Cost Estimating*, 3rd edn, Prentice Hall, Englewood Cliffs, New Jersey.

Palisade Corporation (2014) *@RISK Risk Analysis Software using Monte Carlo Simulation XE "Monte Carlo Simulation" for Excel - at risk - Palisade*. From Palisade: http://www.palisade.com/risk/ (accessed 24 August 2015).

Park, C.S. (2004) *Fundamentals of Engineering Economics*, Pearson-Prentice Hall, Upper Saddle River, NJ.

Parnell, G.B. (2013) *Handbook of Decision Analysis*, John Wiley & Sons, Inc., Hoboken, NJ.

Parnell, G.S., Driscoll, P.J., and Henderson, D.L. (2011) *Decision Making in System Engineering and Management*, Wiley.

Project Management Institute (2013) *A Guide to the Project Management Body of Knowledge (PMBOK® Guide)*, 5th edn, Project Management Institute, Inc., Newtown Square, PA.

Quantitative Module E: Learning Curves (2006) From Learning Curves in Services and Manufacturing: http://wps.prenhall.com/wps/media/objects/2234/2288589/ModE.pdf (accessed 16 July 2015).

Stewart, R.D., Wyskida, R.M., and Johannes, J.D. (eds) (1995) *Cost Estimator's Reference Manual*, 2nd edn, John Wiley & Sons, New York, NY.

Thuesen, H.G. and Fabrycky, W.J. (1989) *Engineering Economy*, 7th edn, Prentice Hall, Englewood Cliffs.

United States Department of Labor (2016) *Consumer Price Index (CPI)*. From U.S. Bureau of Labor Statistics, Division of Consumer Prices and Price Indexes: http://www.bls.gov/cpi (accessed 23 August 2016).

Williams, T.S. and Barber, E. (2011) *DAU Teaching Note - Cost Estimating Methodologies (Feb 2011)*. From ACC Practice Center - Cost Estimating: https://acc.dau.mil/adl/en-US/30373/file/61352/B4_CE_Methodologies_-_Feb%2011_V3.pdf (accessed 16 July 2015).

Wright, T.P. (1936) Factors affecting the cost of airplanes. *Journal of the Aeronautical Sciences*, **3** (4), 122–128.

5

UNDERSTANDING DECISION MANAGEMENT

MATTHEW CILLI

U.S. Army, Armament Research Development and Engineering Center (ARDEC), Systems Analysis Division, Picatinny, NJ, USA

GREGORY S. PARNELL

Department of Industrial Engineering, University of Arkansas, Fayetteville, AR, USA

> Decide what you want, decide what you are willing to exchange for it. Establish your priorities and go to work.
>
> (H. L. Hunt)

5.1 INTRODUCTION[1]

Successful Systems Engineering requires good decision-making. Many systems engineering decisions are difficult decisions in that they include multiple competing objectives, numerous stakeholders, substantial uncertainty, significant consequences, and high accountability. In these cases, good decision-making requires a formal decision management process. The purpose of the decision management process, as defined by ISO/IEC 15288:2015, is "…to provide a structured, analytical framework for identifying, characterizing and evaluating a set of alternatives for a decision at any point in the life-cycle and select the most beneficial course of action."

[1] This chapter is a partial adaptation from Cilli (2015).

This chapter aligns with the structure and principles of the Decision Management Process Section of the INCOSE Systems Engineering Handbook v4.0 (INCOSE SE Handbook Working Group, 2015) and presents the decision management process steps as described therein (written permission from INCOSE Handbook Working Group pending), and it expands on the SEBok section on Decision Management (http://sebokwiki.org/wiki/Decision_Management). Building upon the foundation, this chapter adds a significant amount of text and introduces several illustrations to provide richer discussion and finer clarity.

5.2 DECISION PROCESS CONTEXT

A formal decision management process is the transformation of a broadly stated decision situation (see Chapter 4) into a recommended course of action and associated implementation plan. The process is executed by a resourced decision team that consists of a decision-maker with full responsibility, authority, and accountability for the decision at hand, a decision analyst with a suite of reasoning tools, subject matter experts with performance models, and a representative set of end users and other stakeholders (Parnell et al., 2013). The decision process is executed within the policy and guidelines established by the sponsoring agent. The formal decision management process realizes this transformation through a structured set of activities described later in this chapter. Note the process presented here does not replace the engineering models, performance models, operational models, cost models, and expert opinion prevalent in many enterprises but rather complements such tools by synthesizing their outputs in a way that helps decision-makers thoroughly compare relative merits of each alternative in the presence of competing objectives and uncertainty. (Buede, 2009; Parnell et al., 2011).

Models are central to systems analysis and trade-off analysis. A decision support model is a composite model that integrates outputs of otherwise separate models into a holistic system view mapping critical design choices to consequences relevant to stakeholders. A decision support model helps decision-maker(s) overcome cognitive limits without oversimplifying the problem.

Early in the life cycle, inputs to the decision management process are often little more than broad statements of the decision situation. As such, systems engineers should not expect to receive a well-structured problem statement as input to the decision management process. In later stages of the system life cycle, the inputs usually include models and simulations, test results, and operational data. Opportunities to use a decision management process as part of a systems engineering trade-off analysis throughout the system analysis life cycle are illustrated in Figure 5.3.

The ultimate output of the decision management process should be a recommended course of action and associated implementation plan provided in the form of a high-quality decision report. The decision report should communicate key findings through effective tradespace visualizations underpinned by defendable rationale grounded in analysis results that are repeatable and traceable. As decision-makers seek to understand root causes of top-level observations and build

5.3 DECISION PROCESS ACTIVITIES

The decision analysis process as described in Parnell et al. (2013) and Parnell et al. (2011) can be summarized in 10 process steps: (i) frame decision and tailor process, (ii) develop objectives and measures, (iii) generate creative alternatives, (iv) assess alternatives via deterministic analysis, (v) synthesize results, (vi) identify uncertainty and conduct probabilistic analysis, (vii) assess impact of uncertainty, (viii) improve alternatives, (ix) communicate trade-offs, and (x) present recommendation and implementation plan. An illustration of this 10-step decision process interpretation is provided in Figure 5.1 (http://sebokwiki.org/wiki/Decision_Management, 2015)

Applying this decision process to a new product development context calls for the integration of this process with the systems engineering process. The systems engineering process provides the holistic, structured thinking perspective required for the design and development of complex systems while the analytics-based decision process provides the mathematical rigor needed to properly represent and communicate reasoning and produce meaningful visualization of the tradespace. Figure 5.2 provides a process map of this analytical decision process integrated with

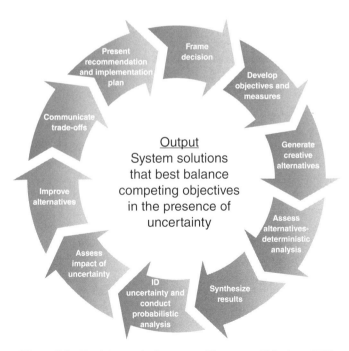

Figure 5.1 Decision analysis process (Courtesy of Matthew Cilli)

Figure 5.2 Integrated Systems Engineering Decision Management (ISEDM) Process Map

six of the systems engineering technical processes, one cross-cutting method, and three technical management processes as described in the INCOSE Systems Engineering Handbook V4. This integrated process will be referred to as the Integrated Systems Engineering Decision Management (ISEDM) Process throughout the rest of this dissertation.

The white text within the outer green ring identifies elements systems engineering processes while the 10 arrows forming the inner ring represent the 10 steps of the decision management process. Interactions between the systems engineering processes and the decision process are represented by the small, dotted green or blue arrows. These interactions are discussed briefly in the subsequent sections of this chapter. (*The reader is referred to the online version of this book for color indication.*)

The focus of the process is to find system solutions that best balance competing objectives in the presence of uncertainty as shown in the center of Figure 5.2. This single focus is important as it can be argued that all systems engineering activities

DECISION PROCESS ACTIVITIES

Table 5.1 Crosswalk Between SE Terms in Figure 5.2 and INCOSE Systems Engineering Handbook V4 and ISO/IEC/IEEE 15288:2015

Systems Engineering Terms in of Figure 5.2	ISO/IEC/IEEE 15288:2015 Section	INCOSE Systems Engineering Handbook V4 Section
Business or mission analysis process	6.4.1	4.1
Stakeholder needs and requirements definition process	6.4.2	4.2
System requirements definition process	6.4.3	4.3
Architecture definition process	6.4.4	4.4
Design definition process	6.4.5	4.5
Measurement process	6.3.7	5.7
System analysis process	6.4.6	4.6
Modeling and simulation	-	9.1
Risk management process	6.3.4	5.4
Project planning process	6.3.1	5.1

should be conducted within the context of supporting good decision-making. If a systems engineering activity cannot point to at least one of the many decisions embedded in a system's lifecycle, one must wonder why the activity is being conducted at all. Positioning decision management as central to systems engineering activity will ensure that the efforts are rightfully interpreted as relevant and meaningful and thus maximize the discipline's value proposition to new product developers and stakeholders.

The decision management process is an iterative process with an openness to change and adapts as understanding of the decision and the tradespace emerges with each activity. The circular shape of the process map is meant to convey the notion of an iterative process with significant interaction between the process steps. The feedback loops seek to capture new information regarding the decision task at any point in the decision process and make appropriate adjustments.

Table 5.1 provides a crosswalk between the Systems Engineering terms used in Figure 5.2, *ISO/IEC/IEEE 15288:2015 Systems and Software Engineering – System Life Cycle Processes* and the section of the *INCOSE Systems Engineering Handbook V4* devoted to the term.

The ISEDM process can be used for trade-off analyses encountered across the system's development life cycle – tailored to the particulars of the decision situation. Figure 5.3 adds the ISEDM process icon several times to the generic life cycle model put forth in the INCOSE Systems Engineering Handbook V4 to illustrate key opportunities to use the process to execute systems engineering trade-off analyses throughout the systems development life cycle.

5.3.1 Frame Decision

The first step of the decision management process is to frame the decision and to tailor the decision process. To help ensure that the decision-makers and stakeholders fully

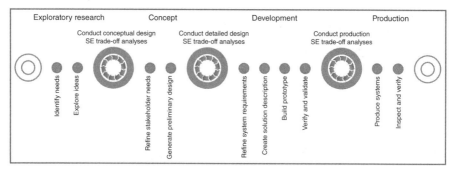

Figure 5.3 Trade-off studies throughout the system's development life cycle

understand the decision context and to enhance the overall traceability of the decision, the systems engineer should capture a description of the system baseline as well as a notion for how the envisioned system will be used (concept of operations) along with system boundaries and anticipated interfaces. Decision context includes such details as the timeframe allotted for the decisions, an explicit list of decision-makers and stakeholders, available resources, and expectations regarding the type of action to be taken as a result of the decision at hand as well as decisions anticipated in the future (Edwards et al. 2007). The best practice is to identify a decision problem statement that defines the decision in terms of the system life cycle. Next, three categories of decisions should be listed: decisions that have been made, decisions to be made now, and subsequent decisions that can be made later in the life cycle. Effort is then focused on the decisions to be made now.

Once the decision at hand is sufficiently framed, systems engineers must select the analytical approach that best fits the frame and structure of the decision problem at hand. For deterministic problems, optimization models can explore the decision space. However, when there are "… clear, important, and discrete events that stand between the implementation of the alternatives and the eventual consequences…" (Edwards et al., 2007), a decision tree is a well-suited analytical approach, especially when the decision structure has only a few decision nodes and chance nodes. As the number of decision nodes and chance nodes grow, the decision tree quickly becomes unwieldy and loses some of its communicative power. However, decision trees and many optimization models require consequences to be expressed in terms of a single number. This is commonly accomplished for decision situations where the potential consequences of alternatives can be readily monetized and end state consequences can be expressed in dollars, euros, yen, and so on. When the potential consequences of alternatives within a decision problem cannot be easily monetized, an objective function can often be formulated to synthesize an alternative's response across multiple, often competing, objectives. A best practice for this type of problem is the multiple objective decision analysis (MODA) approach (Chapter 2).

The decision management method most commonly employed by systems engineers is the trade study and more often than not employs some form of MODA approach. The aim is to define, measure, and assess shareholder and stakeholder values and then synthesize this information to facilitate the decision-maker's search for an alternative that represents the optimally balanced response to often competing objectives. Major system projects often generate large amounts of data from many separate analyses performed at the system, subsystem, component, or technology level by different organizations. Each analysis, however, only delivers one dimension of the decision at hand, one piece of the puzzle that the decision-makers are trying to assemble. These analyses may have varying assumptions and may be reported as standalone documents, from which decision-makers must somehow aggregate system-level data for all alternatives across all dimensions of the tradespace in his or her head. This would prove to be an ill-fated task as all decision-makers and stakeholders have cognitive limits that preclude them from successfully processing this amount of information in their short-term memory (Miller 1956). When faced with a deluge of information that exceeds human cognitive limits, decision-makers may be tempted to oversimplify the tradespace by drastically truncating objectives and/or reducing the set of alternatives under consideration, but such oversimplification runs a high risk of generating decisions that lead to poor outcomes.

By providing techniques to decompose a trade-off decision into logical segments and then synthesize the parts into a coherent whole, a formal decision management process offers an approach that allows the decision-makers to work within human cognitive limits without oversimplifying the problem. In addition, by decomposing the overall decision problem into smaller elements, experts can provide assessments of alternatives as they perform within the objective associated with their area of expertise. Buede and Choisser put it this way,

> These component parts can be subdivided as finely as needed so that the total expertise of the system design team can be focused, in turn, on specific, well-defined issues. The analyses on the component parts can then be combined appropriately to achieve overall results that the decision makers can use confidently. The benefits to the decision maker of using this approach include increased objectivity, less risk of overlooking significant factors and, perhaps most importantly, the ability to reconstruct the selection process in explaining the system recommendation to others. Intuition is not easily reproducible.
> (Buede & Choisser 1992)

MODA approaches generally differ in the techniques used to elicit values from stakeholders, the use of screening techniques, the degree to which an alternative's responses to objectives (and subobjectives) are aggregated, the mathematics used to aggregate such responses, the treatment of uncertainty, the robustness of sensitivity analyses, the search for improved alternatives, and the versatility and quality of tradespace visualization outputs. If time and funding allow, systems engineers may want to conduct trade-off studies using several techniques, compare and contrast results, and reconcile any differences to ensure that the findings are robust. Although there are many possible ways to specifically implement MODA, the discussion contained in the rest of this chapter represents a short summary of best practices.

5.3.1.1 Example of Framing the Decision As an example of decision management process execution, consider the hypothetical small unmanned aerial system (sUAS) case study introduced in the following paragraphs. Note that the lead author of this chapter created a plausible and sufficiently rich example by distilling the technical ideas presented in the-805 page textbook by Dr. Jay Gundlach, Designing Unmanned Aircraft Systems: A Comprehensive Approach. The lead author of this chapter used Gundlach's textbook to inform physical architecture descriptions of the notional UAVs and the stakeholder requirements created for the case study that follows, but no attempt was made to use the mathematical relationships provided in the textbook to generate cost, schedule, and performance estimates for the sUAV concepts within the case study of this chapter. All estimates should be considered illustrative.

We assume that the military is contemplating the start of a new effort to develop the next-generation sUAS and a lead systems engineer has been tasked to conduct a systems engineering trade-off analysis to identify system concepts in order to inform requirements generation. The lead systems engineer is told that the future system will be used primarily in an Intelligence, Surveillance, and Reconnaissance (ISR) mission context as they are now but instead of operating at altitudes of 500–1000 ft, the sUAS will be expected to operate at an altitude of 3000 ft in order to avoid airspace conflicts with military helicopters and to reduce the likelihood of being detected by enemy forces. The lead engineer was also told that the new capability should be operational within 7 years and that the life cycle costs must be affordable.

The lead systems engineer is excited about the opportunity but is a bit anxious about the ambiguity surrounding the problem statement. To curb some of the anxiety, he begins by asking some clarification questions regarding decision timeframe, system boundaries, and expectations regarding affordability. He learns that the final report is due in 12 months with executive-level reviews scheduled every quarter with preliminary study findings expected by the third review. He also learns that the system boundaries include the air vehicle, the ground elements, and the communication links between them. With regard to affordability, he was told not to initially discard concepts on a cost basis but rather collect rough order of magnitude life cycle cost estimates for each concept and show the cost versus performance versus schedule relationship for each. With this information and the information from similar trades being conducted elsewhere in the portfolio, the executive decision board will determine appropriate affordability goals for the next phase of the future sUAS development effort.

Armed with the initial framing of the trade at hand, the lead systems engineer begins wondering how the goodness of system alternatives should be defined. The next section of this chapter addresses the best practices associated with developing objectives and measures and is immediately followed by the continuation of the sUAV case study.

5.3.2 Develop Objectives and Measures

Defining how a decision will be made may seem straightforward, but often becomes an arduous task of seeking clarity amidst a large number of ambiguous stakeholder need statements. The first step is to use the information obtained from the Stakeholder Requirements Definition Process, Requirements Analysis Process, and Requirements Management Processes to develop objectives and measures. If these processes have not been started, then stakeholder analysis is required. Often, this begins with reading documentation on the decision topic followed by a visit to as many decision-makers and stakeholders as reasonable and facilitating discussion about the decision problem. This is best done with interviews and focus groups with subject matter experts and stakeholders.

For systems engineering trade-off analyses, top-level stakeholder value often includes competing objectives of performance, development schedule, life cycle costs, and long-term viability. For corporate decisions, shareholder value would be added to this list. With the top-level objectives set, lower levels of objective hierarchy should be discovered. For performance-related objectives, it is often helpful to work through a functional decomposition (usually done as part of the requirements and architectural design processes) of the system of interest to generate a thorough set of potential objectives. Start by identifying inputs and outputs of the system of interest and craft a succinct top-level functional statement about what the system of interest does, identifying the action performed by the system of interest to transform the inputs into outputs. Test this initial list of fundamental objectives for key properties by checking that each fundamental objective is essential and controllable and that the set of fundamental objectives is complete, nonredundant, concise, specific, and understandable (Edwards et al. 2007). See Figure 5.4 for a list of the key properties for a set of fundamental objectives.

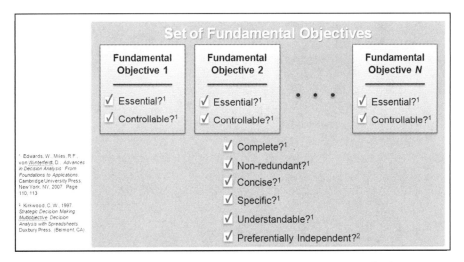

Figure 5.4 Key properties of a high-quality set of fundamental objectives

Beyond these best practices, the creation of fundamental objectives is as much an art as it is a science. This part of the decision process clearly involves subjectivity. It is important to note, however, that a subjective process is not synonymous with an arbitrary or a capricious process. As Keeney points out,

> Subjective aspects are a critical part of decisions. Defining what the decision is and coming up with a list of objectives, based on one's values, and a set of alternatives are by nature subjective processes. You cannot think about a decision, let alone analyze one, without addressing these elements. Hence, one cannot even think about a decision without incorporating subjective aspects
>
> (Keeney 2004)

The output of this process step takes on the form of a fundamental objectives hierarchy as illustrated in Figure 5.5.

For completeness, it is often helpful to build a crosswalk between the objectives hierarchy and any stakeholder need statements or capability gap lists as illustrated in Table 5.2. This activity helps ensure that all stakeholder need statements or capability gaps have been covered by at least one objective. It also aids in identifying objectives that do not directly trace to an expressed need. Such objectives will need additional explanation to justify their inclusion in the hierarchy. It should be noted, however, that it is common to have such objectives included in a hierarchy because stakeholders are often silent about needs that are currently satisfied by the incumbent system but would

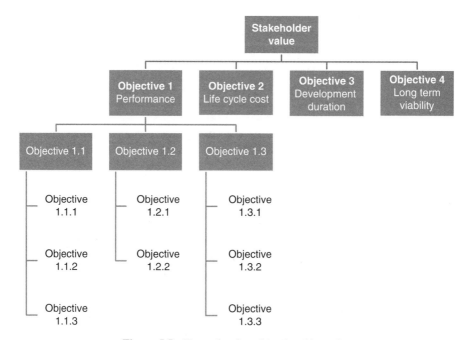

Figure 5.5 Example of an objectives hierarchy

DECISION PROCESS ACTIVITIES

Table 5.2 Crosswalk Between Fundamental Objectives and Stakeholder Need Statements

| | Objective 1 ||||||| | | |
| | OBJ 1.1 ||| OBJ 1.2 || OBJ 1.3 ||| | | |
	OBJ 1.1	OBJ 1.2	OBJ 1.3	OBJ 2.1	OBJ 2.2	OBJ 3.1	OBJ 3.2	OBJ 3.3	Objective 2	Objective 3	Objective 4
Capability Gap 1	x										
Capability Gap 2							x				
Capability Gap 3				x							
Capability Gap 4					x						
Capability Gap 5		x						x			
Business Need 1									x		
Business Need 2										x	
Business Need 3											x

not be happy if, in an effort to fill a perceived need, the new system created a new gap: for example, up-armored vehicles that become unreliable or insufficiently mobile.

For each fundamental objective, a measure (also known as attribute, criterion, and metric) must be established so that alternatives that more fully satisfy the objective receive a better score on the measure than those alternatives that satisfy the objective to a lesser degree. Table 5.3 illustrates this one-to-one mapping of objective and measure.

A measure should be unambiguous, comprehensive, direct, operational, and understandable (Keeney & Gregory 2005). Table 5.4 defines these properties of a high-quality measure.

Keeney has identified three types of measures – natural measure (kilometers, degrees, probability, seconds, etc.), constructed measure (Dow Jones Industrial

Table 5.3 Illustrating the One-to-One Mapping of Objective and Measure

Stakeholder value	Objective 1	Objective 1.1	Objective 1.1.1	Measure 1.1.1
			Objective 1.1.2	Measure 1.1.2
			Objective 1.1.3	Measure 1.1.3
		Objective 1.2	Objective 1.2.1	Measure 1.2.1
			Objective 1.2.2	Measure 1.2.2
		Objective 1.3	Objective 1.3.1	Measure 1.3.1
			Objective 1.3.2	Measure 1.3.2
			Objective 1.3.3	Measure 1.3.3
	Objective 2			Measure 2
	Objective 3			Measure 3
	Objective 4			Measure 4

Table 5.4 Properties of a High-Quality Measure

Property	Definition
Unambiguous	A clear relationship exists between consequences and descriptions of consequences using the measure.
Comprehensive	The attribute levels cover the range of possible consequences for the corresponding objective, and value judgments implicit in the attribute are reasonable.
Direct	The measure levels directly describe the consequences of interest.
Operational	In practice, information to describe consequences can be obtained and value trade-offs can reasonably be made.
Understandable	Consequences and value trade-offs made using the measure can readily be understood and clearly communicated.

Source: Data from Keeney & Gregory 2005.

Average, Heat Index, Consumer Price Index, etc.), and a proxy measure (usually a natural measure of a consequence that is thought to be correlated with the consequence of interest). Keeney recommends natural measures whenever possible since they tend to be commonly used and very understandable. When a natural measure is not available to describe a consequence, he recommends a constructed measure that directly describes the consequence of interest. If neither a natural measure nor a constructed measure is practical, then a proxy measure using a natural scale is often workable although by definition, it is an indirect measure as it does not directly describe the consequence of interest.

A defining feature of Multiobjective Decision Analysis (also called multiattribute value theory) is the transformation from measure space to value space that enables mathematical representation of a composite value score across multiple measures. This transformation is performed through the use of a value function. Value functions describe returns to scale on the measure. In other words, value functions describe the degree of satisfaction stakeholders perceive at each point along the measure scale.

There are several techniques available to elicit value functions and priority weightings from stakeholders. One of the more popular techniques used in marketing circles is Conjoint Analysis (Green et al., 2001) where stakeholders are asked to make a series of pairwise comparisons between hypothetical products. For decisions that involve fewer than eight competing objectives, Conjoint Analysis is a compelling technique for generating representative value schemes. As an objectives hierarchy grows beyond eight objectives, the number of pairwise comparisons required to formulate the representative value schemes balloons to an unreasonable number. Since many complex systems engineering decision tasks tend to involve the balancing of 25 to 35 objectives, the value scheme elicitation approach described in this chapter is a direct interview value function formulation technique coupled with a swing weight matrix methodology for determining priority weightings.

DECISION PROCESS ACTIVITIES

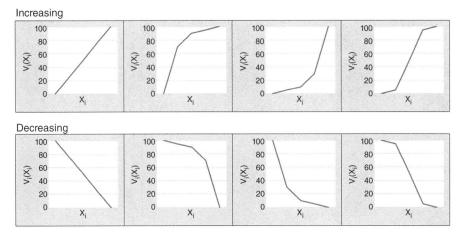

Figure 5.6 Value function examples

When creating a value function, one ascertains whether stakeholders believe there is a walk-away point on the objective measure scale (*x*-axis) and map it to 0 value on the value scale (*y*-axis). A walk-away point is defined as the measure score where regardless of how well an alternative performs in other measures, the decision maker will walk away from the alternative. Working with the stakeholder, find the measure score beyond which an alternative provides no additional value, label it "meaningful limit" (also called ideal), and map it to 100 (1 and 10 are also common scales) on the value scale (*y*-axis). If the returns to scale are linear, connect the walk-away value point to the meaningful value point with a straight line. If there is reason to believe stakeholder value behaves with nonlinear returns to scale, pick appropriate inflection points and draw the curve. The rationale for the shape of the value functions should be documented for traceability and defensibility (Parnell et al., 2011). Figure 5.6 provides examples of some common value function shapes.

Practice suggests that eliciting two end points and three inflection points provides informative value functions without overtaxing the systems engineering or stakeholder.

- Walk Away: Stakeholder will dismiss an alternative if it fails to meet at least this level regardless of how it performs on other value measures (1 point for meeting, 0 points if missed).
- Marginally Acceptable: Stakeholder begins to become interested, and beyond this point the perceived value increases rapidly (10 points).
- Target: Desired level (50 points).
- Stretch Goal: Improving beyond this point is considered gold plating, so there is very little available value between this point and meaningful limit (90 points).
- Meaningful Limit: Theoretical limit or known practical limit beyond which would be considered unrealistic (100 points).

Swing weight matrix	Level of importance			Required relationships
	Defining capability	Critical capability	Enabling/enhancing capability	A > all other cells
High differentiation	A	B2	C3	B1 > C1, C2, D1, D2, E
				B2 > C2, C3, D1, D2, E
Moderate differentiation	B1	C2	D2	C1 > D1, E
				C2 > D1, D2, E
				C3 > D2, E
Low differentiation	C1	D1	E	D1 > E
				D2 > E

(Differentiation in measure range — vertical axis label)

Figure 5.7 Swing weight matrix

In an effort to capture the voice of the customer, systems engineers will often ask a stakeholder focus group to prioritize their requirements. As Keeney puts it,

> Most important decisions involve multiple objectives, and usually with multiple-objective decisions, you can't have it all. You will have to accept less achievement in terms of some objectives in order to achieve more on other objectives. But how much less would you accept to achieve how much more?
>
> (Keeney 2002)

The mathematics of Multiobjective Decision Analysis (MODA) requires that the weights depend on importance of the preferentially independent measure and the range of the measure (walk away to stretch goal or ideal). A useful tool for determining weightings is the swing weight matrix. (Parnell et al., 2011) For each measure, consider its importance by determining if the measure corresponds to a defining capability, a critical capability, or an enabling capability and also consider the variation measure range by considering the gap between the current capability and the desired capability and put the name of the measure in the appropriate cell of the matrix. Swing weights are then assigned to each measure according to the required relationship rules described in Figure 5.7. Swing weights are then converted to measure weights by normalizing such that the set sums to 1. For the purposes of swing weight matrix use, consider a defining capability to be one that directly traces to a verb/noun pair identified at the top-level (level 0) functional definition of the system of interest – the reason why the system exists. Consider enabling capabilities to trace to functions that are clearly not the reason why the system exists but allow the core functions to be executed more fully. Let critical capabilities be those that are more than enabling but not quite defining.

All decisions involve elements of subjectivity, the distinctive feature of formal decision management process is that these subjective elements are rigorously

DECISION PROCESS ACTIVITIES   169

documented so that the consequences can be identified and assessed. Toward this end, it is considered good practice to document the measured, the value function, and the priority weighting along with associated rationale for each fundamental objective.

5.3.2.1 Example of Developing Objectives and Measures Returning to the sUAV case study example, we find that after some quality time with many of the stakeholders and hours digging through white papers, memos, and presentations relevant to the problem statement, the lead systems engineer spearheaded a requirements analysis effort that included a functional decomposition. This exercise helped the systems engineering trade-off study team understand and be able to articulate what the system of interest is expected to "do," which in turn enabled them to construct the functional performance objectives shown in the objectives hierarchy shown in Figure 5.8. Note that stakeholder value is measured in terms of not only functional performance but also life cycle costs and development schedule. (Although not addressed here due to space considerations, the notion of long-term viability fits well within an objectives hierarchy such as this).

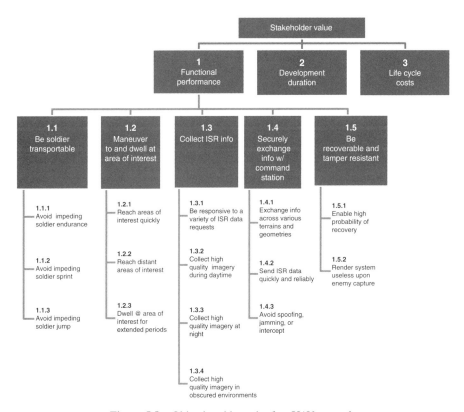

Figure 5.8 Objectives hierarchy for sUAV example

Table 5.5 Measures for sUAV Example

1.1 Be Soldier Transportable	1.1.1 Avoid Impeding Soldier Endurance	Measure: % decrease in sustainable march speed
	1.1.2 Avoid Impeding Soldier Sprint	Measure: % increase in soldier sprint time
	1.1.3 Avoid Impeding Soldier Jump	Measure: % degradation in soldier jump height
1.2 Maneuver to and Dwell at Area of Interest	1.2.1 Reach Areas of Interest Quickly	Measure: Max flight speed (km/hour)
	1.2.2 Reach Distant Areas of Interest	Measure: Maximum operational range (km)
	1.2.3 Dwell @ Area of Interest for Extended Periods	Measure: Operational Endurance (hours)
1.3 Collect ISR Info	1.3.1 Be Responsive to a Variety of ISR Data Requests	Measure: ISR Data Request Responsiveness Index
	1.3.2 Collect High-Quality Imagery During Daytime	Measure: TTP rating per NV-IPM @ 3000m full light
	1.3.3. Collect High-Quality Imagery at Night	Measure: TTP rating per NV-IPM @ 3000m low light
	1.3.4 Collect High-Quality Imagery in Obscured Env.	Measure: TTP rating per NV-IPM @ 3000m w/ smoke
1.4 Securely Exchange Info w/ Command Station	1.4.1 Exchange Info Across Terrains & Geometries	Measure: BLOS comms capable (yes/no)
	1.4.2 Send Large Volumes of Data Quickly & Reliably	Measure: High data rate payload comm link? (Y/N)
	1.4.3 Avoid Spoofing, Jamming, Intercept	Measure: Digital C2 link? (Y/N) Digital Payload Com link? (Y/N)
1.5 Be Recoverable & Tamper Resistant	1.5.1 Enable High Probability of Recovery	Measure: Subjective assessment of landing scheme
	1.7.2 Render System Useless Upon Enemy Capture	Measure: Command self destruct feature?

With the objectives hierarchy in hand, the study team identified measures for each of the functional performance objectives as shown in Table 5.5.

Not shown in Table 5.5 are the measures for the Life Cycle Cost objective and the Development Schedule objective. These two measures are discussed here. The life cycle cost measure for this hypothetical case study is the sum of the rough order of magnitude estimates for development costs, procurement costs, training costs, maintenance costs, and wartime costs. Schedule duration for this exercise is measured as the number of years that the development effort requires estimated

DECISION PROCESS ACTIVITIES   171

at the 80% confidence level after considering the uncertainty associated with the duration estimates for each configuration item to mature from its current state to form, fit, and function tested across temperatures plus the estimate for time required for system-level integration and test.

With a good understanding of how each objective is to be measured, the lead systems engineer worked to understand the degree of satisfaction that each stakeholder perceives at each point along a particular measure scale and then expressed these relationships as a set of value functions. To accomplish this, the lead systems engineer of the sUAV effort worked with a small group of stakeholders to document a walk-away point, a marginally acceptable point, a target point, a stretch goal, and a meaningful limit. The lead systems engineer repeated this process for several stakeholder groups in order to capture any differences among value schemes. The lead systems engineer maintained a record of each set of value functions so that he may use them as part of the sensitivity analysis later in the process. Table 5.6 describes the value functions associated with the 17 measures of this hypothetical sUAV case study as constructed by one of the stakeholder groups. Notice that life cycle cost measures and schedule duration measures are not included in Table 5.6 due to space considerations. Figure 5.9 provides a graphical view of 3 of the 17 value functions.

To complete his understanding of the stakeholder value, the lead systems engineer set out to identify the objectives of which the stakeholders were willing to accept marginal returns in order to achieve high returns on others. By working through a swing weight matrix with each stakeholder group, the lead systems engineer identified weightings for each measure. The normalized weights for one of the stakeholder groups are depicted in Figure 5.10. Weights developed by other stakeholder groups were also documented and maintained for use in the sensitivity analysis of later steps.

With the value schemes of stakeholders having been captured, the lead systems engineer knows how goodness will be measured for each alternative considered within the systems engineering trade-off analysis. The lead systems engineer can now turn his attention to generating sUAV system alternatives. The next section of this chapter describes best practices for generating creative alternatives and is followed by the continuation of the sUAV case study.

5.3.3 Generate Creative Alternatives

For many trade studies, the alternatives will be systems composed of many interrelated subsystems. It is important to establish a meaningful product structure for the system of interest and to apply this product structure consistently throughout the decision process effort in order to aid effectiveness and efficiency of communications about alternatives. The product structure should be a useful decomposition of the physical elements of the system of interest.

Each alternative is composed of specific design choices for each generic product structure element. The ability to quickly communicate the differentiating design features of given alternatives is a core element of the decision-making exercise. Tables 5.8–5.10 provide a template for succinct yet complete system-level alternative

Table 5.6 End and Inflection Points of sUAV Value Functions

Name	Value	Be Transported			Fly			Collect				Communicate			End	
		Avoid Impeding Soldier Endurance	Avoid Impeding Soldier Sprint	Avoid Impeding Soldier Jump	Reach Area of Interest (10 km) Quickly	Reach Distant Areas of Interest	Dwell at Area of Interest	Be Responsive to a Variety of ISR Data Requests	Collect High Quality Imagery During Day	Collect High Quality Imagery During Night	Collect High Quality Imagery In Obscured Environments	Exchange Info Across Various Terrains & Geometries	Send ISR Imagery Quickly and Reliably	Avoid Spoofing, Jamming, or Communicate Intercept	Enable High Probability of Recovery	Render System Useless Upon Enemy Capture
		%	%	%	min	km	hrs	VI	P_d	P_d	P_d	L/B	sec	A/D	%	y/n
Walk-away	1	10	15	20	15	10	1	1	0.1	0.1	0.1	–	90	–	90	–
Marginally Acceptable	10	8	12	16	12	15	2	2	0.2	0.2	0.2	L	30	A	92	N
Target	50	5	7.5	10	8	30	4	5	0.7	0.7	0.7	–	10	–	95	–
Stretch goal	90	2	3	4	4	45	8	8	0.9	0.9	0.9	B	5	D	98	Y
Meaningful limit	100	0	0	0	1	50	16	10	1.0	1.0	1.0	–	1	–	100	–

descriptions. These subsystem design choices have system-level consequences across the objectives hierarchy. Every subsystem design choice will impact system-level cost, system-level development schedule, and system-level performance. It is important to emphasize that these design choices are not fundamental objectives, they are means objectives important only to the degree that they assist in achieving fundamental objectives. It may be useful to think of design choices as the levers used by the system architect to steer the system design toward a solution that best satisfies the all elements of stakeholder value – the full fundamental objectives hierarchy. These levers are very important and care should be given in this step of the process to clearly and completely identify specific design choices for each generic product structure element for every alternative being considered. Incomplete or ambiguous alternative descriptions can lead to incorrect or inconsistent alternative assessments in the process described later. The ability to quickly and accurately communicate

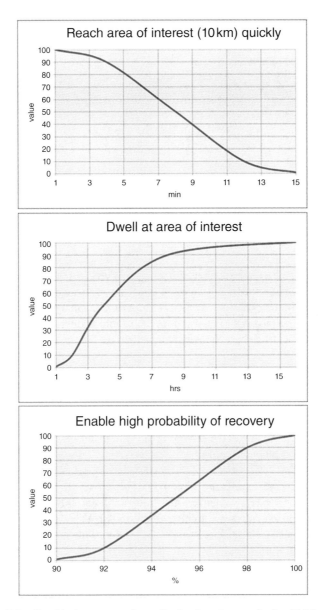

Figure 5.9 Graphical representations of value function graphs for sUAV example

the differentiating design features of given alternatives is a core element of the decision-making exercise.

System characteristics are interesting consequences of subsystem design choices measured at the system level but are not themselves fundamental objectives. For example, the weight of a system alternative is classified as a system characteristic

Objective	Weight
Avoid impeding soldier endurance	0.06
Avoid impeding soldier sprint	0.07
Avoid impeding soldier jump	0.07
Reach area of interest quickly	0.03
Reach distant areas of interest	0.03
Dwell at area of interest	0.03
Be responsive to a variety of ISR requests	0.11
Collect high quality imagery during day	0.16
Collect high quality imagery at night	0.13
Collect high quality imagery in obscured environments	0.11
Exchange info across terrains and geometries	0.06
Send ISR imagery quickly and reliably	0.05
Avoid jamming, spoofing, or comm intercept	0.04
Enable high probability of recovery	0.03
Render system useless upon enemy capture	0.02

Figure 5.10 Weights for sUAV example

in that weight is neither a design decision nor a fundamental objective. Weight is not a design decision but rather a consequence of all the subsystem design choices made throughout the product structure. Weight is not a fundamental objective because it is not inherently good or bad although it does factor into many fundamental objectives measures. Documenting characteristics is considered a best practice for two reasons;

1. if senior stakeholders are known to ask about certain aspects of various alternatives, the lead systems engineer should be prepared to provide an immediate answer, and
2. some system characteristics are part of so many objective measures that it is useful to report them at the same level as system design choices for the sake of clarity and explanatory power.

5.3.3.1 Example of Generating Creative Alternatives Returning to our sUAV example, the lead systems engineer has identified the following four top-level elements of the sUAV physical architecture: the Air Vehicle, the ISR Collecting Payload, the Communication Links, and the Ground Elements. The sUAV Physical Architecture Description in Table 5.7 decomposes the four top level physical elements into generic subelements and also provides a list of specific design choices available for each generic subelement of the physical architecture.

Using the Physical Architecture Description in Table 5.6, the lead systems engineer and his study team developed the 12 system-level sUAV concepts in Tables 5.8–5.10. The team used the table format for describing the alternatives to ensure that each system would be described completely, succinctly, and consistently to aid in the efficiency and effectiveness of communications throughout the trade study.

The lead systems engineer knows that with the alternatives defined, he is ready to start collecting data regarding each system's response to the measures established earlier in the process. The next section of this chapter walks through some best practices

Table 5.7 sUAV Physical Architecture Description

Air Vehicle

Propulsion System	Energy Source	Prop Size & Location	Wing Span	Wing Config.	Fin Config.	Actuators	Airframe Material	Autopilot	Launch	Land
Electric 300W	Li-Ion Battery	18" Rear	4 ft	Conv.	Twin Boom Conv.	Electro-magnetic	Graphite Epoxy	Preprogram, Auto	Hand	Skid and Belly
Electric 600W	Li-S Battery	22" Rear	5 ft	Canard	Inverted V	Hydraulic	Aramid Epoxy	Semiauto	Tensioned Line	Net
Piston Engine 2.5HP	Fuel Cell	26" Rear	6 ft	Tandem Wing	V Tail	MEMS	Boron Epoxy	Remotely Piloted	Gun Launch	Parachute
Piston Engine 4.0HP	Solar	18" Front	7 ft	Three Surface	H Tail		Fiberglass Epoxy			Deep Stall
	JP-8 Fuel	22" Front	8 ft		Cruciform					
		26" Front	9 ft							

(continued overleaf)

Table 5.7 (*Continued*)

ISR Collecting Payload				Communication Links		Ground Elements			
Sensor Actuation	EO Imager	IR Imager		Command & Control Link	Payload Link	Antenna	Computer	User Input Device	Power
Fixed	None	None		Small Fixed Antenna transmits analog data direct to GCS (VHF or UHF)	Small Fixed Antenna transmits analog data direct to GCS (VHF or UHF)	Dipole	Ruggedized Laptop	Keyboard	Generator
Pan–tilt	4 Megapixel Daylight Camera	Cooled 320 × 240 MWIR		Small, Fixed, Nonpointing Antenna transmits digital data to LEO Satellite (L Band)	Small, Fixed, Nonpointing Antenna transmits digital data to LEO Satellite (L Band)	Parabolic Reflector	Wearable Computer	Joystick	Battery + Generator
Roll–tilt	8 Megapixel Daylight Camera	Cooled, 640 × 480 MWIR			Mech. Steerable parabolic dish transmits digital data to GEO Satellite (Ka or Ku Band)		Smartphone	Touchscreen	Battery + Backup Batteries
Pan–tilt–roll		Cooled 1280 × 720 MWIR & LWIR			Electronically Steered Phased Array Antenna transmits digital data to GEO Satellite (Ka or Ku Band)			Stylus	
		Uncooled 1024 × 768 MWIR & LWIR							

Table 5.8 Descriptions for Buzzard I, Buzzard II, Cardinal I, and Cardinal II

Subsystem/Component	1 Buzzard I Design Choice	2 Buzzard II Design Choice	3 Cardinal I Design Choice	4 Cardinal II Design Choice
Air vehicle				
Propulsion System	Electric 300W	Electric 300W	Electric 300W	Electric 300W
Energy Source	Li-Ion Battery	Li-Ion Battery	Li-S Battery	Li-S Battery
Prop Size and Location	18" Rear	18" Rear	20" Rear	20" Rear
Wing Span	5'	5'	6'	6'
Wing Configuration	Canard	Canard	Conventional	Conventional
Fin Configuration	Inverted V	Inverted V	Twin Boom	Twin Boom
Actuators	Electro-magnetic	Electro-magnetic	Electro-magnetic	Electro-magnetic
Airframe Material	Graphite Epoxy	Graphite Epoxy	Graphite Epoxy	Graphite Epoxy
Autopilot	Semiauto	Semiauto	Remotely Piloted	Remotely Piloted
Launch Mechanism	Hand	Hand	Hand	Hand
Landing Mechanism	Belly	Belly	Belly	Belly
ISR Collecting Payload				
Sensor Actuation	Fixed	Fixed	Fixed	Fixed
EO Imager	4 MP	4 MP	4 MP	4 MP
IR Imager	320 × 240 MWIR	640 × 480 MWIR	320 × 240 MWIR	640 × 480 MWIR
Communication Links				
Command and Control Link	Fixed VHF	Fixed VHF	Fixed VHF	Fixed VHF
Payload Data Link	Fixed VHF	Fixed VHF	Fixed VHF	Fixed VHF
Ground Elements				
Antenna	Dipole	Dipole	Dipole	Dipole
Computer	Laptop	Laptop	Smartphone	Smartphone
User Input Device	Keyboard	Keyboard	Joystick	Joystick
Power	Battery + Spare	Battery + Spare	Battery + Spare	Battery + Spare

Table 5.9 Descriptions for Crow I, Crow II, Pigeon I, and Pigeon II

Subsystem/Component	5 Crow I Design Choice	6 Crow II Design Choice	7 Pigeon I Design Choice	8 Pigeon II Design Choice
Air Vehicle				
Propulsion System	Electric 600W	Electric 600W	Electric 600W	Electric 600W
Energy Source	Li-Ion Battery	Li-Ion Battery	Li-S Battery	Li-S Battery
Prop Size and Location	22" Rear	22" Rear	20" Rear	20" Rear
Wing Span	6'	6'	6'	6'
Wing Configuration	Tandem Wing	Tandem Wing	Conventional	Conventional
Fin Configuration	V Tail	V Tail	Twin Boom	Twin Boom
Actuators	MEMS	MEMS	Electromagnetic	Electromagnetic
Airframe Material	Graphite Epoxy	Graphite Epoxy	Graphite Epoxy	Graphite Epoxy
Autopilot	Semiauto	Semiauto	Remotely Piloted	Remotely Piloted
Launch Mechanism	Hand	Hand	Hand	Hand
Landing Mechanism	Belly	Belly	Belly	Belly
ISR Collecting Payload				
Sensor Actuation	Pan–tilt	Pan–tilt	Pan–tilt	Pan–tilt
EO Imager	8 MP	8 MP	8 MP	8 MP
IR Imager	1280×720 MWIR & LWIR cooled	1280×720 MWIR & LWIR uncooled	1280×720 MWIR & LWIR cooled	1280×720 MWIR & LWIR uncooled
Communication Links				
Command and Control Link	Fixed VHF	Fixed VHF	Fixed VHF	Fixed VHF
Payload Data Link	Fixed VHF	Fixed VHF	Phased Array Ka	Phased Array Ka
Ground Elements				
Antenna	Dipole	Dipole	Dipole & Dish	Dipole & Dish
Computer	Laptop	Laptop	Laptop	Laptop
User Input Device	Keyboard	Keyboard	Joystick	Joystick
Power	Battery + Spare	Battery + Spare	Battery + Spare	Battery + Spare

DECISION PROCESS ACTIVITIES 179

Table 5.10 Descriptions for Robin I, Robin II, Dove I, and Dove II

	9 Robin I Design Choice	10 Robin II Design Choice	11 Dove I Design Choice	12 Dove II Design Choice
Subsystem/Component				
Air Vehicle				
Propulsion System	Piston 2.5 HP	Piston 2.5 HP	Piston 4.0 HP	Piston 4.0 HP
Energy Source	JP-8	JP-8	JP-8	JP-8
Prop Size and Location	26" Front	26" Front	28" Front	28" Front
Wing Span	8'	8'	9'	9'
Wing Configuration	Conventional	Conventional	Conventional	Conventional
Fin Configuration	H Tail	H Tail	Cruciform	Cruciform
Actuators	Hydraulic	Hydraulic	Hydraulic	Hydraulic
Airframe Material	Fiberglass Epoxy	Fiberglass Epoxy	Fiberglass Epoxy	Fiberglass Epoxy
Autopilot	Remotely Piloted	Remotely Piloted	Remotely Piloted	Remotely Piloted
Launch Mechanism	Tensioned Line	Tensioned Line	Tensioned Line	Tensioned Line
Landing Mechanism	Net	Net	Net	Net
ISR Collecting Payload				
Sensor Actuation	Pan–tilt	Pan–tilt	Pan–tilt	Pan–tilt
EO Imager	8 MP	8 MP	8 MP	8 MP
IR Imager	1280×720 MWIR & LWIR cooled	1280×720 MWIR & LWIR uncooled	1280×720 MWIR & LWIR cooled	1280×720 MWIR & LWIR uncooled
Communication Links				
Command and Control Link	Fixed VHF	Fixed VHF	Fixed VHF	Fixed VHF
Payload Data Link	Elect. Steered Phased Array Ka	Elect. Steered Phased Array Ka	Mech. Steered Dish Ka	Mech. Steered Dish Ka
Ground Elements				
Antenna	Dipole	Dipole	Dipole and Dish	Dipole and Dish
Computer	Laptop	Laptop	Laptop	Laptop
User Input Device	Keyboard	Keyboard	Joystick	Joystick
Power	Battery + Gen.	Battery + Gen.	Battery + Gen.	Battery + Gen.

with regard to assessing alternatives via deterministic analysis followed by the continuation of this sUAV example.

5.3.4 Assess Alternatives via Deterministic Analysis

With objectives and measures established and alternatives identified and defined, the decision team should engage subject matter experts, ideally equipped with operational data, test data, models, simulations and expert knowledge. Often a mapping between physical architecture elements to fundamental objectives may help identify the types of subject matter expertise needed to fully assess each alternative against a particular objective (Table 5.11 – - Physical Architecture to Fundamental Objective Mapping). These simple maps often grow to more complex flow diagrams showing the interrelationships between physical architecture choices, different levels of intermediate measures, and, finally, the fundamental objectives.

It may be helpful to expand these simple maps into more informative Assessment Flow Diagrams (AFDs) that trace the relationships between physical means, intermediate measures, and fundamental objectives. As an example of such a diagram, consider the sample provided in Figure 5.11. An Assessment Flow Diagram helps individual subject matter experts understand how their area of expertise fits into the larger assessment picture, from where inputs to feed their particular model will be coming and to where their outputs will be consumed. An AFD can be used by the lead systems engineer to organize, manage, and track assessment activities especially when used in conjunction with the consequence scorecard shown in Table 5.12 and Table 5.13.

Table 5.11 Physical Architecture to Fundamental Objective Mapping

	Objective 1										
	OBJ 1.1			OBJ 1.2		OBJ 1.3					
	OBJ 1.1.1	OBJ 1.1.2	OBJ 1.1.3	OBJ 1.2.1	OBJ 1.2.2	OBJ 1.3.1	OBJ 1.3.2	OBJ 1.3.3	Objective 2	Objective 3	Objective 4
Subsystem A	x								x	x	x
Subsystem B							x		x	x	x
Subsystem C				x					x	x	x
Subsystem D					x				x	x	x
Subsystem E			x					x	x	x	x
Subsystem F									x	x	x
Subsystem G									x	x	x
Subsystem H									x	x	x
Subsystem I									x	x	x
Subsystem J		x							x	x	x

DECISION PROCESS ACTIVITIES

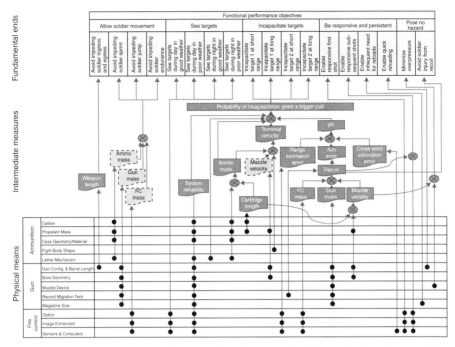

Figure 5.11 Assessment flow diagram (AFD) for a hypothetical gun design choice activity (lead author's original graphic)

Table 5.12 Structured Scoring Sheet for a Given Measure

Detailed Description of Measure:			Assessment				Alternative Description			
			Estimate			Rationale				
ID	Name	Image	Low	Expected	High		Subsystem A	Subsystem B	Subsystem C	Subsystem D
1	Descriptive Name for Alt #1	Illustration for Alt #1								
2	Descriptive Name for Alt #2	Illustration for Alt #2								
3	Descriptive Name for Alt #3	Illustration for Alt #3								
4	Descriptive Name for Alt #4	Illustration for Alt #4								
5	Descriptive Name for Alt #5	Illustration for Alt #5								
6	Descriptive Name for Alt #6	Illustration for Alt #6								

Table 5.13 Consequence Scorecard Structure

ID	Name	Image	Objective 1								Objective 2	Objective 3	Objective 4
			OBJ 1.1			OBJ 1.2			OBJ 1.3				
			OBJ 1.1.1	OBJ 1.1.2	OBJ 1.1.3	OBJ 1.2.1	OBJ 1.2.2	OBJ 1.3.1	OBJ 1.3.2	OBJ 1.3.3			
1	Descriptive Name for Alt #1	Illustration for Alt #1	$x_{1,1.1.1}$	$x_{1,1.1.1}$	$x_{1,1.1.1}$	$x_{1,1.1.1}$	$x_{1,1.1.1}$	$x_{1,1.1.1}$	$x_{1,1.1.1}$	$x_{1,1.1.1}$	$x_{1,2}$	$x_{1,3}$	$x_{1,4}$
2	Descriptive Name for Alt #2	Illustration for Alt #2	$x_{2,1.1.1}$	$x_{2,1.1.1}$	$x_{2,1.1.1}$	$x_{2,1.1.1}$	$x_{2,1.1.1}$	$x_{2,1.1.1}$	$x_{2,1.1.1}$	$x_{2,1.1.1}$	$x_{2,2}$	$x_{2,3}$	$x_{2,4}$
3	Descriptive Name for Alt #3	Illustration for Alt #3	$x_{3,1.1.1}$	$x_{3,1.1.1}$	$x_{3,1.1.1}$	$x_{3,1.1.1}$	$x_{3,1.1.1}$	$x_{3,1.1.1}$	$x_{3,1.1.1}$	$x_{3,1.1.1}$	$x_{3,2}$	$x_{3,3}$	$x_{3,4}$
4	Descriptive Name for Alt #4	Illustration for Alt #4	$x_{4,1.1.1}$	$x_{4,1.1.1}$	$x_{4,1.1.1}$	$x_{4,1.1.1}$	$x_{4,1.1.1}$	$x_{4,1.1.1}$	$x_{4,1.1.1}$	$x_{4,1.1.1}$	$x_{4,2}$	$x_{4,3}$	$x_{4,4}$
5	Descriptive Name for Alt #5	Illustration for Alt #5	$x_{5,1.1.1}$	$x_{5,1.1.1}$	$x_{5,1.1.1}$	$x_{5,1.1.1}$	$x_{5,1.1.1}$	$x_{5,1.1.1}$	$x_{5,1.1.1}$	$x_{5,1.1.1}$	$x_{5,2}$	$x_{5,3}$	$x_{5,4}$
6	Descriptive Name for Alt #6	Illustration for Alt #6	$x_{6,1.1.1}$	$x_{6,1.1.1}$	$x_{6,1.1.1}$	$x_{6,1.1.1}$	$x_{6,1.1.1}$	$x_{6,1.1.1}$	$x_{6,1.1.1}$	$x_{6,1.1.1}$	$x_{6,2}$	$x_{6,3}$	$x_{6,4}$

In addition to the organization and communication benefits, an AFD seems to provide some psychological benefits to the subject matter experts (SMEs) conducting the assessments and to the stakeholders hoping to make use of the results. An AFD gives an SME confidence that their analysis will not be ignored and sends the message that their expertise is important and needed and their results will find their way to the decision table in proper context, as a piece of the whole assessed in terms meaningful to the stakeholder. Similarly, by showing the pedigree of the data feeding the decision support model, an AFD gives the stakeholders confidence that the trade-off analysis rests on a solid foundation and not a product of generalists sitting around the table voting.

The decision team can prepare for subject matter expert engagement by creating structured scoring sheets. Assessments of each concept against each criterion can be captured on separate structured scoring sheets for each alternative/measure combination. Each score sheet contains a summary description of the alternative under examination and a summary of the scoring criteria to which it is being measured. The structured scoring sheet should contain ample room for the evaluator to document the assessed score for the particular concept against the measure followed by clear discussion providing the rationale for the score, noting how design features of the concept under evaluation led to the score as described in the rating criteria. Whenever possible, references to operational data, test data, calculations, models, simulations, analogies, or experience that led to a particular score should be documented.

Creating separate structured scoring sheets for each alternative/measure combination may become somewhat cumbersome for large studies. In practice, a separate structured scoring sheet is often constructed for each measure only and alternatives are identified as separate rows within each measure sheet. Table 5.12 provides a sample format for such a scoring sheet. This approach has the added benefit of reinforcing the notion that each alternative is assessed using the same measure and reducing the risk of inconsistent assessments.

After all the structured scoring sheets have been completed for each alternative/measure combination, it is useful to summarize all the data in tabular form. Each column in such a table would represent a measure and each row would represent a particular alternative. Table 5.13 provides a sample structure identified here as a consequences scorecard.

5.3.4.1 Example of Assessing Alternatives via Deterministic Analysis Continuing the sUAV case study example, the lead systems engineer recruited a team of subject matter experts to assess each alternative against a measure that aligns with their skill set. For instance, human factors experts assessed impact to soldier mobility and endurance, aerospace engineers assessed the measures pertaining to flight, electrical engineers specializing in sensors assessed the ISR data collection measures, communication engineers scored the information transit and receive measures, and mechanical engineers rated each alternative against the recover measures. Each subject matter expert was provided with a scoring sheet for recording their findings. The lead systems engineer took the findings from the scoring sheets and created the consequence scorecard shown in Table 5.14.

With 204 measurements (12 alternatives scored against 17 measures) taken, some would be tempted to claim success and call it a day, but the lead systems engineer for the sUAV trade-off analysis knew there was much to be done in order to fully mine this data for understanding and to communicate the findings and recommendations to the study sponsor in a way that would lead to action. Of course, making 204 measurements and recording them in a well-structured data store is no small task, and it is better than some of the trade-study products he has seen over his career, but the consequence scorecard alone is certainly not conducive to confident decision-making. The next section of this chapter discusses some of the best practices for synthesizing results for rapid and thorough understanding of the trade at hand followed by an application of these best practices to the sUAV case study.

5.3.5 Synthesize Results

At this point in the process, the decision team has generated a large amount of data as summarized in the consequences scorecard. Now it is time to explore the data and display results in a way that facilitates understanding. Transforming the data in the consequences scorecard into a value scorecard is accomplished through the use of the value functions developed in the decision analysis process step described earlier. Table 5.15 shows the structure of a value scorecard. In an effort to enhance speed

184 UNDERSTANDING DECISION MANAGEMENT

Table 5.14 Consequence Scorecard Example for sUAV Case Study

			Functional Performance														
			Be Transported			Fly			Collect				Communicate		End		
			Avoid Impeding Soldier Endurance	Avoid Impeding Soldier Sprint	Avoid Impeding Soldier Jump	Reach Area of Interest (10 km) Quickly	Reach Distant Areas of Interest	Dwell at Area of Interest	Be Responsive to a Variety of ISR Data Requests	Collect High Quality Imagery During Day	Collect High Quality Imagery During Night	Collect High Quality Imagery In Obscured Environments	Exchange Info Across Various Terrains & Geometries	Send ISR Imagery Quickly and Reliably	Avoid Spoofing, Jamming, or Communicate Intercept	Enable High Probability of Recovery	Render System Useless Upon Enemy Capture	Life Cycle Costs	Development Schedule Duration
ID	Name	Image	%	%	%	min	km	hrs	VI	P_d	P_d	P_d	LB	sec	AD	%	y/n	$B	yrs
1	Buzzard I		1	2	3	15	10	1	2	0.3	0.2	0.1	L	30	A	90	90	1	2
2	Buzzard II		1	2	3	15	10	1	4	0.4	0.3	0.2	L	30	A	90	90	1.5	2
3	Cardinal I		2	4	5	12	12	1.5	5	0.5	0.4	0.3	L	90	A	90	90	2.3	3
4	Cardinal II		2	4	5	12	12	1.5	10	0.5	0.4	0.3	L	90	A	90	90	4	3
5	Crow I		4	5	8	10	15	5	10	0.9	0.8	0.7	B	20	D	95	90	5	9
6	Crow II		3	4	7	9	15	6	10	0.8	0.7	0.6	B	30	D	95	90	6	9
7	Pigeon I		5	7	10	8	18	8	10	0.9	0.8	0.7	B	20	D	95	90	6.5	7
8	Pigeon II		4	6	9	7	18	9	10	0.8	0.7	0.6	B	30	D	95	90	7.5	7
9	Robin I		7	12	17	6	22	10	10	0.9	0.8	0.7	B	20	D	98	90	8.3	6
10	Robin II		6	11	16	6	22	11	10	0.8	0.7	0.6	B	30	D	98	90	8.8	6
11	Dove I		10	15	20	4	30	23	10	0.9	0.8	0.7	B	20	D	98	90	9.3	5
12	Dove II		9	14	19	4	30	24	10	0.8	0.7	0.6	B	30	D	98	90	9.9	5

Table 5.15 Value Scorecard Structure

ID	Name	Image	Objective 1					
			OBJ 1.1			OBJ 1.2		OBJ 1.3		

			OBJ 1.1.1	OBJ 1.1.2	OBJ 1.1.3	OBJ 1.2.1	OBJ 1.2.2	OBJ 1.3.1	OBJ 1.3.2	OBJ 1.3.3
1	Descriptive Name for Alt #1	Illustration for Alt #1	$v_{1,1,1}(x_{1,1,1,1})$	$v_{1,1,2}(x_{1,1,1,2})$	$v_{1,1,3}(x_{1,1,1,3})$	$v_{1,2,1}(x_{1,1,2,1})$	$v_{1,2,2}(x_{1,1,2,2})$	$v_{1,3,1}(x_{1,1,3,1})$	$v_{1,3,2}(x_{1,1,3,2})$	$v_{1,3,3}(x_{1,1,3,3})$
2	Descriptive Name for Alt #2	Illustration for Alt #2	$v_{1,1,1}(x_{2,1,1,1})$	$v_{1,1,2}(x_{2,1,1,2})$	$v_{1,1,3}(x_{2,1,1,3})$	$v_{1,2,1}(x_{2,1,2,1})$	$v_{1,2,2}(x_{2,1,2,2})$	$v_{1,3,1}(x_{2,1,3,1})$	$v_{1,3,2}(x_{2,1,3,2})$	$v_{1,3,3}(x_{2,1,3,3})$
3	Descriptive Name for Alt #3	Illustration for Alt #3	$v_{1,1,1}(x_{3,1,1,1})$	$v_{1,1,2}(x_{3,1,1,2})$	$v_{1,1,3}(x_{3,1,1,3})$	$v_{1,2,1}(x_{3,1,2,1})$	$v_{1,2,2}(x_{3,1,2,2})$	$v_{1,3,1}(x_{3,1,3,1})$	$v_{1,3,2}(x_{3,1,3,2})$	$v_{1,3,3}(x_{3,1,3,3})$
4	Descriptive Name for Alt #4	Illustration for Alt #4	$v_{1,1,1}(x_{4,1,1,1})$	$v_{1,1,2}(x_{4,1,1,2})$	$v_{1,1,3}(x_{4,1,1,3})$	$v_{1,2,1}(x_{4,1,2,1})$	$v_{1,2,2}(x_{4,1,2,2})$	$v_{1,3,1}(x_{4,1,3,1})$	$v_{1,3,2}(x_{4,1,3,2})$	$v_{1,3,3}(x_{4,1,3,3})$
5	Descriptive Name for Alt #5	Illustration for Alt #5	$v_{1,1,1}(x_{5,1,1,1})$	$v_{1,1,2}(x_{5,1,1,2})$	$v_{1,1,3}(x_{5,1,1,3})$	$v_{1,2,1}(x_{5,1,2,1})$	$v_{1,2,2}(x_{5,1,2,2})$	$v_{1,3,1}(x_{5,1,3,1})$	$v_{1,3,2}(x_{5,1,3,2})$	$v_{1,3,3}(x_{5,1,3,3})$
6	Descriptive Name for Alt #6	Illustration for Alt #6	$v_{1,1,1}(x_{6,1,1,1})$	$v_{1,1,2}(x_{6,1,1,2})$	$v_{1,1,3}(x_{6,1,1,3})$	$v_{1,2,1}(x_{6,1,2,1})$	$v_{1,2,2}(x_{6,1,2,2})$	$v_{1,3,1}(x_{6,1,3,1})$	$v_{1,3,2}(x_{6,1,3,2})$	$v_{1,3,3}(x_{6,1,3,3})$

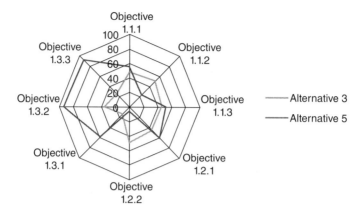

Figure 5.12 Radar value graph structure

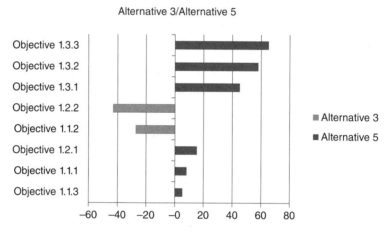

Figure 5.13 Tornado graph structure

and depth of comprehension of the value scorecard, consider associating increments on the value scale with a color according to heat map conventions. This view can be useful when trying to determine which objectives are causing a particular alternative trouble. In addition, one can use this view to quickly see if there are objectives for which no alternative scores well. From this view, the systems engineer can also see if there is at least one alternative that scores above the walk-away point for all objectives. If not, the solution set is empty and the decision team needs to generate additional alternatives or adjust objective measures.

Radar graphs and tornado graphs (Figures 5.12 and 5.13) are popular visualization techniques to show the same value data captured in a Heat-Indexed Value Scorecard discussed earlier but usually for only two alternatives at a time.

5.3.5.1 Example of Synthesizing Results

Returning to the sUAV example, the lead systems engineer has transformed the consequence table into a value scorecard through the use of the value functions created in the second step of this process. Notice how the heat map conditional formatting of the scorecard makes the strengths and weaknesses of each alternative very apparent. It also quickly highlights objectives that are nondiscriminating and objectives that are difficult to achieve for any alternative.

The lead systems engineer was pleased with the value scorecard and rushed to show the emerging results to the study sponsor and several other stakeholders. The feedback he received was very positive and encouraging, but all asked for the cost/schedule/performance trade to be more explicitly shown. The next section in this chapter covers the development of multidimensional value models to create aggregated value visualizations followed by an application of these techniques to the sUAV case study.

5.3.6 Develop Multidimensional Value Model

Beyond the consequence scores for each alternative on each measure, all that was needed to construct the visualizations covered in Table 5.16 were the value functions associated with each objective measure. By introducing the weighting scheme, the systems engineer can create aggregated value visualizations. The first step in assessing an alternative's aggregated value is a prescreen for alternatives that fail to meet a walk-away point for any objective measure and set that alternative's aggregated value to zero regardless of how it performs on other objective measures. For those alternatives that pass the walk-away prescreen, the additive value model[2] uses the following equation to calculate each alternative's aggregated value:

$$v(x) = \sum_{i=1}^{n} w_i v_i(x_i) \qquad (5.1)$$

where $v(x)$ is the alternative's value, $i = 1$ to n is the number of the measure, x_i is the alternative's score on the ith measure, $v_i(x_i) =$ is the single-dimensional value of a score of x_i, w_i is the weight of the ith measure,

$$\sum_{i=1}^{n} w_i = 1 \qquad (5.2)$$

and (all weights sum to 1).

This chapter is devoted to the pragmatic application of the aggregation technique but a thorough treatment of the mathematical foundation for the additive value model is provided in Chapter 2 and by (Keeney 1981; Stewart 1996; Von Winterfeldt et al., 1986).

---

[2] The additive model assumes preferential independence. See Chapter 2. See Keeney and Raiffa (1976), and Kirkwood (1997) for additional models.

Table 5.16 Value Scorecard for sUAV example

			Functional Performance												
			Be Transported			Fly			Collect			Communicate			End
			Avoid Impeding Soldier Endurance	Avoid Impeding Soldier Sprint	Avoid Impeding Soldier Jump	Reach Area of Interest (10km) Quickly	Reach Distant Areas of Interest	Dwell at Area of Interest	Be Responsive to a Variety of ISR Data Requests	Collect High Quality Imagery During Day	Collect High Quality Imagery During Night	Collect High Quality Imagery In Obscured Environments	Exchange Info Across Various Terrains & Geometries	Send ISR Imagery Quickly and Reliably	Avoid Spoofing, Jamming, or Communicate Intercept	Enable High Probability of Recovery	Render System Useless Upon Enemy Capture
ID	Name	Image	%	%	%	min	km	hrs	VI	P_d	P_d	P_d	LB	sec	AD	%	y/n
1	*Buzzard I*		95	93	92	1	1	1	10	18	10	1	10	10	10	1	90
2	*Buzzard II*		95	93	92	1	1	1	37	26	18	10	10	10	10	1	90
3	*Cardinal I*		90	81	83	10	5	6	50	34	26	18	10	1	10	1	90
4	*Cardinal II*		90	81	83	10	5	6	100	34	26	18	10	1	10	1	90
5	*Crow I*		63	72	63	30	10	60	100	90	70	50	90	30	90	50	90
6	*Crow II*		77	81	70	40	10	70	100	70	50	42	90	10	90	50	90
7	*Pigeon I*		50	54	50	50	18	90	100	90	70	50	90	30	90	50	90
8	*Pigeon II*		63	63	57	60	18	91	100	70	50	42	90	10	90	50	90
9	*Robin I*		23	10	8	70	29	92	100	90	70	50	90	30	90	90	90
10	*Robin II*		37	19	10	70	29	94	100	70	50	42	90	10	90	90	90
11	*Dove I*		1	1	1	90	50	100	100	90	70	50	90	30	90	90	90
12	*Dove II*		6	4	3	90	50	100	100	70	50	42	90	10	90	90	90

DECISION PROCESS ACTIVITIES 189

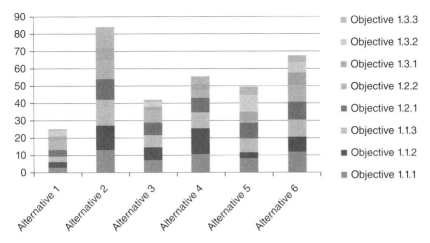

Figure 5.14 Value component graph structure

With the weights in hand, one can construct aggregated visualizations such as the value component graph as shown in Figure 5.14. In a value component graph, each alternative's total value is represented by the total length of a segmented bar. Each bar segment represents the contribution of the value earned by the alternative of interest within a given measure by the weighted value (Parnell et al. 2013). As discussed in Section 5.3.2 and illustrated in Table 5.3, every objective has one measure and only one measure.

The heart of a decision support process for systems engineering trade analysis is the ability to integrate otherwise separate analyses into a coherent, system-level view that traces consequences of design decisions across all dimensions of stakeholder value. The stakeholder value scatterplot illustrated in Figure 5.15 shows in one chart how all system-level alternatives respond in multiple dimensions of stakeholder value.

Figure 5.15 illustrates the structure of a stakeholder value scatterplot showing how the six hypothetical alternatives respond to four dimensions of stakeholder value – performance value, life cycle cost, development schedule, and long-term viability. Each system alternative is represented by a scatterplot marker. An alternative's life cycle cost and performance value are indicated by a marker's x and y positions, respectively. An alternative's development duration is indicated by the color of the marker per heat map conventions shown in the legend, while the long-term viability of a particular alternative is indicated by the shape of the marker as described in the legend.

5.3.6.1 Example of Developing a Multidimensional Value Model Resuming the sUAV case study example, the lead systems engineer is anxious to respond to the study sponsor feedback and provide a more explicit representation of the cost, schedule, and performance trade between the alternatives under consideration. Making use

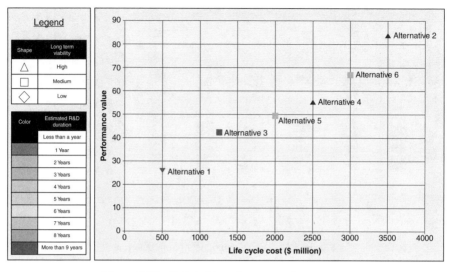

Figure 5.15 Stakeholder value scatterplot structure

of the additive value model and the weighting scheme described in Figure 5.10 along with the value scorecard in Table 5.15 the lead systems engineer is able to create the aggregated value visualizations shown in Figures 5.16 and 5.17.

The lead systems engineer was excited about the value scatterplot and once again hurried to the show this visualization to the study sponsor and several other stakeholders. The feedback he received was again glowing, agreeing that this particular visualization clearly showed the cost/schedule/performance trade. However, this time the study sponsor asked to understand how variations in priority weightings would impact the results. The next section in this chapter covers the best practices associated with identifying uncertainty and conducting probabilistic analysis followed by an application of these techniques to the sUAV case study.

5.3.7 Identify Uncertainty and Conduct Probabilistic Analysis

As part of the assessment, it is important for the subject matter expert to explicitly discuss potential uncertainty surrounding the assessed score and variables that could impact one or more scores. One source of uncertainty that is common within systems engineering trade-off analyses that explore various system architectures is technology immaturity. System design concepts are generally described as a collection of subsystem design choices, but if some of the design choices include technologies that are immature, there may be a lack of detail associated with component-level design decisions that will eventually be made downstream during detailed design. Many times the subject matter expert can assess an upper, nominal, and lower bound measure response by making three separate assessments (i) assuming low performance, (ii) assuming moderate performance, and (iii) assuming high performance.

DECISION PROCESS ACTIVITIES   191

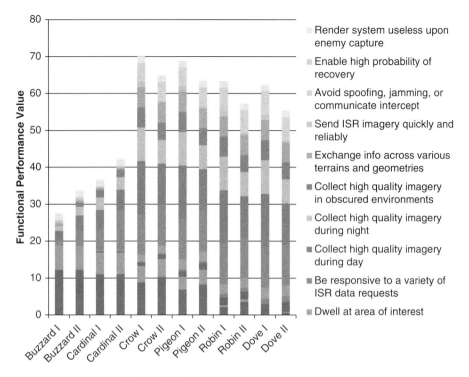

Figure 5.16 Value component chart for sUAV

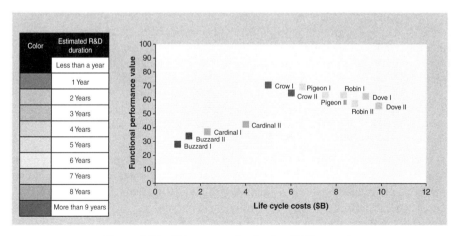

Figure 5.17 Value scatterplot for the sUAV example

Another source of uncertainty has to do with the subjective nature of the value schemes elicited from the stakeholders. Considering that a stakeholder's value scheme is often tied to their forecast of future scenarios and acknowledging that the stakeholder is probably not clairvoyant leads to the conclusion that people can reasonably disagree about things such as priority weightings. One of the common pitfalls of systems engineering trade-off analyses is to collect value scheme information from a small set of like-minded stakeholders and ignore the fact that value schemes likely vary across the full population of stakeholders. The best practice that should be employed here is to collect value scheme information from many different stakeholders and then run a battery of sensitivity analyses pertaining to priority weightings to ensure that a meaningful decision can be made in the presence of such uncertainty. The next section of this chapter applies some of techniques for identifying uncertainty to the sUAV example, and the subsequent section covers techniques used to assess the impact of uncertainty.

5.3.7.1 Example of Identifying Uncertainty and Conducting Probabilistic Analysis Recall the sUAV case study and how the study sponsor asked the lead systems engineer to show how variations in priority weightings would impact the results. Toward this end, the lead systems engineer discussed the composition of the focus group he used to develop the priority weightings with the study sponsor and asked him to provide some recommendations regarding stakeholder representatives that may have a different view on how weightings for this trade should be set. The study sponsor provided the systems engineering lead with a list of contacts that he suspects would offer a somewhat different take on weights associated with this trade. Focus group number 2 was formed from this list, and the lead systems engineer developed Figure 5.18 to highlight the differences in the priority weightings generated by the two groups.

The lead systems engineer noticed the clear differences between the two group's areas of emphasis, group 1 on ISR data collection quality and group 2 on soldier mobility while transporting the sUAV system. He realized that capturing this source of uncertainty is the first step to assessing its impact on the overall decision. The next section of this chapter describes some sensitivity analyses that can be used to gain this understanding followed by a return the sUAV case study.

5.3.8 Assess Impact of Uncertainty

Decision analysis uses many forms of sensitivity analysis including line diagrams, tornado diagrams, and waterfall diagrams and several uncertainty analyses including Monte Carlo Simulation, decision trees, and influence diagrams (Parnell et al., 2013). Due to space limits, only line diagrams of sensitivity to weighting and Monte Carlo Simulation are discussed in this section.

Many decision-makers will want to understand how sensitive a particular recommendation is to weightings and will ask questions regarding the degree to which a particular weighting would need to be changed in order to change in recommended alternative. A common approach to visualizing the impact of measure weighting

DECISION PROCESS ACTIVITIES

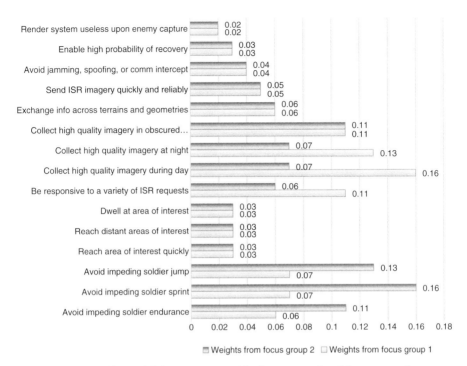

Figure 5.18 Weightings as generated by focus group 1 and focus group 2

on overall value is to sweep each measure's weighting from absolute minimum to absolute maximum while holding the relative relationship between the other measure weightings constant and noting changes to overall score. The output of this type of sensitivity analysis is in the form of a line graph (Parnell et al., 2011). An example of such a graph is provided in Figure 5.19. Note that this particular example shows how sweeping the weight associated with Objective 1.1.2 impacts performance value. The graph in this example shows that the alternative with the highest performance value is alternatives 2 for all cases where priority weighting associated with Objective 1.1.2 is somewhat low but as the weight of Objective 1.1.2 is increased to 0.8 and above, alternative 4 emerges as the high performer.

Sometimes, it is useful to consider all uncertainties at once instead of merely investigating the impact of one particular uncertainty at a time. For this type of view, consider the tradespace visualization in Figure 5.20. Once all the uncertainties have been assessed, Monte Carlo Simulations (Chapter 3) can be executed to identify the uncertainties that impact the decision findings and of the uncertainties that are inconsequential. For example, Figure 5.20 shows that after considering all sources of uncertainty, alternative 1 is less susceptible to changes in stakeholder value than alternative 3. Note, however, that although the stakeholder value of alternative 3 is a bit more volatile in the presence of uncertainty, its stakeholder value never falls below the highest level of alternative 1's stakeholder value. This graph also indicates

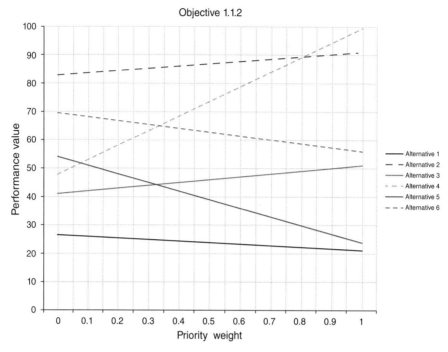

Figure 5.19 Weight sensitivity line graph structure

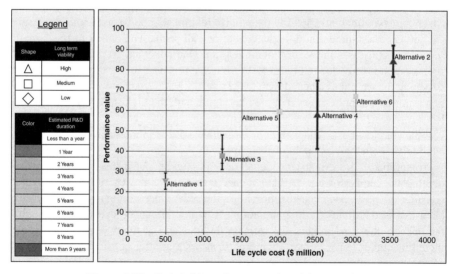

Figure 5.20 Stakeholder value scatterplot with uncertainty

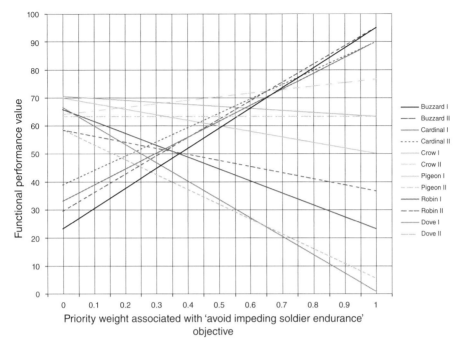

Figure 5.21 sUAV performance value sensitivity to changes in priority weight of "avoid impeding soldier sprint" objectives

that alternative 3 has the edge on long-term viability, whereas development duration differentiation is nil. Consequently, decision-makers may be inclined to pursue alternative 3 over alternative 1 if the differences in life cycle costs were deemed affordable.

The takeaway of this section may be that good decisions can often be made in the presence of high uncertainty. Systems engineers should not let what is not known sabotage a decision-making opportunity if what is known is sufficient.

5.3.8.1 Example of Assessing the Impact of Uncertainty Picking up the sUAV case study, the lead systems engineer applied two sensitivity analysis techniques to assess the impact of the uncertainty surrounding priority weightings. As a first step in his attempt to get his arms around the degree of decision volatility introduced by changes in weighting schemes, he generated a full set of performance value sensitivity line graphs – one graph per measure. Figure 5.21 shows one of the these line graphs, the line graph that shows changes in functional performance value as the priority weight associated with "avoid impeding soldier sprint" objective is swept from 0 to 1. Notice that the top performing alternative only changes twice throughout the entire sweep – from Crow I to Crow II at 0.45 and from Crow II to Buzzard II at 0.75.

Although the set of line graphs were interesting and shed some light on the degree of volatility pertaining to this specific decision, the lead systems engineer feared

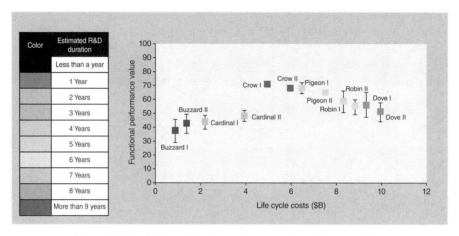

Figure 5.22 sUAV stakeholder value scatterplot with uncertainty

that such graphs would not directly address the question raised by the study sponsor. For this, he decided to generate a stakeholder value scatterplot with uncertainty. Figure 5.22 shows this scatterplot. Notice how this graph clearly shows that Crow I maintains the highest functional performance value under either weight set considered.

The study sponsor was thrilled when the lead systems engineer presented the visualization of Figure 5.22, and the graph formed the focal point for many thoughtful negotiations among the stakeholders and the decision authority.

5.3.9 Improve Alternatives

One could be tempted to end the decision analysis here, highlight the alternative that has the highest total value, and claim success. Such a premature ending, however, would not be considered best practice. Mining the data generated for the first set of alternatives will likely reveal opportunities to modify some subsystem design choices to claim untapped value and reduce risk. Recall the cyclic decision process map and the implied feedback. Taking advantage of this feedback loop and using initial findings to generate new and creative alternatives starts the process of transforming the decision process from "Alternative-Focused Thinking" to "Value-Focused Thinking" (Keeney 1992). To complete the transformation from alternative-focused thinking to value-focused thinking, consider taking additional steps to spark focused creativity to overcome anchoring biases. As Keeney warns,

> Once a few alternatives are stated, they serve to anchor thinking about others. Assumptions implicit in the identified alternatives are accepted, and the generation of new alternatives, if it occurs at all, tends to be limited to a tweaking of the alternatives already identified. Truly creative or different alternatives remain hidden in another part of the mind, unreachable by mere tweaking. Deep and persistent thought is required to jar them into consciousness. (Keeney 1993)

To help generate a creative and comprehensive set of alternatives, consider conducting an alternative generation table (also called a morphological box) (Buede, 2009; Parnell et al., 2011) analysis to generate new alternatives. Chapter 8 presents this and other alternative generation techniques.

5.3.9.1 Example of Improving Alternatives Within the sUAV example, the lead systems engineer worked with the pool of subject matter experts and sUAV design engineers to explore ways to potentially reduce the time required for the development of CROW I from about 8 years to 5 or 6 years without diminishing the functional performance value.

5.3.10 Communicating Trade-Offs

This is the point in the process where the decision team identifies key observations regarding what stakeholders seem to want and what they must be willing to give up in order achieve it. It is here where the decision team can highlight the design decisions that most influence shareholder and stakeholder value and that are inconsequential. In addition, the important uncertainties and risks should also be identified. Observations regarding combination effects of various design decisions are also important products of this process step. Competing objectives that are driving the trade should be explicitly highlighted as well

Beyond the top-level tradespace visualization products, the decision analyst must be able to rapidly drill down to supporting rationale. The decision support tool construct represented in Figure 5.23 will allow the decision team to navigate seamlessly from top-level stakeholder value scatterplot all the way down to any structured scoring sheet so that rationale for any given score is only a click away. Rapid access to rationale associated with the derivation of the value function or priority weightings is also essential for full traceability.

5.3.10.1 Example of Communicating Trade-Offs Concluding the sUAV case study, the lead systems engineer presented the sUAV systems engineering trade-off analysis to the study sponsor and supporting stakeholder senior advisory group. He used Figure 5.22 to summarize the study findings and Figure 5.23 to provide an appreciation for the processed used and the heritage of underpinning data. He ended the talk with a crisp summary of the decision at hand – if an 8-year development time is acceptable, then Crow I offers superior performance at a very attractive life cycle cost point. If the capability is somehow urgent, Dove I can be fully developed within 4 years but comes with about a 15% drop in performance value and about a 90% increase in life cycle costs relative to Crow I.

5.3.11 Present Recommendation and Implementation Plan

It is often helpful to describe the recommendation in the form of clearly worded, actionable task list to increase the likelihood of the decision process leading to some

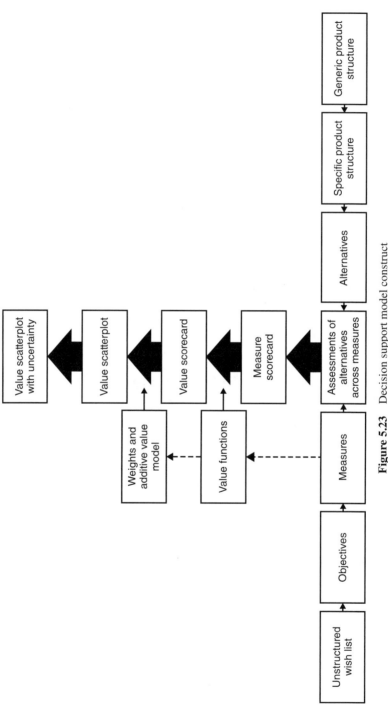

Figure 5.23 Decision support model construct

form of action, thus delivering some tangible value to the sponsor. Reports are important for historical traceability and future decisions. Take the time and effort to create a comprehensive, high-quality report detailing study findings and supporting rationale. Consider static paper reports augmented with dynamic hyperlinked e-reports.

5.4 SUMMARY

The decision management process discussed in this paper integrates decision analysis best practices with systems engineering activities to create a baseline from which the next chapters can explore innovations to further enhance trade-off study quality. The process enables enterprises to develop an in-depth understanding of the complex relationship between requirements, the design choices made to address each requirement, and the system-level consequences of the sum of design choices across the full set of performance requirements as well as other elements of stakeholder value to include cost and schedule. Through data visualization techniques, decision-makers can quickly understand and crisply communicate a complex tradespace and converge on recommendations that are robust in the presence of uncertainty.

The decision management approach is based on several best practices:

(A) Align the decision process with the systems engineering process (Figure 5.2).
(B) Use sound mathematical technique of decision analysis for trade-off studies (Chapter 2).
(C) Develop one master decision model and refine, update, and use it as required for trade-off studies throughout the system development life cycle (Parnell et al, 2013) (Figure 5.3).
(D) Use Value-Focused Thinking (Keeney, 1992) to create better alternatives (Section 3.3.9).
(E) Identify uncertainty and assess risks for each decision (Parnell et al., 2013) (Section 3.3.8).

5.5 KEY TERMS

Integrated Systems Engineering Decision Management (ISEDM) Process: A procedure that combines the holistic perspective required for the design of complex systems with mathematical rigor needed to properly represent and communicate assessments of system level alternatives across all elements of stakeholder value.

Fundamental Objectives: The essential ends that a decision-maker is trying to achieve.

Measures: A scale established to assess the degree to which various alternatives satisfy a fundamental objective.

Value Function: Value functions describe the degree of satisfaction stakeholders perceive at each point along the measure scale.

Swing Weight Matrix: A structured technique to elicit stakeholder perception of the level of importance of each measure and the differentiation in each measure range in order to determine meaningful swing weights.

Assessment Flow Diagram: A mapping technique to trace the relationships between physical means, intermediate measures, and the measures associated with fundamental objectives.

Consequence Scorecard: Assessment results for each alternative/measure combination summarized in tabular form.

Value Scorecard: Assessment results for each alternative/measure combinations transformed into value space and summarized in tabular form.

Value Component Graph: An aggregated value visualization where each alternative's total value is represented by the total length of a segmented bar where each bar segment represents the contribution of the value earned by the alternative of interest within a given measure by the weighted value.

Stakeholder Value Scatterplot: An aggregated value visualization plotting alternatives' assessed values for two dimensions of stakeholder value on Cartesian Coordinates and for third and fourth dimensions of stakeholder value with marker shape and color.

Stakeholder Value Scatterplot with Uncertainty: Markers of the stakeholder value scatterplot are augmented with box-and-whisker plots to express uncertainty associated with assessed value in a particular dimension of stakeholder value.

5.6 EXERCISES

The questions with an * apply only if you are working in an organization performing trade-off analyses.

5.1. Decision management processes
   (a) Why does an organization need a decision management process using decision analysis?
   (b) What is the intended output of the decision analysis process when applied to new product development context?
   (c) List and explain the 10 steps of the decision process as described in this chapter.
   (d) What is the circular shape of the Integrated Systems Engineering Decision Management (ISEDM) process meant to convey?
   (e) (*)Which steps are used in your organization?
   (f) (*)Which of the missing steps would provide the most value to your organization?

5.2. Decision support models using decision analysis
   (a) Why is a decision support model considered to be a composite model?
   (b) What are two key properties of a well-developed fundamental objective?
   (c) What are five key properties of a well-developed set of fundamental objectives?
   (d) What are five key properties of a well-developed measure?
   (e) In terms of value function creation, what is meant by a "walk-away" point? What is meant by a "stretch goal?"
   (f) What is the potential benefit of aggregating several single-dimensional values to create synthesized value visualization?
5.3. Analysis of uncertainty and risk
   (a) List two sources of uncertainty common within systems engineering trade-off analyses.
   (b) Which steps in the ISEDM process explicitly consider uncertainty and risk?
5.4. Tradespace visualization
   (a) List three tradespace visualization techniques discussed in this chapter.
   (b) What insights does each visualization technique provide?
   (c) Explain how the cost versus value plot relates to affordability analysis (described in Chapter 4)?
   (d) Which techniques are used in your organization?
5.5. Perform a trade-off analysis using the ISEDM illustrated in this chapter.

REFERENCES

Buede, D.M. and Choisser, R.W. (1992) Providing an analytic structure for key system design choices. *Journal of Multi-Criteria Decision Analysis*, **1** (1), 17–27.

Buede, D.M. (2009) *The Engineering Design of Systems: Models and Methods*, Wiley.

Cilli, M. (2015) *Improving Defense Acquisition Outcomes Using an Integrated Systems Engineering Decision Management (ISEDM) Approach.* PhD Dissertation. Stevens Institute of Technology, Hoboken, NJ.

Edwards, W., Miles, R.F. Jr., and Von Winterfeldt, D. (2007) *Advances in Decision Analysis: from Foundations to Applications*, Cambridge University Press.

Gundlach, J. (2012) *Designing Unmanned Aircraft Systems: A Comprehensive Approach*, American Institute of Aeronautics and Astronautics.

Green, P.E., Krieger, A.B., and Wind, Y. (2001) Thirty years of conjoint analysis: reflections and prospects. *Interfaces*, **31**, S56–S73.

INCOSE SE Handbook Working Group (2015) Chapter 4: Business or mission analysis, Chapter 5: Technical management processes, Chapter 9: Cross-cutting systems engineering methods, in *Systems Engineering Handbook: A Guide for System Life Cycle Process and Activities*, 4th edn (eds D.D. Walden, G.J. Roedler, K.J. Forsberg *et al.*), International Council on Systems Engineering, Published by John Wiley & Sons, Inc., San Diego, CA.

Keeney, R.L. and Raiffa, H. (1976) *Decisions with Multiple Objectives Preferences and Value Tradeoffs*, Wiley, New York, NY.

Keeney, R.L. (1992) *Value-Focused Thinking: A Path to Creative Decisionmaking*, Harvard University Press, Cambridge, Massachusetts.

Keeney, R.L. (1993) Creativity in MS/OR: value-focused thinking—Creativity directed toward decision making. *Interfaces*, **23** (3), 62–67.

Keeney, R.L. (2004) Making better decision makers. *Decision Analysis*, **1** (4), 193–204.

Keeney, R.L. and Gregory, R.S. (2005) Selecting attributes to measure the achievement of objectives. *Operations Research*, **53** (1), 1–11.

Kirkwood, C.W. (1997) *Strategic Decision Making: Multiobjective Decision Analysis with Spreadsheets*, Duxbury Press, Belmont, CA.

Miller, G.A. (1956) The magical number seven, plus or minus two: some limits on our capacity for processing information. *Psychological Review*, **63** (2), 81.

Parnell, G.S., Driscoll, P.J., and Henderson, D.L. (eds) (2011) *Decision Making for Systems Engineering and Management*, 2nd edn, Wiley & Sons Inc., Wiley Series in Systems Engineering.

Parnell, G., Bresnick, T., Tani, S., and Johnson, E. (2013) *Handbook of Decision Analysis*, Wiley & Sons.

SEBoK authors. "Decision Management." BKCASE Editorial Board. *Guide to the Systems Engineering Body of Knowledge (SEBoK)*, version 1.4, R.D. Adcock (EIC). Hoboken, NJ: The Trustees of the Stevens Institute of Technology ©2015.29 June 2015. Web. 16 Jun 2015, 14:02 <http://sebokwiki.org/w/index.php?title=Decision_Management&oldid=50860>. BKCASE is managed and maintained by the Stevens Institute of Technology Systems Engineering Research Center, the International Council on Systems Engineering, and the Institute of Electrical and Electronics Engineers Computer Society.

Standard, I. (2015) Systems and software engineering-system life cycle processes. *ISO Standard*, **15288**, 2015.

Keeney, R.L. (1981) Analysis of preference dependencies among objectives. *Operations Research*, **29** (6), 1105–1120.

Stewart, T.J. (1996) Robustness of additive value function methods in MCDM. *Journal of Multi-Criteria Decision Analysis*, **5** (4), 301–309.

Von Winterfeldt, D., Edwards, W. *et al.* (1986) *Decision Analysis and Behavioral Research*, Cambridge University Press, Cambridge.

6

IDENTIFYING OPPORTUNITIES

Donna H. Rhodes
Sociotechnical Systems Research Center, Massachusetts Institute of Technology, Cambridge, MA, USA

Simon R. Goerger
Institute for Systems Engineering Research, Information Technology Laboratory (ITL), U.S. Army Engineer Research and Development Center (ERDC), Vicksburg, MS, USA

> Opportunities? They are all around us … There is power lying latent everywhere waiting for the observant eye to discover it.
>
> (Orison Swett Marden)

6.1 INTRODUCTION

This chapter discusses a first step of any trade-off analysis, defining the opportunity space (also called opportunity definition, problem definition, or framing). To analyze the impacts of trades, it is essential to identify and understand the essence of the issue to be resolved. This is achieved through obtaining an understanding of the goals, constraints, and concerns of the decision-makers as well as the impact of and on the stakeholders. This is known as defining the opportunity space. Failure to identify and fully understand the correct opportunity space often results in an expenditure of scare resources to provide a viable solution to an opportunity of limited concern

---

Trade-off Analytics: Creating and Exploring the System Tradespace, First Edition. Edited by Gregory S. Parnell.
© 2017 John Wiley & Sons, Inc. Published 2017 by John Wiley & Sons, Inc.
Companion website: www.wiley.com/go/Parnell/Trade-off_Analytics

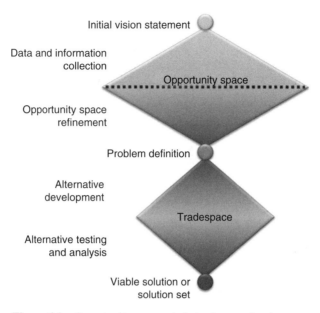

Figure 6.1 Opportunity space role in tradespace development

to the decision-maker(s), while not addressing the key issue (Parnell et al., 2011, pp. 280–281).

The opportunity space provides the foundation for the development of a tradespace to delineate creative, viable system solutions to the opportunity. As such, it is an essential component of the study. Figure 6.1 illustrates how the amount of data and information gathered based on the initial vision statement (opportunity statement) helps to form the opportunity space. The reason for the use of a parallelogram is that the two processes are divergent then convergent processes. As a starting point, the vision statement is an agreement between the decision-maker(s) and key stakeholders on the purpose of the decision. Based on the analysis of the opportunity space, a problem definition (refined vision statement) is developed. The alternative development phase expands the fidelity of the opportunity space to help develop, test, and assess system opportunity. The final result is a viable solution or a set of viable solutions emerging from the tradespace. Implementation plans are developed and executed once the final solution is chosen.

This chapter discusses the categories of knowledge we need to develop the opportunity space, the traps that hinder the data collection and information generation, the tools that are used to collect and analyze data for opportunity space generation, and the techniques for extracting data and information from stakeholders. Using these tools and techniques in a logical and repeatable manner helps to develop the appropriate opportunity space from which the tradespace emerges.

6.2 KNOWLEDGE

Trade-off analysis begins with an initial vision statement, the first attempt to define where the effort will lead, but significant thought and effort are needed to evolve this to an *opportunity definition*. A good starting point is to consider what types of knowledge are needed to formulate the opportunity definition. Effective framing depends upon knowledge drawn from many sources and individuals. Five key types of knowledge are domain knowledge, technical knowledge, business knowledge, expert knowledge, and stakeholder knowledge.

6.2.1 Domain Knowledge

Domain knowledge is the specialized understanding of a specific area or discipline. For today's complex systems, most system opportunities will require domain knowledge from many domains. Consider an opportunity in the unmanned aerial vehicle (UAV); this could involve control systems domain knowledge, remote sensing domain knowledge, and autonomous operations domain knowledge, to name only a few. Scoping the opportunity as commercial or military has domain knowledge implications. Defining the scope to procuring the aircraft platform might reduce the scope of necessary domain knowledge. Keeping the build or buy decision as part of the trade would require expanded domain knowledge.

6.2.2 Technical Knowledge

Technical knowledge can be thought of as deep knowledge within a particular domain. For example, suppose the opportunity space for our UAV necessitates some sort of sensing capability. In this case, domain knowledge of remote sensing technology would be needed and, further, may necessitate specific deep technical knowledge of the Lidar technology. Credibility of the trade-off analysis depends upon sound technical knowledge.

In some cases, the decision-maker(s) may have technical knowledge by climbing up through the ranks. In others, the decision-maker(s) is more focused on the business side of the organization and relies upon the technical staff of scientists, engineers, and others to provide such knowledge. In some ways, it is easier to reconcile conflicting opinions on technical matters than on business matters since technical matters tend to be more factually based and objective. (Parnell et al., 2013, p. 32)

6.2.3 Business Knowledge

Opportunities exist within specific contexts. Business knowledge is an important factor in trade-off analysis, providing a larger context for technical decisions. The analyst may often need to consult other individuals and sources to gather adequate knowledge since there are many facets to consider. Business knowledge includes analysis of the competition/adversaries, analysis of the economic environment, analysis of the legislative environment, and analysis of required rates of return, among other business environment areas. (Parnell et al., 2013, p. 232)

Let us assume that our UAV opportunity is in the commercial space. The target market may have major implications, which must be investigated in defining the opportunity. For example, if the market is local law enforcement, the policies for flying in airspace are likely to be different than if the opportunity was targeted for use in commercial film making.

Trade-off analysts do need to be aware that there is uncertainty in knowledge gathered. Potential shifts over time should also be factored into the framing of the opportunity. In the case of the UAV, the expected time to launching it in the market is to be considered. For example, business knowledge may indicate that policies in use of airspace for commercial purposes may be less restrictive in future years. Since it is uncertain, this type of uncertainty needs to be considered in the analysis. Additionally, there can be omissions, errors, or outdated knowledge that is gathered in defining the opportunity space. As such, it is prudent to involve experts and stakeholders to the extent possible.

6.2.4 Expert Knowledge

Domain knowledge, technical knowledge, and business knowledge is available through many means including books, reports, databases, and websites. Many decisions require very specific and precise knowledge that may not be readily available in such sources. Even when it is available, it is always wise to verify facts and validate their conclusions and assumptions. As part of defining the opportunity space, an analyst may want to seek input from specific individuals or groups of individuals who have unique knowledge, experiences, and/or qualifications. For instance, technical knowledge for a previously developed air vehicle gathered from specifications could be incomplete or out of date. Expert knowledge could be gathered by interviewing the senior engineers who worked on the past system development and then used to validate and supplement the technical knowledge. Given the significant experience it takes to become an expert, such individuals often possess an intuitive understanding of the opportunity space. Experts are particularly important sources of knowledge when there is a lack of empirical data. Unfortunately, experts may have biases (see Chapter 3).

The preferred way to obtain expert knowledge is having direct access to specific individuals and groups, engaging through one of the three techniques we discuss later in this chapter. When necessary, some level of expert knowledge can be gathered from secondary sources, such as a recorded presentations or study panel transcripts.

6.2.5 Stakeholder Knowledge

Trade-off analysts need comprehensive knowledge of the involved and impacted stakeholders within the opportunity space. The simplest taxonomy of stakeholders contains five types: decision authority, client, owner, user, and consumer (Parnell et al., 2011). Knowledge provided by a comprehensive set of stakeholders provides the multiple perspectives necessary for decision analysis, including framing of the opportunity.

Stakeholder inputs are important to accurately specify the opportunity and provide insights into the constraints that bound the opportunity. As the opportunity space is developed, it is prudent to consult key stakeholders to validate the final opportunity statement before continuing in the system decision process.

As analysts begin to involve many different people and organizations, as well as participating in the decision process themselves, it is important to be aware of decision traps. We discuss these in the next section.

6.3 DECISION TRAPS

Decision traps prevent people from making the best decision possible. They are individual and group biases in processing information for decision-making. Our experience provides a set of heuristics that allow us to filter data and make decisions more quickly. Sometimes, these heuristics can also prevent us from making the best decision.

There are numerous decision traps that prevent us from asking questions that will facilitate gathering key information when developing the opportunity space. Others place a filter on information, which limits the fidelity of the opportunity space. The first line of defense against decision traps is to understand and identify them. Although there are numerous decision traps, eight traps that impact development of the opportunity space include (but are not limited to) failure to frame, status quo, anchoring, confirming, groupthink, expert advice, complexity, and numbers.[1,2]

An important initial step taken before defining the question to be decided is to frame the decision (another phrase for defining the opportunity space). This framing process involves understanding the purpose of the decision, identifying the perspective(s) to be taken in making the decision, and clearly delineating the decision scope. Effectively framing the decision is essential to focusing the trade-off analysis on the right areas. Analysts and decision-makers identify key elements of the issue and develop a structure to organize these elements. This creates a framework for placing data and information that will be used in the future to understand the opportunity and assess potential solutions. *Failure to frame* the problem can result in using resources to collect data and information that is outside the scope of effort. It can also result in not collecting information required for a full understanding of the issue or initial solution(s).[3] Using precise language for a vision or opportunity statement agreed

---

[1]Hammond et al. discuss 10 hidden traps in "The Hidden Traps in Decision making." (Hammond et al., 1998).
[2]Parnell et al. define and address 10 decision traps and 7 barriers in "*Handbook of Decision Analysis.*" (Parnell et al., 2013, pp. 33–35).
[3]It is often seen in conjunction with efforts that start collecting data and attempting to solve problems prior to understanding the nature of the issues. Parnell et al. call this "Plunging in." They also break down *failure to frame* traps into to two subcategories: "framing blindness", not defining the mental framework

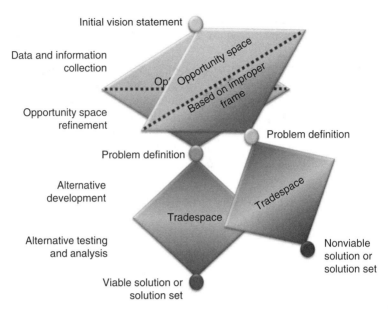

Figure 6.2 Potential impact of Ill-framed opportunity space roll in tradespace development

upon by the decision-maker is one method to frame the effort. The frame should be tested during the study to ensure that it remains valid. Asking questions from different perspectives defined by stakeholder groups and challenging and verifying study assumptions are two means of testing the frame (Henman, pp. 4–5).

Figure 6.2 provides a visual representation of the possible impacts of the ill-framed decision. Although starting out with a poorly framed decision can be detrimental to the effort, quickly realizing that you have fallen into this trap provides the opportunity to correct the mistake and readjust the study in the correct direction. The longer it takes to identify this trap, the more time and effort it will take to correct the mistake.

Other traps deal with methods people use to filter data to come to a decision. An example is when stakeholders prefer to stay with the current system or process and respond to questions in a neutral manner in order to limit the possibility of change. This is known as *status quo* bias. Techniques to avoid this trap include asking if current processes help achieve organization objectives. Analysts should ask if leadership would choose current systems if they were not the status quo. When seeking opinions from stakeholders about changes to systems that affect them, it may be useful to postpone discussion of the cost and effort of change until they understand the need to change.

of the issue; and "lack of frame control," failing to maintain the consistency of the frame by allowing other frames or visions to adversely impact the appropriate frame. The latter is often seen when study teams overextend themselves to resolve issues other than the one defined as the main effort (Parnell et al., 2013, pp. 33–34).

Another common trap is *anchoring* or *first impressions*. This occurs when people give a disproportionate weight to the first information obtained. The initial information can have a surprisingly large impact on the information we seek and evaluations. An example is when a company predicts future product sales of their new mobile device based only on sales of its last device. To avoid this trap, one should seek other sources of information including stakeholder groups that have other perspectives. This is essential if your primary source of data and information comes from internal organizational sources.

Confirming evidence or *confirmation bias* results when we seek information that supports our existing point of view. When a board of directors considers closing mobile phone stores in a specific region of the country and then asks advice from an individual who recently discontinued operating in the same area, it is seeking to gain information to their planned decision. To not fall prey to this trap, analysts must examine all evidence with equal rigor. They should also set up one of the team members as a devil's advocate to argue against potential decisions. When conducting interviews, it is important to seek out independent thinkers and stakeholders with different perspectives.

When dealing with large amounts of data from numerous individuals, this can result in oversimplification by focusing on consensus. This can result in *groupthink*. Groupthink often materializes as a rush to consensus. It often occurs as a result of failure to ask the hard question or challenging statements by dominant members of the group (e.g., the boss). This can be resolved by questioning assumptions and asking challenging question regarding how things can or could be done.

The corollary of groupthink is *over-reliance on expert advice*. This occurs by overly weighting the feedback of a specific individual or group because of their position or experience. An example is limiting data collection to interviewing senior technicians when looking to develop the next mobile devices. Such bias can also emerge when analyzing survey results from the community of interest and focusing on the responses from those with the most experience in the field of interest. Questioning the reason behind an individual's response or recommendation can reduce the impacts of this trap.

The *complexity* trap results from attempting to address too many issues in one solution. Sometimes, the simplest answer is the best answer. Avoid the temptation to develop complex relationships and alternatives as part of the solution set.

The last decision trap deals with the analysis of the information. Over-reliance on *numbers* without the application of statistical analysis is one of these traps. Use of appropriate statistical tests and sensitivity analysis can prevent over-reliance on "worst-case scenarios" and "err on the side of caution". Test to ensure that the probability of these occurring is significant. Reduce the likelihood that decision-makers avoid a decision because of fear of these improbable effects occurring.

Traps can be compounded by two or more working together to adversely impact efforts to collect or assess data. Analysts need to be diligent in seeking to avoid these decision traps. Even when identified and mitigated, traps can reappear if not continuously guarded against.

6.4 TECHNIQUES

There are three common techniques used to obtain information from stakeholders, decision-makers, and subject matter experts (SMEs) to help define the opportunity space. These techniques are interviews, focus groups, and surveys. The appropriate technique depends on five considerations: the time a participant is willing to make available to you, the level of the stakeholder group,[4] the effort required to prepare, the effort required to execute, and the techniques used to analyze the data collected. Table 6.1 provides a brief description of each technique based on the five considerations. The complexity of the opportunity and stakeholder population (the number and diversity of the stakeholders) determine the number of interviews, the size and number of focus groups, and the number of surveys to be performed. The sample size required to ensure the appropriate results of these techniques can be calculated using statistical tests.

6.4.1 Interviews

Interviews are one of the best techniques for obtaining information, especially from senior leaders with limited availability. However, this technique is very time-consuming and requires time to prepare questions, prepare interviewers, schedule and reschedule interviews, conduct interviews, and analyze the results. Often, you have only one opportunity to interview a senior leader; thus, you must be well prepared to leverage every minute of the interview to obtain the key elements of information required to understand their goals, constraints, and concerns. Best practices include asking the appropriate questions and taking good notes that can be used later (Parnell et al., 2011, p. 302).

Depending on the role of the stakeholder, there are several formats one can use to organize questions. These include (Parnell et al., 2013, p. 74) the following:

- "What keeps you up at night?" Useful for identifying major risks.
- "Tell me more" questions are used to expand on the last response or statement of the interviewee.
- Feeling-finding questions are used to collect subjective information such as feelings, opinions, and beliefs.
- Best/least questions are used to collect information on the limits of the wants and requirements (e.g., "What is the most critical factor?").
- Third-party questions are used to gain insight on how the interviewee perceives that others would respond or feel about a topic (e.g., "What do your employees feel is the most important aspect capability of the new mobile phone system?").

---

[4] Levels of stakeholder groups refer to roles and/or ranks within an organization. Three levels of stakeholder groups are used in Table 6.1. These include (i) senior leaders, subject matter experts (SMEs), and key stakeholder representatives; (ii) mid-level to senior stakeholder representatives; and (iii) junior to mid-level stakeholder representatives.

Table 6.1 Stakeholder Analysis Techniques

Technique	Time Commitment of Participants	Ideal Stakeholder Group	Preparation	Execution	Analysis
Interview	30–60 min	Senior leaders, subject matter experts (SMEs), and key stakeholder representatives	Develop interview questionnaire(s) and schedule or reschedule interviews	Interviewer has a conversation with senior leader using questionnaire as a guide. Separate note taker	Note taker types interview. Interviewer reviews typed notes. Team analyzes notes to determine findings, conclusions, and recommendations
Focus groups	Shortest: 60 min Typical: 4–8 h	Mid-level to senior stakeholder representatives	Develop meeting plan, obtain facilities, and plan for recording inputs. May use process-based collaboration software package to record (e.g., ThinkTank©)	At least one facilitator and one recorder. Larger groups may require breakout sessions and multiple facilitators. Include scheduled breaks for long sessions (any session greater than 90 min should have a break every 55–60 min)	Observations must be documented. Analysis determines findings, conclusions, and recommendations
Surveys	5–20 min	Junior to mid-level stakeholder representatives	Develop survey questions, identify survey software, and develop analysis plan with pilot study group(s). Online surveys are useful for distributed groups and digitizing responses	Complete survey questionnaire, solicit surveys, and monitor completion status	Depends on the number of questions and capabilities of the statistical analysis package. Conclusions must be developed from the data

Source: Parnell 2011. Reproduced with permission of John Wiley & Sons.

- "Ruler for a day" questions are used to encourage free thought not limited by current or perceived boundaries (e.g., "If you were designing the next mobile phone system, what new functionality would you include?").

The intent is to gain the insights and level of confidence the interviewees have in these insights. Just because it is a fact, does not mean it is accepted by the leadership or the community at large. This includes identifying possible interviewee biases contained in the responses based on the way the interview was conducted. It can also be done by the interviewer based on how he/she interprets the response to the question. For example, the response "That was a nice piece of work" can be interpreted as "That was a nice piece of work; I like what they produced" or "That was a nice piece of work, let's not produce anything like it again."

When dealing with open-ended questions, the interview analysis team must distinguish between facts, insights, feelings, assumptions, and biases. One technique to accomplish this is to use a combination of question formats. For example, grouping fact-finding or feeling-finding questions with "tell me more" questions can clarify the characteristics of a "nice piece of work."

Whichever question format you choose, there are several practices to ensure that you take advantage of each phase of the interview process: planning, scheduling, conducting, documenting, analyzing, and follow-up. (Parnell et al., 2011, pp. 302–308)

When *planning* for an interview with senior leaders and key stakeholders, it is essential to prepare a series of questions tailored for each individual. This requires extensive background research to identify the several key stakeholder representatives per stakeholder group.

Based on the sensitivity of the topic and the purpose of the study, interviews may need to be without attribution in order to ensure that the stakeholder would be willing to provide detailed responses to questions. For senior leaders, this may be a prerequisite to an interview. Thus, researchers must determine if interview feedback is with or without attribution. This must be conveyed to the potential interviewee when scheduling a session and reinforce this at the start of the interview.

Interviewing key leaders often requires flexibility to *schedule* and reschedule sessions. To optimize the leaders' time, try to schedule individual interviews. This facilitates focused discussion and helps to get buy-in from leaders that you are concerned about and that you seek to understand what they believe is essential to know about the matter. It helps to provide a brief description of the opportunity and a read-ahead packet, but not the questions, when scheduling the interview, so the decision-maker can prepare for the interview. To reduce difficulties with getting on senior leaders' calendars, it is best to use the member of their stakeholder community on your team to help schedule the interview. Ask for 30–60 min for the interview and be prepared for a 15-min or 2-h interview. Although interviews can be done over the phone, it is preferable to meet face-to-face to show your interest in the interviewee's responses and to be able to read their body language.

How you *conduct* the interview can facilitate a positive interaction with the interviewee. Be ahead of time to the interview and limit the number of personnel to the interviewee, the interviewer, and the notetaker if possible. This helps to keep the control of the interview in your hands and not overwhelm the interviewee who may not wish to be recorded or speak openly to a large number of people about the topic. Conduct the interview as a conversation using the list of questions as a guide and taking the questions in the order that seems natural for the interviewee. Confirm with the interviewee if feedback is with or without attribution. Start with an "unfreezing" question to help focus and relax the interviewee. Flexibility in the order of the questions helps you exploit areas that you had not anticipated and keeps the interviewee interested; however, ensure that you bring the interviewee back to the key questions you need addressed. Watch for body language that tells you the interviewee is ready to end the interview (e.g., leafing through the read-a-head packet, looking at the clock or their watch). Manage your questions to ensure that you finish in the allotted time, but be prepared to stay longer or schedule a follow-up if the interviewee wishes to extend the meeting.

Upon completion of the interview, the recorder should *document* the results. Type up the notes and provide them to the interviewer for review. As soon as possible, the notes should be given to the team members responsible for developing the opportunity space. The notes should align with the questions to provide the context for the response. Direct quotes help to reduce interviewer or recorder bias.

Ensure that you *follow-up* with your interviewee to thank them for their participation. This can be done with a note or e-mail. If the interviewer requests a copy of the findings, conclusions, and/or recommendations from their interview, clear this with the sponsor and provide with your thank-you note. This could result in the interviewee request a correction to the record and provide more accurate feedback. If cleared by the sponsor and appropriate, provide the interviewee with a copy of the final report at the end of the project.

To summarize the interview process, Marvin recommended seven best practices to enhance the probabilities of success (Klimack et al., 2011):

1. Establish rapport with the interviewee.
2. Motivate the interviewee by establishing legitimacy and purpose for the interview.
3. Structure the interview using an hour glass approach – go easy early, challenging in the middle, and finish softly and depart.
4. Condition the interviewee to convey an accurate perspective on the issue.
5. Be flexible with the order of the question and follow up on important information with "tell me more" questions.
6. Verify key responses with follow-up question (e.g., "Did I hear correctly that … ") or statement (e.g., "I understand you to say that … ").
7. Finish on time.

6.4.2 Focus Groups

We can use Focus Groups to obtain insights from groups of individuals. These groups are often homogeneous with regard to stakeholder categories (e.g., budget team, assembly line workers, soldiers that deployed together as part of a single unit) but do not have to be. Focus groups create insights through discussion between group members. The number of focus groups will depend of the structure of the organization and the vision statement. For example, if you want to produce a product more efficiently and at a lower cost, you could run one or more focus groups consisting of assembly line workers, floor managers, and quality control managers. Large focus groups are difficult to manage and do not allow for everyone to participate; thus, it is best to keep the size of groups between 6 and 12 persons. Focus group sessions can last from 60 min to 8 h. Similar to preparing for an interview, the process can be broken into three phases: planning, execution, and analysis of data. (Parnell et al., 2011, p. 309)

In preparation (*planning*) for a focus group, it is essential to develop a set of goals and objectives for each group that will support the opportunity space vision. Although this is an essential step, it is insufficient to ensure a successful focus group. Some best practices for preparing a focus group session are listed as follows. (Greenbaum, 1997) It may be appropriate to clear the group membership with the study sponsor.

- Develop a clear purpose statement and goals for each focus group.
- Develop participant profiles for each focus group.
- Select the pool of viable participants for each focus group.
- Select and prepare session moderators to facilitate discussion and reduce the impacts of bias (traps). Parnell et al. outline the skills required to be an effective moderator (Parnell et al., 2013, pp. 81–86).
- Schedule the time and location for conduct of the focus group. If using a software product to collect data and facilitate discussion, ensure you have sufficient IT support for the session.
- Prepare a series of questions tailored for each focus group. To facilitate discussion, use open-ended questions (e.g., "What is the most commonly used function on a mobile phone?").

The most important elements of *executing* a focus group session are the moderator and the data collection plan. McNamara has several recommendations to help facilitate the execution of a focus group session (McNamara, 2008). Below is a summary of eight recommendations.

- Moderators must review focus group session goal(s) and objective(s), provide an agenda to participants, and present a plan for recording session feedback.
- When asking questions, moderators should allow participants to discuss the topic for a few minutes. In order to prevent one or more individuals from dominating the discussion, moderators must ensure all participants are afforded an opportunity to speak.
- Moderators can leverage technology such as ThinkTank© (ThinkTank Business Process Collaboration Software, 2014) to collect data and facilitate discussion. This is a means to reduce the impacts of overdominating focus group members.

They also assist in recording participant responses in a timely and accurate manner for later analysis.
- To facilitate accurate note-taking, record these sessions or use multiple notetakers.
- Moderators may leverage participant discussion to explore topics germane to the goals and objectives of the session that are exposed during open discussion.
- At the end of the session, inform participants that they will receive a copy of the session record to review their comments for accuracy and clarity.
- After focus sessions are complete, send participants thank-you notes or e-mails.

Focus groups can generate vast amounts of qualitative data for *analysis*. The study team must first verify the data collected by consolidating notes for review and eliciting clarification by the participants. The approved records will then be binned by goal and objective and analyzed to produce findings, conclusions, and study recommendations for the opportunity space.

6.4.3 Surveys

Surveys are a technique of gathering insights from large groups of mid-level stakeholders who are dispersed by time or location. Surveys are often used to collect qualitative and/or quantitative data on a system's status (perceived or actual), understanding of organizational goals, identification of tradespace uncertainties, and possible solutions. Surveys can be conducted using hard copy forms in person or via mail, leveraging the Internet with survey collection tools (e.g., SurveyGizmo© (SurveyGizmo|Professional Online Survey Software; Form Builder), SurveyMonkey® (SurveyMonkey: Free online survey software & questionnaire tool), and Wufoo© (Online Form Builder with Cloud Storage Database|Wufoo)), and e-mail. Parnell et al. describe three phrases of a survey: planning, execution, and analysis.[5,6] This section reviews seven steps included in these three phases to provide a set of best practices for the use of surveys to collect trade data. The seven steps include the following:

- (Planning) Establish the goals of the survey.
- (Planning) Determine the target population of stakeholder from which you will present the survey and the number of respondents required to provide significant results.
- (Planning) Select the format of and the means to distribution and collection responses for the survey.

---

[5]There are numerous ways to bin the steps of the survey process. Kulzy and Ficker explain an alternative set of phases in an article entitled, "The Survey Process: With Emphasis on Survey Data Analysis." (Kulzy & Fricker, 2015).

[6]These are found in (Parnell et al., 2011, pp. 310–313), (Parnell et al., 2013), and (Parnell et al., 2013, pp. 78–80).

- (Execution) Develop survey questions.
- (Execution) Test the survey for appropriateness for the community of interest and time to execute.
- (Execution) Distribute the survey to target population of stakeholders and collect responses.
- (Analysis) Analyze survey responses.

The *planning phase* consists of three key steps. For best results, it is best to execute these in order. This helps to ensure the appropriateness of the survey to address its goals. Defining the goals of the survey helps to frame its purpose, length, and target population. Well-defined and articulated goals are used to focus the questions as well as the respondents. The start of the survey should include them to provide respondents with the proper context for their responses. The goals help to identify the information requirements and the likely stakeholders to provide this type of information. This may result in multiple target groups of stakeholders who are asked different sets of questions. An example is a survey to understand user requirements for a mobile communications device and a second survey of developers to identify the current and forecasted capabilities of the mobile devices they produce. The first survey may result in large population of stakeholders (e.g., hundreds to thousands) while the second may be a much smaller pool of respondents (e.g., tens to hundreds). The survey of users may consist of a series of standard response questions that would require a large number of responses to gain statistical significance. An example question might be, "Do you need a calculator in your mobile device (Yes/No)?" Feeling-finding and "tell me more" questions may be text-based and provide insight but not statistically significant responses.

The required sample sizes can be calculated using one of numerous methods described in most statistics books. Online survey software can also provide tools to calculate these numbers. Three examples of online survey size calculators are hosted by Creative Research Systems© (Systems©, 2012), SurveyMonkey® (Sample Size Calculator), and SurveyStatz© (Sample Size Calculator – SurveyStatz). Regardless of the number of respondents required or the stakeholder group(s) to be surveyed, researchers should work with the project sponsor to ensure that the appropriate stakeholders are engaged for the survey.

Before developing the questions, select the survey method (e.g., online, e-mail, mail, phone, door-to-door, or in person) and tool. The method employed and the capabilities of online tools may limit the type and number of questions that can be used. Table 6.2 provides a list of advantages and disadvantages of four survey methods. One of the most prominent methods of surveying in recent years is the online survey. The ability to reach out to large numbers of individuals and electronically collect responses into a database makes this method attractive to many analysts. However, the survey is limited to individuals who have access to computer systems and respondents may choose not to respond or forget to respond to the survey based on the lack of a physical paper survey. To maximize the number of target audience responses, researchers must provide several reminders to survey recipients to complete the survey.

Table 6.2 Advantages and Disadvantages of Popular Survey Methods

Survey Method	Advantages	Disadvantages
In person (group survey)	• Easy to check compliance and conduct follow-up with respondents • Consistency in the environment and information provided to the respondents • Captive pool of respondents • Can include extensive supporting graphics • Respondents have flexibility in completing the survey	• Requires a physical space to conduct the survey (e.g., auditorium) • Requires additional coordination to get respondents and survey monitors • Limited to no ability for respondents to research their response • Response data will have to be transformed by the analysis team into a format for analysis
Mail	• Can include extensive supporting graphics • Respondents have flexibility in completing the survey	• Takes a great deal of time • Hard to check compliance and conduct follow-up with respondents • Response data will have to be transformed by the analysis team into a format for analysis • Postage for delivery and return could be cost prohibitive
Electronic mail (e-mail)	• Fast to distribute and receive responses • Low cost • Easy to check compliance and conduct follow-up	• Need to obtain e-mail addresses for survey sample • Cannot program automatic logic into the survey (e.g., "Skip over the next set of questions if your answer is 'No' to this question") • Respondent e-mail programs may limit the type of information that can be sent in the survey • Response data will have to be transformed by the analysis team into a format for analysis
Internet web survey	• Extremely fast • Can include special graphics and formatting • Can collect responses in a database to facilitate analysis • Can be opened to the general public	• May be hard to control who and number of responses to the survey due to worldwide Internet access • Respondents can provide partial responses to the survey

Source: Parnell 2011. Reproduced with permission of John Wiley & Sons.

The *execution phase* of a survey also consists of three steps: developing survey questions, pilot testing the survey, and distributing the survey to the sample population. Developing a well-formed survey can ensure that the data required to meet the survey goal is collected and can reduce the time required for the final phase, analysis. To increase the likelihood of responses, the use of focused unambiguous questions that are well worded is encouraged. Questions should be worded from the survey population's perspective, using their terminology.

Fowler presented general principles for survey question development that are applicable to all survey methods (Fowler, 1995):

- Ask survey respondents about their firsthand experiences to help them provide informed and focused responses.
- Ask only one question at a time.
- In wording questions, ensure that the respondent understands the question. If the question includes terms that can be interpreted in more than one way, provide definitions to these terms before you ask the question to resolve possible ambiguities.
- Articulate to respondents the type of response you expect to the question. For objective questions, a scale can be provided with multiple choice answers from a rating scale or level-of-agreement scale. Open-ended text response questions should be worded to encourage respondents to provide information germane to the *survey goals*. The final question in the survey should offer respondents an opportunity to provide additional information relevant to the survey goal(s).
- Format the survey so it is easy for respondents to read the question, follow instructions, and provide answers. For example, responses to scales should follow a consistent pattern (e.g., the least desirable outcome to the most desirable outcome).
- Orient the respondent to the survey in a consistent manner. This is often accomplished by providing a list of instructions that describe the goals of the survey, the method for completing responses, and the means to submit the completed survey.

To provide context and focus the audience, the survey should start with a short statement outlining the purpose and major goal of the survey. It should also include the number of questions and likely amount of time required to complete the survey. The opening and completion of the survey should include a "thank you" to the respondent for participating in the survey. In some cases, it may be appropriate to include a statement of who is sponsoring and who is conducting the survey.

Once the survey is drafted, conduct a pilot study with a small group of individuals who did not draft the survey. Have them use the same survey method as the survey sample pool will use (e.g., online, e-mail, mail, phone, door-to-door, or in person). Ask them to provide feedback on the clarity and appropriateness of the instructions, wording of the questions, answer format, and scales. When using a web survey, assess the method for collecting data (e.g., open-ended text, radio buttons, and checkbox responses). Use the feedback from the pilot survey to enhance the survey.

Develop the plan for *distributing the survey* to target population of stakeholders and collect responses. The plan should include a response rate monitoring plan and a means of reminding those who have not completed the survey of the pending submission deadline. The survey team should also have a standard format for thanking respondents for the time and effort to complete the survey (e.g., thank-you note or e-mail). When applicable and desired, the thank-you response should inform survey respondents when and how they can obtain the survey findings.

Analyze survey responses includes the formatting of the data collected. Use of a web survey facilitates this process by the preformatting of the database where the responses will be stored. Placing the data in a standardized database for the survey allows team members to quickly perform data scrubbing and initial statistical analysis on objective-type questions. The goal of the effort should be to identify data and generate information that is relevant to the survey goals. Responses should be grouped by question for independent analysis. Analysis of the bins will lead to conclusions, which inform the survey team about the tradespace development, and to recommendations for the opportunity space.

6.5 TOOLS

There are numerous tools that a trade-off analyst can use to scope and bound the opportunity space and elaborate the framing from initial vision statement toward a full definition of the opportunity. We introduce five tools in this section, although there are many others in use by analysts.

6.5.1 Concept Map

Concept maps are graphical tools used for organizing and representing relationships between concepts indicated by a connecting line linking the concepts (Novak, 2006). Concept maps can be constructed by an individual constructing the map from concepts identified during the mapping process or constructed from a set of concepts agreed upon by a team. An example of a concept map is shown in Figure 9.1.

For opportunity space definition involving a team, concept mapping can be a useful tool to obtain a shared mental model. DeFranco, 2011 described the Cognitive Collaborative Model (CCM) as a framework to facilitate critical thinking and effective problem-solving among a group of collaborators. In their investigation of mental models, they used concept mapping to elicit and represent team member mental models, with pathfinder network analysis to determine degree of commonality and similarity of models within the team (DeFranco, 2011).

Concept maps have proven to be effective for stimulating ideas and for organizing and communicating complex concepts. There are various ways to construct a concept map. A common set of steps includes beginning with the topic, collecting and listing the key concepts, arranging the concepts in a pattern best suited to the information, ordering the concepts, linking the concepts, and reviewing the whole map. There is a large literature available on constructing concept maps.

6.5.2 System Boundary

A system boundary is a physical or conceptual boundary that contains all the essential elements, subsystems, and interactions necessary to address a decision opportunity. The boundary effectively and completely isolates the system under study from its external environment except for the inputs and outputs that are allowed to move across the system boundary (Parnell et al., 2011).

The system boundary is a key tool to define and understand the opportunity space, resulting in a boundary description artifact of some type, often a combination of illustrative graphic and descriptive text. There are several types of boundary conditions that must be addressed. The boundary denotes what is inside and outside the opportunity area and, accordingly, what is relevant in the external environment given the boundary that is chosen. Various constraints in regard to feasibility and practicality also drive the boundaries, yet some openness to what is possible is also necessary.

6.5.3 Decision Hierarchy

The decision hierarchy is a decision analysis conceptual tool for categorizing what is in and out of scope given the opportunity space, as this is refined. Figure 6.3 shows the decision hierarchy as a three-level pyramid structure. High-level context-setting

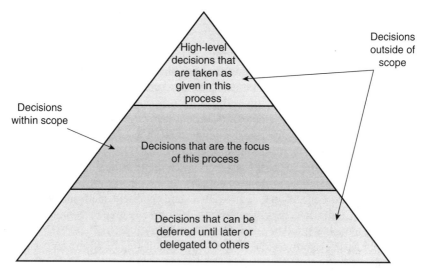

Figure 6.3 Format of a decision hierarchy (Source: Parnell et al. 2013. Reproduced with permission of John Wiley & Sons)

choices are assigned at the top level, and these are choices that are assumed to have been already made. The middle level has choices that are within the scope of the decision under consideration, and the combinations of these choices will be defined as alternatives. Choices allocated to the lowest level of the decision hierarchy are those that are out of scope of the decision because they are deferred to a later time or delegated to others outside the opportunity space (Parnell et al., 2013).

The process of creating the decision hierarchy engages the various stakeholders in an activity to place the decisions into one of the levels. Once the hierarchy is complete and agreed upon, the decision choices in the middle tier are those that are within the opportunity space.

6.5.4 Issues List

A simple but effective tool that contributes to defining the opportunity space is the issues list. Developing an issues list is an effective way to elicit the numerous perspectives of the involved individuals in regard to how they view the opportunity. The process of raising and discussing issues generally leads to convergence on a key set of opportunity space concerns that are at a roughly equivalent level. The process of raising issues can be enhanced by the use of a designated facilitator.

Issues should be categorized, using affinity diagramming or binning, in order to add structure and simplify their use going forward. Four useful categories are (Parnell et al., 2013, pp. 117–118) as follows:

- **Decisions:** issues suggesting choices that can be made as part of the decision
- **Uncertainties:** issues suggesting uncertainties to be taken into account when making decisions
- **Values:** issues that refer to measures with which decisions should be compared
- **Other:** issues not belonging to the other categories, such as those referring to the decision process itself

Use cases are a mechanism for enriching the discussion of issues and have proven to be an effective tool (Parnell et al., 2013, pp. 116–118). Use cases prepared in advance to portray possible futures that could strongly impact the outcome of the decision can lead to a more extensive discussion of the opportunity space. For example, the commercial aerial vehicle decision analyst might benefit from generating an issues list for use cases with restrictive airspace policy and less restrictive policy.

6.5.5 Vision Statement

A vision statement is sometimes referred to as the statement of purpose, the opportunity description, or the opportunity definition. Developing a vision statement is an effective contributor to reaching agreement and shared purpose among the decision-maker(s) and other key stakeholders. This tool provides a compelling statement of the motivation, answering three key questions (Parnell et al., 2011): What are we going to do? Why are we doing this? How will we know we have succeeded?

> We will design and develop an effective and affordable commercial unmanned air vehicle to serve the needs of local law enforcement. The product line will be available within 18 months, and will provide capabilities for surveillance that are presently not available in the commercial market. We will measure our success through direct feedback from the law enforcement agencies based on their perceived usability of and impact of our system.

Figure 6.4 Example of a vision statement

An example vision statement is shown in Figure 6.4. The statement may be obvious in the end; however, the process of articulating the statement and gaining the consensus of the group serves an important goal of getting everyone "on the same page." As part of the opportunity space activity, or later in the trade-off analysis process, further details can be elaborated to enrich the understanding of the opportunity and how it will be addressed.

6.5.6 Influence Diagram

Influence diagrams (sometimes referred to as decision networks, value maps, or decision diagrams) show the interrelationships between uncertain variables and decisions

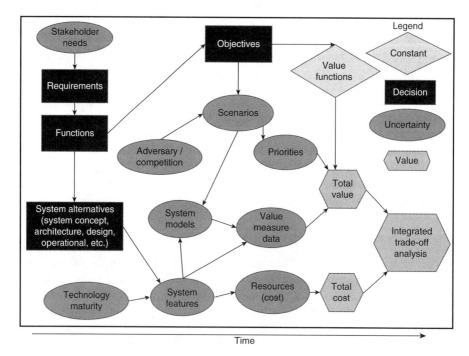

Figure 6.5 Influence diagram – same as Figure 2.1

through three levels of specification: relation, function, and number (Howard, 1984). The relational (or graphical) level shows relationships through a graphical representation consisting of various node types and directed arcs. The functional level defines the conditional probability distribution of each uncertain chance node and the set of alternatives associated with each decision node. The numerical level specifies actual numbers associated with probability functions and utility values (Diehl, 2004). The relational level diagram is particularly useful while defining the opportunity space.

Influence diagrams provide an easily understood approach for structuring and sharing information as the opportunity space is being explored. Figure 6.5 shows four elements (constant, decision, uncertainty, value) employed in a relational-level influence diagram as a compact graphical representation and is further explained in Chapter 2.

The influence diagram can be an effective tool for displaying the structure of a decision opportunity, providing an effective communications tool to reveal the various elements that are important for a given decision situation and their relationships. More importantly, during the effort to characterize the opportunity space, the activity of developing influence diagrams provides a means to analyze and think about various aspects of a complex opportunity situation (see Chapter 9).

6.5.7 Selecting Appropriate Tools and Techniques

As we look at the general process of opportunity space development, we need not apply all the techniques or tools. No matter the stakeholder analysis or questioning technique(s) used, one must take into account the decision traps. Framing the topic and developing questions that are resilient to bias (e.g., what is the most important thing you do each day?) are essential to maximizing the time spent with stakeholders. It helps to reduce the amount of post data collection clean-up and analysis required to gain insights for the development of the opportunity space. It also ensures that the opportunity space created is as complete as possible for creation of an appropriate tradespace.

6.6 ILLUSTRATIVE EXAMPLES

This section includes two illustrative examples that show the application of the tools and techniques used to obtain knowledge of the opportunity and defining the opportunity space. The section contains an example of developing the operational space for a commercial drone, describing knowledge gathering, and illustrating the decision hierarchy techniques and opportunity statement. The second example is about developing the operational space for a defense UAV illustrating how three techniques and four tools were applied to collect four types of knowledge.

6.6.1 Commercial

This section presents an illustrative case for developing the opportunity space for a small drone to be used in performing community-based forest monitoring. Adapted from insights in a recent study (Paneque-Galvez et al., 2014), this simplified example discusses an opportunity that is at the intersection of the commercial domain and the sustainable forest management domain. Community-based forestry monitoring strategies are an important part of the overall community-based forest management strategies, which are central to many development and conservation projects across the globe. Community-based forestry has the potential to be improved through the use of technology, and this opportunity is specifically targeted to developing a specialized new small drone. The capability provided by the drone will be drone-assisted community-based forestry monitoring, as part of a larger forestry management system.

Gathering domain knowledge, technical knowledge, expert knowledge, and stakeholder knowledge has been an important first step in developing the opportunity space for a new drone. First of all, there is a need to have commercial domain knowledge and general knowledge of the forestry management domain. This was satisfied through including team members representing both domains. Specialized technical knowledge needed by the team included autonomous operations, remote sensing, control systems, monitoring systems, forest environment characteristics, and other

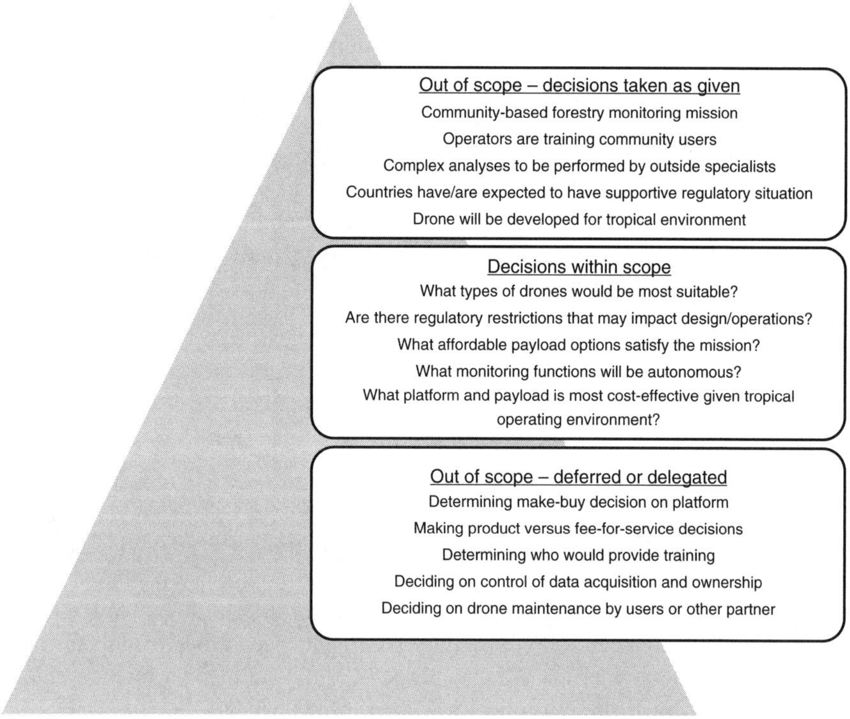

<u>Out of scope – decisions taken as given</u>
Community-based forestry monitoring mission
Operators are training community users
Complex analyses to be performed by outside specialists
Countries have/are expected to have supportive regulatory situation
Drone will be developed for tropical environment

<u>Decisions within scope</u>
What types of drones would be most suitable?
Are there regulatory restrictions that may impact design/operations?
What affordable payload options satisfy the mission?
What monitoring functions will be autonomous?
What platform and payload is most cost-effective given tropical operating environment?

<u>Out of scope – deferred or delegated</u>
Determining make-buy decision on platform
Making product versus fee-for-service decisions
Determining who would provide training
Deciding on control of data acquisition and ownership
Deciding on drone maintenance by users or other partner

Figure 6.6 Commercial drone decision hierarchy example

topics. This knowledge was gathered through multiple sources including academic and nonacademic literature, forestry management reports, recent environmental studies, and recent technology studies. Expert knowledge was acquired in several ways. First of all, some members of the team had expert knowledge from personal experience, including forest monitoring, training monitors, and operating small drones. Second, the team gathered knowledge through discussions with experts on technical areas such as image generation, data acquisition, and spatial resolution. Third, experts were also consulted on topics relevant to the opportunity context including environmental topics, regulatory concerns, ethical issues, safety and security, and suppliers (funders, partners, training providers). Stakeholder knowledge gathering included interviews with selected individual observers in community-based forest monitoring, secondary source data from recent surveys, and group interviews with environmental agencies and partner agencies involved in community-based forestry management.

The system boundary for this opportunity area is important to recognize and understand relevant to the larger forest management system. The team determined that the drone platform, payload, and operational process are all within the opportunity area. External to this area are the community monitors (operators), technology suppliers, geospatial analysis providers, partner organization, forest data end users, regulatory agencies, and the forest management enterprise.

Once the team had completed its initial knowledge gathering and scoped the system boundary, they created a decision hierarchy to categorize the in-scope and out-of-scope opportunities for the drone opportunity. Figure 6.6 shows a partial list of the categorized decisions.

The process of completing the decision hierarchy helped the team to clarify the decisions relevant to the opportunity. They determined where they needed to delegate some decisions to others who would be involved in decisions at a later point in time (e.g., the decision on payload suppliers would need to be made once the decision on platform and payload was made). The team consciously avoided the complexity decision trap, by constraining the decision analysis to the drone itself.

Once the team completed its efforts, they produced an opportunity statement regarding the purpose of the study, which was agreed upon by the involved decision-makers and stakeholders.

> We will develop the system requirements for a small drone that provides a user-operated autonomous capability for community-based forest monitoring in a tropical environment. The system will meet/exceed current technical and operational requirements for community-based monitoring, at less cost and with added environmental benefit. The small drone product will be available and operating in the field within 3 years. We will measure system success through the number of drone units delivered and fielded, feedback from the community-based forest management leadership on the impact to the monitoring mission, and usability feedback from monitoring participants who operate and maintain the drones.

6.6.2 Defense

This section describes the process for developing an opportunity space for a new UAV to support dismounted soldiers while on patrol. The example will provide notional data and results to help illustrate the process. As part of the process, four types of knowledge were sought (Domain, Technology, Expert, and Stakeholder). To collect this knowledge, a collection of techniques and tools was used. This section outlines why certain techniques and tools were used to collect each of these types of knowledge and how that knowledge informed the opportunity space generated for this effort.

The first task to be accomplished was defining an initial vision statement based on the system decision-maker's guidance. After holding conversations with the requirements development community, it was determined that the initial vision statement (opportunity statement) would read:

> Develop the requirements for a tactical system that would provide users with situational awareness of the environment from an elevated position. The product line will be available within 5 years and provide capabilities for surveillance at the platoon level (24 infantry soldiers). We will measure system success through the Network Integration Evaluation assessments at Fort Bliss.

To generate domain knowledge, the team reviewed a series of books on situational awareness, after action reports on recent dismounted operations conducted in urban and open terrain, Army and Marine Corps doctrine and training manuals for dismounted operations, and Infantry and Army AL&T magazine articles published over the last 10 years. Synthesis of this information helped to define the system boundary and generate the influence diagram (see Figure 6.7) that provided insight into the complexities to the environment for which the system had to be designed, built, tested, procured, deployed, and maintained.

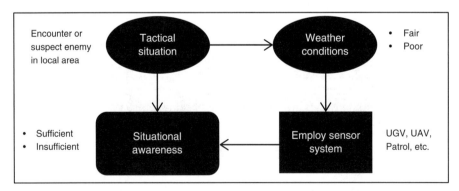

Figure 6.7 Example influence diagram for desired capability

To obtain technical knowledge, the team reviewed technical reports from six Department of Defense communities of interest: control systems, remote sensing, autonomous operations, information assurance, dismounted military operations, and logistics. The information obtained from this effort was used to frame the vision, goals, and objectives of the surveys designed to collect expert knowledge.

To obtain expert knowledge, the team conducted a survey of the technical community with the goal of understanding the current capabilities available from government programs or industry. A second goal of understanding the possible within the next 5, 10, and 15 years was also achieved using these surveys. The surveys consisted of quantitative questions and were conducted via web-based surveys on government research networks to protect the sensitivity of the information. For information too sensitive for the network data collection system, study team members scheduled and conducted follow-up sessions with stakeholders to collect sensitive information. The number of participants sent questions was calculated using an online survey size calculator and approved by the study sponsor. The results of the survey helped to verify the influence diagrams and refine the issue list.

To generate stakeholder knowledge, the team developed a series of surveys and follow-up interview questions for use in collecting information from the decision authority, client (study sponsor), and program owner (Program Executive Office). Two focus group sessions were conducted to gather information from the user and consumer community of system operators and logistics personnel. The focus groups were able to leverage some of the questions used during the interview session, but most of the questions were specific for the individual focus groups. The number of participants per sessions was 12 with the individuals selected from a cross section of each community. The final list of participants for the interviews and focus groups were reviewed with the sponsor for concurrence. As with the other data collection efforts, the knowledge generation efforts, data was collected, binned, cleaned, and analyzed for insights on elements to be placed in the issue list as well as to produce findings, conclusions, and study recommendations for the opportunity space.

The information gathered in these efforts resulted in an opportunity space peppered with an abundance of possible insights based on the initial vision statement. Analysis of the information allowed the team to conduct opportunity space refinement, and from this, they generated a refined and better framed vision statement:

> Develop the requirements for a tactical system that will provide users with a Nonconsumable Expendable tool to deliver secure local situational awareness of the environment from an elevated position for at least 45 min. The product line will be available within 10 years and will provide capabilities for surveillance that are presently not available at the platoon level (24 infantry soldiers). We will measure system success through the Network Integration Evaluation assessments at Fort Bliss, TX and direct feedback from the units using these systems in deployed environments based on user perceived usability and system impacts on mission success.

Opportunity space refinement also produced a refined set of requirements that would be used to develop viable alternatives to accomplish the vision statement. These included the following:

- System must be Nonconsumable Expendable Property (Army Regulation 735-5, 2013).
- Per unit price must be less than $1000.
- System transport must be via standard Army light cargo carriers.
- System must be man-portable (max weight of system and any power sources for operation cannot exceed 5 pounds).
- System must be operational within 60 s of decision to employ.
- System must be maintainable, deployable, operable, and recoverable by the average infantry soldier with less than 30 min of training by a fellow soldier.
- System must be able to be maintained by a single individual with basic soldier skills.
- Spare parts must be obtained via the Army supply system within 24 h.
- Consumable items (e.g., fuel, batteries, and lubricants) must be part of the current Army supply system.
- Data transfer must be real time to operator via Secure File Transfer Protocol (SFTP).
- System must be able to be operated by one soldier and provide information to at least 10 designated personnel within 1 km of the system operator.
- System must be able to maintain stable observation of an area no smaller than 10 m by 10 m and no larger than 100 m by 100 m area for at least 30 min.
- System must be able to stay on station for no less than 45 min.

6.7 KEY TERMS

Business Knowledge: An understanding of the broader industry, competition, economic environment, legislative environment, rates of return, and other factors related to a client or company's culture and business processes (Parnell et al., 2013, p. 32 and 44).

Concept Map: An illustration of the ideas and objects that link major concepts (Parnell et al., 2011, p. 2).

Decision: An irrevocable allocation of resources.

Decision Hierarchy: The "primary conceptual tool for defining the scope of a decision. The decision hierarchy is portrayed as a pyramid structure with three levels which summarize the decisions that have been made, the decision made in this decision, and the subsequent decisions." (Parnell et al., 2011, p. 118 and 126).

Decision Traps: Factors that prevent people from making appropriate decisions. Examples include failure to frame the opportunity and status quo.

KEY TERMS

Domain Knowledge: An understanding of specific area or specialized discipline.

Expert Knowledge: Authority extensive knowledge of a specific area or specialized discipline.

Focus Groups: A "group of 6–12 people used to solicit qualitative opinions and attitudes about how satisfied users and customers of a service or product have been. They are also used to gather data on new or proposed products, services, or ideas." (Parnell et al., 2013, p. 86 and 89).

Influence Diagram: "A compact graphical representation of conducting relationships among uncertainties and decisions in a perspective on a decision situation" (Parnell et al., 2011, pp. 169–170, 223).

Interviews: "The one-on-one process used to elicit information from decision makers and SMEs." Although structured to maximize the productivity of the interaction, interviewers modify questions to leverage the previous responses of the interviewee. "Interviews can be in person, by telephone, or by electronic means." (Parnell et al., 2013, pp. 73–77, 89).

Initial Vision Statement: The first attempt to define where the effort will lead to. It is refined into a problem definition or vision statement after the opportunity space has been identified.

Issue List: Consists of the issue(s) as well as a description of the issue. It can be a prioritized or an unprioritized list.

Opportunity Space: The area from which a viable solution can be found. It is framed by the issue, constraints, and assumptions.

Stakeholder Knowledge: Information from a person with an interest or concern in the system under assessment, improvement, or development.

Subject Matter Experts (SMEs): "Someone with credible substantive knowledge about the decision". Quantitative means of designating an SME include but are not limited to 10,000 h of experience in this domain or certification via standard test in the field (e.g., Principles and Practice of Engineering (PE) exam and Certified Modeling & Simulation Professional) (Parnell et al., 2013, p. 90).

Surveys: A structured means used to elicit quantitative or qualitative information from large groups of decision-makers and SMEs. It is a series of written questions, which can be distributed and collected in person or via mail, e-mail, or the Internet (Parnell et al., 2013, pp. 78–80).

System Boundary: A means of delineating the elements, interactions, and subsystem the decision analyst believes should be part of a system definition from elements, interactions, and subsystem, which may influence or be influenced by the system but lie outside the influence of the system decision of the analysis (Parnell et al., 2011, pp. 35–38).

Technical Knowledge: A substantive knowledge of a specific domain of interest (Parnell et al., 2013, p. 32 and 44).

Use Case: A description of system or process that defines interactions between external actors and the system to attain particular goals (Janssen).

Vision Statement (Opportunity Statement): An effective way to illustrate an agreement between the decision-maker(s) and key stakeholders on the purpose of the decision. "The vision statement answers three questions: What are we going to do? Why are we doing this? How will we know that we have succeeded?" (Parnell et al., 2011, p. 115 and 126).

6.8 EXERCISES

6.1. Distinguish between an opportunity space and a tradespace.

6.2. Identify and describe the types of knowledge that systems engineers use in defining the opportunity space.

6.3. List and describe three decision traps that you have observed in a professional (or family) setting.

6.4. Write an initial vision statement to describe the opportunity for a pizza company seeking to expand their business through extending delivery of their product beyond local Hoboken, NJ, to customers on Manhattan Island.

6.5. Develop an influence diagram for how a professor might determine student grades based on data from multiple inputs (e.g., such as homework, quizzes, projects, exams, class attendance, and class participation) for a distance learning course.

6.6. Consider a trade-off analysis study that you have done or might do in the future. Develop each of the following.
  (a) List of stakeholders and decision-makers
  (b) Types of knowledge that would be relevant
  (c) Who you would interview and five sample interview questions
  (d) Purpose of a relevant focus group and desired list of stakeholders
  (e) Who you would survey and five questions you might ask
  (f) System boundary
  (g) Vision statement
  (h) Decision hierarchy
  (i) Influence diagram

REFERENCES

Army Regulation 735-5 (2013) *Property Accountability Policies*, Headquarters Department of the Army, Washington, DC.

DeFranco, J.N. (2011) A cognitive collaboration model to improve performance in engineering teams - a study of team outcomes and mental model sharing. *Systems Engineering*, **14** (3), 267–278.

Diehl, M. (2004) Influence diagrams with multiple objectives and tradeoff analysis. *IEEE Transactions on Systems, Man, and Cybernetics —Part A:Systems and Humans*, **34** (3), 293–304.

Fowler, F.J. (1995) *Improving Survey Questions: Design and Evaluation*, 1st edn, vol. **38**, SAGE Publications, Inc., Thousand Oaks, CA.

Greenbaum, T. L. (1997). *Focus Groups: A Help or a Waste of Time?*. From Groups Plus: http://www.groupsplus.com/pages/pmt0797.htm (accessed 7 June 2015).

Hammond, J.S., Keeney, R.L., and Raiffa, H. (1998) The hidden traps in decision making. *Harvard Business Review*, **76** (5), 3–9.

Henman, L. D. (n.d.). *How to Avoid the Hidden Traps of Decision Making*, from Articles by Linda - Henman Performance Group: http://www.henmanperformancegroup.com/articles/Avoid-Hidden-Traps.pdf (accessed 2 May 2015).

Howard, R.M. (1984) Influence diagrams, in *The Principles and Applications of Decision Analysis* Vol II (ed. R.a. Howard), Strategic Decisions Group, Menlo Park, CA, pp. 719–762.

Janssen, C. (n.d.). *What is a Use Case - Definition from Techopedia*. From Techopedia: http://www.techopedia.com/definition/25813/use-case (accessed 12 June 2015).

Klimack, W., Marvin, F., Paul, W. et al. (2011) *Soft Skills Workshop: Real World Skills for Decision Analysis and OR/MS Professionals*, Wiley, Vienna, VA.

Kulzy, W.W. and Fricker, R.D. (2015) The survey process: with emphasis on survey data analysis. *PHALANX*, **47**, 32–37.

McNamara, C. (2008). *Basics of Conducting Focus Groups (C)*. From Free Management Library TM: http://managementhelp.org/businessresearch/focus-groups.htm (accessed 24 May 2015).

Novak, J. (2006). The theory underlying concept maps and how to construct them. *Technical Report IHMC Cmap Tools 2006-1*. Florida Institute for Human and Machine Cognition.

Online Form Builder with Cloud Storage Database|Wufoo. (n.d.). From Wufoo: http://www.wufoo.com/ (accessed 24 May 2015).

Paneque-Galvez, J., McCall, M.K., Napoletano, B.M. *et al.* (2014) Small drones for community-based forest monitoring: an assessment of their feasibility and potential in tropical areas. *Forests*, **5**, 1481–1507.

Parnell, G., Bresnick, T.A., Tani, S.N., and Johnson, E.R. (2013) *Handbook of Decision Analysis*, John Wiley & Sons, Inc., Hoboken, NJ.

Parnell, G.S., Driscoll, P.J., and Henderson, D.L. (2011) *Decision Making in System Engineering and Management*, Wiley, Hoboken, NJ.

Sample Size Calculator. (n.d.). From SurveyMonkey®: https://www.surveymonkey.com/mp/sample-size-calculator/ (accessed 24 May 2015).

Sample Size Calculator - SurveyStatz. (n.d.). From SurveyStatz©: http://www.surveystatz.com/Sample-size-calculator.aspx (accessed 24 May 2015).

SurveyGizmo | Professional Online Survey Software; Form Builder. (n.d.). From SurveyGizmo: http://www.surveygizmo.com/ (accessed 24 May 2015).

SurveyMonkey: Free Online Survey Software & Questionnaire Tool. (n.d.). From SurveyMonkey®: https://www.surveymonkey.com/ (accessed 24 May 2015).

Systems©, C. R. (2012). *Sample Size Calculator - Confidence Level, Confidence Interval, Sample Size, Population Size, Relevant Population - Creative Research Systems*. From The Survey System: http://www.surveysystem.com/sscalc.htm (accessed 24 May 2015).

ThinkTank Business Process Collaboration Software. (2014), From ThinkTank©: http://thinktank.net/ (accessed 24 May 2015).

7

IDENTIFYING OBJECTIVES AND VALUE MEASURES*

GREGORY S. PARNELL
Department of Industrial Engineering, University of Arkansas, Fayetteville, AR, USA

WILLIAM D. MILLER
Innovative Decisions, Inc., Vienna, VA, USA; School of Systems and Enterprises, Stevens Institute of Technology, Hoboken, NJ, USA

> If you don't know where you are going, any road will get you there.
>
> (Lewis Carroll)

7.1 INTRODUCTION

The decision opportunity and our values determine the decision objectives. The objectives define the goals that we are trying to achieve. In systems engineering, the objective space includes business/mission objectives, stakeholder objectives, and system objectives. Business/mission objectives are derived from organizational and customer needs. Stakeholder objectives include the goals of other important stakeholders in the system life cycle. Finally, system objectives include the technical objectives necessary for the system to meet business/mission and stakeholder objectives throughout the system life cycle. The systems engineering objectives space

---

*This chapter draws on material from Chapters 3 and 7 of Parnell, G. S., Bresnick, T. A., Tani, S.N., and Johnson, E. R., *Handbook of Decision Analysis*, Wiley Operations Research/Management Science Handbook Series, Wiley & Sons, 2013.

Trade-off Analytics: Creating and Exploring the System Tradespace, First Edition. Edited by Gregory S. Parnell.
© 2017 John Wiley & Sons, Inc. Published 2017 by John Wiley & Sons, Inc.
Companion website: www.wiley.com/go/Parnell/Trade-off_Analytics

spans the life cycle of the system's products and services in both commercial and government enterprises including the business/mission need, concept, requirements, architecture, design, integration, verification, validation, production, deployment, operations, support, and retirement phases of the life cycle.

Defining the value of the system's products and services is one of the most critical tasks of systems engineering. In the previous chapter, we discuss the critical role of understanding the decision opportunity or problem. With this understanding, which may include a partial list of objectives, we next strive to identify the full list of objectives for the system. Of course, the identification of a significant new objective can change our understanding of the opportunity. In addition to improving the opportunity space, systems engineers use objectives as the foundation for developing value measure(s) used to evaluate alternatives and select the best alternative(s).

We believe it is good practice in any trade-off study to consider values and objectives before determining the full set of alternatives. Our experience is that studies focusing on alternative development first, with measures used only to distinguish between those alternatives, often miss several important objectives. This results in a missed opportunity to create a broader and more comprehensive set of alternatives with the potential to create more value.

To identify objectives, it is generally not sufficient to interact with only the decision-maker(s) because, for complex decisions, they may not have a complete and well-articulated list of objectives. Instead, significant effort may be required to identify, define, and perhaps even carefully craft the objectives based on many interactions with multiple decision-makers, diverse stakeholders, and recognized subject matter experts.

The chapter is organized as follows. In Section 7.2, we introduce the concept of Value-Focused Thinking. In Section 7.3, we describe shareholder and stakeholder value as the basis for the decision objectives. In Section 7.4, we describe why the identification of objectives is challenging. In Section 7.5, we list some key questions we can use to identify decision objectives and four major techniques for identifying decision objectives: research, interviews, focus groups, and surveys. In Section 7.6, we discuss key considerations for the financial objective of private companies and cost objectives of public organizations. In Section 7.7, we discuss key principles for developing value measures to measure, at the time of the decision, how well an alternative could achieve an objective. In Section 7.8, we describe the structuring of multiple objectives including objective and functional value hierarchies; four techniques for structuring objectives through the use of platinum, gold, silver, and combined standards; best practices; and provide cautions about risk and cost objectives. In Section 7.9, we describe the diverse approaches used to craft objectives for two illustrative problems. We conclude with a summary of the chapter in Section 7.10.

7.2 VALUE-FOCUSED THINKING

An important philosophical approach to creating value was introduced by Ralph Keeney in his book entitled "Value-Focused Thinking" (VFT) (Keeney, 1992). To be

VALUE-FOCUSED THINKING   235

successful, VFT must interactively involve decision-maker(s), stakeholders, subject matter experts, and analysts to create higher shareholder and/or stakeholder value.

7.2.1 Four Major VFT Ideas

VFT has four major ideas: start first with our values, use our values to generate better alternatives, create decision opportunities, and use our values to evaluate the alternatives.

7.2.1.1 Start First with Values VFT starts with values and objectives before identifying alternatives. Keeney describes the contrasting approach as Alternative-Focused Thinking (AFT), which starts with the alternatives and seeks to differentiate them. He lists three disadvantages of AFT. First, if we start with known alternatives, this will limit our decision frame (our understanding of the opportunity or problem, see Chapter 6). Second, if we try to evaluate only known alternatives, we may not identify important new values and objectives that are relevant for future solutions. Finally, since we may not have a full understanding of the opportunity and the decision frame, we may not have a sound basis for the generation of alternatives.

7.2.1.2 Generate Better Alternatives Our decision can only be as good as the best alternative that we identify. If we have several poor alternatives, the best analysis will only identify a poor alternative! Once we have identified the values and objectives for the decision problem, we can use them to generate better alternatives. We do this by qualitatively and quantitatively defining the value gaps. See Chapter 8.

7.2.1.3 Create Decision Opportunities Many of us have been taught to look for potential problems, carefully define the problems, and then look for solutions. Whereas problem definition is reactive, opportunity definition is proactive. Keeney encourages us not to wait for problems to appear, but instead to focus our energy and creativity on identifying decision opportunities that may result in added value for an organization's shareholders and stakeholders. See Chapter 6.

7.2.1.4 Use Values to Evaluate Alternatives Finally, we should use our values to qualitatively and quantitatively evaluate our alternatives. The mathematics of single and multiple objective decision analysis can be used to evaluate the alternatives, and once completed, we can use this information to improve existing alternatives or develop new alternatives. See Chapter 8.

7.2.2 Benefits of VFT

Keeney has identified several benefits of VFT:

1. Since strategy is a choice among alternatives based on strategic objectives, VFT can support the organization's strategic planning by helping to identify strategic objectives and creative strategies for evaluation in strategic planning.

2. VFT is broadly applicable to both decision opportunities and decision problems.
3. Since VFT takes a broad perspective of the decision opportunity, it can help identify hidden objectives. Objectives may be hidden because they result from new opportunities and challenges faced by the organization.
4. VFT can identify creative new alternatives with potential to achieve value for shareholders and stakeholders. This is a critical part of any decision analysis, since the quality of the alternatives limits the potential value that can be identified.
5. Since VFT identifies important objectives and new alternatives, it can focus information collection on what is important and NOT just on what is available. The prioritization of objectives and measures is critical to establishing what is important.
6. The focus on stakeholder participation to identify their objectives and measures is important to achieve their involvement in analysis, decision-making, and strategy implementation. When stakeholders see that their objectives are included, they are more willing to participate.
7. Definition of objectives and measures can improve evaluation of alternatives. By using the mathematics of decision analysis, we can develop a value model directly from the objectives and measures that can be used to evaluate alternatives.
8. A clear understanding of the organization's values, objectives, and value measures can improve communications between stakeholders. This understanding is critical for development of alternatives, analysis of alternatives, and implementation of the selected alternative(s).
9. Better alternatives and better evaluation can lead to better decisions. The identification of better alternatives offers the potential for higher value outcomes and better analysis provides insights and decision clarity for decision-makers.
10. Early stakeholder involvement and better communication can increase the probability of successful solution implementation, which increases the likelihood of converting potential value to actual value.

7.3 SHAREHOLDER AND STAKEHOLDER VALUE

Decision objectives should be based on shareholder and stakeholder value. Value can be created at multiple levels within an organization. An organization creates value for shareholders and stakeholders by performing its mission (which defines the customers); providing products and services to customers; improving the effectiveness of its products and services to provide better value for customers; and improving the efficiency of its operations to reduce the resources required to provide the value to customers. Defining value is challenging for both private and public organizations. Private companies must balance shareholder value with being good corporate citizens (sometimes called stakeholder value). Management literature includes discussion of stakeholder value versus shareholder value for private companies (Charreaux & Desbrieres, 2001). Public organizations, which operate without profit incentives, focus on stakeholder value, with management and employees being key stakeholders.

Next, we consider shareholder and stakeholder objectives in a private company example and stakeholder objectives in a public organization.

7.3.1 Private Company Example

Consider a large publicly owned communications company operating a cellular network that provides voice and data services for customers. Stakeholders include the shareholders, the board of directors, the leadership team, the employees (managers, designers, developers, operators, maintainers, and business process personnel), the customers, the cell phone manufacturers, the companies that sell the cell phone services, and the communities in which the company operates. Competition includes other cellular communications companies and companies that provide similar products and services using other technologies (satellite, cable, etc.).

Since the company is publicly owned, the board of directors' primary objective is to increase shareholder value. Stakeholders have many complementary and conflicting objectives. For example, the board may want to increase revenues and profits; the leadership may want to increase executive compensation; the sales department may want to increase the number of subscribers; the network operators want to increase availability and reduce dropped calls; the safety office may want to decrease accidents; the technologists may want to develop and deploy the latest generation of network communications; the cell phone manufacturers may want to sell improved technology cell phones; companies that sell the cell phones and services may want to increase their profit margins; operations managers want to reduce the cost of operations; the human resources department may want to increase diversity; and the employees may want to increase their pay and benefits.

7.3.2 Government Agency Example

Next, consider a government agency that operates a large military communications network involving many organizations. There are no shareholders in this public system that provides communications to support military operations. However, similarly to the private company example, there are many stakeholders. Some of the key stakeholders are the Department of Defense (DoD) office that establishes the communications requirements and submits the budget; Congress that approves the annual budgets for the network; the agency that acquires the network; the agency that manages the network; contractor personnel who manufacture and assemble the network; contractor, civilian, and/or military personnel who operate the network; information assurance personnel who maintain the security of the network; mission commanders who need the network to command and control their forces; and military personnel whose lives may depend on the availability of the network during a conflict. Instead of business competitors, the network operators face determined adversaries who would like to penetrate the network to gain intelligence data in peacetime or to disrupt the network during a conflict.

The stakeholders have many complementary and conflicting objectives. For example, the DoD network management office wants the best network for the budget; Congress wants an affordable communications network to support national security; the acquiring agency wants to deliver a network that meets the requirements, on time and on budget; the network management agency wants to ensure an adequate budget; contractors want to maximize their profits and obtain future work; network operators want to increase network capabilities and maximize availability of the network; information assurance personnel want to maximize network security; mission commanders want to maximize the probability of mission success; and military personnel who use the network want to maximize availability and bandwidth.

7.4 CHALLENGES IN IDENTIFYING OBJECTIVES

The identification of objectives is more art than science. The four major challenges are (i) identifying a full set of values and objectives, (ii) obtaining access to key decision-makers and stakeholders, (iii) differentiating fundamental and means objectives, and (iv) structuring a comprehensive set of fundamental objectives for validation by the decision-maker(s) and stakeholders.

In a complex decision, especially if it is a new opportunity for the organization, the identification of objectives can be challenging. In a research paper (Bond et al., 2008), the authors concluded that "in three empirical studies, participants consistently omitted nearly half of the objectives that they later identified as personally important. More surprisingly, omitted objectives were as important as the objectives generated by the participants on their own. These empirical results were replicated in a real-world case study of decision-making at a high-tech firm. Decision-makers are considerably deficient in utilizing personal knowledge and values to form objectives for the decisions they face." To meet this challenge, we must have good techniques to obtain the decision objectives.

A second challenge is obtaining access to a diverse set of decision-makers (DMs), stakeholders (SHs), and subject matter experts (SMEs). Sometimes, clients are reluctant to provide the trade-off analysis team access to senior decision-makers and diverse stakeholders who have the responsibilities and breadth of experience essential to providing a full understanding of decision objectives. In addition, it can be difficult to obtain access to recognized experts instead of individuals who have more limited experience. Many times, the best experts resist meetings that take their focus away from their primary responsibilities. In addition, even if we have access, we may not have the time in our analysis schedule to access all the key individuals. To be successful, the analysis team must obtain access to as many of these key individuals as possible in the time allocated for the study.

The third challenge is the differentiation of fundamental and means objectives (Keeney, 1992). Fundamental objectives are what we ultimately care about in the decision. Means objectives describe how we achieve our fundamental objectives.

An automobile safety example helps to clarify the difference. The fundamental objectives may be to reduce the number of casualties due to highway accidents and to minimize cost. The means objectives may include to increase safety features in the automobile (e.g., airbags and seat belts), to improve automobile performance in adverse weather (e.g., antilock brakes), and to reduce the number of alcohol-impaired drivers (e.g., stricter enforcement). The mathematical considerations of multiple objective decision analysis discussed in Chapter 2 require the use of fundamental objectives in the value model.

The fourth challenge is structuring the knowledge about fundamental objectives and value measures. This structure should enable decision-makers, stakeholders, and experts to validate that the set of objectives and value measures are both necessary and sufficient to evaluate the alternatives.

7.5 IDENTIFYING THE DECISION OBJECTIVES

7.5.1 Questions to Help Identify Decision Objectives

The key to identifying decision objectives is asking the right questions, to the right people, in the right setting. Keeney identified 10 categories of questions that can be asked to help identify decision objectives (Keeney, 1994). These questions should be tailored to the problem and to the individual being interviewed, the group being facilitated, or the survey being designed.[1] For example, the strategic objectives question may be posed to the senior decision-maker in an interview while the consequences question may be posed to key stakeholders in a facilitated group.

1. **Strategic objectives:** What are your ultimate or long-range objectives? What are your core fundamental values? What is your strategy to achieve these objectives?
2. **A wish list:** What do you want? What do you value? What should you want? What are you trying to achieve? If money was not an obstacle, what would you do?
3. **Alternatives:** What is a perfect alternative, a terrible alternative, and a reasonable alternative? What is good or bad about each?
4. **Problems and shortcomings:** What is wrong or right with your organization or enterprise? What needs fixing? What are the capability, product, or service gaps that exist?
5. **Consequences:** What has occurred that was good or bad? What might occur that you care about? What are the potential risks you face? What are the best or worst consequences that could occur? What could cause these?
6. **Goals, constraints, and guidelines:** What are your goals or aspirations? What limitations are placed upon you? Are there any legal, organizational, technological, social, or political constraints?

---

[1] Our focus with these questions is the decision objectives. However, the answers may provide valuable insights on issues, alternatives, uncertainties, and constraints that can be used in later phases of the analysis.

7. **Different perspectives:** What would your competitor or your constituency be concerned about? At some time in the future, what would concern you? What do your stakeholders want? What do your customers want? What do your adversaries want?
8. **Generic fundamental objectives:** What objectives do you have for your customers, your employees, your shareholders, and yourself? What environmental, social, economic, or health and safety objectives are important?
9. **Structuring objectives:** Follow means–ends relationships: why is that objective important, how can you achieve it? Use specification: what do you mean by this objective?
10. **Quantifying objectives:** How do you measure achievement of this objective? Which objective is the most important? Which objective is least important? How much more important is any one objective to another objective? Why is any one objective more important than another objective?

7.5.2 How to Get Answers to the Questions

Four techniques to obtain answers to these questions and help identify objectives and value measures are research, interviews, focus groups, and surveys. See Chapter 6 for more details on interviews, focus groups, and surveys. We discuss the use of these techniques to identify functions, objectives, and value measures. The amount of research, the number of interviews, the number and size of focus groups, and the number of surveys we use depend on the scope and importance of the problem, the number of decision levels, the diversity of the stakeholders, the number of experts, and the time allocated to defining objectives and identifying value measures. Best practices for these techniques are described in (Parnell et al., 2011) and (Parnell et al., 2013).

7.5.2.1 Research Research is an important technique to understand the problem domain; to identify potential objectives to discuss with decision-makers, stakeholders, and experts; and to understand suggested objectives. See Chapter 6 for discussion on knowledge. The amount of research depends on the analysts' prior understanding of the problem domain, knowledge of key terminology, and amount of domain knowledge expected of the analyst in the decision process. The primary research sources include the problem domain (including work done on functional and requirements analysis) and the analysis literature. Research should be done throughout the objective identification process. Many times, information obtained with the other three techniques requires research to fully understand or validate the objective recommendation.

7.5.2.2 Interviews Senior leaders, stakeholders, and "world class" experts can identify important value and objectives. Interviews are the best technique for obtaining objectives from senior decision-maker(s), senior stakeholders, and "world class" experts since they typically do not have the time to attend a longer focus group nor the interest to complete a survey. However, interviews are time-consuming for the interviewer due to preparation, execution, and analysis.

7.5.2.3 Focus Groups Focus groups are another useful technique for identifying decision objectives. We usually think of focus groups for problem/opportunity identification and product market research; however, they can also be useful for identifying decision objectives. While interviews typically generate a two-way flow of information, focus groups create information through discussion and interaction between all the group members. As a general rule, focus groups should comprise between 6 and 12 individuals. Too few may lead to too narrow a perspective, while too many may not provide all attendees the opportunity to provide meaningful input. It is very important to have a good facilitator to keep the focus group on track and to make sure that a few members do not dominate the discussion.

7.5.2.4 Surveys In our experience, surveys are not used as frequently to identify potential decision objectives as interviews and focus groups. However, surveys are a useful technique for collecting decision objectives from a large group of individuals in different locations. Surveys are especially good for obtaining general public values. Surveys are more appropriate for junior- to mid-level stakeholders and dispersed experts. We can use surveys to gather qualitative and quantitative data on the decision objectives. A great deal of research exists on techniques and best practices for designing effective surveys.

7.6 THE FINANCIAL OR COST OBJECTIVE

The financial or cost objective may be the only objective, or it may be one of the multiple objectives. Shareholder value is an important objective in any firm. In many decisions of private companies, the financial objective may be the only objective or the primary objective. Firms employ three fundamental financial statements to track value: balance sheet, income statement, and cash flow statement. In addition, firms commonly use discounted cash flow (net present value) to analyze the potential financial benefits of their alternatives. For public company decisions, the cost objective is usually a major consideration. The cost can be the full life cycle cost or only a portion of the costs of the alternative.

7.6.1 Financial Objectives for Private Companies

In order to understand the financial objectives for private companies, we begin with the three financial statements. Next, we consider the conversion of cash flows to net present value.

7.6.1.1 Balance Sheet Statement A balance sheet is developed using standard accounting procedures to report an approximation of the value of a firm (called the net book value) at a point in time. The approximation comes from adding up the assets and then subtracting out the liabilities. We must bear in mind that the valuation in a balance sheet is calculated using generally accepted accounting principles, which make it reproducible and verifiable, but it will differ from the valuation we would develop taking account of the future prospects of the firm.

7.6.1.2 Income Statement The income statement describes the changes to the net book value through time. As such, it shares the strengths (use of generally accepted accounting procedures and widespread acceptance) and weaknesses (inability to address future prospects) of the balance sheet.

7.6.1.3 Cash Flow Statement A cash flow statement describes the changes in the net cash position through time. One of the value measures reported in a cash flow statement is "free cash flow" (which considers investment and operating costs and revenues, but not financial actions such as issuing or retiring debt or equity). Many clients are comfortable with this viewpoint; hence, a projected cash flow statement is the cornerstone of many financial analyses.

7.6.1.4 Net Present Value To boil a cash flow time pattern down to a one-dimensional value measure, companies usually discount the free cash flow to a "present value" using a discount rate. The result is called the net present value, or NPV, of the cash flow. We usually contemplate various possible alternatives, each with its own cash flow stream and NPV cash flow. In many decision situations, clients are comfortable with using this as their fundamental value measure. See Chapters 4 and 14.

7.6.2 Cost Objective for Public Organizations

For many public organizations, minimizing the cost is a major decision objective. Depending on the decision, different cost objectives may be appropriate. The most general cost objective (subject to acceptable performance levels) is to minimize life cycle cost, the full cost over all stages of the life cycle: concept development, design, production, operations, and retirement. However, costs that have already been spent (sunk costs) should never be included in the analysis. Since sunk costs should not be considered and some costs may be approximately the same across alternatives, in practice, many analysts consider the delta life cycle costs among the alternatives. In public organizations, the budget is specified by year and may or may not be fungible between years. When multiple years are analyzed, a government inflation rate is used to calculate net present cost.

See Chapter 4 for further discussion on cost analysis and affordability analysis.

7.7 DEVELOPING VALUE MEASURES

In order to quantitatively use the decision objectives in the evaluation of the alternatives, we must develop value measures for each objective that measure the a priori potential value, that is, before the alternative is selected. The identification of the value measures can be as challenging as the identification of the decision objectives. We can identify value measures by research, interviews, and group meetings with decision-makers, stakeholders, and subject matter experts. Access to stakeholders and experts with detailed knowledge of the problem domain is key to developing good value measures.

Kirkwood (1997) identifies two useful dimensions for value measures: alignment with the objective and type of measure. Alignment with the objective can be direct or proxy. A direct measure focuses on attaining the full objective, such as net present value for shareholder value. A proxy measure focuses on attaining an associated objective that is only partially related to the objective (e.g., reduce production costs for shareholder value). The type of measure can be natural or constructed. A natural measure is in general use and commonly interpreted, such as dollars. We have to develop a constructed measure, such as a five-star scale for automobile safety. Constructed measures are very useful but require careful definition of the measurement scales to ensure that we have an interval or ratio scale (see Chapter 2). In our view, the use of an undefined scale, for example, 1–7, is not appropriate for trade-off analysis since the measures do not define value and scoring is not repeatable.

Table 7.1 reflects our preferences for types of value measures. Our first preference is to select a value measure that directly aligns with the objective and has a natural scale. Our second preference is direct and constructed. We prefer direct and constructed to proxy and natural for two reasons. First, alignment with the objective is more important than the type of scale. Second, one direct and constructed measure can replace many natural and proxy measures. When value models grow too large, the source is usually the overuse of natural and proxy measures.

7.8 STRUCTURING MULTIPLE OBJECTIVES

Not all problems have only the financial or the cost objective. In many public and business decisions, there are multiple objectives and many value measures. Once we have

Table 7.1 Preference for Types of Value Measure

Type of Scale	Direct Alignment	Proxy Alignment
Natural	1	3
Constructed	2	4

a list of the preliminary objectives and value measures, our next step is to organize the objectives and value measures (typically called structuring) to remove overlaps and identify gaps. For complex decisions, this can be quite challenging. In this section, we introduce the techniques for identifying and structuring, using hierarchies. The decision analysis literature uses several names: value hierarchies, objectives hierarchies, value trees, objective trees, functional value hierarchy, and qualitative value model.

7.8.1 Value Hierarchies

The primary purpose of the objectives hierarchy is to identify the objectives and the value measures so we understand what is important in the problem and can do a much better job of qualitatively and quantitatively evaluating the alternatives. This may be the result of an enterprise-level business case analysis. Most decision analysis books recommend beginning with identifying the objectives and using the objectives to develop the value measures. For complex systems, decisions we have found that it is very useful to first identify the system functions that create value (Parnell et al., 2011). Functional analysis, central to both enterprise business processes and systems engineering, is the starting point for modeling the functional value hierarchy. For each function, we identify the fundamental objectives we want to achieve for that function. For each objective, we identify the value measures that can be used to assess the potential to achieve the objectives. In each application, we use the client's preferred terminology. For example, functions can be called missions, capabilities, activities, services, tasks, or other terms appropriate to the level of the decision analysis. Similarly, objectives can be called criteria, evaluation considerations, or other terms. Value measures can be called any of the previously mentioned terms (see Chapter 2).

The terms objectives hierarchy and functional value hierarchy are used in this book to make a distinction between the two approaches. The functional value hierarchy is a combination of the functional hierarchy from systems engineering and the value hierarchy from decision analysis (Parnell et al., 2011). In decisions where the functions of the alternatives are the same, or are not relevant, it may be useful to group the objectives by categories to help in structuring the objectives.

Both hierarchies begin with a statement of the primary decision objective as the first node in the hierarchy. An objectives hierarchy begins with the objectives in the first tier of the hierarchy, (sometimes) subobjectives as the second tier, and value measures as the final tier of the hierarchy. A functional value hierarchy uses functions as the first tier, (sometimes) subfunctions as the second tier, objectives as the next tier, and values measures as the final tier of the hierarchy. The value hierarchy integrates the enterprise business case analysis objectives hierarchy with the business process or systems engineering functional analysis.

Consider Figure 7.1. When the randomly ordered objectives hierarchy is logically organized by functions, the objectives and measures make more sense to the decision-makers and, many times, we identify the missing objectives (and value measures). The benefit of identifying the functions is threefold. First, a logical structure of the functional objectives hierarchy is easier for the decision analyst to develop.

STRUCTURING MULTIPLE OBJECTIVES

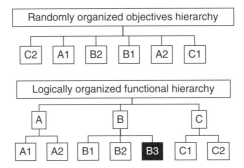

Figure 7.1 Comparison of objectives and functional value hierarchy

Second, it may help identify additional objectives (and value measures) that might be missed in the objectives hierarchy. Third, the logical order helps the decision-makers and stakeholders understand the hierarchy and provide suggestions for improvement.

In either the objectives hierarchy or functional hierarchy, we can create the structure from the top-down or from the bottom-up. Top-down structuring starts with listing the fundamental objectives on top and then "decomposing" into subobjectives until we are at a point where value measures can be defined. Top-down structuring has the advantage of being more closely focused on the fundamental objectives, but often, we initially overlook important subobjectives. Bottom-up structuring starts by discussing at the subobjective or subfunctional level, grouping similar things together, and defining the titles at a higher level for the grouped categories. Value measures are then added at the bottom of the hierarchy. Bottom-up structuring has the advantage of discussing issues at a more concrete and understandable level, but if we are not careful, we may drift from the fundamental objectives in an attempt to be comprehensive. In theory, both approaches should produce the same hierarchy. In practice, this rarely happens.

7.8.2 Techniques for Developing Value Hierarchies

The credibility of the qualitative value model is very important since it is the basis of multiple decision-maker and stakeholder reviews. If decision-makers do not accept the qualitative value model, they will not (and should not!) accept the quantitative analysis. We discuss here four techniques for developing objectives: the platinum, gold, silver, and combined standards (Parnell et al., 2013).

7.8.2.1 Platinum Standard A platinum standard value model is based primarily on information from interviews with senior decision-makers and key stakeholders. Decision analysts should always strive to interview the senior leaders (decision-makers and stakeholders) who make and influence the decisions. As preparation for these interviews, they should research potential key problem domain documents and talk to decision-maker and stakeholder representatives. Affinity diagrams (Parnell et al., 2011) can be used to group similar functions and objectives into logical, mutually exclusive, and collectively exhaustive categories. For example, interviews with senior decision-makers and stakeholders were used to develop a value model for the Army's 2005 Base Realignment and Closure value model (Ewing et al., 2006).

7.8.2.2 Gold Standard When we cannot get direct access to senior decision-makers and stakeholders, we look for other approaches. One approach is to use a "gold standard" document approved by senior decision-makers. A gold standard value model is developed based on an approved policy, strategy, or planning document. Many military acquisition programs use capability documents as a gold standard since the documents define system missions, functions, and key performance parameters (Parnell et al., 2001). A systems engineering functional analysis serves as a "gold standard" artifact in the development of systems products and services. Many times, the gold standard document has many of the functions, objectives, and some of the value measures. If the value measures are missing, we work with stakeholder representatives to identify appropriate value measures for each objective. It is important to remember that changes in the environment and leadership may cause a gold standard document to no longer reflect leadership values. Before using a gold standard document, confirm that the document still reflects leadership values.

7.8.2.3 Silver Standard Sometimes, the gold standard documents are not adequate (not current or not complete) and we are not able to interview a significant number of senior decision-makers and key stakeholders. As an alternative, the silver standard value model uses data from the many stakeholder representatives. Again, we use affinity diagrams to group the functions and objectives into mutually exclusive and collectively exhaustive categories. For example, inputs from about 200 stakeholders' representatives were used to develop the Air Force 2025 value model (Parnell et al., 1998). This technique has the advantage of developing new functions and objectives that are not included in the existing gold standard documents. For example, at the time of the study, the Air Force Vision was Global Reach, Global Power. The Air Force 2025 value model identified the function, Global Awareness (later changed to Global Vigilance), which was subsequently added to the new Air Force Vision of Global Vigilance, Reach, and Power.

7.8.2.4 Combined Standard Since it is sometimes difficult to obtain access to interview senior leaders, and many times, key documents are not sufficient to completely specify a value model, the most common technique is the combined standard. First, we research the key gold standard documents. Second, we conduct as many interviews with senior leaders as we can. Third, we meet with stakeholder representatives, in groups or individually, to obtain additional perspectives. Finally, we combine the results of our review of several documents with findings from interviews with some senior decision-makers and key stakeholders, and data from multiple meetings with stakeholder representatives. This technique was used to develop a space

STRUCTURING MULTIPLE OBJECTIVES 247

technology value model for the Air Force Research Laboratory Space Technology R&D Portfolio (Parnell et al., 2004).

7.8.3 Value Hierarchy Best Practices

The following are some recommended best practices for developing value hierarchies for systems engineering trade-off analyses.

7.8.3.1 Put the Opportunity or Problem Statement at the Top of the Hierarchy Use the language of the decision-maker and key stakeholders. A clear problem statement is a very important tool to communicate the purpose of the system to the decision-maker(s), senior stakeholders, and the trade-off study team.

7.8.3.2 Use Terms from the Domain Select terms (e.g., functions, objectives, and value measures) used in the problem domain and from functional analysis in the engineering domain. This improves understanding by the users of the model.

7.8.3.3 Develop Functions, Objectives, and Value Measures from Functional Analysis, Research, and Stakeholder Analysis Functional analysis encompasses structure, behavior, and performance of products and services. Structure includes the functional decomposition. Behavioral includes the control flow and the functional interfaces, that is, data, material, and energy flows. Performance includes the timeline analysis as well as throughputs and latencies.

7.8.3.4 Define Functions and Objectives with Verbs and Objects Avoid buzzwords. This improves the understanding of the function or objective. Data, material, and energy are defined with nouns. Physical entities are also defined as nouns.

7.8.3.5 Logically Sequence the Functions (e.g., Temporally) in the Requirements Analysis Phase and the Hierarchy This provides a framework for helping decision-makers and stakeholders understand the value hierarchy and identify missing functions. In the architecture phase, functions are initially allocated to the generic physical entities of the system or service. (Software is considered a physical entity.) The instantiated physical architecture results from decisions as to the selection of specific physical entities and their spatial distribution.

7.8.3.6 Use Fundamental Objectives and Not Means Objectives in the Hierarchies Fundamental objectives are about why, what, when, where, and how well. These go in the functional/value hierarchy. Means objectives are about how. Means objectives are related to the alternatives.

7.8.3.7 Use the Value Measure Preferences in Table 7.1 Use value measures that are direct measures of the objectives and not proxy measures. If no natural measure exists, consider decomposing the objective or developing a constructed scale for attainment of the objective. Proxy measures result in more measures and increased data collection for measures that are only partially related to the objectives.

7.8.3.8 Vet the Value Hierarchy with Decision-Makers, Stakeholders, and Engineers The hierarchy needs to include business/mission objectives, stakeholders' objectives, and system technical objectives. Developing a good functional value hierarchy for a major system development is very difficult.

7.8.4 Cautions about Cost and Risk Objectives

Two commonly used objectives require special consideration: cost and risk objectives.

7.8.4.1 Cost Objective Mathematically, minimizing cost can be one of the objectives in the value hierarchy and cost can be a value measure in the value model. However, for many multiple objective trade-off studies, it is useful to treat cost separately and show the amount of value per unit cost. In our experience, this is the approach that is the most useful for decision-makers who have a budget that they might be able to increase or might have to accept a decrease.

7.8.4.2 Risk Objective Risk is a common decision-making concern, and it is tempting to add minimization of risk to the set of objectives in the value hierarchy. However, this is not a sound practice. A common example is helpful to explain why we do not recommend this approach. Suppose there are three objectives: maximize performance, minimize cost, and minimize time to complete the schedule. It may be tempting to add minimize risk as a fourth objective, but what type of risk are we minimizing and what is causing the risk? The risk could be performance risk; cost risk; schedule risk, performance and cost risk; cost and schedule risk; performance and schedule risk; or performance, cost, and schedule risk. In addition, there could be one or more uncertainties that drive these risks. In Chapter 8, we use probabilistic modeling to model the sources of risk and their impact on the value measures and the objectives. We believe this is a much more sound approach than the use of a vague risk objective in the value hierarchy. Furthermore, if utility (see Chapter 2) is used, we can model the risk preference directly.

7.9 ILLUSTRATIVE EXAMPLES

In this section, we provide military and homeland security examples of value models for trade-off analyses.

7.9.1 Military Illustrative Example

The following functional value hierarchy was developed for an illustrative trade-off analysis study. The example is also used in Chapter 9. The Army is interested in the Squad of the future. The purpose of the study is to have the future Army squad achieve overmatch against enemies in complex environments. Five functions are identified

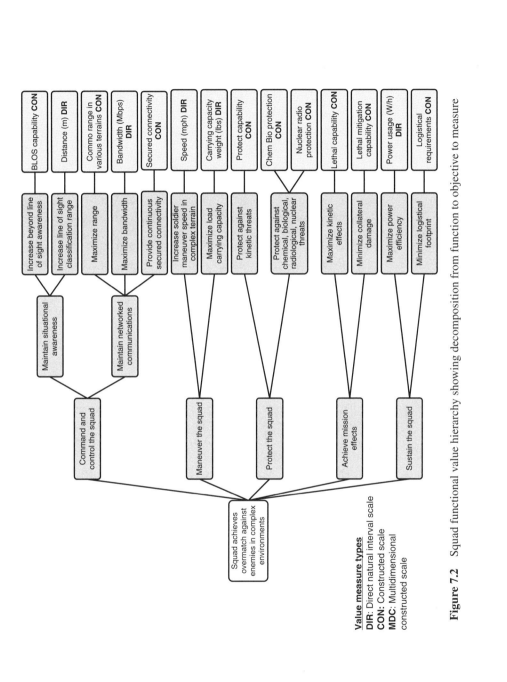

Figure 7.2 Squad functional value hierarchy showing decomposition from function to objective to measure

and presented in a logical time sequence in Figure 7.2. Two subfunctions are identified for the Command and Control Squad.

Figure 7.2 also identifies the objectives for each function or subfunction if the function is divided into subfunctions. For each objective, at least one value measure is identified. Two types of value measures are used: direct natural measures on an interval scale (e.g., speed in mph) and constructed interval scales (e.g., secured connectivity).

7.9.2 Homeland Security Illustrative Example

The Domestic Nuclear Detection Office (DNDO) of the Department of Homeland Security was created to increase the United States' ability to detect radiological and nuclear (RN) material that could be obtained and then used by terrorists. The office coordinates the Global Nuclear Detection Architecture (GNDA), an international and interagency strategy for detecting, analyzing, and reporting RN materials outside of regulatory control. In 2012, the Government Accountability Office expressed concern about the prioritization of GNDA resources as well as the documentation of GNDA improvements over time. As a result, the DNDO asked the National Research Council (NRC) for advice on how to develop performance measures and metrics to quantitatively assess the GNDA's effectiveness. The result of the NRC study was a report titled "Performance Metrics for the Global Nuclear Detection Architecture." In the report, the committee created a notional strategic planning framework for evaluating the performance of the GNDA. Using the data from the public report, multiobjective decision analysis techniques, and notional data from research, the NRC framework was expanded to a complete value model. The value model was used to demonstrate that it is possible to evaluate the potential performance of the GNDA over time and to evaluate the cost-effectiveness of potential improvements (Hilliard et al., 2015).

The vision of the study was "For U.S. citizens to live free from the fear of nuclear or radiological terrorism." The mission of the GNDA was to "Protect the nations from terrorist attacks that use radiological or nuclear materials." Table 7.2 shows the goals, objectives, and value measures that were used in the analysis. The objectives were numbered to align with the goals and the value measures were numbered to align with the objectives. Twenty-five value measures were used in the study.

7.10 SUMMARY

Decision objectives are based on shareholder and stakeholder value. Crafting objectives and value measures is a critical step in the trade-off analyst's support to the decision-maker and helps qualitatively define the value we hope to achieve with the decision. We described the differences between Value-Focused Thinking and Alternative-Focused Thinking. We recommend the use of Value-Focused Thinking for trade-off analyses. It is not easy to identify a comprehensive set of objectives and value measures for a complex decision. We describe the four techniques for identifying decision objectives: research, interviews, focus groups, and surveys. Research is

SUMMARY

Table 7.2 Value Model Structure

Goals	Objective	Measure
1. Reduce the threat	O 1.1 Deter terrorists' RN attacks by demonstrating high likelihood of failure	M 1.1 Probability of deterrence by denial (1)
	O 1.2 Identify terrorist plans for RN attacks	M 1.2 Probability of identifying plans (2)
2. Reduce the vulnerability	O 2.1 Detect RN materials out of regulatory control outside of the United States O 2.1.1 At foreign ports of departure	M 2.1.1 Probability of detection (PD) at foreign seaports (3)
	O 2.1.2 In route to the United States	M 2.1.2.1 PD at foreign airports (4) M 2.1.2.2 Percent of vessels in US coastal regions that are searched (5)
	O 2.2 Detect at US Borders O 2.2.1 Detect at US POEs	M 2.2.1.1 PD major seaports (6) M 2.2.1.2 PD at minor seaports (7) M 2.2.1.3 PD at airports (8) M 2.2.1.4 PD at vehicle entries (9) M 2.2.1.5 PD at airports (10)
	O 2.2.2 Detect between US POEs	M 2.2.2.1 PD on land between POEs (11) M 2.2.2.2 PD on coasts between seaports (12)
	O 2.3 Detect inside the US O 2.3.1 PD inside US	M 2.3.1.1 Train (13) M 2.3.1.2 Truck (14) M 2.3.1.3 Inland waterways (15) M 2.3.1.4 Primary airports (16) M 2.3.1.5 Other airports (17)
	O 2.3.2 PD around target vicinity	M 2.3.2.1 Number of major urban areas (18) M 2.3.2.2 Number of critical infrastructure (19)
3. Reduce the consequences of a successful attack	O 3.1 Divert nuclear attacks to lower consequence targets O 3.1.1 Urban areas O 3.1.2 Critical infrastructure (CI)	M 3.1.1 Percent of population covered by Urban Security Initiative (20) M 3.1.2 Percent of CI targets protected (21)
	O 3.2 Provide early warning of RN attacks	M 3.2 Detection alert times (22)

(continued overleaf)

Table 7.2 (*Continued*)

Goals	Objective	Measure
4. Reduce unintended side effects	O 4.1 Minimize impacts on privacy and civil liberties	M 4.1 Number of complaints from civil liberties groups (23)
	O 4.2 Minimize impacts on flow of commerce and the economy	M 4.2 Level of impact (24)
	O 4.3 Avoid transfer of RN risks to other nations	M 4.3 Number of countries spending money on GNDA efforts (25)

The bold items are in the NRC report. (RN = radiological and nuclear, PD = probability of detection, POE = point of entry/embarkation, CI = Critical infrastructure).

essential to understand the problem domain and to determine applicability of different decision analysis modeling techniques. Interviews are especially useful for senior leaders. Focus groups work well for stakeholder representatives. Surveys are especially useful to obtain public opinion. We note that the financial or cost objective is almost always an important objective. Next, we describe the important role of hierarchies in structuring objectives and providing a format that is easy for decision-makers and stakeholders to review and provide feedback. We present objectives and functional hierarchies. We recommend functional value hierarchies for complex system decisions. We define four standard techniques for structuring objectives: Platinum (senior leader interviews), Gold (documents), Silver (meetings with stakeholder representatives), and Combined (using all three). The combined standard is the most common. Two examples are provided illustrating approaches to identifying objectives and value measures.

7.11 KEY TERMS

Combined Standard: An objective identification process that uses the Platinum (senior leader interviews), Gold (documents), and Silver (meetings with stakeholder representatives) standards.

Focus Groups: Decision objectives are obtained through a facilitated discussion and interaction of 6 to 12 group members.

Fundamental Objectives: The most basic objectives of the system. For example, reduce driving deaths.

Functional Value Hierarchy: A hierarchy displaying the decision purpose, functions to be met by the system, fundamental objectives for each function, and value measure(s) for each objective that will be used to evaluate the alternatives.

Gold Standard: An objective identification process that uses an approved document.

Interviews: A meeting with a leader or subject matter expert using a series of planned questions and follow-on discussion to identify the decision objectives.

Life Cycle Cost: The total cost of an alternative across all stages of the alternative life cycle. Sunk costs should not be included when evaluating alternatives.

Means Objective: Describes how we can achieve a fundamental objective. For example, "increase use of safety belts" is a means objective for "reduce driving deaths."

Net Present Value: The calculation of the present worth from a sequence of cash flows using a discount rate.

Objective Identification: The process of identifying potential decision objectives.

Objective Structuring: The process of organizing decision objectives and value measures into a hierarchical structure.

Objectives Hierarchy: A hierarchy displaying the decision purpose, the decision objectives, and the value measures that will be used to evaluate the alternatives.

Platinum Standard: An objective identification process that uses interviews with senior leaders.

Risk: The likelihood of an undesired outcome and the consequence of the undesired outcome.

Shareholder Value: Value is defined in financial terms understandable to shareholders.

Stakeholder Value: Value is defined by meeting the objectives of the stakeholders.

Silver Standard: An objective identification process that uses meetings with stakeholder representatives.

Survey: A prepared list of questions that is sent to stakeholders to help identify functions, objectives, and/or value measures.

7.12 EXERCISES

7.1. Distinguish between Value-Focused Thinking (VFT) and Alternative-Focused Thinking. Why do the authors recommend the use of VFT in systems engineering trade-off analyses?

7.2. Shareholder and stakeholder values. Consider an organization that you have worked in. If you have never worked, consider the university you attend.
   (a) List the shareholder values, if appropriate.
   (b) List the stakeholder values.

7.3. Fundamental versus means objectives.
   (a) Distinguish between the two types of objectives.
   (b) Provide an example of each type of objective.
   (c) Which type of objective should we use in value hierarchies? Why?

7.4. The chapter presented four techniques for stakeholder analysis to develop a value hierarchy: research, interviews, focus groups, and surveys.
   (a) Discuss the advantages and disadvantages of each technique.

(b) Identify the types of information you would try to obtain using each technique.

(c) Identify when you would use each technique.

7.5. Categorize the value measures below based on the four categories in Table 7.1.

Objective	Value Measure(s)	Category
Maximize fuel efficiency	Miles per gallon	
Minimize impact on environment	Miles per gallon	
Maximize safety in crash	National Highway Traffic Safety Administration (NHTSA) 5-star crash test rating	
Maximize automobile safety	National Highway Traffic Safety Administration (NHTSA) 5-star crash test rating	
Maximize vehicle safety	Number of seat belts Vehicle stopping distance Depth of tire tread remaining	

7.6. Consider the Deepwater Horizon drilling rig accident in April 2010 that caused the largest oil spill in US waters.

(a) Identify the stakeholders involved in drilling oil in the Gulf of Mexico.

(b) Identify the shareholder and stakeholder values that became evident for British Petroleum and the other companies involved in the accident.

(c) Develop a functional value hierarchy for a trade-off analysis for oil rig concepts. Identify business, stakeholder, and system objectives.

7.7. Identify a systems engineering trade-off analysis that you will be performing.

(a) What life cycle phase is the system in?

(b) Write a one sentence description of the decision opportunity.

(c) Will you use an objectives hierarchy or a functional value hierarchy? Why?

(d) Which of Keeney's 10 questions would be the most useful?

(e) Which of the four standards would you use to develop the value hierarchy?

(f) Develop a hierarchy for your trade-off study.

REFERENCES

Bond, S., Carlson, D., and Keeney, R.L. (2008) Generating objectives: can decision makers articulate what they want? *Management Science*, **54** (1), 56–70.

Charreaux, G. and Desbrieres, P. (2001) Corporate governance: stakeholder value versus shareholder value. *Journal of Management and Governance*, **5** (7), 107–128.

Ewing, P., Tarantino, W., and Parnell, G. (2006) Use of decision analysis is the Army Base Realignment and Closure (BRAC) 2005 military value analysis. *Decision Analysis*, **3** (1), 33–49.

Hilliard, H., Parnell, G., and Pohl, E. (2015). Evaluating the effectiveness of the global nuclear detection architecture using multiobjective decision analysis. *Systems Engineering*, **18** (5), 441–452.

Keeney, R.L. (1992) *Value-Focused Thinking: A Path to Creative Decisionmaking*, Harvard University Press, Cambridge, MA.

Keeney, R.L. (1994) Creativity in decision making with value-focused thinking. *Sloan Management Review*, **35** (4), 33–41.

Kirkwood, C. (1997) *Strategic Decision Making with Multiobjective Decision Analysis with Spreadsheets*, Duxbury Press, Belmont, CA.

Parnell, G.S., Bresnick, T.A., Tani, S.N., and Johnson, E.R. (2013) *Handbook of Decision Analysis*, Wiley & Sons.

Parnell, G.S., Driscoll, P.J., and Henderson, D.L. (eds) (2011) *Decision Making for Systems Engineering and Management*, 2nd edn, Wiley & Sons.

Parnell, G.S., Metzger, R.E., Merrick, J., and Eilers, R. (2001) Multiobjective decision analysis of theater missile defence architectures. *Systems Engineering*, **4** (1), 24–34.

Parnell, G., Burk, R., Schulman, A. *et al.* (2004) Air Force Research Laboratory Space Technology value model: creating capabilities for future customers. *Military Operations Research*, **9** (1), 5–17.

Parnell, G., Conley, H., Jackson, J. *et al.* (1998) Foundations 2025: a framework for evaluating air and space forces. *Management Science*, **44** (10), 1336–1350.

8

DEVELOPING AND EVALUATING ALTERNATIVES

C. ROBERT KENLEY
School of Industrial Engineering, Purdue University, West Lafayette, IN, USA

CLIFFORD WHITCOMB
Systems Engineering Department, Naval Postgraduate School in Monterey, Monterey, CA, USA

GREGORY S. PARNELL
Department of Industrial Engineering, University of Arkansas, Fayetteville, AR, USA

> Truly successful decision making relies on a balance between deliberate and instinctive thinking.
> **Malcolm Gladwell**, *Blink: The Power of Thinking Without Thinking, 2005*

8.1 INTRODUCTION

As discussed in Chapter 1, value identification requires that we create feasible alternatives. We also need techniques for evaluating alternatives that measure the value of each alternative and identify the alternatives that provide the most potential value, as presented in Chapter 2. This chapter reviews techniques for creating and evaluating alternatives, provides references for the techniques, and provides an assessment of the techniques. When using the methods in this chapter, always keep in mind the decision traps discussed in Section 6.3.

Trade-off Analytics: Creating and Exploring the System Tradespace, First Edition. Edited by Gregory S. Parnell.
© 2017 John Wiley & Sons, Inc. Published 2017 by John Wiley & Sons, Inc.
Companion website: www.wiley.com/go/Parnell/Trade-off_Analytics

This chapter is organized as follows: Section 8.2 presents an overview of decision-making, creativity, and teams, and Section 8.3 presents techniques for creating alternatives. These sections describe divergent and convergent phases for alternative development, and when techniques are embedded in what is customarily proposed as a complete process for creating and evaluating alternatives, it identifies and describes the embedded alternative development techniques. Section 8.4 assesses the alternative creation techniques with respect to when they are best applied during the life cycle and their contribution to the divergent and convergent phases of alternative development. Section 8.5 presents alternative evaluation techniques, including the techniques embedded in complete processes for alternative development and evaluation. Section 8.6 assesses the alternative evaluation techniques with respect to best practices for making good decisions (Howard & Abbas, 2016) along the dimensions of information, preferences, and logic. The chapter concludes with key terms, exercises, and references.

8.2 OVERVIEW OF DECISION-MAKING, CREATIVITY, AND TEAMS

The section provides an overview of three important areas that are relevant to developing alternatives.

First, decision-making approaches vary according to the decision opportunity. It is important to understand the characteristics of the decision opportunity and approaches that are applicable to the situation before committing to complete a trade study when perhaps another approach may be more appropriate.

Second, research in engineering design has identified the cognitive methods that are the most effective for creating alternatives. The research results are not surprising to those who have experience with approaches that are employed by highly skilled system engineers. We believe it is critically important to create innovative alternatives before taking the deep dive into decision calculus that provides analytic insights into the trade-offs between the alternatives.

Finally, due to multiple stakeholder objectives, uncertainties about competition and adversaries, and the complexity of systems, we need expertise in many areas to perform a quality trade-off analysis – trade studies are a team sport. Research in the psychology of teams has identified personality-based suggestions for improving the performance of individuals in a team environment, the roles that must be balanced when building a team, and the tasks that are more appropriate for individuals versus being completed as a team exercise.

8.2.1 Approaches to Decision-Making

Table 8.1 summarizes three modes of decision-making from Watson & Buede (1987, 123–124) that characterize different approaches to making decisions that are appropriate for different decision opportunities. The "Choosing" mode employs decision analysis methods and is the principal approach used for performing trade-off analyses. The "Allocating" mode captures decision-making approaches that apply optimization techniques that distribute scarce resource(s) among competing projects. This mode is typical for operations research and multidisciplinary design optimization. Also included in the allocating mode is the determination of constraints

Table 8.1 The Modes for Making Decisions

Mode	Definition	Distinctive Traits of the Decision Opportunity	Common Methods
Choosing	Selecting one alternative from a list	• Finite number (usually 5–15) of discrete alternatives that may need to be developed • Finite number of sequential decisions • Constraints on resources not treated explicitly • Uncertainties important • Single decision-maker • Value trade-offs not well-established	• Uncertainty modeling with influence diagrams, decision trees, or Monte Carlo simulation • Value modeling using single or multiple objective value functions
Allocating	Distributing scarce resource(s) among competing projects	• Alternatives are discrete or continuous variables • Sequential decisions under certain assumptions • Constraints on resources treated explicitly • Constraints on based on physics and laws of probability • Uncertainties with certain assumptions • Single decision-maker • Known parameters (e.g., cost, weight, effort, benefit–cost, profit)	• Optimization methods such as linear, nonlinear, dynamic, and stochastic programming using an objective function • Trade-off analysis based on physics and probability to determine the constraints on the decisions
Negotiating	Defining an agreement with one or more actors	• Discrete or continuous alternatives • Possible sequential decisions • Constraints on resources not treated explicitly • Multiple interacting decision-makers • Often highly dynamic, unstructured environment	• Game theory in structured cases • Descriptive, empirical, and heuristic methods possibly informed by game theory, decision analysis, and optimization

based on physics (Slepian 1963) and on the laws of probability (Kenley & Coffman, 1998). The "Negotiating" mode captures decision-making approaches that apply game theory and agent-based modeling to understand the dynamics of multiple interacting decision-makers. Negotiating includes mechanism design, which is the design of the rules that provide for a situation that is efficient, effective, and fair for all parties.

8.2.2 Cognitive Methods for Creating Alternatives

The quality of any decision-making, no matter what mode is employed, is dependent on generating a range of decision alternatives that provide multiple means for achieving the objectives of the decision-makers and stakeholders. Jones (1992) indicates that the range and structure of alternatives to be searched at the system level change form based on assumptions made and the willingness of others to adopt the design solutions. The decision alternatives are characterized by variables that parameterize the discrete and continuous choices that are available and by the sequential ordering of the choices, which may include choices to gather additional information about contextual or environmental factors. This additional information may result in an update to the assumptions, and it may increase the willingness of decision-makers to commit to taking action.

Results from research on design methods highlight the importance of two key methods for effectively generating alternatives: visualization and organizing information according to functionality. Cross (2000) states that the ability to design depends on both internal (in the mind's eye) and external (on the whiteboard) visualizations such as drawings and that designers employ these visualizations to structure the design space and generate alternatives. Cross also notes that visualization enables designers to work simultaneously with the overall concept and lower-level details that have a critical influence on the overall design. Ullman (2010) explains that experienced designers use functional groupings as an indexing mechanism to generate information about design options and that the information typically is a graphical representation of the structure of the options rather than the values of specific physical parameters. Fortunately, for trade studies performed within the context of an overall systems engineering approach, visual and functional representations of the problem space and the design space are commonly used, and a trade study team should make every effort to take advantage of the visual and functional modeling artifacts that the system design team has developed.

8.2.3 Key Concepts for Building and Operating Teams

Complex system designs require a large number of designers and systems engineers. These individuals usually work in teams within and across organizational boundaries.

Extensive research (McGrath, 1984) indicates that individuals working separately generate many more ideas than do groups. Two-part group methods such as the nominal group technique that allow for initial generation of concepts by individuals and follow-up in group sessions to enhance concepts were found to be very effective in taking advantage of the concept generation capabilities of individuals. These two-part methods provide the proper balance between generating alternatives and achieving the group buy-in needed to avoid down-selecting alternatives prematurely. In addition, decision-makers who participate in the group sessions should be more willing to commit to recommendations from the decision-making process than if they were left out of the process altogether.

Belbin (2010) describes nine team roles for team members by identifying the role name, the characteristics of individuals in the role, and the key activities the individual performs as shown in Table 8.2.

Belbin indicates that balancing these roles among team members is crucial to achieving successful outcomes and that these team roles are distinct from the functional roles that team members fulfill in accordance with their formal job descriptions. Belbin provides guidance for building and operating teams to ensure that the team roles effectively balanced.

Ullman (2010) describes five problem-solving dimensions of individuals that are relevant to their performance on teams and provides suggestions for the most effective approaches to contributing to team success along these dimensions:

1. Problem-solving style (extroverted vs introverted)
2. Information management style (preference to work with facts vs with possibilities)
3. Information language (verbal vs visual)
4. Deliberation style (objective vs subjective)
5. Decision closure style (decisive vs flexible).

Ullman provides suggestions on maximizing the productivity of individuals in a team environment organized along these dimensions that are summarized in Table 8.3.

Table 8.2 Belbin's Nine Roles for Team Members

Role	Characteristics	Key Activities
Plant	Creative, imaginative, unorthodox	Solves difficult problems
Resource investigator	Extrovert, enthusiastic, communicative	Explores opportunities. Develops contacts
Coordinator	Mature, confident, a good chairperson	Clarifies goals, promotes decision-making, and delegates well
Shaper	Challenging, dynamic, thrives on pressure	Has the drive and courage to overcome obstacles
Monitor-Evaluator	Sober, strategic, discerning	Sees all options. Judges accurately
Team Worker	Cooperative, mild, perceptive, diplomatic	Listens, builds, averts friction, and calms the waters
Implementer	Disciplined, reliable, conservative, efficient	Turns ideas into practical actions
Completer-Finisher	Painstaking, conscientious, anxious	Searches out errors and omissions. Delivers on time
Specialist	Single-minded, self-starting, dedicated	Provides knowledge and skills in rare supply

Table 8.3 Ullman's Suggestions for Increasing Team Performance

Problem-Solving Dimension	Suggestions for Increasing Team Performance
Problem-solving style	Externals need to • Allow others time to think • Practice listening to the ideas and suggestions of others • Recap what has been said to make sure they have been heard • Realize that silence does not always mean consent Internals need to • Share more than their final response • Have an equal say in selecting ideas and plans • Develop nonverbal signals understood by other team members that indicate assent or dissent • Restate their ideas • Push externals for more clarity and meaning
Information management style	Fact-oriented members need to • Fantasize, think wildly, and allow others to think wildly • Allow the team to set goals rather than dive right into the problem and tackle the details Possibility-oriented team members need to • Deal with details • Be specific and avoid generalities • Stick to the issues
Information language	Team leaders need to • Identify information that needs to be communicated, regardless of language • Identify differences in team members' mental models • Encourage extra effort by both visual and verbal people to communicate clearly with other members • Try a diagram or picture if words and equations are not working • Try words and equations if a picture is not working
Deliberation style	Objective team members need to • Pay attention to feelings of others • Understand that how the team functions is as important as what is accomplished • Understand that not everyone likes to discuss a topic merely for the sake of argument • Express how they feel about the outcome once in a while

(*continued overleaf*)

Table 8.3 (*Continued*)

Problem-Solving Dimension	Suggestions for Increasing Team Performance
	Subjective team members need to
	• Realize that it is acceptable to disagree and argue
	• Realize that not every resolved issue will satisfy everyone even though consensus may have been reached
	• Understand that discussions about ideas are not personal attacks
Decision closure style	Flexible team members need to
	• Receive plans in advance so that they can think on their own time
	• Receive acknowledgement of their contribution as a step toward moving to closure
	• Receive clear decision deadlines in advance
	• Provide feedback to the team
	• Settle on something and live with it for a period of time
	Decisive team members need to
	• Understand that most problems need to be subdivided into smaller problems
	• Organize the data collection and review process
	• Utilize techniques, such as brainstorming, that suppress judgment
	• Understand that they are not always right

8.3 ALTERNATIVE DEVELOPMENT TECHNIQUES

Alternative development should be a two-phase process. The first phase is divergent thinking that results in creating many ideas. The second phase is convergent thinking that requires organized, analytic thinking to identify a smaller number of well-crafted, feasible alternatives for the initial evaluation. Once the initial evaluation is complete, the process should be repeated to refine and improve the alternatives (Figure 8.1).

8.3.1 Structured Creativity Methods

Creativity is a critical aspect for successfully executing the divergent phase when engineering new products and systems. It is "the ability to transcend traditional ideas, rules, patterns, relationships, or the like, and to create meaningful new ideas, forms, methods, interpretations, and so on; originality, progressiveness, or imagination" (Dictionary.com). Structured methods provide useful approaches for an individual or a system development team to think about problems and to create solutions. The concepts for building and operating teams described in Section 8.2.3 should be applied when using structured methods to create an environment that generates many

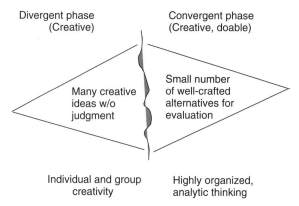

Figure 8.1 Two phases of alternative development (Source: Parnell et al. 2013. Reproduced with permission of John Wiley & Sons)

alternative viewpoints and does not yield to pressures toward uniformity, which lead to seeking consensus and eliminating alternatives too early in the process.

Several representative methods are outlined, although many others are available. The methods covered are brainstorming, Reversal, SCAMPER, Six Thinking Hats, Gallery, and Force Field.

1. Brainstorming. This method can be either an individual exercise or a group exercise. In groups, the objective of this method is to generate within a given time limit as many ideas as possible for solving a problem. Having ground rules and posting them where all participants can see them is useful to keep everyone aware of the rules before and during the process. All members of the group should participate and, if someone has not participated, someone in the group should ask that person to give an idea. Ideas are recorded with no criticism or evaluation of ideas from the other participants. Encourage participants to augment and enhance others' ideas. Do not let single participants do too much explaining or digressing or giving all the idea inputs. Enthusiastic participation by everyone keeps the process moving. If the process reaches a lull, then try to get new ideas by kick-starting the conversation. When the time limit is reached, or the ideas stop coming, stop the process. Make sure to record the ideas, either on a computer that can be displayed on a video screen or using a white board or flip chart, so that everyone can see them, consider what has been offered, and use them to generate new ideas. Once documented, the ideas can then be used in various ways to see how they might apply to the problem, with constructive feedback and making sure that the process moves to convergence on one or several possible best ways forward.

2. Gallery. This method is implemented by having team members write down ideas about the system problem at hand, either on computer or flip chart paper or on sticky notes, which are then placed on a sheet of flip chart paper or poster board – or even individually partitioned parts of a white board. Set a time limit for participants to generate their ideas. Post the resulting posters or flip chart papers on a wall, and have the group of participants walk around the room individually or in informal groups to view the posters as if they were in a gallery. Having participants place their own notes on desired posters will allow everyone to add their comments. The person who created each poster can then present their summary to the group. This method has the advantage of allowing those who do not like to present their ideas in a group setting to get their ideas out in front of everyone through a more private idea generation process. The group gets to review everything and has a chance to comment before the group discussions.
3. Reversal. This method has participants take contrary or opposite positions by responding to questions about what you would not like to have happen to your product or system. By taking the opposite position, ways to prevent this outcome may become clear, so you can make sure that the system does not let these things happen. The method has the advantage of allowing participants to state openly some of the ways a system can fail in a noncritical environment so they can be considered, especially during the system conceptualization and design stages.
4. SCAMPER. Bob Eberle (1996) created this method that uses a set of techniques that define different ways to approach thinking about a problem. The use of the various techniques forces thinkers to apply methods, including ones that they might not have considered for application to the problem at hand. A key point is to force a response from each perspective and, thus, provide many possibilities to expand thinking. Some responses may end up not being feasible and should be eliminated. For SCAMPER, the seven ways to think about a system problem are as follows:
   - Substitute – Remove something from a solution or alternative, and substitute it with something else.
   - Combine – Put together two or more system aspects in ways that yield a new solution or possibility.
   - Adapt – Change something about the system, so it does something it could not before the change.
   - Modify – For the attributes of the system, make changes, even arbitrarily, and study the results or possibilities.
   - Put to Other Use – Alter the intention of the system. Challenge assumptions about the system, and suggest new purposes.

- Eliminate – Arbitrarily take out some or all parts of the system or solution to simplify the solution or to use as a basis to find a way to proceed without what was eliminated.
- Reverse – Change the orientation (turn it upside down, invert it, etc.) or direction (go backward, against typical direction, etc.) of the system or components.

5. Six Thinking Hats. This method (De Bono, 1985) forces participants to take on different perspectives about a problem based on the hat they wear, substituting that perspective for one that they might have used themselves:
   - The White Hat calls for information known or needed based on facts.
   - The Yellow Hat symbolizes brightness and optimism and explores the positives and probes for value and benefit.
   - The Black Hat uses judgment to articulate why something may not work.
   - The Red Hat signifies feelings, hunches, and intuition and calls for expressing emotions and feelings and sharing fears, likes, dislikes, loves, and hates.
   - The Green Hat focuses on creativity to generate possibilities, alternatives, and new ideas and allows for expressing new concepts and new perceptions.
   - The Blue Hat manages the thinking process and ensures that the overall process guidelines are observed.

6. Force Field. This method was developed by Lewin (1943) to consider all the forces that come to play on a system. The process begins with a brief problem statement, and then participants add all the forces that might either help or hinder the process or system from going forward toward a solution. Each idea is categorized as either "help" or "hinder." Arrows are shaped around each idea, with the direction indicating help or hinder and size indicating the relative magnitude of the force. The overall process outcome is to identify the forces and combination of forces that exist and to then strengthen the helping (positive) ones and to weaken the hindering (negative) ones.

8.3.2 Morphological Box

Morphology has been used in many fields of science, such as anatomy, geology, botany, and biology, to designate research on structural interrelations. Fritz Zwicky generalized and systematized morphological research and applied it to abstract structural interrelations among phenomena, concepts, and ideas for the purpose of discovery, invention, research, and construction of approaches to deal with all situations in life more effectively (Zwicky, 1969, 273). Zwicky's principal tool is the morphological box, chart, or table that is constructed using different dimensions that define a multivariate decision space. A morphological box provides a structure for divergent thinking that generates multiple decision alternatives and for first-order convergent thinking that eliminates unreasonable and unachievable alternatives.

ALTERNATIVE DEVELOPMENT TECHNIQUES

There are three variants of the morphological box described in this section: the functional allocation box, the physical architecture definition box, and the strategy table.

8.3.2.1 The Functional Allocation Box Ullman (2010) describes a technique for designing with function that constructs a morphological box to create the alternatives by defining physical means for accomplishing the system's functions as shown in Figure 8.2. The alternatives are all possible combinations of the means for all of the functions. Assuming that the last three functions listed in Figure 8.2 are constrained to be achieved by the same means, there are $4 \times 1 \times 4 \times 2 \times 5 \times 4 \times 1 \times 1 = 640$ alternatives that can be created from this morphological box. Of course, not all 640 combinations are reasonable or even physically achievable. For example, if a cantilever is used to transmit the vertical force to the body, then it is reasonable to use it to transmit the horizontal force rather than adding a truss to transmit the horizontal force. In addition, the means to achieve each function should be carefully evaluated for completeness before finalizing the list of options. For example, a rearward-facing fork end that typically is used on track bicycles could be added to the list of means for transmitting vertical and horizontal forces to the suspension system.

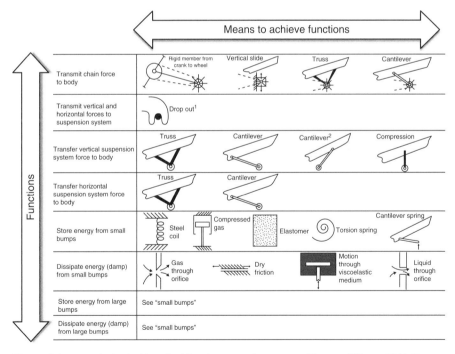

Figure 8.2 Morphological box for bicycle suspension system (Source: Ullman 2010. Reproduced with permission of McGraw-Hill Education).

System Functions		Architectures for Unguided Interceptors				Architectures for Guided Interceptors			
		U1	U2	U3	U4	G1	G2	G3	G4
	Detect, Acquire & Declare	Passive Cuer	Passive Cuer / Coarse Tracker	Passive Cuer	Active Cuer / Tracker	Passive Cuer	Passive Cuer / Coarse Tracker	Passive Cuer	Active Cuer / Tracker
	Classify			Passive or Active Coarse Tracker				Passive or Active Coarse Tracker	
	Coarse Track	Active Tracker				Active Tracker			
	Initial Slew / Tube Selection	Launcher	Launcher	Launcher	Launcher	Launcher	Launcher	Launcher	Launcher
	Fine Track	Active Tracker	Active Fine Tracker	Active Fine Tracker	Active Cuer / Tracker	Active Tracker	Active Fine Tracker	Active Fine Tracker	Active Cuer / Tracker
	Final Slew / Tube Selection & Fire Control	Launcher	Launcher	Launcher	Launcher	Launcher	Launcher	Launcher	Launcher
	In-Flight Track	None	None	None	None	Active Tracker	Active Fine Tracker	Active Fine Tracker	Active Cuer / Tracker
	In-Flight Guidance					Guided Interceptor	Guided Interceptor	Guided Interceptor	Guided Interceptor
	Terminal Track	Unguided Interceptor	Unguided Interceptor	Unguided Interceptor	Unguided Interceptor	Active Tracker	Active Fine Tracker	Active Fine Tracker	Active Cuer / Tracker
	Terminal Guidance & Fuze					Guided Interceptor	Guided Interceptor	Guided Interceptor	Guided Interceptor
	Warhead Effect								

Figure 8.3 Allocation of functions to physical components for interceptor system architecture alternatives (Adapted with permission of Salvatore F. (2008). The Value of Architecture. NDIA 11th Annual Systems Engineering Conference. San Diego, US-CA)

In the context of a systems engineering process, the selection of means to achieve each function is equivalent to the functional-to-physical allocation that defines candidate allocated architectures. Figure 8.3 shows this quite explicitly; it was developed using a morphological box and presents a feasible set of alternatives by listing the physical means to achieve each of the system functions for each of the alternative architectures (Salvatore, 2008).

8.3.2.2 The Physical Architecture Definition Box Buede (2009) employs morphological boxes to define alternatives for physical architectures. As an example, Buede considers a physical device such as a hammer, which is comprised of a handle and a head, which can be further subdivided into the face, neck, cheek, and claw. For developing the morphological box, we will consider a generic physical architecture for a hammer that is defined by the parameters of the hammer components identified in Figure 8.4. The morphological box for the hammer in Figure 8.5 defines the parameters as handle length, handle material, face size surface, head weight, and claw curvature and is used to generate different alternatives by instantiating the parameters through selection of one of the allowable settings listed for each of parameters. The hammer depicted in Figure 8.4 has a 22-in. wood handle, a 1.25-in. grooved face, a 20-ounce head, and 60° claw curvature. A different instantiation constitutes an 8-in. graphite/rubber handle, a 1-in. flat face, a 12-ounce head, and straight claw.

8.3.2.3 The Strategy Table Howard & Abbas (2016) use the term "strategy table" for their version of a morphological box that was introduced to the decision analysis community by Howard (1988). An example of a strategy table that was used by General Motors in 1988 to define alternative approaches for the fifth-generation Corvette is shown in Figure 8.6. Note that the parameters that are varied in a strategy table are the key decisions, and they are not necessarily structured by systems engineering concepts such as allocating components of the functional architecture to physical architectures (Salvatore, 2008; Ullman, 2010) or instantiating the parameters of a

ALTERNATIVE DEVELOPMENT TECHNIQUES

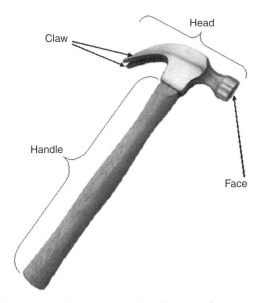

Figure 8.4 Generic physical architecture of a hammer

	Parameters that Define Components of the Generic Physical Architecture				
	Handle Length	Handle Material	Face Size-Surface	Head Weight	Claw Curvature
Possible Parameter Settings	8 in.	Fiberglass/ rubber grip	1 in. – grooved	12 oz.	Straight steel claw
	22 in.	Graphite/ rubber grip	1 in. – flat	16 oz.	60° claw
		Steel/ rubber grip	1.25 in. – grooved	20 oz.	
		I-beam	1.25 inch - flat	24 oz.	
		Wood			

Figure 8.5 Morphological box used to instantiate architectures (Source: Buede 2009. Reproduced with permission of John Wiley & Sons)

generic physical architecture (Buede, 2009). For the Corvette example, the decision parameters are a system performance requirement (brake horsepower), a physical architecture design constraint (extent of changes), and a multidimensional generic physical architecture parameter (door sill location, foot room, and trunk space). An important feature of the strategy table is to provide a descriptive title for each alternative that captures the theme that highlights their costs and benefits. This may be useful

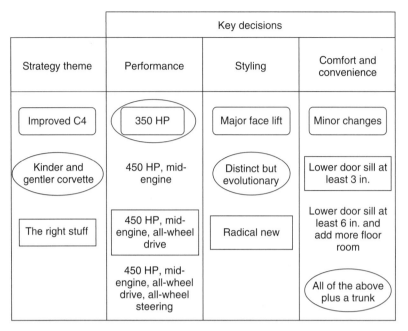

Figure 8.6 Strategy table for fifth-generation Corvette (Adapted from Barrager 2001)

for helping participants recall the alternative under consideration; however, care must be taken that the theme titles are adequately descriptive and at the same time do not lead to premature elimination of alternatives.

8.3.3 Pugh Method for Alternative Generation

The Pugh method of controlled convergence (Pugh, 1991) uses the nominal group technique to allow individuals to generate ideas that are traceable to product expectations or the voice of the customer prior to convening group meetings and to transition these individually generated alternatives to group-owned alternatives. This transition from individually owned concepts to group-owned concepts reduces the political negotiation that can dominate team interactions, and it achieves group buy-in for the alternative set under consideration as well as allowing creative interaction to take advantage of the wide range of experience of the team. The method is proposed as a complete method for selecting a single concept that has successive divergent and convergent phases. It addresses three principles of value-focused thinking (Keeney, 1992): it incorporates information from opportunity investigation to identify the values of an organization's shareholders and stakeholders, it uses values to generate better alternatives, and it uses values to evaluate alternatives. As we will discuss later, using the Pugh method for evaluating alternatives has its mathematical faults; however, when properly applied, it avoids (explicit) numerical scoring and instead focuses on generating alternative concepts that improve the richness of the set of alternatives.

Clausing (1994) lays out a 10-step process for the Pugh method shown in Table 8.4 that avoids (explicit) numerical scoring and instead focuses on group buy-in and generating new alternatives. If this process is followed, the group can retain the alternatives that are generated and perform analysis that explicitly models uncertainty and the value that stakeholders derive from achieving various level of satisfaction of the decision criteria.

8.3.4 TRIZ for Alternative Development

TRIZ is an acronym for the Russian phrase теория решения изобретательских задач, which transliterates as *teoriya resheniya izobretatelskikh zadatch* and is translated into English as the theory of inventive problem solving (TIPS). Altshuller (1998) developed this problem-solving approach based on his analysis of global patent literature that identified means to classify problems and relate the abstract problem classification to innovative solutions in patent literature. TRIZ can be used as both an alternative generation technique and an alternative evaluation technique (see Section 8.5.4). The TRIZ method has been applied to improvements in existing technologies and technological forecasting as a problem-solving method in product engineering (Clausing & Fey, 2004; Fey & Rivin, 2005).

Shanmugaraja (2012) describes the framework or process used by Altshuller as Define–Model–Abstract–Solve–Implement (DMASI). The first step of the DMASI framework defines the problem by a set of functional requirements (FRs). It then models the problem by defining a set of resources that are components of the physical architecture and the environment, which play a role in achieving the FRs or detract from achieving the FRs. Hipple (2012) lists the following types of resources used for this step:

1. Substances and materials
2. Time
3. Space
4. Fields (mechanical, thermal, acoustic, chemical, electronic, electromagnetic, and optical) and field/functional conversions
5. Information
6. People and their skills.

Abstracting the solution applies substance-field (Su-Field) modeling to the each of the FRs to identify which solutions among a set of 76 standard solutions, or high-level concepts, can be applied to meet each requirement. Silverstein et al. (2007) lists the

Table 8.4 Ten-Step Process for Controlled Convergence

Step	Activity	Description
1	Choose criteria	Criteria chosen by group and derived from product expectations or voice of the customer, for example, usability, reliability, cost, and operational life. Avoid engineering requirements that are test- or inspection-oriented and typically do not impact concept selection
2	Form the matrix	Rows are criteria. Columns are alternative concepts brought to the group meeting by individuals or distinct groups. Quite often, this is a large sheet of paper on a wall, for example, 2 m high by 5 m long
3	Clarify the concepts	Achieve common understanding and move from individual ownership of design concepts to group ownership Different points of view enhance each design concept Increased group ownership reduces defensiveness
4	Choose the datum concept	Group chooses reference concept to compare against other concepts. Important to choose one of better concepts: progress toward convergence and increased team insight
5	Run the matrix	Three-level rating scale for each criterion: better (+), worse (−), or same (S) relative to datum. Do not add more levels: it reduces insights gained that result from looking across the alternatives instead, for example, "maybe we can combine features from Concept 1 with Concept 3 to produce a better concept"
6	Evaluate the ratings	For each concept, add up number of each rating (+, −, S) Three separate sums are not combined: combining hides positive and negative features of each design that are to be evaluated
7	Attack the negatives and enhance the positives	Team attacks negatives, especially for most promising concepts Often, a concept can be easily changed to overcome a negative Team reinforces positives, by applying strong positive features to other concepts Can lead to hybrid concepts to be added to matrix before the next evaluation Can drop worst concepts, especially after they have been studied for positive features that might be incorporated to other concepts

(continued overleaf)

ALTERNATIVE DEVELOPMENT TECHNIQUES

Table 8.4 (*Continued*)

Step	Activity	Description
8	Select new datum and rerun the matrix	Next datum can be concept with "best" rating or a concept that was formed by enhancing or hybridizing an original concept Intent is not to verify that datum is the "best" concept Intent is to gain additional insights for further creative work New datum gives different perspective and reveals any distinctions that team was tempted to capture by using more than three levels in first run of matrix
9	Plan further work	Gathering more information, conducting analyses, performing experiments, and recruiting help, especially from people who can provide support in critical areas where the team has little or no expertise
10	Iterate to arrive at the winning concept	Team returns armed with very relevant information not available in the first session Team members may bring in new concepts or hybrids or extensions Matrix is run again and team continues to iterate until it has converged to the consensus-dominant concept Opportunity to involve customers at this stage

standard solutions organized into the following categories (with the total number in each category shown in parentheses):

1. Improving the System with Little or No Change (13)
2. Improving the System by Changing the System (23)
3. System Transitions (6)
4. Detection and Measurement (17)
5. Strategies for Simplification and Improvement (17).

Abstracting then proceeds with evaluation of technical and physical contradictions. Altshuller developed a set of technical contradictions defined by pairs from a list of 39 problem parameters that he discovered from his analysis, and he provides a matrix that maps each possible pair of contradictions to a subset of no more than 4 principles from a set of 40 inventive principles; these principles are suggested as high-level design concepts. Altshuller also developed four classes of physical contradictions that Hipple (2012) describes as:

1. Separation in space
2. Separation in time

3. Separation between parts and whole
4. Separation upon condition where a physical property of a system changes in response to an external condition.

Altshuller provides a matrix mapping from the four physical contradictions to identify a subset of no more than 7 principles from the set of 40 inventive principles that are suggested as high-level design concepts.

After completing the abstracting, a narrowly focused description of a set of alternative concepts for solving the problem at hand is available. Developing specific alternatives from the TRIZ-provided generic solution alternatives does require the ability to reason using analogy and experience. For example, the first inventive principles is segmentation, and Table 8.5 shows three generic solutions for the first inventive principle from which specific alternatives can be created (Silverstein et al., 2007). The original inventive principles were developed based on patented design solutions for

Table 8.5 Example of Generic Solutions and Specific Alternatives Using the First TRIZ Inventive Principle

Generic Solutions	Specific Alternatives
Divide an object into independent parts	Replace mainframe computers by personal computers Replace a large truck by a truck and a trailer Use a work breakdown structure for a large project
Make an object easy to disassemble	Modular furniture Quick-disconnect joints in plumbing
Increase the degree of fragmentation or segmentation	Replace solid shades with Venetian blinds Use powdered welding metal instead of foil or rod to get better penetration of the joint

Table 8.6 First TRIZ Inventive Principle Interpreted for Marketing, Sales, and Advertising

Generic Solutions	Specific Alternatives
Divide an object into independent parts	Market segmentation: clustering prospective buyers into groups that have common needs Sales splitting between customers Autonomous sales region centers Division and sorting advertisements by categories
Increase the degree of fragmentation or segmentation	Mass customization: Each customer is a market Stratified sampling for heterogeneous customer population Product advertisement minikits
Transition to microlevel	Description of product function in advertisement on molecular level (e.g., drugs, cosmetics, and food)

engineered systems. Subsequently, they have been interpreted for other applications areas such as marketing, sales, and advertising, as shown in Table 8.6 (Silverstein et al., 2007).

8.4 ASSESSMENT OF ALTERNATIVE DEVELOPMENT TECHNIQUES

Table 8.7 presents the assessment of the alternative development techniques by identifying the life cycle stage appropriate for their use and their potential contributions in the divergent and convergent phases of alternative development.

All of the techniques are dependent on the expertise, diversity, and creativity of individual team members, on the effectiveness of the team leadership, and on the processes used.

Table 8.7 Assessment of Alternative Development Techniques

Technique	Recommended Life Cycle Stage	Divergent Phase	Convergent Phase
Structured Creativity methods	All stages	Develop many ideas	Do not focus on convergence
Morphological box	All stages	Develop many ideas (can be functional-based)	Allows for elimination of unreasonable or physically unachievable alternatives
Pugh method	Concept and design stages	Nominal group technique for individual reflection Improves concepts in group setting by attacking the negatives and improving the positives, which can lead to hybrid concepts	Allows for eliminating worst concepts, especially after they have been studied for positive features that might be incorporated to other concepts
TRIZ	Concept and design stages	Defines the problem by a set of functional requirements. Models the problem by defining a set of resources that play a role in achieving the functional requirements or detract from achieving the functional requirements	Focuses search for alternative to a set of solutions based on historical patent data. Identifies which solutions among a set of 76 standard solutions, or high-level concepts, can be applied to meet each requirement. Uses 40 inventive and 4 separation principles to further reduce the number of alternatives

8.5 ALTERNATIVE EVALUATION TECHNIQUES

Once alternatives have been developed, evaluation techniques help determine which of the alternatives best meets the needs. Several alternative evaluation techniques are described.

8.5.1 Decision-Theory-Based Approaches

By decision-theory-based approaches, we mean the approaches that follow the integrated model for trade-off analyses as presented in Chapter 2. They use value measures to assess how well the alternatives achieve objectives, where value can be single-dimensional (e.g., monetary value) or multidimensional using many value measures derived from the objectives. In the case of multidimensional value measures, they capture the trade-offs among the different objectives using multidimensional value models. Decision-theory-based approaches also account for uncertainty and use either expected value in risk-neutral situations or expected utility in other situations.

8.5.2 Pugh Method for Alternative Evaluation

Pugh developed a method for alternative evaluation using a basic comparison between alternative attributes as better (+), same (S), or worse (−) (Pugh, 1991). Many practitioners (Tague, 2009) convert the Pugh evaluation criteria of better, same, or worse to numerical scores to evaluate the alternatives and select the best alternative. Hazelrigg (2012, 496–499) describes a scenario in which underlying quantitative measures of value are encoded as Pugh evaluation criteria. Subsequently, the Pugh criteria are converted to numeric measures of value (+1, 0, −1) and a total score is calculated to determine the ranking of alternatives. Table 8.8 presents the underlying quantitative measures of value for four of the alternatives from Hazelrigg's example. The value measures range from 0 to 1, where 0 is the least valuable and 1 is the most valuable. Table 8.9 encodes the quantitative measures of value as Pugh evaluation criteria and calculates a total score using alternative 1 as the datum (reference concept) against which the alternatives are scored. It shows the rank ordering of alternatives from most preferred to least preferred as alternative as {4, 3, 2, 1}, with 4 as preferred alternative.

If the choice of datum is changed, as shown in Table 8.10, the ranking of alternatives changes. It shows the rank ordering of alternatives from most preferred

Table 8.8 Initial Quantitative Measures of Value for Alternatives

Attribute	Alternatives			
	1	2	3	4
A	0	0	0	0.81
B	0	0	0.9	0.81
C	0	1	0.9	0.81

ALTERNATIVE EVALUATION TECHNIQUES

Table 8.9 Initial Scoring of Pugh Matrix

Attribute	Alternatives			
	1	2	3	4
A	Datum	S	S	+
B		S	+	+
C		+	+	+
Scores	0	1	2	3

Table 8.10 Updated Scoring of Pugh Matrix with Different Datum

Attribute	Alternatives			
	1	2	3	4
A	S	S	Datum	+
B	−	−		−
C	−	+		−
Scores	−2	0	0	−1

to least preferred alternative as {(2, 3), 4, 1}, with alternatives 2 and 3 being tied as the preferred alternative. This possible outcome indicates that it is not advisable to convert Pugh evaluation results to numeric measures of value and use the results for selecting alternatives.

8.5.3 Axiomatic Approach to Design (AAD)

Suh (2001) developed the Axiomatic Approach to Design (AAD) as a formalized decomposition method to organize design into manageable subsystems through systematic decomposition. We include it in this section, because it guides the evaluation of design alternatives based on two axioms, and it does not provide specific guidance on developing alternative designs. Whitcomb and Szatkowski applied AAD to the design of a naval combatant ship (Szatkowski, 2000; Whitcomb & Szatkowski, 2000).

Using decomposition, a system can be broken down into any number of logical subsystems arranged in a hierarchy that defines the interconnections among the subsystems. The hierarchy maintains the structure of the system through subsystem interconnections. The hierarchy can be useful in studying both the analysis of a large system and the working organization of the engineering teams performing a system design. Any hierarchy created by decomposing a system depends on the perspective taken by the viewer, and subsequently, any number of decomposed hierarchies can be defined for the same set of systems. When viewing a system, the designer defines a desired perspective, focusing on an aspect, and then decomposes that aspect into subsystems in order to create a logical structure with bounded subsystems that can be more easily analyzed and engineered.

As a more formally developed method to approach engineering design, AAD decomposes the design process into four separate domains: the customer domain, the functional domain, the physical domain, and the process domain. A specified vector type characterizes each domain. Mapping enables the designer to logically progress through the design process by first determining **what** is required in each domain and then specifying **how** these requirements are satisfied in the next successive domain (Szatkowski, 2000). The entire process advances by "zigzagging" between adjacent domains, thereby producing a hierarchical decomposition as the design is defined in increasing detail. The AAD framework provides the basis for systems engineers to define a design process driven by stakeholder need with feasible design solutions mapped to physical form through functional allocation.

The needs are used to determine the customer attributes (CAs). In turn, the CAs further map to determine the functional requirements (FRs) and the overall constraints placed on the design process. Constraints can easily be implemented to limit the designer's available choices of design parameters (DPs), as desired for the task at hand. Effectiveness of a design is based on its ability to satisfy the specified FRs (Szatkowski, 2000). Beyond the operational characteristics defined by the FRs, DPs in the physical domain are fulfilled by process variables (PVs) in the process domain. PVs are the production and manufacturing resources needed to physically construct the required DPs. In the context of systems design, the production tools and techniques used to construct each portion of the system comprise the possible PVs, allowing manufacturing or product or service realization considerations to be integrated into the process framework.

Suh's first axiom is the Independence Axiom. A good design maintains the independence of the FRs according to the Independence Axiom. The design process does not continue to the next level of decomposition until the Independence Axiom is satisfied (Szatkowski, 2000). It is this independence that allows subsystem engineers to continue their designs in their own discipline, since interfaces have been accounted for in the decomposition. Independence is achieved by either an uncoupled or a decoupled design. An uncoupled design is one in which only one DP satisfies each FR. A decoupled design is one in which the independence of FRs is satisfied if and only if the DPs are changed in the proper sequence (Szatkowski, 2000).

A coupled design does not satisfy the Independence Axiom. This type of design signifies the need for iteration because successive DPs are not necessarily fixed as FRs are sequentially satisfied. In other words, a DP may require modification to satisfy one or more additional FRs. Once this modification occurs, the fulfillment of the original FR (in part by the subject DP) must again be verified. If fulfillment is not achieved, the subject DP must once again be altered initiating the iteration process. A design matrix with elements populating both sides of the diagonal characterizes a coupled design (Szatkowski, 2000).

A decoupled design allows the designer to concentrate all efforts in a logical sequence, thereby eliminating the iteration process and allowing independent design. Once a portion of the design is complete in sequence, it theoretically does not require further modification upon completion of another aspect of the design (Szatkowski, 2000).

ALTERNATIVE EVALUATION TECHNIQUES

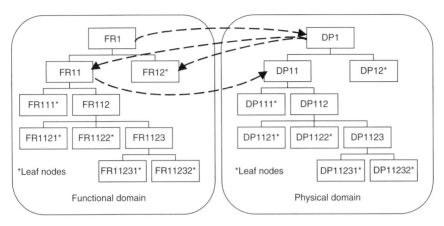

Figure 8.7 Zigzagging between domains from (Szatkowski, 2000)

The design questions become "what FRs must be provided" and "how is each specified requirement fulfilled by use of DPs." A "zigzagging" process from FR to DP, then back to the next lower level of FR, enables the designer to logically decompose the design, thereby developing FR and DP hierarchies. Figure 8.7 illustrates this process. First, the designer selects a DP to satisfy a particular FR. Then a determination regarding further decomposition is made. If the selected DP is a well-established component or system that does not require redesign, the decomposition stops. On the other hand, if the chosen DP is not a well-understood legacy component or system, decomposition is required. The designer decomposes the DP by determining the FRs it fulfills. Then, each of these FRs is satisfied with a suitable DP. Once again, a determination regarding the status of the lower level DP decomposition is made using the stated criteria. The designer "zigzags" between the two domains in this fashion until all the lowest level DPs do not require redesign. This lowest level of decomposition is referred to as the *leaf level*. The DPs at this level are called *leaf nodes* (Szatkowski, 2000).

Suh's second axiom is the Information Axiom, which states that a superior design is one that minimizes information content. This measures the information content of a design using an entropy measure that is consistent with Shannon's approach (Shannon, 2001). Each functional requirement, FR_i, has a design range that is defined on a continuous scale with a lower and an upper bound of acceptable performance. Similarly, the system range for each functional requirement, SR_i, is defined on a continuous scale and as the lower and upper bound of capability that the system design is able to provide. For each functional requirement, FR_i, the percentage of this design range that is covered by SR_i is treated as an approximation to the probability, P_i, of meeting the FR. If the design is decoupled, the FRs are independent and the entropy measure for the total system

$$I_{sys} = -\sum_i \log_2(P_i) = -\log_2\left(\prod_i P_i\right) \quad (8.1)$$

For coupled systems, Suh develops expressions that are analogous to conditional probabilities to calculate the entropy measure based on the design matrix.

Hazelrigg (2012, 506) describes AAD as a constrained optimization framework, where the independence axiom defines constraints on the structure of the FRs, and the information axiom defines an objective function based on entropy that is to be minimized. Hazelrigg notes that objectives such as maximizing profit or expected utility are not included in the AAD objective function. The system entropy is a function of the product of the fraction of each performance goal that is achieved. This could be considered a utility function only in the instance that (i) the utility of achieving each FR is identical for all requirements, (ii) the utility is identical for all outcomes that fall within the lower and upper bounds of acceptable performance, and (iii) the utility is zero for all outcomes that fall outside the lower and upper bounds of acceptable performance. The percentage of this design range that is covered by FRs are valid probabilities only for the very limited case of uniform probability distributions across the system range for the capability to achieve each FR. As a consequence, AAD does not properly account for the full range of uncertainties that could exist when measuring the value provided by the system's capabilities. Furthermore, AAD does not account for uncertainties in other important attributes of the design, such as cost.

8.5.4 TRIZ for Alternative Evaluation

The strength of TRIZ is in assisting a design team in developing focused set of alternatives and a description of the technical merits of the alternatives. Silverstein et al. (2007, 44) provides a means of evaluating alternatives when using the TRIZ method by linking its top-level ideality objective to axiomatic design. This linkage provides an analytic approach that assigns increased value to decoupling in a design and decreased value to coupling and costs. The mathematics of the approach is not rigorous and is not well documented. Other attributes of the design are not included in the alternative evaluation method, and it does not account for uncertainty.

8.5.5 Design of Experiments (DOE)

Design of experiments (DOE) uses statistics to analyze the measures of quality for a design across a predefined set of levels for variables that describe the design. Often, DOE is used to develop a surrogate model based on the important factors to predict the model response for factor-level combinations, including those that optimize the response.

The method uses a linear statistical model known as analysis of variance (ANOVA) that relates the design variables, which the statistical literature calls inputs or factors, to the measures of quality, which the statistical literature calls the response or effect variables (Montgomery & Runger, 2014). It is an efficient, reliable method that allows engineers to run a minimum number of experiments to determine the relationships between design variables and measures of quality. The relationships subsequently are used to determine the levels of the design variables that optimize a product or process design with respect to the measures of quality.

DOE starts by identifying the input variables and the output response to be measured. For each input variable, the engineer defines a number of levels that represent the range for which they want to know the effect of that variable. An experimental plan is produced that tells where to set each input parameter for each run of the experiment. Then the response is measured for each run. Any differences in outputs among the groups of inputs are then attributed to *single effects* (input variables acting alone) or *interactions* (input variables acting in combination).

The most common DOE method is to select an array (a matrix of numbers in rows and columns) to experiment over a wide variety of factor settings. Each row represents a given experiment with the settings of the factors shown across the columns of the row. Each additional experiment is added as a new row in the matrix with its own setting of the columns. When the row describing a new experiment is the same as one of the previous rows, it is known as a replicate of the previous experiment and indicates that the identical settings of the input factors are to be used for the new experiment.

Many DOE approaches use orthogonal arrays, where each set of levels has to occur equally across columns when going down the rows. Until the advent of digital computers, virtually every experiment design used an orthogonal array, because the matrix that is inverted in the least squares formulation of the ANOVA is a diagonal matrix, which makes it is easy to calculate the effect of changing any factor without extensive calculations. This also results in the estimated effect of any factor being statistically independent of the estimated effect of any other factor. Although the effect of each individual factor, known as the main effect for each factor, is independent of each other, the effects of two-factor interactions may not be independent, so orthogonal arrays should be used with caution.

The following are the basics steps in a designed experiment:

- Identify the factors to be evaluated.
- Define the levels of the factors to be tested.
- Create an array of experimental combinations.
- Conduct the experiment under prescribed conditions.
- Evaluate the results and conclusions.

DOE, if used judiciously, helps to identify the key attributes and shorten design development time. A DOE analysis can achieve several objectives. First, it can be used to create mathematical models that can be used in the design process to optimize the design for a given measure of quality. Second, it provides insight into the design levels that can be used to create a robust design, which is a product or process design that is insensitive to conditions to use or operation. Third, it identifies which experimental factors are so influential on the key performance metrics that the system will benefit from monitoring them.

The DOE method is best implemented through the use of a small set of predefined arrays, using Taguchi, Full Factorial, Fractional Factorial, Box–Behnken, or Central

Composite designs. Custom design spaces can be created using statistical analysis software, such as JMP, and Minitab. Section 8.5.6 summarizes the Taguchi approach to creating a robust design and provides additional references.

8.5.6 Taguchi Approach

Roy (1990) presents the Taguchi approach to evaluating designs as a two-step process:

(a) Optimizing the design of the product or process (system approach)
(b) Making the design insensitive to the influence of uncontrollable factors (robustness).

The first step uses a DOE approach that employs a set of predefined orthogonal arrays developed by Taguchi. The Taguchi arrays do allow for estimation of some, but not all, of the interactions that may be present. The analysis of experimental results and judgment of the engineering team are needed to determine if there are interactions present and to determine if another DOE approach is necessary to account for the interactions. Once experimental runs are completed, results are analyzed to determine the design levels that optimize the measures of quality. Yurkovich (1994) describes applying the Taguchi optimization approach to the problem of minimizing weight as the measure of quality for an aircraft wing design, and Olds & Walberg (1993) describe applying the approach to the problem of minimizing both dry weight and gross weight as the measures of quality for a launch vehicle design.

The second step uses a quadratic quality loss function to develop a signal-to-noise ratio to evaluate design alternatives for robustness. The quality loss function is defined

$$L_{ij} = k(y_{ij} - m)^2 \tag{8.2}$$

where y_{ij} is the measure of quality for design i that is observed when experiment j is performed, m is the target value that the design is trying to hit, and k is a proportionality constant that can be established by assessing the value of the loss function at specific value of y. If robustness were not being evaluated, the evaluation function for design where n total experiments have been performed

$$E(L_i) = k \sum_{j=1}^{n} (y_{ij} - m)^2 \tag{8.3}$$

The design that minimizes $E(L_i)$ is the optimum design; however; when considering the robustness of the design, the Taguchi approach uses the signal-to-noise ratio as the figure of merit that is to be maximized. When hitting the target value m of the quality measure that is desired (known as nominal is best), the signal-to-noise ratio that is to be maximized

$$SN_i = 10 \log_{10} \left(\frac{\left[\frac{1}{n} \sum_{j=1}^{n} (y_{ij} - m) \right]^2}{\frac{\sum_{j=1}^{n} (y_{ij} - m)^2}{n-1}} \right) \tag{8.4}$$

ALTERNATIVE EVALUATION TECHNIQUES   283

Note that the $y_{ij} - m$ values in the numerator are assumed to be a (radial) distance from the target, which implies that these terms are all nonnegative. Taguchi et al. (2007, 264–269) describe other forms for the single-to-noise ratio when

1. the design target is the nominal value m and the $y_{ij} - m$ are allowed to be both negative and nonnegative,
2. the design target is zero and deviations can only be positive (known as smaller is better), and
3. the design target is infinity and all deviations from the target are positive (known as larger is better).

Although Taguchi's robust design method does account for uncertainty, the value function is limited to these four functional forms for a single-dimensional design target. It is further restricted by the assumption that the variability of the design response around the target and the average deviation of the response from the target are of equal importance in forming the figure of merit that is to be optimized. Under utility theory, the relative importance of these two measures should be allowed to vary according to the risk tolerance of the decision-maker.

8.5.7 Quality Function Deployment (QFD)

The Quality Function Deployment (QFD) method was first developed by Yoji Akao in Japan for the Mitsubishi Heavy Industries Kobe Shipyard to provide a way to trace ship development aspects from the voice of the customer through definition of metrics, requirements, functions, and physical designs, including manufacturing processes. Hauser and Clausing introduced the method to the United States in a widely cited article in *Harvard Business Review* (1988). Clausing describes applications and enhancements to applying QFD to product development (Clausing, 1994). Cohen provides very good resource on the details of applying QFD (Cohen, 1995). Ficalora and Cohen describe QFD implementation in a Six Sigma context (Ficalora & Cohen, 2009).

The QFD method is well established and consists of the following main characteristics:

- Prioritizes explicit and implicit customer wants, needs, and requirements
- Translates requirements into technical system or product characteristics
- Provides traceability from customer (stakeholder) requirements to product technical characteristics
- Builds a quality product focused on stakeholder satisfaction.

The main purpose was originally to improve customer satisfaction. The structured approach implemented in QFD allows engineers and designers to better organize the information needed to trace from the customer need down to the final manufactured product. This allows engineers to be able to ensure that the way the system is realized

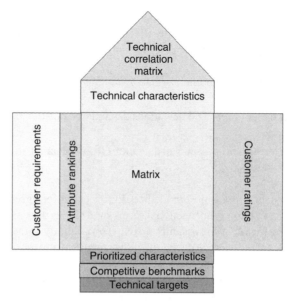

Figure 8.8 Typical QFD house of quality matrix

aligns with the original customer intent, and if problems arise, they can be traced to throughout the process through the use of a "House of Quality" (HOQ) interconnected set of matrices. The use of these standardized matrices not only allows the engineer to understand the context within which a design exists, but allows the communication of aspects to all members of the design team consistently. Many HOQ implementations also allow for benchmarking a system against competing or alternative designs and documenting and tracking customer ratings of various system or product attributes. Figure 8.8 shows a typical QFD HOQ matrix for use near the beginning of a product development, when customer (or stakeholder) needs or requirements are compared to the potential Technical Characteristics of the eventual product or system.

The HOQ are created between various pairs of attributes. In Figure 8.8, Customer Requirements are paired against system Technical Characteristics. The Customer Requirements are determined from the system needs analysis and requirements definition processes. The Attribute Rankings are relative-importance weights that capture the relative importance of each of the Customer Requirements. The Technical Characteristics are determined by reviewing each Customer Requirement and determining what aspect of the product or system would have some impact on the ability of the product or system to meet a customer requirement or need. The name of each Technical Characteristic is entered into the column heading along that row near the top of the matrix. The Technical Correlation Matrix, in the shape of a triangle at the top of the HOQ, is where correlations between Technical Characteristics are captured. It is this part of the HOQ that gives it the shape that makes it look like a house, thus the name HOQ. The entries are none (left blank), positive (plus sign), or negative (minus sign).

ALTERNATIVE EVALUATION TECHNIQUES 285

The intersection of each technical characteristic column with each other column is where the comparative correlation is recorded. This keeps track of possible interconnections that might need to be considered when developing technical characteristics of the resulting product or system. The Customer Ratings keep track of any customer survey ratings of requirements as found in competitors' products that are currently available to meet the requirements. They can also be used to record levels of existing ratings of systems that exist or have been proposed that attempt to meet the customer requirements and for which the HOQ is being used to determine a new replacement, or upgraded, system. The customer satisfaction levels recorded in this area are typically presented as line charts to allow a visual interpretation of the information.

The Matrix has an entry in each cell that indicates the impact of each column header to each row header item. Each matrix cell then records the impact of a specific technical characteristic on a specific customer requirement. The matrix is typically best kept somewhat sparse, to keep characteristics aligned to requirements keeping track of only the most important impacts. The matrix entries either can be symbols, typically indicating high, medium, or low impact, or can have number values entered. The use of numeric entries is useful for spreadsheet implementations. The use of symbols is sometimes preferred so that the recording of impact when facilitating the group development of the entries for the Matrix does not imply a definite quantity – which probably better represents that the impact is a subjective assessment and probably does not have an objective value behind it. In other words, when facilitating the development of the entries in the Matrix, symbols are used to indicate the impact, rather than using numerical values. Once the entries in the Matrix have been recorded, numerical values can be applied to compute the Prioritized Characteristics in a spreadsheet.

Once the matrix is filled out after reviewing the mapping of the technical characteristics and the customer requirements, the value for each cell entry is multiplied by the Attribute Ranking to yield a value for each entry in the Prioritized Characteristics row, which are importance scores for the Technical Characteristics. Competitive Benchmarks can be included to compare the HOQ product prioritized characteristic levels with those that existing product levels achieve against the same customer requirements. Finally, Technical Targets can be included to document the desired levels of that the product development team is hoping to achieve.

HOQ are linked together by next using some rows and transposing them into columns in succeeding matrices. For a system development, the next HOQ might use the Prioritized Characteristics row and transpose it so that it becomes the left-hand column. The Prioritized Characteristics becomes the Ranking column for this second HOQ. This HOQ maps Technical Characteristics against Functions that a system needs to perform as measured by the technical characteristics in the first HOQ. The Matrix of this second HOQ is then filled in to obtain the Prioritized Functions along the bottom row. Figure 8.9 shows a diagram of the transposition from the first HOQ to the second in our example.

A third HOQ would then be formed to map Functions to Physical Characteristics, such as components or subsystems. The Functions row headings are transposed to form the first column for the third HOQ, mapping Physical Characteristics to the

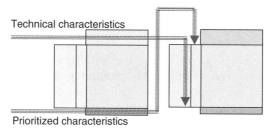

Figure 8.9 Typical QFD house of quality matrix mapping by transposing previous HOQ header row to the next HOQ column

Functions, with Prioritized Functions being used as the attribute ranking column. By linking these matrices together, the flow down from Customer requirements can be done to the system Functions and Physical Components. This provides traceability from the customer needs all the way through product design, development, implementation, and operation as various HOQ are created and linked for the various aspects.

Various design alternatives are compared via a weighted scoring approach that uses importance weights contained in the QFD matrix. Delano et al. (2000) provide an example of using QFD to evaluate alternative aircraft designs. They calculate a scalar score for how well each design alternative satisfies the total set of requirements by numerically evaluating how well each design performs on each Technical Characteristic and calculating a weighted sum of the numerical evaluations that uses the Prioritized Characteristics row as the importance weights for the weighted sum. They separately develop importance weights for a set of program go-ahead cost attributes and a set of political impact attributes, numerically evaluate how well each design performs on the cost and political attributes, and calculate separate weighted sums for cost and political impact. Finally, they combine the requirements, cost, and political impact scores into a single score for each design alternative by weighting these three top-level CAs.

Hauser & Clausing (1988, 66) describe how Attribute Rankings (relative-importance weights) typically are defined: "Weightings are based on team members' direct experience with customers or on surveys. Some innovative businesses are using statistical techniques that allow customers to state their preferences with respect to existing and hypothetical products. Other companies use' revealed preference techniques,' which judge consumer tastes by their actions as well as by their words—an approach that is more expensive and difficult to perform but yields more accurate answers." As was discussed in Chapter 2, weights used in constructing value trade-offs should be based on a ratio scale, which are implemented by assessing swing weights instead of assessing importance weights. Typically, Attribute Rankings and other rankings, such as the weights for costs and political impact in the aircraft example from Delano et al. (2000), are based on assessing importance weights and not on assessing swing weights, although the statistical and revealed preference techniques alluded to by Hauser and Clausing may provide a means to develop swing weights. The numerical entries in the Technical Correlation

Matrix are essentially a measure of the partial correlation coefficients and, therefore, are scale-invariant. This implies that if the Attribute Rankings are importance weights, then the weights in Prioritized Characteristics row of the QFD matrix are also importance weights; if the Attribute Rankings are ratio-scale weights, the Prioritized Characteristics row contains ratio-scale weights.

8.5.8 Analytic Hierarchy Process AHP

The Analytic Hierarchy Process (AHP) is a method for multiattribute analysis that enables engineers to explicitly capture stakeholder needs rankings for both tangible and intangible factors against each other in order to establish priorities. The method was developed by Saaty and has been utilized extensively in many applications both in engineering and in many other multiattribute decision-making situations (Saaty, 2000, 2012).

AHP is based on determining the relative importance of the attributes of the needs, comparing each one against each of the others. All options are paired separately for each criterion, for comparison. The pairing comparison is accomplished by stakeholders using paper survey forms, spreadsheets, or specialized AHP software. Then, some simple calculations determine a weight that will be assigned to each criterion. The weights are calculated and normalized to be between 0 and 1 with the resultant weight values of all attributes totaling to 1.

Stakeholders for systems design are often encouraged to put forth their reasons for their comparisons through verbal dialog that express relative merit or interest. Then, a group of stakeholders can negotiate as to how they rank the relative attributes, providing a more consistent outcome.

Barzilai & Golany (1994) describe situations in which preference reversals can occur when adding a new alternative that ranks lower on most of the attributes under consideration. This is because "the correct interpretation of weight in the context of the AHP's normalisation procedures is the relative contribution of an average score

Table 8.11 Best Practices in Forming a Basis for Good Decisions

Element of Decision Basis	Associated Best Practice
Alternatives (What you can do)	A rich set of alternatives that include alternatives for immediate action and for postponing action after some of the uncertainties have been resolved
Information (What you know)	Information must link the alternatives to what will ultimately happen in terms of value delivered. Because outcomes are uncertain, models used to link alternatives to outcomes must account for uncertainty
Preferences (What you want)	Preferences must account for the trade-offs among different attributes that measure value delivered, the relative value delivered across time, and the relative value of different levels of risk

Table 8.12 Assessment of Alternative Evaluation Techniques

Technique	Information (Uncertainty)	Preferences	Logic
		Best Practice	
Decision-theory-based	Explicitly accounts for uncertainty	Interaction among preferences accounted for via multiattribute value modeling or using independent value measures. Can incorporate time and risk preferences	Mathematically consistent
Pugh Method	Does not account for uncertainty	Value preferences built from bottom up and may not account for interactions. Does not incorporate time and risk preferences	Ignores relative importance of performance measures
Axiomatic Approach to Design (AAD)	Accounts for uncertainty in a very limited way	Seeks to maintain independence of requirements and does handle interactions for entropy value measure. Entropy value measure does not link to cost, performance, schedule, and other system-wide measures. Does not incorporate time and risk preferences	"Information Axiom" makes assumptions about performance trade-offs
TRIZ	Does not account for uncertainty	Does not incorporate time and risk preferences	Mathematics of the approach is not consistent and is not well documented
Design of Experiments (DOE)	Captures uncertainty via experiments and analysis of variance models	Typically does not focus on value modeling and/or on time and risk preferences, although it is possible to use DOE to do so. Does focus on developing statistical models that can link alternatives to scalar value measures	Mathematically consistent

Taguchi Approach (Step 1)	Captures uncertainty via experiments and analysis of variance models	Typically does not focus on value modeling and/or on time and risk preferences, although it is possible to use DOE to do so. Does focus on developing statistical models that can link alternatives to scalar value measures. Limited treatment of interactions among design input variables	Mathematically consistent
Taguchi Approach (Step 2)	Captures uncertainty via experiments and analysis of variance models	Limited to scalar design target value measure. Signal-to-noise measure does not link to cost, performance, schedule, and other system-wide measures. Does not incorporate time and risk preferences	Mathematically consistent
Quality Function Deployment (QFD)	Does not account for uncertainty	Typically uses importance weights. Requires additional effort to use swing weights. Does not incorporate time and risk preferences	Mathematically consistent
Analytic Hierarchy Process (AHP)	Does not account for uncertainty	Requires additional effort to use swing weights. Does not incorporate time and risk preferences	Bases weights on average of alternatives

(average over the options under consideration) on each criterion--in contrast to the relative contribution of a unit as specified by the MAV approach" (Belton, 1986). When the lower ranking alternative is added, the average shifts far enough to affect the ranking of the alternatives.

8.6 ASSESSMENT OF ALTERNATIVE EVALUATION TECHNIQUES

Howard & Abbas (2016) define three elements that form the basis for making good decisions and the associated best practices for forming a good basis for decision-making (see Table 8.11). With a good basis, the best practice for the logic that evaluates alternatives should follow a normative framework that is mathematically correct and invariant to the order in which preferences are measured.

In Table 8.12, we assess the alternative evaluation techniques with respect to the best practices for information, preferences, and logic.

The decision-theory-based techniques use an integrative, top-down approach to measuring and modeling values; however, the other approaches tend to be bottom-up in their approach to measuring and modeling values. As a result, they can suffer what Saari & Sieberg (2004) call loss of information due to separation; for example, customers evaluating car designs may prefer body style A over body style B, and engine C over engine D when each is considered separately, but they may not prefer the combination (A, C) over (A, D) if mounting engine C on body style A results in unacceptable handling characteristics.

8.7 KEY TERMS

Convergent Thinking: Cognitive approach to problem-solving focusing on a limited number of alternatives, eventually converging on a solution. The alternatives considered are typically predetermined or limited to a known set. The end result is a choice from among the limited alternatives (Bernhard, 2013).

Divergent Thinking: Cognitive approach to problem-solving looking for possibilities from which to define alternatives. The alternatives are developed in different directions to explore ways of looking at a problem that may not have been readily apparent at first (Bernhard, 2013).

Creativity: The use of imagination to develop new concepts or ideas.

Alternative: One of a set of possible objects or ideas under consideration for decision-making.

8.8 EXERCISES

8.1. Decision-making.
  (a) Describe the three modes of decision-making presented in this chapter.

EXERCISES   291

   (b) Provide an example of each from your organization, university, or family.
   (c) Which mode do you think is the most frequently used in trade-off decision-making? Why?

8.2. Creating alternatives.
   (a) Why is creating alternatives important?
   (b) How have you generated alternatives in your past professional and personal life?
   (c) What ideas and techniques from this chapter will help you develop better alternative in the future? Explain.

8.3. Role of teams in alternative generation.
   (a) Why does a chapter on alternative generating present material on teams?
   (b) How does your organization or university use teams to generate alternatives?
   (c) How could your organization or university improve their use of teams to generate better alternatives?

8.4. Alternative development techniques.
   (a) What are the two stages and why is each stage required?
   (b) Select a systems engineering journal article that uses one of the techniques described in this chapter to develop alternatives. How effective was their use of the technique?
   (c) Which technique is the most appropriate for developing alternatives in early life cycle stages? Why?
   (d) Which alternative is the most appropriate for developing alternatives in later life cycle stages? Why?

8.5. Alternative evaluation techniques.
   (a) This chapter compares the techniques using three elements of the basis for good decisions and associated best practices. Identify and briefly describe the three elements.
   (b) Select a systems engineering journal article that uses one of the techniques described in this chapter to evaluate alternatives. How effective was their use of the technique?
   (c) Which element of the basis is the most important in early life cycle stages? Why?
   (d) Which element is the most important in later life cycle stages? Why?

8.6. Morphological Box – Functional Allocation Box.
   (a) Consider an autonomous vehicle system for personal transportation. Brainstorm a set of functions that such a system should have in order to operate successfully. Give yourself 10–15 min to generate the list.
   (b) Review your list, and edit each item, so it is in the form of a functional description (verb or verb phrase) rather than a physical description (noun or noun phrase).

(c) Select from 5 to 9 functions to use as row entries for developing a morphological box as a table.

(d) For each function, enter as many ways as you can think of as a means to accomplish that function with a physical system or component. Use a noun or noun phrase to describe the means. Use a sketch, too, if desired.

(e) How many possible system alternative configurations are possible to create based on the number of functions and the number of physical means you have in your table?

(f) Create a set of alternative system configurations with each alternative consisting of one means from each function. Do not include combinations that are not physically feasible. For example, if you consider an electric propulsion subsystem that is recharged using an electric outlet, combining that with an energy storage subsystem using gasoline or diesel fuel would not be a feasible system alternative.

(g) How many feasible alternatives do you have for future consideration? How might you proceed in a design process to focus efforts to develop one or more alternatives from this conceptual design stage to a more detailed design stage?

8.7. Morphological Box – Strategy Table.

(a) Consider you are a member of a design team for a major manufacturer developing a self-driving car for mass production. Brainstorm a list of Strategy Themes that you think might be key for developing a successful product. Give yourself 10–15 min to generate the list.

(b) Select from 3 to 5 Strategy Themes and enter those as row headings in a table.

(c) Brainstorm a list of Key Decisions that you think would be important to potential customers. Give yourself 10–15 min to generate the list.

(d) Select from 3 to 5 Key Decisions and enter those as column headings in your table.

(e) Create Strategy Theme alternatives by selecting one Key Decision from each column for each Strategy Theme. Provide a brief description of each Strategy Theme that captures the benefits and costs of each.

(f) Summarize the resulting Strategy Themes to persuade the company decision-makers to consider the alternatives you have created in the development of the new product.

8.8. Pugh Method for Alternative Evaluation.

(a) Consider that you thinking about purchasing a vehicle. Brainstorm a list of characteristics that are important to you. Give yourself 10–15 min to generate the list.

(b) Organize the list with features that are most important at the top with items of lower importance toward the bottom. Select between 5 and 9 items to use as attributes for alternative evaluation.

(c) Review the top attributes you selected. Identify a metric, in terms of something that can be measured, for each attribute. For example, passenger capacity, cargo capacity, gas mileage, safety rating, quality rating, and similar attributes have available metrics published. For attributes such as comfort, color, and style, you will have to create your own subjective scale to be able to discriminate among the alternatives. Create a scale for these subjective attributes so you can compare attribute levels for different alternative vehicles.

(d) Create a Pugh Table with the rows labeled with the attributes, and the columns labeled with the alternatives, a number of vehicles you would like to evaluate.

(e) Select one of the vehicles as the datum or reference. For each attribute, score each alternative as plus (+), minus (−), or same (S) against the datum vehicle. Sum the scores across the alternatives.

(f) Select your highest rated alternative. Reflect on the outcome. Do you think you would be satisfied with this choice based on this analysis? Why or why not?

REFERENCES

Altshuller, G.S. (1998) *Creativity as an Exact Science: The Theory of the Solution of Inventive Problems*, Gordon and Breach Publishers, Amsterdam.

Barrager, S.H. (2001). Design of the Corvette. Presentation to Committee on Theoretical Foundations for Decision Making in Engineering Design of the National Research Council Board on Manufacturing and Engineering Design.

Barzilai, J. and Golany, B. (1994) AHP rank reversal, normalization and aggregation rules. *INFOR*, **32** (2), 57–64.

Belbin, R.M. (2010) *Team Roles at Work*, 2nd edn, Butterworth-Heinemann, Oxford.

Belton, V. (1986) A comparison of the analytic hierarchy process and a simple multi-attribute value function. *European Journal of Operational Research*, **26** (1), 7–21.

Bernhard, T. (2013) What Type of Thinker are You?. Psychology Today. https://www.psychologytoday.com/blog/turning-straw-gold/201302/what-type-thinker-are-you (accessed June 2016).

Buede, D.M. (2009) *The Engineering Design of Systems: Models and Methods*, 2nd edn, John Wiley & Sons, Hoboken, NJ.

Clausing, D. (1994) *Total Quality Development: A Step-by-Step Guide to World Class Concurrent Engineering*, ASME Press, New York.

Clausing, D. and Fey, V. (2004) *Effective Innovation: The Development of Winning Technologies*, American Society of Mechanical Engineers. ISBN-13: 978-0791802038. ISBN-10: 0791802035.

Cohen, L. (1995) *Quality Function Deployment: How to Make QFD Work for You*, Prentice-Hall. ISBN-10: 0201633302. ISBN-13: 978-0201633306.

Cross, N. (2000) *Engineering Design Methods: Strategies for Product Design*, 3rd edn, John Wiley & Sons, New York.

De Bono, E. (1985) *Six Thinking Hats: An Essential Approach to Business Management*, 1st edn, Little Brown and Company, Boston.

Delano, G., Parnell, G.S., Smith, C., and Vance, M. (2000) Quality function development and decision analysis: a R&D case study. *International Journal of Operations & Production Management*, **20** (5), 591–609.

Dictionary.com. (nd). *Unabridged*, http://dictionary.reference.com/browse/creativity (accessed 22 July 2015).

Eberle, B. (1996) *Scamper: Games for Imagination Development*, Prufrock Press, Waco, TX.

Fey, V. and Rivin, E. (2005) *Innovation on Demand: New Product Development Using TRIZ*, Cambridge University Press. ISBN-13: 978-0521826204. ISBN-10: 0521826209.

Ficalora, J. and Cohen, L. (2009) *Quality Function Deployment and Six Sigma: A QFD Handbook*, Second edn, Prentice-Hall. ISBN-10: 0133364437. ISBN-13: 978-0133364439.

Hauser, J.R. and Clausing, D. (1988) The house of quality. *Harvard Business Review*, **66** (3), 63–73.

Hazelrigg, G. (2012) *Fundamentals of Decision Making for Engineering Design and Systems Engineering*, Worldcat.org. ISBN 978-0-9849976-0-2.

Hipple, J. (2012) *The Ideal Result: What it is and How to Get it*, Springer.

Howard, R.A. (1988) Decision analysis: practice and promise. *Management Science*, **34** (6), 679–695.

Howard, R.A. and Abbas, A.E. (2016) *Foundations of Decision Analysis*, Pearson, Boston, MA.

Jones, J.C. (1992) *Design Methods*, 2nd edn, John Wiliey & Sons, New York.

Keeney, R.L. (1992) *Value-Focused Thinking: A Path to Creative Decisionmaking*, Harvard University Press, Cambridge, MA.

Kenley, C.R. and Coffman, R.B. (1998) The error budget process: an example from environmental remote sensing. *Systems Engineering*, **1** (4), 303–313.

Lewin, K. (1943) Defining the 'field at a given time'. *Psychological Review*, **50** (3), 292–310.

McGrath, J.E. (1984) *Groups: Interaction and Performance*, Prentice-Hall, Englewood Cliffs, NJ.

Montgomery, D.C. and Runger, G.C. (2014) *Applied Statistics and Probability for Engineers*, 6th edn, John Wiliey & Sons, Hoboken, NJ.

Olds, J., & Walberg, G. (1993) Multidisciplinary Design of a Rocket-Based Combined-Cycle SSTO Launch Vehicle Using Taguchi Methods. *AIAA/AHS/ASEE Aerospace Design Conference*. Irvine, CA.

Parnell, G., Bresnick, T., Tani, S., and Johnson, E. (2013) *Handbook of Decision Analysis*, John Wiliey & Sons, Honoken, NJ.

Pugh, S. (1991) *Total Design: Integrated Methods for Succesful Product Engineering*, Addison-Wiley, Reading, MA.

Roy, R.K. (1990) *A Primer on the Taguchi Method*, Van Nostrand Reinhold, New York.

Saari, D.G. and Sieberg, K.K. (2004) Are partwise comparisons reliable? *Research in Engineering Design*, **15** (1), 62–71.

Saaty, T. (2000) *Fundamentals of Decision Making and Priority Theory with the Analytic Hierarchy Process*, RWS Publications. ISBN-10: 0962031763. ISBN-13: 978-0962031762.

Saaty, T. (2012) *Decision Making for Leaders: The Analytic Hierarchy Process for Decisions in a Complex World*, 3rd Revised edn, RWS Publications. ISBN-10: 096203178X. ISBN-13: 978-0962031786.

REFERENCES

Salvatore, F. (2008) The Value of Architecture. *NDIA 11th Annual Systems Engineering Conference*. San Diego, CA.

Shanmugaraja, M. (2012) Quality improvement through the integration of Six Sigma QFD and TRIZ in manufacturing and service organizations. Dissertation, Faculty of Mechanical Engineering, Anna University, Chennai, India.

Shannon, C.E. (2001) A mathematical theory of communication. *SIGMOBILE Mobile Computing Communications Review*, **5** (1), 3–55.

Silverstein, D., DeCarlo, N., and Slocum, M. (2007) *Insourcing Innovation: How to Achieve Competitive Excellence Using TRIZ*, Taylor & Francis, Hoboken, NJ.

Slepian, D. (1963) Bounds on communication. *Bell System Technical Journal*, **42** (3), 681–707.

Suh, N. (2001) *Axiomatic Design : Advances and Applications*, Oxford University Press, New York.

Szatkowski, J. (2000) Manning and automation of naval surface combatants: a functional allocation approach using axiomatic design theory. Masters Thesis. MIT, Cambridge, MA.

Taguchi, G., Chowdhury, S., and Wu, Y. (2007) *Taguchi's Quality Engineering Handbook*, John Wiliey & Sons, Hoboken, NJ.

Tague, N.R. (2009) *The Quality Toolbox*, 2nd edn, ASQ, Milwaukee.

Ullman, D.G. (2010) *The Mechanical Design Process*, 4th edn, McGraw-Hill, Boston, US-MA.

Watson, S.R. and Buede, D.M. (1987) *Decision Synthesis: The Principles and Practice of Decision Analysis*, Cambridge University Press, Cambridge, England.

Whitcomb, C. and Szatkowski, J. (2000). Concept level naval surface combatant design in the axiomatic approach to design framework. ICAD 2000, *First International Conference on Axiomatic Design*. MIT, Cambridge, MA.

Yurkovich, R.. (1994) The use of Taguchi techniques with the ASTROS code for optimum wing structural design. 35th Structures, Structural Dynamics, and Materials Conference. American Institute of Aeronautics and Astronautics.

Zwicky, F. (1969) *Discovery, Invention, Research Through the Morphological Approach*, New York Macmillan, New York, NY.

9

AN INTEGRATED MODEL FOR TRADE-OFF ANALYSIS

ALEXANDER D. MACCALMAN
Department of Systems Engineering, United States Military Academy, West Point, NY, United States

GREGORY S. PARNELL
Department of Industrial Engineering, University of Arkansas, Fayetteville, AR, United States

SAM SAVAGE
School of Engineering, Stanford University, Stanford, CA, United States

> Interactive Simulation Connects the Seat of the Intellect to the Seat of the Pants
> (Sam Savage, Author of The Flaw of Averages)

9.1 INTRODUCTION

System engineers often use value modeling to capture a composite perspective of multiple stakeholders with conflicting objectives to help understand value trade-offs among several system alternatives. There are many uncertainties involved with designing a system, and we believe they must be considered in all system decisions throughout the life cycle. These uncertainties include stakeholder needs, technological maturity, adversary and competition actions, scenarios, costs, schedules, and many more. Uncertainty and risk analyses are often performed independent of the value model. As a result, our decisions become biased toward deterministic

---

Trade-off Analytics: Creating and Exploring the System Tradespace, First Edition. Edited by Gregory S. Parnell.
© 2017 John Wiley & Sons, Inc. Published 2017 by John Wiley & Sons, Inc.
Companion website: www.wiley.com/go/Parnell/Trade-off_Analytics

solutions, and system decision-makers may not understand the key uncertainties and risks. To eliminate this cognitive bias, we propose an approach that integrates uncertainty modeling with value and cost modeling in order to help understand value and risk while we analyze alternatives. Chapter 7 introduced value modeling, Chapter 4 introduced a number of cost modeling methods, while Chapter 3 introduced uncertainty modeling. In this chapter, we use these methods to model the uncertainties associated with both value and cost in order to demonstrate our integrated approach. By propagating the uncertainties of the independent system variables through the value and cost models, we can examine stochastic Pareto charts and cumulative distribution functions (cdfs) and identify dominant solutions. In addition, we can use tornado diagrams to identify which value measures and cost components explain the majority of the alternatives' value and cost variations. Figure 9.1 shows the many elements that contribute to performing a trade study and the key relationships between them; these concepts will be discussed in detail in this chapter.

To demonstrate our approach, we use a notional illustrative example to explain the types of trade-offs and uncertainties system engineers typically face when designing a system. Our next section will provide a brief description of this example; Section 9.3 will introduce an influence diagram with nodes that represent the types of decisions, uncertainties, and values involved in a system trade-off study; we will explain each node in detail as it applies to the system decision and provide examples from our notional trade study. Section 9.4 will discuss other types of trade-off analysis that can be performed using our approach. Section 9.5 will discuss the types of Monte Carlo simulation tools available, and Section 9.6 will summarize the chapter.

9.2 CONCEPTUAL DESIGN EXAMPLE

Our conceptual design example is a defense system design problem that involves the development of new Infantry squad technologies that will enhance their effectiveness (see Section 7.9.1). An Infantry squad is a nine-person organization that consists of a squad leader and two teams of four persons each. Each team has a team leader, a rifleman, an automatic rifleman, and a grenadier. The Infantry squad technologies enhancement problem provides an opportunity to invest in new systems that will increase the squad's capability to overmatch the current and future adversaries in complex environments. The system is the collection of integrated technologies that consists of soldier, sensors, weapons, exoskeletons, body armor, radios, unmanned aerial vehicles (UAV), and robots. We use the squad enhancement design example to highlight trade-offs across multiple types of costs, performance, schedule, risk, and scenario considerations. In addition, it allows for a wide variety of alternatives that are comprised of different combination of six system technology components.

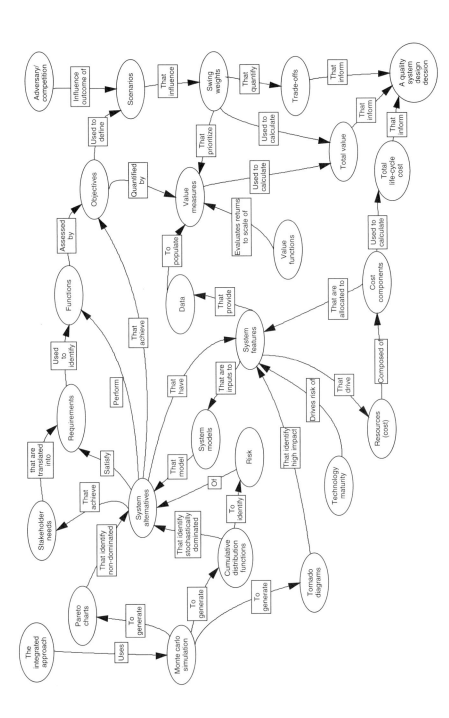

Figure 9.1 Concept diagram for the integrated trade-off analysis

9.3 INTEGRATED APPROACH INFLUENCE DIAGRAM

We use an influence diagram, a decision analysis technique (Buede, 2000) to present our integrated model for trade-off analysis. We chose the influence diagram because it is offers a probabilistic decision analysis model that identifies the major system variables and their probabilistic dependencies. This influence diagram assumes a multiple-objective model using the additive value model that has a functional layer above the objectives. If net present value is used, the diagram could be easily modified. Figure 9.2 shows the integrated approach influence diagram that displays many of the system decision elements and how they influence each other. The subsequent sections will define each node and the arrows that represent how the nodes influence other nodes.

9.3.1 Decision Nodes

A decision node is a rectangle that represents the set of choices the decision-maker must make. For a system design problem, the decisions include establishing the system functions, the objectives, the requirements, and system alternatives. Adding the functions, objectives, and requirements as systems engineering decisions is very important, especially in the system concept, architecting, and design decisions. In later life cycle stages, once decisions are made, the functions, objectives, and requirements may be known constraints or constants.

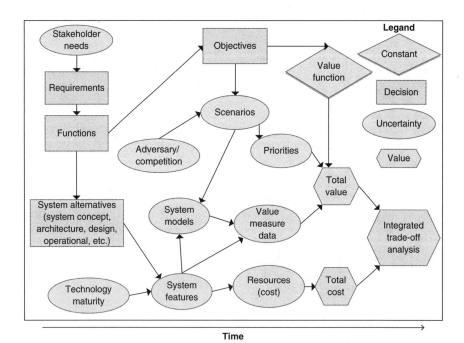

Figure 9.2 Integrated approach influence diagram

9.3.1.1 Requirements Requirements describe the technical capabilities and levels of performance that a system must have and any constraints on the system or system design. System engineers translate stakeholder needs into clear requirement statements specific enough for domain engineers to implement and test in order to verify that they are satisfied. Requirement engineering is a critical specialty that involves extensive analysis and management, especially when there are requirement changes in later stages of the life cycle. Requirement changes typically have the highest impact on cost and schedule due to the redesign and rework needed to implement the change. In order to mitigate a change's impact, it is important to establish a logical requirement structure with auditable records that can be traced to functions, objectives, and system features. Understanding which functions, objectives, and system features satisfy requirements will reduce the impact of a requirements change when they occur. The Model-Based Systems Engineering (MBSE) approach is a new paradigm that supports the specification, analysis, design, and verification of a complex system using an integrated system model with a dedicated tool. The integrated system model effectively manages the auditable records of a system design by defining a system element once to be used throughout the model. As a result, once a change is made to an element in the integrated system model, the dedicated tool will instantly identify how the change will impact the system. The MBSE approach is gaining popularity and is expected to become a common state of practice in the near future (National Defense Industrial Association, 2011).

Each requirement can be classified as either a desired capability or a constraint that must be met (Parnell et al., 2011). The arrow from the Requirements decision node to the Functions decision node in Figure 9.2 represents how the desired capabilities inform what functions the system must perform in order to achieve the objectives. Requirements influence the value functions indirectly through the Functions and Objectives decision nodes in two ways. First, the desired capabilities inform the threshold and objective values used to define the shape of the value function. Second, the constraints are those that define the screening criteria, minimum acceptable value, or walk-away point that eliminates alternatives from the decision.

9.3.1.2 Functions A function is an action that transforms inputs and generates outputs, involving data, materials, and/or energies (SEBoK, 2015). The INCOSE Systems Engineering Handbook (SEH) defines a function as a characteristic task, action, or activity that must be performed to achieve a desired outcome (SEBoK, 2015). Functional analysis is a key systems engineering task that identifies the system functions and system element interfaces required to perform the performance objectives (Parnell et al., 2011). The arrow from the Functions decision node to the Objectives decision node in Figure 9.2 represents how functions are used to identify the fundamental objectives we want to achieve; see Section 7.8 for a discussion on how to develop a functional hierarchy.

Functions are allocated to structural elements of the system alternative that performs them. Depending on the system decision, these system elements may be subsystems, components, or parts. Each system element has system features that define

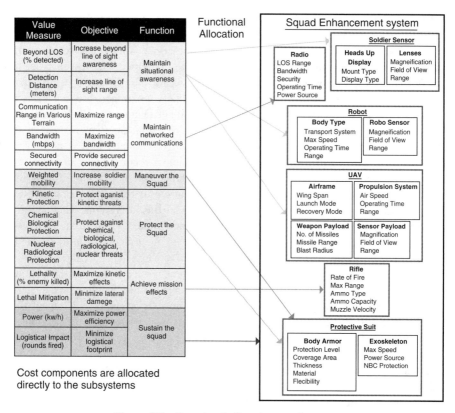

Figure 9.3 Functional allocation to subsystems

the element's characteristics. System engineers use models and simulations, subject matter expertise (SME), operational testing results, and data from legacy systems to decide which functions will be allocated to each system feature. The arrow from the Functions decision node to the System Alternatives decision node in Figure 9.2 represents the functional allocation to the system features; these allocations depend heavily on the perspective of how functions will be satisfied. In Figure 9.3, we show an example of a functional allocation to subsystems from the squad enhancement design example.

9.3.1.3 Objectives An objective is a statement of something that we desire to achieve (Keeney, 1992). Identifying the objectives we use to evaluate our system alternative is the most critical step in the decision process; see Chapter 7 for a detailed discussion on the development of the objectives. Figure 9.2 shows two outgoing arrows from the Objective decision node: one to the Scenarios uncertainty node that represents the objective's influence on the scenario context and another to the Value Function constant node. Each objective has one or more value functions that assess a system alternative's achievement of that objective.

9.3.1.4 System Alternatives This decision node involves the selection of alternatives within the decision space that will be considered in the system decision. The decision space involves the exploration of creative alternatives that span the opportunity space as much as possible; Chapter 8 will discuss in detail methods system engineers can use to generate these creative alternatives. The alternatives within the decision space are what the SEBoK refers to as the physical architecture. System alternatives each have a collection of elements with system features that define the characteristics of the alternative. The settings of the system features distinguish one alternative from another. The arrow in Figure 9.2 from the System Alternative decision node to the System Feature uncertainty node represents how the alternative has system features that define its characteristics. For the squad enhancement design example, we have seven alternatives. Table 9.1 shows each alternative for our example, along with their system features that define their unique characteristics; the types of UAVs are from the example in Chapter 5.

9.3.2 Uncertainty Nodes

Uncertainty nodes are ovals that represent uncertain information relevant to the decision; they could be a single probability value, a random variable, or a vector of data. For the system design problem, major uncertainties include the following: stakeholder needs, scenarios, priorities, adversary actions, competition actions, technological maturity, system features, system models, data, value measures (system performance), and resources (cost). An outcome space or a probability distribution can be assigned to the independent uncertain variables. The uncertainty can then be propagated through the value and cost models by Monte Carlo simulation.

9.3.2.1 Stakeholder Needs In the early stages of the conceptual design, stakeholders develop need statements that express what they think the system should be able to do from their perspectives. The arrow in Figure 9.2 from the Stakeholder needs uncertainty node to the Requirements decision node represents the need's refinement into a system requirement. The uncertainty associated with stakeholder needs is the result of an unclear problem statement within the opportunity space and ill-defined values in the objective space. To address this challenge, we use a value hierarchy that captures a composite perspective of multiple stakeholders with conflicting objectives. For the squad enhancement design example, the key stakeholders include the decision-makers responsible for the system decisions, the acquisition personnel who manage the development, the technologists that develop the technology, the specialty engineers who design the system, the contractors that build the systems, the logistical personnel who distribute and maintain the system, and the soldiers who operate the system.

9.3.2.2 Adversary/Competition This type of uncertainty node involves the external factors that influence the outcome of a scenario; this relationship is represented by the arrow in Figure 9.2 from the Adversary/Competition uncertainty node to the Scenarios uncertainty node. In defense, information security, and homeland defense

Table 9.1 Squad Enhancement Alternatives and System Features

Subsystem	System Feature	Baseline Design Choice	Sustainable Design Choice	Attack Design Choice	LongRange Design Choice	Survivable Design Choice	Defendable Design Choice	Performance Design Choice
Heads Up Display	Mount Type	NA	Helmet	Goggle	Helmet	Goggle	Integrated	Integrated
	Display Type	NA	2D Small	3D Small	2D Large	3D Medium	2D Medium	3D Large
	Magnification	0X	0.5X	1X	3X	1X	2X	4X
Lenses	Field of View	50 deg	60 deg	80 deg	60 deg	90 deg	100 deg	150 deg
	Range	300 m	900 m	800 m	1300 m	1000 m	1100 m	1400 m
	LOS Range	1000 m	1200 m	1500 m	800 m	1400 m	1300 m	1600 m
Radio	Bandwidth	5 mbps	10 mbps	9 mbps	13 mbps	12 mbps	6 mbbs	12 mbps
	Security	Freq Hop	Freq Hop	TK400	TK650	KLM40	PX33	FH98
	Operating Time	12 hrs	14 hrs	10 hrs	24 hrs	16 hrs	10 hrs	18 hrs
	Power Source	5 watts	10 watts	40 watts	30 watts	38 watts	20 watts	25 watts
	Rate of Fire	700 rpm	700 rpm	1300 rpm	600 rpm	1100 rpm	800 rpm	1500 rpm
	Max Range	500 m	600 m	700 m	1200 m	800 m	1200 m	1500 m
Rifle	Ammo Type	5.56 mm	7.62 mm	12.7 mm	6.5 mm	10 mm	8 mm	12.7 mm
	Ammo Capacity	30 rds	50 rds	150 rds	50 rds	200 rds	300 rds	500 rds
	Muzzle Velocity	880 m/s	1000 m/s	1500 m/s	900 m/s	1100 m/s	950 m/s	1500 m/s
	Max Speed	4 mph	4 mph	5 mph	5 mph	5.5 mph	6 mph	8 mph
Exoskeleton	Power Source	NA	BA90	BA75	Tablet	BA550	Integrated	Integrated
	Frame Size	NA	Small	Medium	Medium	Medium	Large	Large
	Protection Level	Type III	Type III	Type III	Type IV	Type III	Type V	Type V
	Coverage Area	Vital Only	Vital & Extreme	Full	Full	Full	Full	Full
Body Armor	Thickness	2 in	1 in	2.5 in	3 in	2.5 in	3 in	3.5 in
	2.5 in	Kevlar	Kevlar	MX500	P4	LM500	Kevlar III	Telex
	Flexibility	Hard	Soft	Medium	Hard	Hard	Soft	Medium
UAV	Model	NA	Cardinal	Buzzard	Crow	Pigeon	Robin	Dove
	Transport System	NA	Track	Track	Quad	Track	Mixed	Quad
Robot Type	Max Speed	NA	2 mph	2.3 mph	4 mph	2.7 mph	5 mph	8 mph
	Operating Time	NA	24 hrs	8 hrs	20 hrs	5.5 hrs	10 hrs	18 hrs
	Range	NA	600 m	500 m	700 m	300 m	700 m	1000 m
	Magnification	NA	1X	2X	1.5X	3X	2.5X	2X
Robot Sensor	Field of View	NA	50 deg	360 deg	180 deg	80 deg	100 deg	75 deg
	Range	NA	300 m	400 m	200 m	500 m	4050 m	500 m

type problems, we typically have adversaries while in the private sector, we have competitors that influence the outcome of a scenario. For the squad enhancement design example, the adversaries are the enemy forces that fight against the squad.

9.3.2.3 Scenarios Scenarios involve political–military situations, missions, and the environment. Scenarios allow systems engineers to understand how a system will operate in different environmental conditions. In order to evaluate the system, system engineers can leverage operational simulations that model the system performing its intended purpose within different scenarios. Because there are many situational outcomes for each scenario, we should use stochastic models that simulate multiple scenarios or run trials. In Figure 9.2, there are two outgoing arrows from the Scenarios uncertainty node to the Priorities and System Models uncertainty nodes; these arrows represent the following two influences: first, each scenario typically has different priorities that depend on the scenario's context; second, scenarios influence the development of system simulation models by specifying what should occur during the conduct of the simulation. In the squad enhancement design example, we use two scenarios that represent the assault and defend missions.

In the assault scenario, the squad conducts an assault to seize terrain and destroy the enemy. The squad moves toward the enemy with their robot in front and UAV flying overhead. The squad calls for indirect fire as they approach the enemy. The enemy is positioned on high ground with Improvised Explosive Devices around their perimeter. The squad establishes a support by fire position with the Automatic Weapons and Grenadiers while the Rifleman assaults the enemy. The enemy calls for another enemy element to reinforce their position. The squad calls indirect fire on the reinforcements once they identify their location.

In the defensive scenario, the squad is in a defensive position in a combat outpost with 10 ft walls, two gate entry points, and fighting positions around their perimeter. Robots are positions outside the perimeter to act as forward sensors. Six enemy insurgents approach the combat outpost wearing suicide vests and attempt to detonate at the gate to breach into the combat outpost, make entry, and detonate additional suicide vest within the perimeter. The enemy then calls indirect fire from a mortar position that is beyond line of friendly sight. Enemy crew serve weapons open fire on the combat outpost from a long range distance while the enemy approaches the combat outpost from three different directions. The squad calls for indirect fire once they identify enemy force locations beyond line of sight. A UAV will loiter over the area of operations in order to provide early warning to the squad.

When developing a scenario, we must ensure that the entire system has an opportunity to perform all of its intended functions in order to properly evaluate each alternative. For example, in the assault and defense scenario, the UAV and robot have an opportunity to identify the enemy and provide earlier warning to the squad beyond line of sight. In addition, both scenarios have enough hostile enemies that fire at the squad in order to evaluate the system effectiveness with regard to the protection function.

9.3.2.4 Priorities In a multiple-objective decision model, priorities are instantiated using swing weights (Chapter 2). If we have one scenario, the priorities would be constant. In general, when there are multiple scenarios, the priorities will be different for each of them (Parnell et al., 1999). The swing weight matrix is a useful way to assess swing weights (Parnell et al., 2013). In the squad enhancement design example, we evaluate the system with respect to the assault and defense scenarios; Tables 9.2 and 9.3 show the swing weight matrices for each scenario, respectively. Within each matrix, the value measures are placed in different importance columns depending on whether they are mission-critical, enable, or enhance the system's capabilities. The row of the matrix classifies the value measure's impact on capability and is based on the range of the value measure scale (large, medium, or small capability gap between the walk-away and the ideal levels). Placing the value measures within the columns and rows of the swing weight matrix provides both a subjective importance criteria and an objective criteria based on the impact of the value measure scale's range. The scenario swing weights quantify the trade-offs between the value measures differently, which have an impact on our decision.

It is important to understand how changes in the swing weight assignments impact the results of the alternative values. A common approach to visualizing the impact of swing weight changes is with a line chart that varies the weight from 0 to 1; see Section 5.3.15.

9.3.2.5 Technology Maturity Technological maturity is one of the most difficult uncertainty nodes to assess and typically has the highest impact on each alternative's risk. The Department of Defense has established nine Technical Readiness Level criteria (TRL 1–9) that describe categories of system readiness acquisition managers use to classify system technological maturity (Mankins, 1995). Each subsystem, component, or part has system features with their own unique level of maturity. The technology drives the uncertainty in the system features and the data used to populate the value measures. Existing system alternatives that are already in use may have features with high levels of technical maturity while nonexistent, conceptual systems may have low levels of maturity. Low levels of system feature maturity can cause higher risk of achieving system value, cost, and schedule. System integration is another challenge that we can classify within the Technology Maturity uncertainty node. Functional analysis identifies the component and part integration requirements that are often the leading cause of system redesign and rework during the life cycle. Depending on the technological maturity of the component or part, there may be significant uncertainty in its ability to integrate with other components or parts in the system. These integration challenges add an additional layer of complexity that increases the risk of achieving system value, cost, and schedule. The arrow in Figure 9.2 from the Technology Maturity uncertainty node to the System Features uncertainty node represents the risk technology imposes on the system feature settings that define each alternative.

9.3.2.6 System Features System features are the characteristics or design parameters that define each alternative. The types of features coincide with the type of

Table 9.2 Swing Weight Matrix for the Assault Scenario

Capability Impact	Mission-Critical			Enables Capability			Enhances Capability		
		Matrix Weight	Swing Weight		Matrix Weight	Swing Weight		Matrix Weight	Swing Weight
Significant impact	Lethality	100	0.14	Weighted mobility	70	0.10	Power	20	0.03
	Lethal mitigation	90	0.13				Logistical impact	20	0.03
	Beyond LOS	90	0.13						
	Kinetic protection	70	0.10						
Medium impact				Bandwidth	50	0.07	Chemical bio protection	15	0.02
	Secured connectivity	65	0.09						
	Communication range	60	0.09	Detection distance	50	0.06			
Minimal impact							Nuclear radio protection	5	0.01

Table 9.3 Swing Weight Matrix for the Defense Scenario

Capability Impact	Mission-Critical			Enables Capability			Enhances Capability		
		Matrix Weight	Swing Weight		Matrix Weight	Swing Weight		Matrix Weight	Swing Weight
Significant impact	Kinetic protection	100	0.15	Beyond LOS	70	0.11	Lethality	30	0.05
	Lethal mitigation	90	0.14				Power	20	0.03
	Detection distance	90	0.14				Weighted mobility	20	0.03
							Logistical impact	20	0.03
Medium impact	Chemical bio protection	65	0.10	Secured connectivity	50	0.08	Communication range	10	0.02
				Bandwidth	40	0.06			
Minimal impact	Nuclear radio protection	60	0.09						

system decision within each phase of the life cycle (concept, architecture, design, or operations). A system feature is analogous to what is known as a local property within physical architecture: a property that is local to a single system element (SEBoK, 2015). In Figure 9.2, the System Features uncertainty node influences the System Models, Value Measure Data, and Resource (Cost) uncertainty nodes. System Models model alternatives within a scenario by setting the model inputs so that the model run represents the alternative of interest. As a result, the system features that define the characteristics of an alternative become the inputs to the simulation model(s). When models are not used to evaluate the whole system or a subset of the system, we acquire value measure data from subject matter experts, development testing, operational testing, and legacy system architectures. System feature settings drive the cost of the alternative. Typically, costs are decomposed into different components that are allocated to subsets of system features. As a result, we can attribute the cost drivers to system features of the alternative. Table 9.1 has examples of system feature settings for each of the seven alternatives in the squad enhancement design example.

9.3.2.7 System Models System models are critical to understanding the behavior of a system, especially in the early stages of the life cycle. There are many types of models that represent different domains and perspectives; they include operational simulations that model a system performing a mission, models that estimate life cycle costs, physics-based computational models that help understand system feasibility, and many more. Because models are abstractions of reality, they must be verified and validated so that they can accurately inform the system decision. The arrow in Figure 9.2 from the System Models uncertainty node to the Value Measure Data uncertainty node represents how the model outputs generate the data used to evaluate an alternative. For each alternative in the squad enhancement design example, we modeled the assault and defense scenarios using an agent-based simulation; for each scenario, we executed 100 trails. The value model used four model outputs to populate the data for the following value measures: Beyond Line of Sight Awareness, Line of Sight Distance, Protection, and Logistical Impact. Because the model is stochastic, each alternative has 100 different outcomes for each of the four outputs.

9.3.2.8 Value Measure Data Data, information, and knowledge along with values and alternatives are what drive decisions. Data is for the systems engineer what electricity is for the electrical engineer and construction material is for the civil engineer; it is the underlying commodity to base all system engineering methods, processes, and activities. We use data to populate value measures for each alternative. Value measures quantify the objectives we use to evaluate our alternatives. See Section 7.7 for details on how to develop value measures. Generally, we obtain data from models and simulations, subject matter experts, development testing, operational testing, and legacy system architectures. The data we use are either known (deterministic) or uncertain (stochastic). Chapter 3 describes a number of ways to handle uncertainty. In this chapter, we demonstrate the use of discrete probability elicitations for the constructed scale value measures, SME estimation of triangular distribution parameters, and agent-based simulation output data for the natural scale value measures.

Table 9.4 shows the value measures used for the squad enhancement design example. Additionally, the table indicates the functions and objectives the value measure evaluates, their types, minimal acceptable levels, ideal levels, and type of uncertainty data.

Table 9.5 shows the assault scenario alternative data for each value measure. The value measure data with a single number in the table are deterministic, while the data with a distribution are stochastic; the mean is shown in the upper right corner. The distributions in Table 9.5 are the actual data distributions used in the model. We note the importance in considering the variability rather than relying on the mean only; the input distributions shown in Table 9.5 indicate that there is significant uncertainty in the alternatives' performance. A companion Excel file is provided on the Wiley website for this book.

9.3.2.9 Resources (Cost) Systems require a number of resources in order to bring the system into being. The most common type of resource that will generally always be present is cost. Life cycle cost estimation is a challenging endeavor, especially for unprecedented systems. Chapter 4 reviews a number of different cost estimating methods. For the squad enhancement design example, the cost model decomposes the system into cost components that align with the major subsystems of the squad system; this alignment serves as the allocation of cost components to subsystems. The cost components are the soldier sensor, radio, rifle, protective suit, the UAV, and robot. In order to capture the life cycle costs, the model incorporates four types of costs: unit costs, training, maintenance, and disposal. The problem assumes that the deterministic costs are derived from the learning curve, analogy, or the cost estimating relationship methods described in Chapter 4. The component costs that are stochastic are derived from a triangular distribution with parameters elicited from subject matter experts.

9.3.3 Constant Node

A constant node is a number or a function that remains constant and is represented by a diamond. There is one constant node in Figure 9.2 that represents the value functions. The value functions are constant with respect to the decision opportunity and the requirements, functions, and objectives decision nodes; when any of these change, the value functions will change.

9.3.3.1 Value Functions Value functions show the returns to scale of the value measures. These functions translate the raw value measure data into a scale between 0 and 100 (0 and 1 or 0 and 10 can also be used). The value function range should include the walk-away point or minimal acceptable value, the threshold, the objective, and the ideal values. The shape of the value function is often dictated by the requirements defined by the systems engineer as either a desired capability or a constraint. The constraints are what set the minimal acceptable level or walk-away point, while the desired capabilities help determine the marginally acceptable, target, stretch goal, and meaningful limit levels; see Section 5.3.2. Table 9.6 shows the value functions for the squad enhancement design example.

Table 9.4 Value Measures for the Squad Enhancement Design Example

Function	Objective	Value Measure	Type	Minimal Acceptable Value	Ideal Value	Uncertainty Type
Maintain situational awareness	Increase beyond line of sight awareness	Beyond LOS (% detected)	Natural	0	1	Simulation output
	Increase line of sight range	Detection distance (meters)	Natural	300	1500	Simulation output
Maintain networked communications	Maximize range	Communication range in various terrain	Constructed	1	5	Probability elicitation
	Maximize bandwidth	Bandwidth (mbps)	Natural	3	15	Triangular distribution
	Provide secured connectivity	Secured connectivity	Constructed	1	8	Probability elicitation
Maneuver the squad	Increase soldier mobility	Weighted mobility	Multi-dimensional constructed	1	8	Probability elicitation
Protect the squad	Protect against kinetic threats	Kinetic protection	Constructed	1	9	Probability elicitation
	Protect against chemical, biological, radiological, nuclear threats	Chemical biological protection	Constructed	1	9	Probability elicitation
		Nuclear radiological protection	Constructed	1	7	Probability elicitation
Achieve mission effects	Maximize kinetic effects	Lethality (% enemy killed)	Natural	0	1	Simulation output
	Minimize lateral damage	Lethal mitigation	Constructed	1	5	Probability elicitation
Sustain the squad	Maximize power efficiency	Power (kw/h)	Natural	300	100	Triangular distribution
	Minimize logistical footprint	Logistical impact (rounds fired)	Natural	6000	1000	Simulation output

Table 9.5 Squad Scores on Each Value Measure

	Beyond LOS	Detection Distance	Commo Range Various Terrain	Bandwidth	Secured Connectivity	Weighted Mobility	Kinetic Protection	Chem Bio Protection	Nuclear Radio Protection	Lethality	Lethal Mitigation	Power	Logistical Impact
Baseline	0	300	1	5	2	1	1	1	1	1	1	280	5500
LongRange	0.50	1300	4	13	7	7	6	4	4	0.41	4	110	4042
Attack	0.80	983	4	9	5	7	5	8	3	0.39	2	120	4278
Sustainable	0.84	1063	2	10	3	3	3	2	1	0.38	3	130	4213
Survivable	0.55	957	4	12	8	4	8	7	7	0.30	4	150	3695
Defendable	0.68	1084	3	6	6	5	7	7	3	0.74	4	110	3704
Performance	0.57	1400	5	12	7	8	9	5	4	0.69	3	110	2728

Table 9.6 Squad Enhancement Design Example Value Functions

Value Measure	Value Function	Value Measure	Value Function	Value Measure	Value Function
Beyond LOS (% detected)		Weighted Mobility		Lethality (% enemy killed)	
Detection Distance (meters)		Kinetic Protection		Lethal Mitigation	
Communication Range in Various Terrain		Chemical Biological Protection		Power (kw/h)	
Bandwidth (mbps)		Nuclear Radiological Protection		Logistical Impact (rounds fired)	
Secured Connectivity					

9.3.4 Value Nodes

The value node is a hexagon representing either the total value or total life cycle cost of an alternative.

9.3.4.1 Total Value The decision total value can be calculated based on decision, the value measure score, the value functions, and the priorities (swing weights) using the additive value model (see equations 2.1 and 2.2). One way to understand how each alternative achieves each of the objectives in the value hierarchy is to use a value component chart. Figure 9.4 shows the value components of each alternative as a stacked bar chart. To the right, we see the *Ideal* alternative that represents the maximum possible value score, derived from the swing weights for each value measure. The *Hypothetical Best* alternative is the maximum value achieved for each value measure in the set of alternatives. The difference between the *Ideal* and *Hypothetical Best* is the value gap that the set of alternatives cannot achieve with existing technologies. Depending on its size, we may want to consider including other new alternatives that close the gap. In addition, we may have an opportunity to combine components from the existing alternatives to create new alternative. When we reconfigure system components to create new alternatives, we must consider the system integration

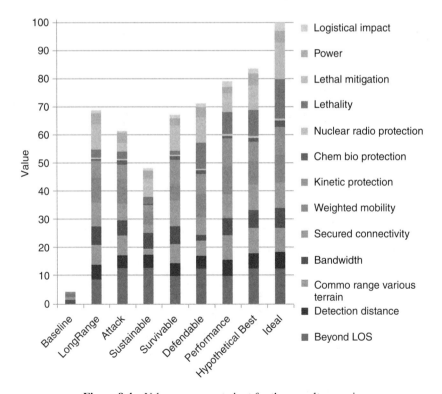

Figure 9.4 Value component chart for the assault scenario

challenges that may result as described earlier in the Technology Maturity uncertainty node section.

9.3.4.2 Total Cost The total cost value node is the estimated total life cycle cost for each alternative. Because life cycle cost is such an important aspect of any systems decision, we often exclude cost as a value measure and treat it as an independent variable. An effective way to understand the trade-offs between value and cost is to use a Pareto chart. The Pareto chart is a scatter plot with cost on the horizontal axis and value on the vertical axis; each dot represents an alternative's total life cycle cost and total value score. Figure 9.5 shows a deterministic cost versus value Pareto chart using the average costs and value scores from the squad enhancement design example. We used a notional cost spreadsheet model described in Section 9.3.2 to calculate each alternative's life cycle cost. We can see that the *Sustainable* and *Survivable* alternatives are deterministically dominated by all the others. It makes no sense to select a dominated alternative when we can select nondominated alternative with a higher value for less cost. The set of nondominated alternatives is known as the Pareto Frontier.

However, Figure 9.5 does not consider the uncertainty associated with each alternative's value and cost and does not show the risk or the probability of a lower value. The majority of value trade-off studies in the literature are deterministic. Eliminating deterministically dominated alternatives without considering risk may lead to the wrong decision. Cost estimation techniques already incorporate uncertainties using

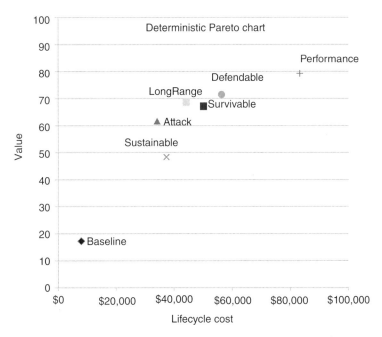

Figure 9.5 Deterministic Pareto chart for the assault scenario

Monte Carlo simulations and other methods. A major contribution of our approach is that we simultaneously integrate value trade-off uncertainties and cost uncertainties in order to facilitate better value and risk identification. If we do not consider how much variation there is in the consequences of our decision, we may end up making the wrong decision.

9.3.4.3 Integrated Trade-Off Analysis

The integrated trade-off analysis value node represents our approach that simultaneously models value and cost uncertainties to better identify value and risk. Figure 9.6 illustrates the integrated approach by showing how we use a variety of uncertainty modeling methods to propagate uncertainty through the value and life cycle cost models in order to analyze value and risk. After performing a Monte Carlo simulation, we have a collection of value and cost vector data for each alternative.

When faced with an uncertain system decision, we can leverage three types of analytical charts that help the systems engineer understand the risk associated with a decision and identify what drives the uncertainties in each alternative. First, stochastic Pareto charts identify nonstochastically dominated alternatives with respect to value and cost; second, cdf charts (*S*-curves) compare the alternative risk profiles; and third, tornado charts identify the value measures and cost components that have the highest impact on the alternative uncertainties. We will now describe each of these charts separately and use the squad enhancement design example to demonstrate the insights we can obtain from them with respect to value and cost.

9.3.4.4 Stochastic Pareto Chart

In order to address the limitations of the deterministic Pareto chart, we create a stochastic Pareto chart, shown in Figure 9.7, by displaying a two-dimensional box plot for each alternative's cost and value. The boxes along each axis represents the second and third quantiles while the lines represent the first and fourth quantiles of the vector output data from the value and cost models. We can create these box plots in Microsoft Excel using a scatter plot with a combination of four data series for each alternative, two for the cost axis and two for the value axis. The lines, otherwise known as whiskers, are created from the vector data using the maximum and minimum data points. The boxes are created using the third quantile, mean, and first quantile; to create the box, increase the line style width of the data series. The box plots allow us to understand the uncertainty associated with each alternative's cost and value simultaneously.

The stochastic Pareto chart allows us to consider value, risk, and dominance simultaneously. In addition, it provides important information for affordability analyses (see Chapter 3). We can see in Figure 9.7 that *Sustainable* is stochastically dominated by all other alternatives and can be eliminated from consideration. If *Performance* and *Defendable* are affordable we then focus on understanding what drives their uncertainty to mitigate risk. If *Defendable* is not affordable, we then consider either *LongRange* or *Survivable*. If we used the deterministic Pareto chart from Figure 9.5 to eliminate the *Survivable* alternative as a dominated solution, we would have missed an important trade-off consideration. We can see in Figure 9.7 that *LongRange* has a higher risk in value (probability of lower value) compared to *Survivable*. We may

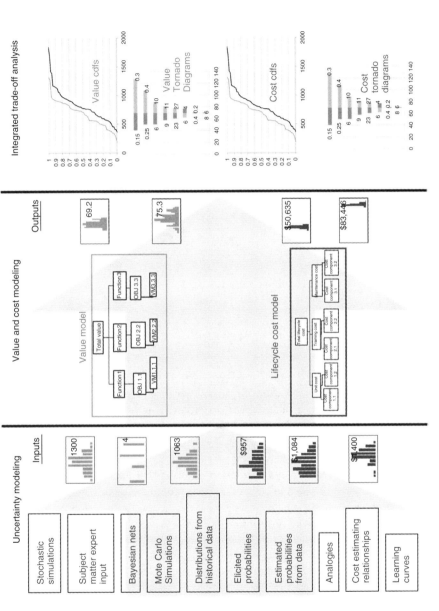

Figure 9.6 The integrated approach

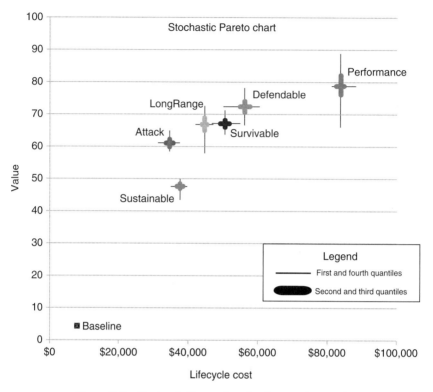

Figure 9.7   Stochastic Pareto chart for the assault scenario

want to accept a higher cost by choosing *Survivable* to mitigate the risk associated with *LongRange*. Of course, an alternative is to choose *LongRange* to reduce the risk. In order to better understand the risk implications, we can use cdf charts to compare alternative risk profiles.

9.3.4.5 Cumulative Distribution Function (S-Curve) Charts
The cdf chart displays an alternative's potential outcomes by accumulating the area under the outcome's probability mass functions for discrete data and the probability density functions for continuous data. Typically, the shape of the line in the chart is an S-curve that depicts the probability that the outcome will be at or below a given value. The horizontal axis has the outcome scale, either value or cost, while the vertical axis has the probability. Figure 9.8 shows a cdf chart with six S-curves that represent the uncertain alternative value outcomes, otherwise known as the risk profiles. *Sustainable* is deterministically dominated by the other five alternatives. *Attack* is deterministically dominated by *Survivable*, *Defendable*, and *Performance*. *Survivable* is deterministically dominated by *Defendable* and *Performance*. The *Performance* and *Defendable* alternatives stochastically dominate all others because their S-curves are positioned completely to the right of all others. The risk profiles

INTEGRATED APPROACH INFLUENCE DIAGRAM

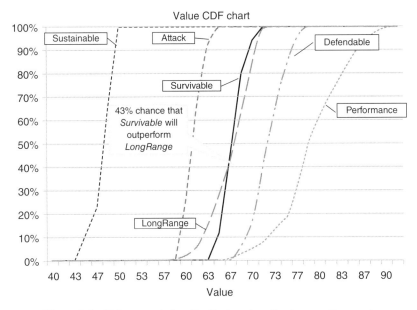

Figure 9.8 Value cumulative distribution chart for the assault scenario

of *Survivable* and *LongRange* value cross indicating that there is no clear winner between the two; we may want to accept a higher cost by choosing *Survivable* to mitigate the risk associated with *LongRange*. The cdf chart tells us that there is a 43% chance *Survivable* will outperform *LongRange* and that *Survivable* has less risk due to its steeper risk profile. In general, when the risk profiles of alternatives cross, we then consider risk preference (risk-averse, risk-neutral, or risk-taking) during our system decision. A risk-averse decision would spend more for *Survivable* to guarantee a higher value, while a risk-taking decision would select *LongRange* to save money with the risk of achieving less value.

When we want to understand how to best mitigate an alternative's risk, we can use tornado diagrams to identify the value measures and cost components that have the highest impact on value and cost, respectively.

9.3.4.6 Tornado Diagrams An effective way to identify the impact of uncertainty is to perform sensitivity analysis using tornado diagrams. Tornado diagrams allow us to compare the relative importance of each uncertain input variable with horizontal bars; the longer the bar, the higher the impact on the output variable's variation. The bars are sorted so that the longest bars are at the top; sorting the bars in this way makes the diagram look like a tornado. The length of the bars depends on the type of tornado diagram. Deterministic tornado diagrams vary each input variable using low, base, and high settings while all other input variables are held constant. A stochastic tornado diagram uses the vectors of input and output variable trials from a Monte Carlo simulation (Parnell et al., 2013). The low end of the bar is the average output variable

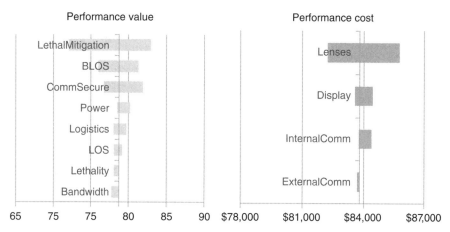

Figure 9.9 Value and cost stochastic tornado diagrams for the *performance* alternative from the assault scenario

from the subset of trials where the input is less than a specified lower percentile. Similarly, the high end of the bar is the average output variable from the subset of trials where the input is greater than a specified higher percentile. For the squad enhancement design example, we use stochastic tornado diagrams with a low percentile of 0.3 and a high percentile of 0.7. Figure 9.9 shows the value and cost tornado diagrams for the *Performance* alternative. The input variables for value are the value measures and the input variables for cost are the cost components. The horizontal axis shows each input variable's impact on the total variation for the value and cost. We can see in Figure 9.9 that the LethalMitigation value measure has the highest impact on the value's variation while the Lenses cost component has the highest impact on the cost's variation.

Prior to performing the integrated trade-off analysis, we allocated functions and cost components to subsystems (see Figure 9.3 for an example). Our value hierarchy contains objectives that assess the performance of functions and value measures that define how well an alternative achieves the objectives. As a result, the value measures are indirectly allocated to system features through the objectives and functions. Cost components are generally allocated directly to subsystems. Because of these indirect and direct allocations, we can use tornado diagrams to identify the system features that have the highest impact on the system decision. Figure 9.10 illustrates how we use tornado diagrams to indirectly trace high impact value measures and directly trace cost components to system features.

As we learned from our stochastic Pareto chart and cdf charts, we do not have a clear winner between the *Survivable* and *LongRange* alternatives. To help understand how these alternative uncertainties impact the system decision, we can use tornado diagrams to identify the system features that are driving risk.

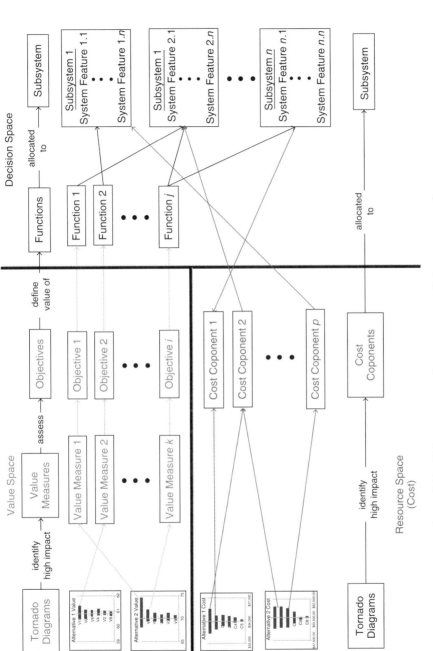

Figure 9.10 Value measure and cost component linkage to system features

We can see from the top bar of the tornado diagrams in Figure 9.11 that the radio drives the majority of the risk for the *LongRange* value while soldier sensor drives the majority of its cost. The protective suit drives the majority of the risk for the *Survivable* value while the soldier sensor drives the majority of its cost. These subsystems and the system features that characterize them have the highest impact on the system decision. These insights provide a clearer understanding of what drives the alternatives' risk and how to prioritize system feature refinements. For example, the systems engineer can reduce the risk of the *LongRange* alternative by investing more resources into improving the radio's range, security, and bandwidth features.

9.4 OTHER TYPES OF TRADE-OFF ANALYSIS

There are a number of other types of trade-off analysis we can perform using the integrated approach. The approach can be applied to evaluate any type of system decision encountered throughout the life cycle; the key application differences are the types of data, information, models, value measures, and system features used during the decision. Chapters 10–14 will explain in more detail the concept, architecture, design, sustainment, and programmatic decisions, respectively. Often times, we must compare alternatives across different scenarios in order to capture their value differences and consider them in the systems decision. When we can assume that only one system component or subsystem affects only one value measure, we can perform system component optimization to identify a solution that maximizes value under a set of constraints (Parnell et al., 2011). When our value component chart reveals an unacceptable value gap, we can evaluate new technologies that will achieve higher value. When the system design specifications are frozen during development or while the system is in the operational stage, we can use the approach to prioritize system modifications; these modifications typically add new components as new technologies emerge. We can examine the tornado diagrams and the lower end of the *S*-curve in the cdf chart to identify the source of risk and support risk management programs during all stages of the life cycle.

9.5 SIMULATION TOOLS

In order to effectively deal with the multiple sources of uncertainty, we need software that can facilitate Monte Carlo simulations. We prefer tools that are Microsoft Excel Add-Ins so that we can build our value and cost models in the same environment. There are a number of software features that every tool should offer; these include the capability to run thousands of trials, model correlated uncertain input variables, identify the uncertain input variables that have the highest impact on the model's variation, and clearly display the results of the output distributions. Our next two sections discuss some of the Microsoft Excel Add-Ins available to perform Monte Carlo simulations.

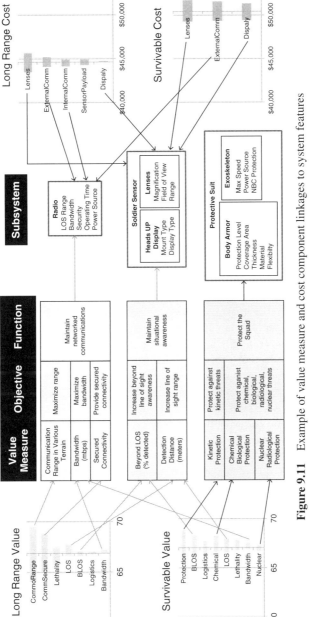

Figure 9.11 Example of value measure and cost component linkages to system features

9.5.1 Monte Carlo Simulation Proprietary Add-Ins

There are a variety of tools used to conduct Monte Carlo simulation analysis. Three of the leading proprietary software tools are Crystal Ball (www.crytalball.com), @Risk (www.palisade.com), and Risk Solver (www.solver.com/risksolver.htm). All three of these tools provide an intuitive user interface that facilitates all the software features mentioned earlier. The key advantage of using any of these proprietary software tools is the relative ease of use and accessibility of the output analysis charts they provide. The disadvantages are the cost, the need to create distributions data from other simulations, and that they create static output charts requiring the user to rerun the simulation after changing the input variables.

9.5.2 The Discipline of Probability Management

ProbabilityManagement.org (http://probabilitymanagement.org/), a 501(c)(3) nonprofit, has pioneered an approach to simulation based on an open data standard for storing Monte Carlo trials. The new data standard, known as the Stochastic Information Packet (SIP), ushers in a new category of data, which simulates the future instead of recording the past. The SIP makes the abstract concept of a probability distribution **actionable**, **additive**, and **auditable**. The discipline of probability management was formalized by Savage et al. (2006) and further developed by Savage (2012). Traditional simulation illuminates uncertainty by generating random variates and running them through an analytical model in a single application. The discipline of probability management allows random variates to be generated in one application for use in other applications. As an analogy, substituting electricity for random variates, the former is like a generator with a light bulb attached, while the latter is like a collection of generating plants, light bulbs, and appliances connected by a power grid. The SIPs are the electricity, which allow the results of simulations to be aggregated across such platforms as Crystal Ball, @RISK, Risk Solver, Matlab, and R. A coherent collection of SIPs that preserve statistical dependence is known as a Stochastic Library Unit with Relationships Preserved (SLURP).

Computers are now fast enough to perform interactive simulation, in which thousands of trials are processed in real time while the user adjusts parameters of the model. This provides an experiential understanding not possible with command driven simulation. Just as light bulbs may be used by those with no knowledge of how the electricity was generated, probability management enables stochastic dashboards for managers with little understanding of how the random variates were generated. Nonprofit probabilitymanagement.org developed and maintains the SIPmath™ open cross-platform standard for SIP libraries, which may be easily generated by such software stored as XML, CSV, or XLSX files. It also provides a suite of tools to facilitate the generation of SIP libraries and SIPmath models.

9.5.3 SIPmath™ Tool in Native Excel

The open SIPmath™ standard is platform agnostic, but fortunately, the Data Table function in native Microsoft Excel is now powerful enough to run thousands of trials

SIMULATION TOOLS   325

through a model extremely quickly http://viewer.zmags.com/publication/90ffcc6b#/90ffcc6b/29 (Savage, 2012). Not only is this approach free to any Excel user, but it is fully interactive. Because setting up the Data Table for this purpose can be time-consuming, the nonprofit offers a free add-in to automate this. Their SIPmath™ Modeler Tools create simulations in Excel, which do not require the tool to rerun once the inputs are changed. When the user changes inputs, the outputs are dynamically updated after each change; this feature allows the user to build compelling, dynamic, custom-made dashboards in the Microsoft Excel environment without the need for macros. The SIPmath Modeler Tools operate in two modes: Random and SIP Library Modes.

9.5.3.1 Random Mode In this mode, random inputs are generated with built-in Excel generators. This can quickly create interactive Monte Carlo simulations in Excel, in which thousands of trials are run before the user's finger leaves the <Enter> key. Since inputs are produced dynamically by Excel, the resulting output distributions will vary from run to run.

9.5.3.2 SIP Library Mode Stochastic library mode utilizes precompiled Monte Carlo trials (SIPS) to represent the uncertainty in the model inputs. This guarantees repeatability of results across multiple users and trials and allows results to be aggregated across applications. A SIP Library may be either built into the model workbook or linked to as an external file, for example, in the cloud, so that many users share a common set of trials.

We used the SIP Library Mode for the squad enhancement design example model. This model does contain macros for data manipulation, but the simulation is entirely performed within the Data Table. Within our squad enhancement model, there are three types of uncertain data: simulation output data, elicited discrete probability distributions, and triangular distributions. The model used the INDEX function and the Data Table functionality to generate the random variates for each of the discrete and triangular distributions. We used the SIPmath™ Modeler Tool Add-in to define cells as inputs for the simulation data SIPs. We defined the cells that needed the discrete and triangular distributions as output cells in order to generate a SIP of random variates from each of these distributions. Finally, we defined the cells that contained the total value scores and life cycle costs for each alternative as outputs to generate their SIPs of random variates. The SIPmath™ Modeler Tool Add-in uses the Microsoft Excel sparkline feature to display the input and output data distributions within a cell (see Table 9.5 for an example).

9.5.4 Model Building Steps

This section outlines the steps we used to build our value and cost models using the SIPmath™ Modeler Tool Add-In. As a reminder, there are two major differences between SIPmath models in Excel and traditional Monte Carlo simulation. First, SIPmath is interactive in native Excel, in that a full simulation is run with each keystroke using the Excel Data Table. Thus, the simulation will not require macros and may be stored in an .xlsx file without the need for additional software. Second, the random trials may be run in advance and stored as auditable data in a SIP Library in the open SIPmath format. Among the tools at probabilitymanagement.org are macros that allow both the @RISK and Crystal Ball simulation packages to write their results directly into this format. The SIP libraries generated may be used in two ways: either built into the model workbook or linked as external files. The squad model has the library built in, so that it may be widely distributed without linking to other files. In a setting in which a new SIP Library is published periodically to reflect the latest probabilistic estimates, it makes more sense to have all users link to the same external file.

Before we discuss each step, we want to note that we should not use functions defined using Visual Basic for Applications (VBA). We can have macros that perform procedures in Excel, but they cannot be user-defined VBA functions because they will slow down the Data Table calculations considerably. To create our value functions, we used the INDEX and MATCH functions for the natural scales and the VLOOKUP function for the constructed scales; see the Excel model that accompanies this book for the exact formula. We also note that the INDIRECT formulas should be used sparingly in models as it also slows down the Data Table. For more on building SIPmath models in Excel, visit the Tools page of Probability Management.org for videos, tutorials, and documentation.

Prior to using the SIPmath^{TM} Modeler Tool Add-In, it is important to organize the model so that each value measure and cost component for each alternative has a unique name specified in a cell. We organized these names above the input data and to the side of our output data so that they are easily identified when defining our inputs and outputs. Figure 9.12 shows a screenshot of our model organization with the input names positioned above the input data cell entry area and the output names alongside the output data cells. At the lower right of Figure 9.12, there is a screenshot of the SIPmath^{TM} Modeler Tools ribbon.

Step 1: Initialize. Before we initialize, we must create a named range called "PM_Trials" that contains the number of trials in each of our SIPs; an appropriate place to name this range is at the top of our worksheet that contains the input distribution column data. We then press the "Initialize" button in the SIPmath ribbon and select the "In current workbook" radio button, specify the default number of bins we want to display in our graphics, and press ok. The "In current workbook" tells the tool that the column data reside in the model workbook. You should now notice that there is a "PMTable" and a "SIP Chart Data" worksheet created for you. The "PMTable" worksheet will contain a Data Table that creates the columns of outputs. In cell A1, there is a named range called "PM_Index." The "PM_Index" is the row index used for the Data Table row input cell. Next, we must name each input distribution column as a named range. An effective way to name a collection of column data with the column header at the top row is to use the "Create from Selection" Excel feature found in the "Formulas" ribbon tab under the "Defined Names" group.

Step 2: Define inputs. Our next step defines input distributions in each of the appropriate empty cells in Figure 9.12. These cells represent an alternative's value measure that has input distributions that will be defined as a SIP. We can define multiple inputs simultaneously if they are positioned contiguously, otherwise we must define them separately. First, select the cell(s) in the alternative data entry area that need SIPs assigned to them. In the SIPmath ribbon, we click the "Define Inputs" button and designate the starting cells for the input names located above the entry area in Figure 9.12. These input ranges can be arranged as rows or columns; our data is arranged as columns. Under the window titled "Select Input," select the named ranges that contain the input SIP data located in the "SIP Library" worksheet. Within the cells we selected, we should now see sparklines that show the input data distributions they were assigned.

Step 3: Define outputs. We now will define outputs for two types of data. For the first type, we must generate columns of data that result from our triangular distributions and discrete probability distributions. The second type will be for the actual value and life cycle cost outputs for each alternative.

9.5.4.1 Triangular and Discrete Probability Distribution Output Creation
Within our model, we have a worksheet named "Uniform Library" that has a collection of columns with 100 trials from a uniform distribution, the 101st row contains the mean, the 102nd row contains the 5th percentile, and the 103rd row contains the 95th percentile. We created these columns using the RAND Excel function and copied and pasted special values; we create a unique named range for each of them in the same way we did for the input distributions. We must ensure that we create the same number of trials for the uniform distributions as we do for the input distributions defined earlier. In another area of our workbook, we have triangular distribution formulas with the minimum, maximum, and mode parameter entry cells for each value measure and cost component that use them. For the discrete probability distributions, we use a probability table that contains a row for each constructed scale category and an ascending cumulative table that adds the probabilities for each category. The VLOOKUP formula has the last parameter set to "True" so that the function looks up the closes matching random number value (between 0 and 1) within the cumulative table and returns the category number. For each distribution, the cell assigned for the random number parameter uses an INDEX function with a named range that represents a unique uniform random number column as the array parameter and the "PM_Index" as the row index parameter. In the "PMTable" worksheet, if we enter 101 as the "PM_Index," we can show the mean value for all inputs and outputs in our model. For each cell in Figure 9.12 that uses the triangular and discrete distributions, we set them equal to the cells that provide the output for each of the distributions. We then select each cell in the alternative data entry area shown in Figure 9.12, click "Define Outputs" in the SIPmath ribbon, assign the appropriate output named range, and click ok;

	Squad achieves overmatch aganist enemies in complex environments													
Funciton	Command and control the Squad					Maneuver the squad	Protect the Squad			Achieve Mission Effects		Sustain the Squad	
Sub-funciton	Maintain situational awareness		Maintain networked communications										
Objectives	Incr. beyond line of sight awareness	Incr. line of sight range	Max. range	Max. bandwidth	Provide secured connectivity	Incr. soldier mobility	Protect aganist kinetic threats	Protect aganist chemical, biological, radiological, nuclear threats		Maximize kinetic effects	Min. lateral damage	Max. power efficiency	Min. logistical footprint
Value Measures	Beyond LOS	Detection Distance	Commo Range Various Terrain	Bandwidth	Secured Connectivity	Weighted Mobility	Kinetic Protection	Chem Bio Protection	Nuclear Radio Protection	Lethality	Lethal Mitigation	Power	Logistical Impact
Short Names for named range inputs	B BLOS	B Detect	B Commo	B Bandwidth	B Secure	B Mobility	B Protect	B Chem	B Nuc	B Lethal	B LethalMitigate	B Power	B Log
	ATK_BLOS_LR	ATK_LOS_LR	LR_Commo	LR_Bandwidth	LR_Secure	LR_Mobility	LR_Protect	LR_Chem	LR_Nuc	ATK_LethalLR	LR_LethalMitigate	LR_Power	ATK_Sustain_LR
	ATK_BLOS_A	ATK_LOS_A	A_Commo	A_Bandwidth	A_Secure	A_Mobility	A_Protect	A_Chem	A_Nuc	ATK_Lethal_A	A_LethalMitigate	A_Power	ATK_Sustain_A
	ATK_BLOS_Sus	ATK_LOS_Sus	Sus_Commo	Sus_Bandwidth	Sus_Secure	Sus_Mobility	Sus_Protect	Sus_Chem	Sus_Nuc	ATK_Lethal_Sus	Sus_LethalMitigate	Sus_Power	ATK_Sustain_Sus
	ATK_BLOS_Sur	ATK_LOS_Sur	Sur_Commo	Sur_Bandwidth	Sur_Secure	Sur_Mobility	Sur_Protect	Sur_Chem	Sur_Nuc	ATK_Lethal_Sur	Sur_LethalMitigate	Sur_Power	ATK_Sustain_Sur
	ATK_BLOS_D	ATK_LOS_D	D Commo	D Bandwidth	D Secure	D Mobility	D Protect	D Chem	D Nuc	ATK_Lethal D	D LethalMitigate	D Power	ATK_Sustain_D
	ATK_BLOS_P	ATK_LOS_P	P_Commo	P_Bandwidth	P_Secure	P_Mobility	P_Protect	P_Chem	P_Nuc	ATK_Lethal_P	P_LethalMitigate	P_Power	ATK_Sustain_P
Baseline													
LongRange													
Attack													
Sustainable													
Survivable													
Defendable													
Performance													

Value Output Names	**Value**	Cost Output Names	**Cost**
V_Baseline		C_Baseline	
V_LongRange		C_LongRange	
V_Attack		C_Attack	
V_Sustainable		C_Sustainable	
V_Survivable		C_Survivable	
V_Defendable		C_Defendable	
V_Performance		C_Performance	

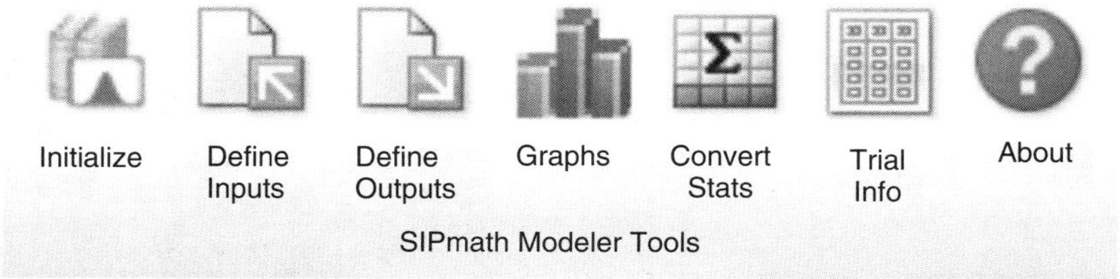
Initialize
Define Inputs
Define Outputs
Graphs
Convert Stats
Trial Info
About
SIPmath Modeler Tools

Figure 9.12 Value model named range data entry setup and SIPmath ribbon

we should now see sparklines that show the output distributions of the 100 trials of uniform random variates that were passed through each of the triangular and discrete distribution functions. You will now see the output data columns from these distributions in the "PMTable" worksheet.

9.5.4.2 Value and Cost Output Creation In order to create these output columns, we simple select the cells that contain the value and cost outputs, click the "Define Outputs" in the SIPmath ribbon, and designate the output named ranges. We should now see sparklines for the resulting distributions. Table 9.5 shows a screenshot of the squad model input data entry area with the sparklines for each uncertain input.

9.6 SUMMARY

This chapter demonstrated the integrated trade-off approach that simultaneously models value and cost in order to better identify value and risk. Trade-off decisions are present throughout the system life cycle; therefore, the trade-off analysis techniques we use have a high impact on the quality of our system decisions. The majority of value trade-off studies are deterministic without considering the types of uncertainties associated with a system decision. We used an influence diagram to express the types of decisions and uncertainties that influence system value and cost. Chapters 3 and 6 reviewed a variety of methods to model uncertainty and cost. In this chapter, we demonstrated how to use these methods to propagate uncertainties through the value and cost models with Monte Carlo simulations. We then used stochastic Pareto charts to visualize the Pareto Frontier, cumulative distributions function charts to display alternative risk profiles, and tornado diagrams to identify high impact value measures and cost components that are indirectly and directly allocated to system features. We learned that deterministic analysis does not tell the whole story and can mislead system decisions when we do not consider the types of uncertainties shown in our influence diagram. More often than not, it is unclear which alternative is the true winner in terms of value and cost because of the system decision uncertainties. The alternative risk profiles that cross in the cdf charts highlight the unclear winners that need further investigation. To better understand how to address alternative risk, we can trace the allocation of the high impact value measures and cost components identified by the tornado diagrams to system features. We can then invest more resources to improve these system features in order to reduce risk. Finally, we mentioned a few Monte Carlo software packages that facilitate the integrated approach, discuss the advantages and disadvantages of them, and implement our approach using a freely available software tool by Probability Management.org.

System decisions involve several stakeholders, multiple conflicting objectives, and a variety of different uncertainties. Our integrated trade-off approach incorporates all three of these aspects in order for the wider community to make better quality system trade-off decisions. We do this by providing the decision-makers all the critical trade-offs and risks associated with each alternative. Generally, the higher the value, the higher the risk; understanding where these risk reside facilitates a better quality system design decision.

9.7 KEY TERMS

Constant Node: A number or a function that remains constant and is represented by a diamond within an influence diagram.

Decision Node: A rectangle within an influence diagram that represents the set of choices the decision-maker must make.

Functional Allocation: The allocation of functions to system components, parts, and system features that will perform them.

Integrated Trade-Off Analysis: An approach that integrates uncertainty modeling with value and cost modeling in order to help understand value and risk while we analyze alternatives.

Pareto Chart: A scatter plot with cost on the horizontal axis and value on the vertical axis; each dot represents an alternative's total life cycle cost and total value score.

Pareto Frontier: The set of nondominating alternatives such that value or cost cannot be improved without degrading the other.

Probability Management: A nonprofit organization that uses computer technology to address the Flaw of Averages through improvements in communication, calculations, and credibility of uncertainty estimates.

Scenario: A sequence of events used to evaluate a system's performance in different environmental conditions.

Stochastic Information Packet (SIP): Data arrays used to communicate uncertainties.

Stochastic Tornado Diagrams: Compare the relative importance of each uncertain input variable with horizontal bars; the longer the bar, the higher the impact on the output variable's variation. The bars are sorted so that the longest bars are at the top; sorting the bars in this way makes the diagram look like a tornado. The low end of the bar is the average output variable from the subset of trials where the input is less than a specified lower percentile. Similarly, the high end of the bar is the average output variable from the subset of trials where the input is greater than a specified higher percentile.

System Feature: The characteristics or design parameters that define each alternative. A system feature is analogous to what is known as a local property within physical architecture, a property that is local to a single system element. The settings of the system features define each system alternative.

Technology Maturity: The technical readiness level of a system feature, component, or part that often drives the system design risk. Unprecedented systems with advanced system features typically have a low level of technological maturity.

EXERCISES   331

Uncertainty Node: Ovals within an influence diagram that represent uncertain information relevant to the decision; they could be a single probability value, a random variable, or a vector of data.

Value Node: A hexagon within the influence diagram representing either the total value or total life cycle cost of an alternative.

9.8 EXERCISES

9.1. This chapter is an example of the Decision Management process presented in Chapter 5.
  (a) Identify the techniques that were used in the chapter for each of the 10 steps in the Decision Management process.
  (b) Were any of the 10 steps not included in this chapter?

9.2. Integrated trade-off analysis model.
  (a) Use Figure 9.6 to describe an integrated trade-off analysis model.
  (b) Why does this chapter advocate an integrated trade-off analysis approach?

9.3. The squad enhancement model illustrated in this chapter used three types of uncertainty data: probability elicitation, distribution (triangular), and simulation output.
  (a) Are there other types of uncertainty data that could be used in the integrated approach?
  (b) Briefly describe the criteria a systems engineer should use to determine the most appropriate type of uncertainty data to use in the integration approach.

9.4. Briefly explain how to construct and how to interpret the results of the following three outputs of an integrated trade-off analysis model.
  (a) Stochastic Pareto charts
  (b) cdf charts (*S*-curves)
  (c) Stochastic tornado charts

9.5. Briefly describe the advantages and disadvantages of using the SIPmath approach versus a proprietary Excel add-in to perform Monte Carlo simulation

9.6. The following exercises walk you step by step through an example that builds a Monte Carlo simulation using the SIPmath Modeler Tool add-in for a simple value and cost model that uses the integrated approach discussed in this chapter. Note that when using any other Excel templates provided with the book for new work, be sure to delete any existing named ranges using Named Range Manager feature.
  (a) In Excel, create three value functions charts for the following three value measures using the raw data and value measure scores given. For the natural scale, use a line chart, and for the constructed scale, use a column bar chart.

Natural Scale Measure		Constructed Scale Measure		Natural Scale Measure	
Maximum Vehicle Speed (mph)		Vehicle Safety Star Rating Category		Miles per Gallon (mpg)	
Raw Data (X)	Value Scores	Raw Data (X)	Value Scores	Raw Data (X)	Value Scores
45	0	1	0	8	0
50	30	2	20	20	20
60	60	3	50	30	65
100	80	4	90	40	90
130.001	100	5	100	50.001	100

(b) You are given three vehicle alternatives that you want to assess using the value functions defined in problem 1 called *Baseline*, *Frontier*, and *Starlight*. *Baseline* has a maximum vehicle speed of 60 mph, a vehicle safety star rating of 3, and 18 mpg. *Frontier* has a maximum vehicle speed of 90 mph, a vehicle safety star rating of 3, and 42 mpg. *Starlight* has a maximum vehicle speed of 100 mph, a vehicle safety star rating of 4, and 22 mpg. In Excel, create the value function formulas for the natural value measure using the INDEX and MATCH Excel functions and the constructed scale value measure using a VLOOKUP Excel function. Note that we cannot use a macro-enabled function while using the SIP Math modeling tool. Calculate the value measure scores for each alternative. In addition, calculate the *Hypothetical Best* alternative using the MAX or MIN Excel functions depending on if more is better or less is better. Finally, calculate the *Ideal* alternative.

Use the following procedure to create the natural value measure function. For the Maximum Vehicle Speed natural measure, assume that the raw data column has a named range of **RawSpeed**, the value scores data column has a named range of **ValueSpeed**, and the maximum vehicle speed for the *Baseline* alternative has a named range of **BaseSpeed**. To calculate the value for the natural measure, use the following Excel function: =**INDEX (ValueSpeed,MATCH(BaseSpeed,RawSpeed))+(BaseSpeed-INDEX (RawSpeed,MATCH(BaseSpeed, RawSpeed)))*(INDEX(ValueSpeed, MATCH(BaseSpeed, RawSpeed)+1)-INDEX(ValueSpeed, MATCH (BaseSpeed, RawSpeed)))/(INDEX(RawSpeed, MATCH(BaseSpeed, RawSpeed)+1)-INDEX(RawSpeed,MATCH(BaseSpeed,RawSpeed)))**. Note that we do not have to name the data ranges for this formula to work. We assume that they are named as defined earlier in order show the Excel function with the appropriate cell ranges. Also note that the bottom-most value in the table has 0.001 added to it. We must add a small amount to the last value in order for the INDEX and MATCH functions to work properly.

(c) Calculate the total value score for each alternative using the matrix weights for the value measures shown in the following table and create a value component chart using a stacked bar chart.

Value Measure	Matrix Weight
Maximum vehicle speed (mph)	100
Vehicle safety star rating	65
Miles per gallon (mpg)	40

(d) Assign each alternative with the life cycle costs shown in the following table and create a deterministic Pareto chart using a scatter plot with cost on the x-axis and total value on the y-axis.

Alternative	Life Cycle Cost ($1000s)
Baseline	70
Frontier	150
Starlight	110

(e) For each value measure, perform a swing weight sensitivity analysis. Use an Excel data table to vary the value measure's matrix weight at 0, 20, 40, 60, 80, and 100, and calculate the resultant total value score for each alternative. Graph the results of the data table using a line graph; each alternative should have its own data line series showing the total value as the matrix weight varies from 0 to 100.

(f) Perform a Monte Carlo simulation using the SIP Math Modeler Tool add-in; ensure that the add-in is present within Excel. Start with the Excel file titled "Problem 6.xlsx." Notice that there is a new worksheet named "SIP Library." This worksheet contains the data used to perform the Monte Carlo simulation and a text box with a detailed description of the worksheet structure. We can perform a simulation for a set number of trials simply by using the Excel data table feature and INDEX function; see the SIP Math Modeler Tool tutorials and users guide on the Tools page of ProbbilityManagement.org for a detailed explanation on how to do this.

To complete this exercise, follow the instructions in Steps 1–3 from Section 9.5.4. First, initialize the model using the instructions in Step 1; once complete, you should see two additional worksheets called "PMTable" and "SIPMath Chart Data." Next, define the inputs for the *Frontier* and *Starlight* alternative vehicle speed raw data scores using the instructions in Step 2; these input cells are highlighted in green in the "Value Model" worksheet. Remember that the data is arranged in columns. You should see green sparklines in these cells once this step is complete. (*The reader is referred to the online version of this book for color indication.*)

Next, each cell that contains the uncertain alternative raw data score in the "Value Model" worksheet should have a reference (set equal) to the cell under the "Output" titles in the "SIP Library" worksheet in columns D through H; these are the cells that use the discrete and triangular probability distributions. Next, follow the instructions in Step 3 to define outputs for each of the cells using the discrete and triangular probability distributions as well as the total value for the *Frontier* and *Starlight* alternatives; these output cells are highlighted in blue in the "Value Model" worksheet. You should see blue sparklines in these cells once this step is complete. In addition, you will see new named ranges and a data table within the "PMTable" worksheet and data used for chart construction in the "SIPmath Chart Data" worksheet. (*The reader is referred to the online version of this book for color indication.*)

(g) Create a stochastic Pareto chart using the template found in the Excel file titled "Problem 6.xlsx." within the worksheet titled "Stochastic Cost vs. Value Chart." Follow the instructions within the worksheet. Once complete, copy the chart and paste it into the "Value Model" worksheet and format appropriately.

(h) Create a cumulative distribution chart for the *Frontier* and *Starlight* alternative total value scores. Select the two cells that contain the total value sparklines for the *Frontier* and *Starlight* alternatives. In the SIPmath Modeler Tools ribbon, select "Graphs." For the Cumulative Chart Starting Location, select and area in the "SIPmath Chart Data" worksheet. Once complete, there will be two lines series charts shown. Copy one data series after clicking a line in one line chart and paste it into the other chart. You should now have one line chart with two data series. Delete the chart with only one data series. At the top of the table in the "SIPmath Chart Data" worksheet, adjust the chart data to the desired settings; ensure that each alternative has the same values, especially the minimum values. Format the chart as needed and copy and paste the chart into the "Value Model" worksheet.

(i) Create a stochastic tornado diagram for the *Frontier* and *Starlight* total value scores using the template found in the Excel files titled "Problem 9.xlsm." Creating the tornado diagrams in this template requires an extensive use of the Excel INDIRECT function. Because the INDIRECT function significantly slows the performance of the Excel file when used extensively, the template provides a macro to apply the INDIRECT function where needed and paste special values to obtain the percentiles used to create the tornado diagrams. Follow the instructions within the worksheet titled "Tornados." Set the lower percentile to be 0.3 and the higher percentile to be 0.7. Copy the stochastic tornado diagrams into the "Value Model" worksheet.

(j) After completing exercises a–k, you should have a working model with the results of the Monte Carlo simulation. Answer the following questions once you have verified your model.

(1) What insights can you identify from the value component chart? Provide some example of how to address the value gap.

(2) What can you conclude from looking at the deterministic component chart?

(3) How sensitive is each of the value model swing weights?

(4) After performing the Monte Carlo simulation, what impact does the value measure uncertainty have on the overall assessment of each alternative?

(5) Use the stochastic tornado diagrams to determine how we can best mitigate the risk in value of the *Frontier* and *Starlight* alternatives?

(6) Change the matrix weight of the vehicle safety measure from 100 to 0. How do the answers to question j (1) to (5) change?

9.7. Perform a Monte Carlo simulation of problem 5 using an Excel add-in. Compare the results of both Monte Carlo analyses using the following three outputs: Stochastic Pareto charts, cdf charts (*S*-curves), and Stochastic tornado charts.

REFERENCES

Buede, D.M. (2000) *The Engineering Design of Systems: Models and Methods*, Wiley Inter-Science.

Keeney, R.L. (1992) *Value-Focused Thinking: A Path to Creative Decision Making*, Harvard University Press, Cambridge, MA.

Mankins, J.C. (1995). Technology readiness levels. White Paper, April, 6.

Parnell, G.S., Bresnick, T.A., Tani, S.N., and Johnson, E.R. (2013) *Handbook of Decision Analysis*, Wiley & Sons.

Parnell, G.S., Driscoll, P.J., and Henderson, D.L. (2011) *Decision Making for Systems Engineering and Management*, 2nd edn, Wiley and Sons, Hoboken, NJ.

Parnell, G., Jackson, J., Lehmkul, L., and Engelbrecht, J. (1999) R&D concept decision analysis: using alternate futures for sensitivity analysis. *Journal of Multi-Criteria Decision Analysis*, **8**, 119–127.

National Defense Industrial Association (2011) Systems Engineering Division. Final Report of the Model Based Engineering (MBE) Subcommittee. Arlington, VA. From http://www.ndia.org/Divisions/Divisions/SystemsEngineering/Documents/Committees/M_S %20Committee/Reports/MBE_Final_Report_Document_(2011-04-22)_Marked_Final_ Draft.pdf (accessed 23 September 2016).

Savage, S.L., Scholtes, S., and Zweidler, D. (2006) Probability Management, OR/MS Today, Volume 33 Number 1. From http://viewer.zmags.com/publication/90ffcc6b#/90ffcc6b/29 (accessed 23 Sep 2016).

Savage, S.L. (2009) *The Flaw of Averages*, John Wiley and Sons, Hoboken, NJ.

Savage, S.L. (2012) Distribution Processing and the Arithmetic of Uncertainty. Analytics Magazine, November/December 2012.

SEBoK (2015) BKCASE Editorial Board. Guide to the Systems Engineering Body of Knowledge (SEBoK), version 1.4, R.D. Adcock (EIC). Hoboken, NJ: The Trustees of the Stevens Institute of Technology ©2015.29 June 2015. Web. 16 Jun 2015, 14:02 http://sebokwiki.org/w/index.php?title=Decision_Management&oldid=50860. BKCASE is managed and maintained by the Stevens Institute of Technology Systems Engineering Research Center, the International Council on Systems Engineering, and the Institute of Electrical and Electronics Engineers Computer Society.

10

EXPLORING CONCEPT TRADE-OFFS

AZAD M. MADNI

Systems Architecting and Engineering and Astronautical Engineering, Viterbi School of Engineering, University of Southern California, Los Angeles, CA, USA

ADAM M. ROSS

Systems Engineering Advancement Research Initiative (SEAri), Massachusetts Institute of Technology (MIT), Cambridge, MA, USA

> There is nothing worse than a sharp image of a fuzzy concept.
>
> Ansel Adams

10.1 INTRODUCTION

Systems engineering is undergoing a transformation motivated by both challenges and opportunities. The challenges are posed by increasing system scale and complexity. The opportunities are afforded by advances in computing and communication and social media (Madni, 2015d). Specifically, a major shift is underway that requires systems engineering to go beyond the "how" to the "why." In particular, systems engineering is going beyond traditional process prescriptions, with checkbox steps, to the creation of a framework that supports decision space definition, expansion, exploration, and explanation. This is a major shift in that today's SE process-driven approach comes with implied assumptions and "shallow" thinking. When things go awry, the process is invariably blamed. In sharp contrast, an "explore and explain" approach seeks to formalize systems engineering while grounding it in theoretical foundations. At the heart of today's thinking about systems engineering is identification of the opportunity space (Chapter 6), definition and expansion

---

Trade-off Analytics: Creating and Exploring the System Tradespace, First Edition. Edited by Gregory S. Parnell.
© 2017 John Wiley & Sons, Inc. Published 2017 by John Wiley & Sons, Inc.
Companion website: www.wiley.com/go/Parnell/Trade-off_Analytics

of the decision space, and exploration of the tradespace. The intent is on gaining insights, uncovering previously unknown interactions, uncovering constraints, and discussing regions of "good" and "bad" possibilities. This approach develops an appreciation of epistemic and aleatoric (i.e., known and unknown) uncertainties, as well as contingencies and conditions. The explanation aspect communicates insights and articulates the rationale behind decisions. Example insights include determining context-dependence of stakeholder needs; value-driving design decisions; value-limiting constraints and sensitivities to their removal; important risks, impact of multiple stakeholder preferences on cost, performance, and schedule of potential concept solution. Concept trade-off analysis is an essential activity in this transformation.

Concept trade-off analysis is a key activity in the conceptual design phase of systems engineering and is part of the overall systems engineering trade-off analysis process (Madni et al. 2014c). It consists of a set of concepts that need to be traded off against each other to find the best balance among them. The conceptual design phase is characterized by a high degree of uncertainty as customer needs are understood and then incrementally translated into requirements, which are then transformed into system design concepts and associated concepts of operation (CONOPS). Trade-offs performed in this phase are critical to conceptual design decisions. The quality of conceptual design is determined in large part by the importance of the opportunity space, having concept alternatives that span the decision space, the completeness of the concept design variables, and the effective and full exploration of the conceptual design tradespace.

Figure 10.1 illustrates the layered hierarchy relationship between concept, architecture, design, and system representations. Depending on the nature of the system being developed, there may be more than one instance of lower levels of the hierarchy. For example, the concept of an iPhone has more than one set with it over time

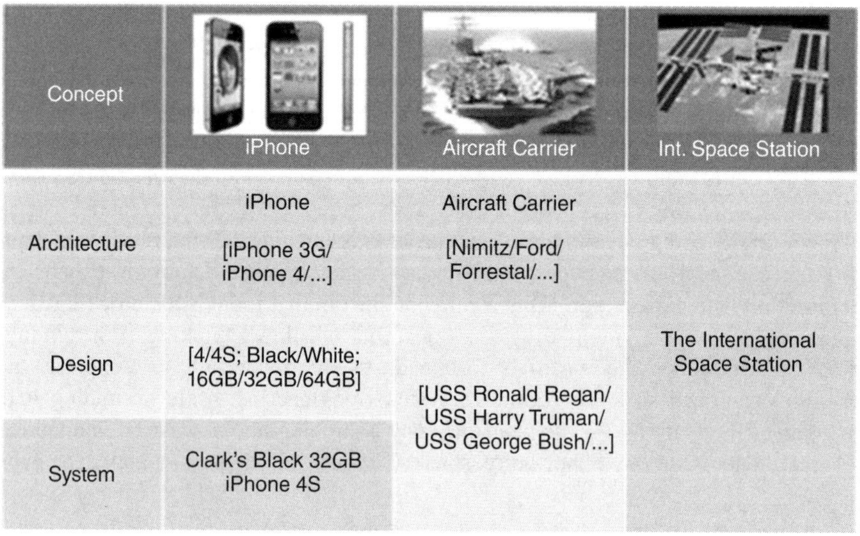

Concept	iPhone	Aircraft Carrier	Int. Space Station
Architecture	iPhone [iPhone 3G/ iPhone 4/...]	Aircraft Carrier [Nimitz/Ford/ Forrestal/...]	
Design	[4/4S; Black/White; 16GB/32GB/64GB]	[USS Ronald Regan/ USS Harry Truman/ USS George Bush/...]	The International Space Station
System	Clark's Black 32GB iPhone 4S		

Figure 10.1 Concept, architecture, design, and system abstraction layer examples

INTRODUCTION

(e.g., iPhone 3G, 4, 5, and 6). Associated with each architecture are particular design specifications that adhere to that architecture (e.g., the iPhone 4, Black, 16GB), which may allow for multiple design variants within the same architecture. Many individual systems may be manufactured for each specified design. For low production or unique systems, there may be only one architecture, or one design specified (e.g., USS Ronald Reagan aircraft carrier), or even one system developed (e.g., the International Space Station). The specification of the concept constrains the possible architectures, designs, and ultimately the systems that may be produced.

Conceptual design is where the mapping of function to form occurs. The resulting allocated architecture, in large part, determines most of the performance, cost, and schedule of subsequent development. The decisions made at this stage are the key determinants of the achievable performance of a system, the costs to produce and operate a system, and the time needed to produce deliver an operational system. The output of conceptual design is a design concept and high-level system specification that are used by preliminary design. The design concept is created and selected from the design tradespace with the intent of minimizing risks of costly downstream changes while simultaneously maximizing value for stakeholders. In this stage, it is important to not prematurely eliminate design alternatives because doing so, intentionally or inadvertently, can result in loss of valuable information that reduces the odds of realizing high-value, robust systems.

Trade-off Space (or Tradespace, for short) is defined by an enumerated set of design variables. When these design variables are assigned different levels, candidate design options are generated. Thus, it can be said that the tradespace is the space defined by the set of alternative designs that can be generated by all of combinations of the levels that are contained in the enumerated set of design variables. A tradespace can be expanded through novel combinations of existing design variables or generation of new design variables. Through the use of modeling and simulation, the entire tradespace (i.e., the total set of design alternatives) can be evaluated in terms of their benefits and costs to stakeholders. Here the benefits include all of the factors that particular decision-maker desires from the system (typically performance- or outcome-oriented), and costs include all of the resources the decision-maker cares about in order to achieve the desired benefits (typically these include monetary costs, but could also include schedule).The benefit–cost scatterplot is referred to as the tradespace because it offers an actionable representation for making "best" value trade-off decisions for the system (see Figure 10.2, for example).

The concept of value in this chapter is defined as the benefit for cost realized by a particular decision-maker, taking into account the importance and scarcity of the outcome. Because of the nature of systems engineering decision-making and the desire to help make explicit cost versus benefit trade-offs during early stage design, the representation of value in this chapter intentionally keeps the costs and benefits distinct (i.e., not aggregated together into a single function). It is often the case for complex systems that those who benefit and those who pay are different entities (e.g., government-funded science missions for scientist users and Air Force acquisition office for military end users) Keeping the cost and benefit functions separate is useful to foster more explicit conversations about benefits achievable for different levels of

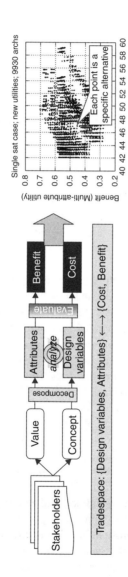

Figure 10.2 Example tradespace reflecting designer-controlled parameterized concepts in terms of stakeholder value metrics (Source: Ross, Massachusetts Institute of Technology, 2006. Reproduced with permission of Ross)

INTRODUCTION

resources. This chapter will use "utility" as a functional representation of perceived benefit under uncertainty of an outcome derived from a particular system alternative.

This chapter is organized as follows. First key terms such as concept trade-offs and concept exploration are defined. Then key considerations that come into play in exploring the concept space are discussed. The ongoing transformation in systems engineering is discussed as the field is beginning to emphasize the "why," beyond its traditional focus on the "how." (Madni, 2015d) Systems engineering decisions and potential outcomes are discussed, and it is emphasized that a good decision does not guarantee a good outcome because of the element of uncertainty and unpredictability that characterizes the behavior of complex systems. Then trade-off analysis frameworks are discussed and contextualized within the broader context of systems engineering frameworks. The importance of conceptual trade-offs is discussed from a life cycle perspective. Then value-based multiattribute tradespace analysis methods are presented in terms of their key features and limitations. Thereafter, conceptual tradespace analysis is discussed in terms of tradespace exploration, sensitivity analysis, and system quality attributes (e.g., reliability, agility) that comprise the tradespace. Tradespace exploration is also presented within the context of spiral development and compared to traditional optimization and decision analysis. Finally, the dynamic nature of concept design tradespace is discussed with storytelling-based simulations advanced as an effective way to uncover previously unknown interactions that lead to an expansion of the design variables that contribute to conceptual trade-offs (Madni et al., 2014a). Storytelling is also shown to be a viable approach for dynamic trade-offs analysis (Madni, 2015a,b). The chapter concludes with a summary of key points made and a brief discussion of the outlook for the future.

10.1.1 Key Concepts, Concept Trade-Offs, and Concept Exploration

Systems engineering trade-offs tend to be poorly defined and not well understood (Chapter 1). Not surprisingly, they are underemphasized in technical project management and often become a significant source of conflict among stakeholders as the project progresses. The inevitable outcome of ignoring trade-off analysis is selection of a concept that has low probability of meeting the needs and has high probability of cost overruns that sometimes lead to outright program failures.

At its core, trade-off analysis (or tradespace analysis as it is often referred to in systems engineering) is an analytical method for evaluating and comparing system designs based on stakeholder-defined criteria. Trade-off analysis requires measurable system attributes (also called measures of performance (MOPs) and measures of effectiveness (MOEs)). An example of MOP is "speed in engineering units," while an example of MOE is "probability of mission accomplishments" in terms that directly reflect stakeholder value. These are quantifiable criteria that are useful in characterizing how a stakeholder values important system attributes (e.g., performance, cost, and ilities). "Ilities" are quality attributes or properties of an engineered system that contribute to system value once the system is put into use. Examples of ilities are resilience, interoperability, and flexibility. Ilities are nonfunctional system requirements that are concerned with broader system impacts with respect to time and

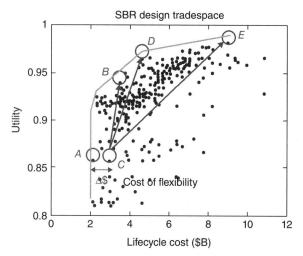

Figure 10.3 Concept of flexibility in tradespaces (Source: Ross 2005. Reproduced with permission of John Wiley & Sons)

stakeholders. While system attributes and decision-makers' utility functions capture the system need to a large extent, there are other sources of unarticulated or poorly characterized value. These tend to be quality attributes such as flexibility, extensibility, and resilience. Attributes such as flexibility and resilience are not easily defined, quantified, or costed. Flexibility offers the ability to respond to expected change. Adaptability provides the ability to respond to unexpected change. Resilience is the ability to respond to disruptions (both expected and unexpected). Therefore, flexibility and adaptability are enablers of resilience. In this regard, tradespace exploration techniques are likely to provide insights by affording the system designer a larger system view that includes relationships between architectural and design choices. The tradespace shifts as a result of changes in stakeholder preferences. This shift can potentially produce a change in the ranking of candidate design options. Comparing the relative shift in the ranking of candidate designs along an isocost can provide insights into how a candidate design may or may not be susceptible to such changes.

Figure 10.3 is a depiction of the tradespace for a Space-Based Radar (Spaulding, 2003). This tradespace is analyzed to determine the best design (Roberts, 2003; Shah, 2004). Typical analysis methods choose from among candidate designs that lie on the Pareto Frontier. The choice is invariably based on the option that has the lowest cost. However, when we consider other attributes, this may not always be the case.

With respect to Figure 10.3, a closer examination tells a somewhat different story if flexibility is deemed an important consideration. Flexibility implies that a system can be readily transformed into a different but related system at a modest cost. If flexibility is desirable, then Pareto Frontier solutions may not be the best. Roberts (2003) and Shah (2004) argue that these points in a sense are optimized for best value at a point in time and do not account for changes that might occur in the future. The latter, in fact, is the key motivation for wanting flexibility. While it may be appealing to

INTRODUCTION

incorporate flexibility into the utility model, one can potentially run into mathematical challenges (McManus et al., 2007). Therefore, treating quality attributes as separate criteria to be considered in addition to benefit and cost may provide more decision insights. Of course, all quality attributes including flexibility tend to come at a cost (Neches & Madni, 2013). So, the question that needs to be answered is whether or not the solutions on the Pareto Frontier are sufficiently flexible and, if so, what does flexibility cost. With respect to Figure 10.3, candidate designs A, B, C, D, and E all lie on the Pareto Front. However, close examination reveals that they are somewhat rigid in that they are not easily transitioned to other solutions on the Pareto Front. However, candidate option C, which has the same utility as option A but costs more than option A, can be "transitioned" to solution points B, D, or E more easily than option A. In this case, the cost differential between option C (the dominated point) and option A can be considered the cost of flexibility to move up the Pareto frontier. The generalized qualification of path dependence in transitioning between candidate solution options in a tradespace is an area of ongoing research (Ross, 2006; Ross & Hastings, 2006; Fitzgerald et al., 2012). It is intended to explicate the costs and benefits of including system quality attributes such as flexibility into an architecture or design. The rigid designs are those that have fewer design options to transition to than flexible designs do.

In the context of trade-off studies, measurable system attributes are used to determine how to balance stakeholder preferences (e.g., functions, objectives, measures, and requirements) to identify the system design that most closely satisfies stakeholders' objectives.

With respect to systems engineering trade-off analysis, a trade-off hyperspace is created to address the multiple trade-off factors (i.e., system attributes) for a system. Figure 10.4 presents a partial trade-off hyperspace for an unmanned vehicle. In other words, the trade-off hyperspace (i.e., tradespace) shown in this figure is meant to be a representative, not exhaustive. Different potential unmanned vehicles will display different trade-offs among these and other factors. Design features are represented in the design tradespace, while performance factors such as system performance, mission effectiveness, cost, schedule, and risk are represented in the performance tradespace.

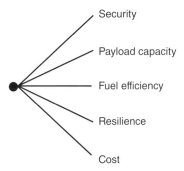

Figure 10.4 A trade-off hyperspace for an unmanned vehicle

As can be expected, systems engineering trade-off analysis is complicated by several factors. These include the following: multiple stakeholders with conflicting objectives and differing preferences for system functions and attribute level; uncertainty surrounding technology maturity and the future actions of adversaries/competitors, interactions among hardware, software, and human elements in the system; and the need to address trade-offs at multiple levels of abstraction and spanning multiple domains (conceptual, physical, social). It is important to note that Trade-off analysis in the concept space needs to be employed in several systems engineering contexts:

- When creating an unprecedented system
- When integrating legacy components into a new system
- When developing and establishing feasibility of system design and CONOPS

Trade-off analysis is further complicated by the number of system attributes in the tradespace, number of stakeholders, percentage of nonengineer stakeholders, presence of uncertainty in key attributes and the future environment, and whether or not the trade-offs vary with context. The concept space is typically explored in the conceptual design stage (also called upfront engineering). During conceptual design, the mapping from function to form occurs. As noted earlier, the physical form (i.e., implementation) determines most of the performance potential, cost, and schedule of subsequent development in the life cycle. A suboptimal decision in conceptual design can potentially result in cost and schedule overruns, as changes become increasingly more difficult to make later in the system development life cycle. The design concept and high-level specification are the outputs of this stage. These outputs serve as inputs to the preliminary design stage. The design trade-offs space (or tradespace, for short) is the space that is spanned by the fully enumerated set of design variables. Given a set of design variables, the tradespace is the space of possible design options. It is from this space that the initial system design concept is created/selected. The system design concept that is selected from the conceptual design tradespace is one that mitigates the likelihood of costly change later in the system life cycle and that maximizes stakeholder value. In this stage, intentional or inadvertent premature reduction of the design space can result in loss of potential future value (Chapter 1). Such premature pruning of the design options tradespace can preclude the realization of high value, risk-mitigated system.

A fundamentally important activity is the linking of the concept trade-offs to specific systems engineering life cycle phases and appropriate stakeholders. These links ensure that the right factors and considerations are addressed in the early stages of systems engineering, that is, conceptual design or preliminary design. These links serve to inform stakeholders about the applicable trade-offs that address variables of interest to them. This timely communication of trade-offs and subsequent decisions are essential for stakeholder understanding and, ultimately, stakeholder buy-in.

The creation of this link can be aided by defining a system ontology that spans system life cycle phases, tradespace attributes/variables, stakeholders and their interests in specific attributes, the weights assigned to the various attributes, and the supporting rationale.

Concept trade-offs apply to both system design and operation. Concept trade-offs during operation are often dynamic and can vary with changes in context (Madni & Madni, 2004). In the spirit of true concurrent design, system CONOPS need to be defined at the same time that the system is specified. This is key to assessing the feasibility of the design and CONOPS that result from a systematized analysis of the tradespace and sensitivity analysis of key design variables and their impact on key MOPs and MOEs. It is important to realize that at this stage, there is a high degree of uncertainty and ambiguity that is gradually reduced through modeling and simulation-based analysis. At this stage, great care needs to be exercised to resist premature elimination of options through untimely imposition of constraints during conceptual trade-offs analysis.

The outcome of conceptual trade-off analysis is an assessment in terms of feasibility, short-term impact, and long-term viability including affordability (Chapter 4). A variety of trade-off analysis techniques can be employed as the system moves through a series of architectural abstractions, that is, concept, function, form, and fit. Finally, it is imperative to go beyond the obvious technical and programmatic trade-offs to include the enterprise, social, and human cognitive dimensions and their impact on technical and programmatic trade-offs.

10.2 DEFINING THE CONCEPT SPACE AND SYSTEM CONCEPT OF OPERATIONS

As systems continue to grow in scale and complexity, upfront engineering becomes a critical activity with significant downstream impact. Specifically, initial specification of the system concept becomes a key driver of system performance and effectiveness.

Tradespace typically defines the region within which an optimum balance needs to be struck among factors such as performance, cost, and schedule. However, the tradespace defined solely by considering these three variables may not be sufficiently encompassing to strike an ideal balance among system attributes. This is where the system concept of operations (CONOPS) comes in. In fact, CONOPS needs to be a key variable in the tradespace (Mekdeci et al., 2011). For a CONOPS to be part of the tradespace, the CONOPS must perform its intended role in the development process and abide by the specific constraints that govern the system. Context comprising distinct but related elements serves to bound the CONOPS tradespace. These elements are the mission, system capability, and operational scenarios. These relationships can be captured within an ontology that encompasses the key concepts and relationships associated with a system CONOPS (Figure 10.5).

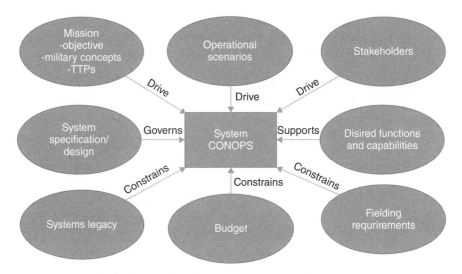

Figure 10.5 System CONOPS ontology (Adapted from Madni 2015c,d)

10.3 EXPLORING THE CONCEPT SPACE

Systematic exploration of the concept tradespace is essential to identifying superior and novel options (Madni et al., 1985; Madni, 2014). Key considerations during exploration are the success criteria, constraints, and system attributes. There are other factors that also bear on concept exploration, including functions, objectives, information availability, expertise availability, data availability, ambiguity and uncertainty in these factors, imposed inheritance and constraints, schedule and budget pressures, as well as cultural and organizational factors. Exploration can be aided by systematically applying and relaxing different constraints at the right time to assess the impact of these actions on the tradespace and to determine viable options and their sensitivities to constraints.

10.3.1 Storytelling-Enabled Tradespace Exploration

Storytelling is a powerful instrument for uncovering variables that should become part of the tradespace and for performing dynamic trade-offs (Madni, 2015a). Storytelling brings the benefit of illuminating changing contexts over time. Stories are created from a combination of mission description, use cases, and domain expert knowledge elicitation (Madni et al., 2014a). Interactive stories, that is, stories that allow users to interact with the system and thereby shape story evolution, are especially useful in uncovering previously unknown interactions as well as "hotspots." Stories allow simplification (e.g., through abstraction and selective timeline compression) and thereby allow the user to uncover interesting interactions and "surprises" (both good and bad outcomes) with fewer runs than possible with traditional simulation (Madni et al., 2014b). The challenge is to ensure that the story is rich enough (i.e., complicated enough) to be interesting and insightful and simple enough to find short-term answers of interest to the different stakeholders. Because of these properties, stories add temporal, spatial, and human dimensions to the tradespace.

10.3.2 Decisions and Outcomes

Ward Edwards said that "a good decision cannot guarantee a good outcome" (Edwards, 1962). Real decisions invariably involve uncertainty. A decision is, therefore, a bet, and evaluating whether it is good or bad depends on the stakes and the odds, not on the outcome. Also, the goodness of a decision is with respect to the criteria one defines. The possible results from the resolution of an uncertain event are called outcomes. With some uncertain events, there are only a few possible outcomes. In other cases, the outcome is a value within a specified range. The outcome of an uncertain event comes from a range of possible values and may fall anywhere within that range. Multiple uncertain events can potentially be considered in a decision situation. However, only a subset might be relevant. For an uncertain event to be deemed relevant, it must have some impact on at least one of the objectives. An example of an uncertain event in systems engineering is the time for a technology to mature from TRL3 to TRL7. This uncertain event has an impact on system objectives such as system performance, system affordability, system scalability, and possibly other system quality attributes (also called ilities). The complexity of a decision situation depends on the number of relevant uncertain events. The larger the number of uncertain relevant events in a particular context, the more complex the decision. Early in the system life cycle, such as conceptual design, the number of uncertain events is at its largest; some of these events will resolve themselves before the system enters operations (e.g., system test results, while others will resolve during operations (e.g., adversary/competitor actions). Also, some relevant uncertain events may depend on other such events. This dependence further adds to the complexity of the decision. Decisions made during the concept stage then are not only far-reaching, but also made complex by the sheer number and variety of uncertain events.

10.3.3 Contingent Decision-Making

In systems engineering, proactively managing the impact of uncertainty is critical. This capability is especially crucial when exploring the concept space and performing conceptual trade-offs analysis. One method of managing uncertainty is to create flexible opportunities that allow for deferring some decisions until uncertainties are resolved. Exploring the concept space and performing conceptual trade-offs analysis provide mechanisms to identify opportunities for deferring decisions and to define what have been described as "real options" (Myers, 1977)

Figure 10.6 illustrates some of the key concepts of a real option, including implementation of the option (real option enabled), the period during which the option enables the "right but not obligation" to execute a decision (at some cost), followed by the end of the option (real option expires).

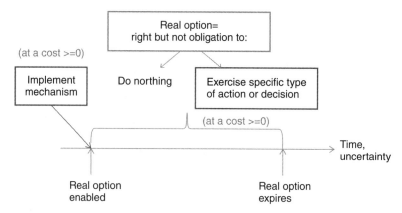

Figure 10.6 Anatomy of a real option (Source: Mikaelian et al. 2008. Reproduced with permission of Donna H. Rhodes)

In systems engineering, real options have been exploited to hedge against downside risk while promoting upside opportunity. However, such flexibility, similarly to all quality attributes, comes at a cost. The cost in this case is increased system complexity, additional financial outlay, and potential delay in schedule. The challenge is where to embed the real options to derive maximum leverage and enable the ability to satisfy the appropriate objectives at the appropriate time. Maximum leverage is derived by having key options available when they are needed (often as late as possible in the system life cycle, when cost of change is at its highest). During the option-enabled period, there may be associated carrying costs (e.g., storage costs for spare satellites, which enable the option to launch replacements to an existing constellation). Having such options allows for contingent decision-making where decisions can be deferred and executed contingent on emergent information. The challenge for designers is determining which options to embed in the system in order to respond to both anticipated and unanticipated uncertainties (Ricci et al., 2013).

10.4 TRADE-OFF ANALYSIS FRAMEWORKS

The purpose of a trade-off analysis framework is to identify and organize the key components comprising the trade-off hyperspace (i.e., tradespace for short) to facilitate subsequent analysis. Frameworks provide a formal construct for: (i) analyzing trade-offs associated with designing a new system or modifying an existing one using stakeholders' values; (ii) optimizing system design and operation in light of stakeholders' values; and (iii) analyzing sensitivities of stakeholders' values and system outputs to system perturbations introduced as "what-if" changes. Frameworks need to be generalizable and capable of supporting the analysis of tradespaces for both system design and system operation.

In systems engineering, the trade-off analysis framework is an essential part of the system engineering framework. The systems engineering framework informs and guides system design and incorporates feedback from system operation. Systems engineering frameworks tend to be process-oriented in the sense that they implicitly adopt a life cycle (i.e., beginning to end) perspective of a system. They also tend to be holistic in the sense that they focus on the key activities involved in developing, manufacturing, testing, deploying, and operating a system, rather than on specific methods for executing the steps in the systems engineering process. The V-model is an example of a traditional systems engineering framework. A trade-off analysis framework would have to fit within such a framework.

There are several trade-off analysis frameworks in use today. These frameworks have several features in common: (i) multiple criteria (i.e., trade-off dimensions); (ii) multiple stakeholders, some with conflicting preferences; and (iii) system-specific or general-purpose use. Examples of trade-off analysis frameworks are provided in Table 10.1. (Chapter 8 provides an expanded list and assessment of trade-off frameworks for developing and evaluation alternatives.)

There are multiple sources of uncertainty in systems engineering trade-offs analysis: (i) system-related; (ii) external-environment-related; (iii) valuation-method-related; and (iv) model-related.

System-related uncertainty arises from partial observability of system behavior. External-environment-related uncertainty stems from a lack of perfect information about various entities in the environment. These include the characteristics of an adaptive, asymmetric adversary, the time frame for an immature (e.g., TRL 3) but promising technology to mature, pending new policies and regulations, new standards that developers have to comply with, and potential competitors working on similar products. Valuation-method-related uncertainty arises from the inability of stakeholders to accurately convey preferences. The latter includes differences in the manifestations of preferences and time-dependent or context-dependent assumptions used in the valuation methods. Model-related uncertainty originates from approximations and abstractions made to simplify the model. All these uncertainties have to be accounted for when conducting systems engineering trade-offs analysis.

There is also a multistakeholder alignment challenge. The diversity of stakeholder preferences grows with increase in number of stakeholders causing a preference alignment challenge. There are several methods proposed in the decision analysis literature to deal with the stakeholder alignment challenge. These include the following: supraobjective function, negotiations, analytical methods, and minimum acceptable threshold.

10.5 TRADESPACE AND SYSTEM DESIGN LIFE CYCLE

Design efforts have typically focused on generating design options and evaluating their feasibility for various missions. Practically speaking, despite impressive advances in computing, evaluating design alternatives exhaustively remains time-consuming and expensive. This is why some engineers begin a design with a preferred, previously developed baseline concept and then systematically perturb the baseline to generate alternatives. Occasionally, concept trade-offs are performed on

Table 10.1 Example Trade-Off Analysis Frameworks

Name	Key Characteristics
Decision-based design (DBD) (Hazelrigg, 1996)	Evaluates systems/products that have a market, and associated demand and supply such that revenues are generated from selling the product at a particular price; implicitly considers trade-offs based on the rationale that design with the highest expected utility (EU) is the preferred candidate; alternatives are ranked using expected utility
Multiattribute tradespace exploration (MATE) (Ross et al., 2004)	Explores design/configuration options and then evaluate them in a benefit–cost space; iteratively utility aggregates benefit attributes (or trade-off dimensions) of interest for a given design that benefit relevant stakeholders and positions the design options relative to cost(s) of design, thereby capturing dominance of the designs in terms of benefit-to-cost trade-offs designs with most efficient trade-offs being preferred
Change propagation analysis	Tracks changes in complex systems with goal of evaluating the potential impact of specific changes on the system. Impact may implicitly contain multiattribute trade-offs based on stakeholder input (similarly to decision-based design and MATE); relies on design structure matrices (DSM) to characterize system and system interdependencies (Eppinger & Tyson, 2012) Changes in system can be identified, tracked, and quantified and form the basis for deriving measures of change type and magnitude in a system; change in DSM density is a proxy for magnitude of change in a system; change propagation index (CPI) metric is for determining the level of change in a system (Bartolomei et al., 2012)
Hazard and risk analysis	Addresses how hazards lead to risks in the deployed system by evaluating how design decisions affect reliability issues or risk posed by a system; used to analyze a system to mitigate residual risk in a system before it is fielded; for example, failure mode, effects, and critically analysis (FMECA) and systems-theoretic accident modeling and processes (STAMP); Marais framework, an improvement on STAMP (Marais, 2005, Macharis et al., 2009)
Multiactor, Multicriteria Analysis (MAMCA) (Macharis et al., 2009) Negotiations	Analyzes trade-off hyperspaces associated with design and operation of a system while accommodating needs and preferences of multiple stakeholders; broadest, most generally applicable for analyzing a system; often not overly specific to a particular system or method Facilitate consensus among stakeholders when there are competing preferences; many formal approaches exist to facilitate alignment (e.g., Game Theory)

a larger scale using lower fidelity models. Chapter 8 discusses holistic techniques to develop a creative set of alternatives that span the concept space. The system design life cycle begins with collaborative and iterative specification of requirements and concept development and concludes with system operation, periodic upgrades, and ultimately disposal.

Keeping the tradespace of potential design options open as late as possible is essential for ultimately identifying an effective candidate solution. Unfortunately, the tradespace is often prematurely and inadvertently reduced when customers specify requirements in a manner that implicitly implies a solution direction (e.g., choice of a particular technology or legacy component). In such cases, design creativity is stifled and design options are limited to a subset of potential options.

The process of design can be viewed simply as a series of system designer decisions, where each decision effectively prunes the set of available alternatives before arriving at the final preferred alternative (Madni, 2014). While narrowing the tradespace is both useful and desirable at the proper time, premature elimination of potential options can limit the value delivered to stakeholders (Madni, 2014).

10.6 FROM POINT TRADE-OFFS TO TRADESPACE EXPLORATION

Ross and Hastings (2005) present four types of trade-offs that represent increasing scope and effort: local point solution; Pareto frontier subset solution points; Pareto frontier solution sets; and full tradespace exploration. Figure 10.7 illustrates these four. A local point solution represents the least effort and value because of incomplete knowledge of the bigger picture needed to arrive at higher value solutions. The Pareto frontier subset solution begins to acknowledge key value trade-offs that exist in the tradespace. Multiple design options are considered with no "best" solution clearly identified. The complete Pareto frontier explicitly identifies the key cost–benefit trade-offs for each design option. New candidate design options can be rapidly assessed in terms of their distance from the "optimal" trade-off curve. Full tradespace exploration takes into account both dominated solutions and the Pareto Front. The inclusion of dominated solutions in the analysis acknowledges the possible existence of uncaptured value metrics (i.e., the requirements may be incomplete or wrong), recognizes that uncertainty may exist in the location of particular points (e.g., external-environment-related or model-related uncertainty), and allows for more detailed, dynamic analysis of the tradespace boundary.

10.7 VALUE-BASED MULTIATTRIBUTE TRADESPACE ANALYSIS

Chapter 2 presents the decision analysis underpinnings for trade-off analysis. Attributes are the key system performance/quality metrics of importance to stakeholders and decision-makers. They reflect how well the objectives specified by stakeholders/decision-makers are met. A system is characterized by multiple

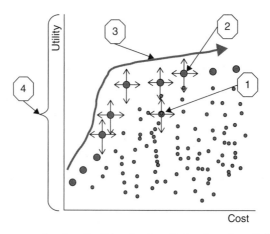

Figure 10.7 Four types of trade-offs: 1) local points, 2) frontier points, 3) frontier sets, and 4) full tradespace exploration (Source: Ross 2005. Reproduced with permission of John Wiley & Sons)

Table 10.2 Example Value Functions

Value Function Name	Function Type	Value Representation
Requirement	Binary	Yes/No
Functionality measure	Continuous	Amount of benefit provided
Single-attribute value (SAV)	Discrete or continuous	Desirability of a given level of a particular attribute
Single-attribute utility (SAU)	Discrete or continuous	Perceived SAV under uncertainty
Multiattribute value (MAV)	Discrete or continuous	Aggregate value of independent attributes
Multiattribute utility (MAU)	Discrete or continuous	Perceived MAV under uncertainty

attributes. The attribute set needs to be complete, minimal, nonredundant, measurable, decomposable, and independent. Practically speaking, an attribute set, at a given level of a goal hierarchy, should be limited to seven or fewer attributes (Keeney & Raiffa, 1993). This reflects individual cognitive limitations for perceiving value (i.e., an individual has difficulty thinking about more than seven items simultaneously, typically closer to three or five).

Value functions aggregate attributes into a single metric that reflects the collective preferences of decision-makers/stakeholders. The value function is used to compare candidate system designs to arrive at the "best" solution. In principle, there are a variety of value functions that have been used in evaluating candidate system designs (Table 10.2).

These value functions (Table 10.2) range from least difficult to most difficult to generate. The more difficult to generate, the more useful is the function in tradespace exploration because it captures the key preferences of decision-makers/stakeholders in greater detail, thereby allowing system developers to assess an expanded tradespace. Value function attributes can range from "least desirable" to "most desirable." In the limit, when the value function acceptability range converges to a point, it becomes a requirement (Ross & Hastings, 2005).

While some debate exists in the academic community about the role of uncertainty (vs ambiguity) in early phase design, this chapter will focus on using utility functions, rather than the more general value functions. This is done in order to reflect both the role of uncertainty in the mind of the decision-maker (and the ability to accurately predict the outcomes of particular designs) and the desire to have the concept of value reflect explicit trade-offs between benefit and cost. Utility will be constrained to represent only the benefit criteria under uncertainty, while cost will represent the resource criteria. The concept of "value" will be determined as an interpretation of desirable utility for cost and will vary from decision-maker to decision-maker.

10.7.1 Tradespace Exploration and Sensitivity Analysis

Tradespaces can be represented using a variety of visualization formats, including Kiviat Charts, bar charts, and utility–cost scatterplots. The utility–cost plot representation of the tradespace is a concise representation that highlights two critical decision metrics. The value-oriented focus of the tradespace plot is that utility, a key decision metric, is defined by the decision-maker who ultimately chooses from among the available alternatives. It is important to recognize that with designer-specified metrics such as speed or power; utility has to be inferred because these engineering characteristics used by designers do not directly measure the value or benefit that accrues to owners of the system when it is operated to achieve mission objectives.

The value-centric representation of a tradespace allows for the comparison of different concepts on the same basis (e.g., utility and cost). Figure 10.8 illustrates a tradespace for an operationally responsive disaster surveillance system, with different concepts dominating different regions for the Owner stakeholder. In this case, the satellite concept delivers less benefit and more cost than many of the alternatives. Surprisingly, the sensor swarm concept appears most promising at low cost and high benefit (however, further inspection reveals that the swarm is difficult to deploy in various potential mission areas, making the concept infeasible. This example illustrates the importance of iterative exploration of tradespaces, especially when utilizing low fidelity models for evaluating costs and benefits. Assumptions made during concept development may not be compatible with new information discovered later in the life cycle; therefore, it is essential to explore the uncertainties and sensitivities in tradespace results.

When a requirement or a stakeholder's preference changes, the impact of that change on the tradespace can be rapidly assessed. Ross & Hastings (2005) show a representative tradespace impact of a change in the utility function for a system after the decision-maker reviewed the initially proposed design (Figure 10.9). Another use

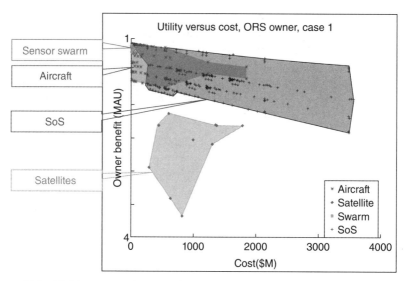

Figure 10.8 Multiconcept tradespace with sensor swarms, aircraft, satellites, and systems of systems (SoS) composed of pairs of assets (Source: Chattopadhyay, Massachusetts Institute of Technology, 2009. Reproduced with permission of Ross)

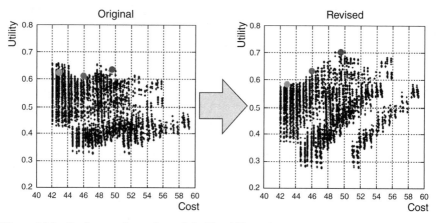

Figure 10.9 Preference change by stakeholder shifts tradespace (Source: Ross 2005. Reproduced with permission of John Wiley & Sons)

of the tradespace plot is to compare point designs in a macro-tradespace sense, that is, in terms of high-level alternatives. Designs that are dominated (i.e., inferior to the Pareto front) can be readily identified and examined.

10.7.2 Tradespace Exploration and Uncertainty

Uncertainty complicates tradespace exploration and alternative selection (Chapter 9). Various approaches have been pursued to include uncertainty in tradespace

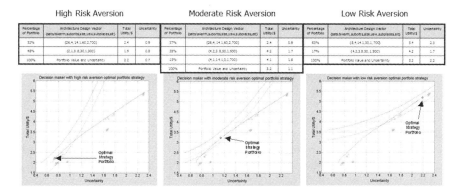

Figure 10.10 Example tradespace: uncertainty mitigation through system portfolios (Source: Walton 2002. Reproduced with permission of the Massachusetts Institute of Technology)

representation and exploration. Walton (2002), Walton & Hastings (2004) employs Monte Carlo simulations to generate probability distributions associated with each point in the tradespace. In this case, uncertainty is quantified as the variance of the value metric. Once uncertainty is quantified, system designers can exploit both the upside and downside of uncertainty (Walton & Hastings, 2001). For example, portfolio of system designs that combine options with anticorrelated uncertainties will have lower uncertainty than the uncertainty in individual designs (Walton, 2002; Walton & Hastings, 2004). Figure 10.10 presents an example of uncertainty-value space from which portfolios are constructed.

Each point in the figure is a point in the tradespace. As with financial portfolio theory, the efficient frontier of maximum value for a given level of uncertainty represents various combinations (or portfolios) of options. For financial instruments such as stocks and bonds, the uncertainty, or variance, is equated with "risk" and the expected value is called "return." Portfolios with best risk–return ratios are constructed based on the risk tolerance of the portfolio holder (Howard & Abbas, 2015, pp. 622–624). In Figure 10.10, three types of decision-makers are shown: highly risk-averse, moderately risk-averse, and minimally risk-averse. The optimal investment strategy for each type of decision-maker is determined by the intersection of the isovalue contours of the decision-maker with the efficient frontier. The analogy is made between partial ownership of financial instruments (i.e., stocks, bonds) and the partial investment in the development of physical systems. As uncertainties are resolved with further development of the system, the portfolio is modified accordingly (Walton, 2002). The key idea behind the uncertainty-oriented view of the tradespace is to develop an understanding of high leverage points for technology insertion or investment. Technology development that results in more anticorrelated design options can potentially mitigate uncertainty and help system engineers and program managers protect against program failure that is more likely with correlated technology development.

One high payoff application of the uncertainty-focused interpretation of the tradespace is in understanding high leverage points for technology insertion or investment. Technology development that results in more anticorrelated design options has the potential of reducing uncertainty and helping the system developer hedge against the risk of program failure due to correlated technology development. Of course, this hypothesis provides the basis for a whole new area of research.

10.7.3 Tradespace Exploration with Spiral Development

With spiral development, successive generations of system architecture and design result from each iteration of the spiral. Each iteration produces a learning opportunity that results in the refinement of the system architecture and design. A possible way to understand designs resulting from each spiral is through understanding the tradespace associated with each spiral (Derleth, 2003). Each spiral pass has a Pareto frontier, with some changes occurring in the Pareto Front with each spiral as new needs impact expectations around new capabilities. Thus, alternatives that were on the Pareto Front of the first spiral may not all be in the Pareto Front of the second spiral. In other words, some designs continue to offer the highest utility for a given cost, whereas others move to lower utility. This finding implies that designs are best chosen within the dynamic context of spirals. Sensitivities to hypothesized learning from one spiral to the next can be investigated in the tradespace context, resulting in superior architecture choices, especially where prior spiral decisions can limit future achievable capabilities.

10.7.4 Tradespace Exploration in Relation to Optimization and Decision Theory

For complex systems, value-driven tradespace exploration paradigm recognizes that concept design is a dynamic and complex process, and exploration is key to developing solutions that offer greater value to stakeholders. There are several reasons why system tradespace exploration is a preferred approach to optimization methods and decision-theory-only-based approaches such as utility functions (Table 10.3). While tradespace exploration can leverage both optimization and decision theory, the exploration paradigm encourages investigation of changing problem formulations and potential solutions and engagement with soft factors in order to iterate toward deeper understanding around the potential value of different alternatives.

Trade-offs between design choices, perceived utility, and costs of alternatives are conveniently performed today using tradespace exploration. Tradespace exploration and visualization can support a variety of analysis and can be a source of much needed insights during the conceptual stage of system design. Tradespace analysis is complicated by changing needs and contexts over time. Today, tradespace exploration is being exploited: to identify promising designs; understand strengths and weaknesses of each design alternative across the tradespace; identify requirements and constraints that limit the ability to uncover less expensive solution; understand design sensitivities to changes in needs and context; and assess differential impacts of uncertainty on different alternative designs with opportunities for risk mitigation.

System architects and designers have the daunting challenge of evaluating multiple candidate design options, in terms of satisfying the needs of multiple stakeholders with potentially conflicting objectives while also contending with changes in stakeholders needs, emergence of new, relevant promising technologies, and uncertain

Table 10.3 Comparison of Tradespace Exploration with Optimization and Decision-Theoretic Approaches

Comparison Criteria \ Methods	Tradespace Exploration	Optimization	Decision Theory
Objective function	Can handle changing, poorly defined objectives; can leverage decision theory and optimization	Require well-defined objectives	Employ utility functions to account for uncertainty in outcomes
Decision-maker preferences	Can handle changing preferences	Reflected in well-defined objective function	Can handle decision-maker preferences; changing preferences require revisiting decisions
Handling of "Soft" issues	Policy issues; system quality attributes such as flexibility and resilience	Not equipped to handle soft issues	Limited handling of soft issues
Uncertainty management and mitigation	Uncertainty–cost tradespace; uncertainty, mitigation through real options and decision rules (engineering view)	Limited uncertainty handling (stochastic optimization)	Uncertainty handling through real options (normative view) and expected value of perfect and imperfect information; use of expected utility; Bayesian networks
Visualization	Risk presentations; Kiviat charts; utility–cost functions; uncertainty cost functions; Pareto Front representation and change	Traditional visualization formats used by optimization methods	Utility curves; risk–benefit curves; sensitivity curves

Source: Adapted from Madni 2015c,d.

geopolitical futures (Madni, 2015a). Fortunately, increases in computing power and decreasing cost of computing resources and wide availability of flexible tools have made it possible to more extensively explore the tradespace, generate more design options and CONOPS, and assess their feasibility.

In light of the forgoing, the first step is generating a tradespace for exploration. Figure 10.11 provides the overall concept for tradespace generation and exploration.

Figure 10.11 presents the tradespace generation and exploration system concept and functional architecture of a future system. As this figure shows, in the database generation phase, tradespace analysis is used to generate a large, representative number of candidates designs, concepts of operations, and their preanalyzed performance.

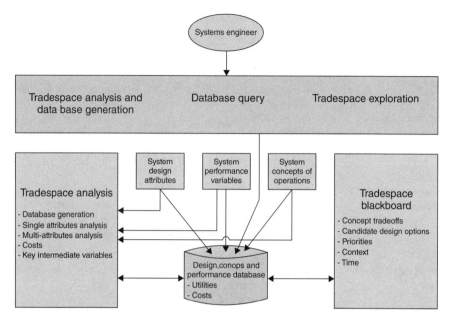

Figure 10.11 Tradespace generation and exploration

The design and CONOPS are parameterized and expressed as parametric curves. System performance is expressed in terms of attributes and metrics of interest to the stakeholders. Computational and graphical tools allow the system engineers to query the database. Tradespace analysis employs utility–cost curves to concisely portray cost–benefit for each candidate design and CONOPS. Some of the key benefits of using this approach are as follows:

- Visualizing the immediate effects of changing needs
- Comparing candidate design alternatives (point designs) in a tradespace under a variety of contexts and conditions
- Readily identifying both functional and performance constraints on feasible solution set
- Visualizing different impacts of uncertainty across alternatives from tradespace
- Discerning alternatives in terms of their ability to change state (flexibility)
- Informing spiral acquisition
- Assessing policy robustness of alternative solutions

Tradespace exploration, a concept borrowed from decision analysis, is domain-independent. It explicitly links value propositions to alternatives, which may be obscured by politics, time, and misrepresentation. Tradespaces are constructed using concept-independent criteria (i.e., perceived benefits and costs) that allow comparison of distinct concepts based on the same tradespace.

Tradespaces can change over time. Therefore, tradespace exploration approaches need to be able to handle a dynamic tradespace. To this end, methods are needed that relax assumptions about static performance and constraints while adding the ability to take into account changing contexts and needs (epochs) over the short and long run (eras) using, for example, Epoch-Era Analysis (Ross, 2006; Ross & Rhodes, 2008; Roberts et al., 2009). In addition, there is a need to quantify the ability of the system to change ("changeability") in the tradespace (Ross et al., 2008). This capability allows for the evaluation of a system's ability to reconfigure or be altered to respond to changing definitions of utility and cost over time. Several metrics have been defined to identify "good" designs in a time-varying context. These metrics fall into specific categories to achieve system value robustness, that is, maintain system value despite changing contexts and needs. These include highly changeable designs (filtered outdegree) and highly versatile or passively value robust designs (Pareto Trace, normalized Pareto trace, fuzzy normalized Pareto trace) (Ross et al., 2009a, 2009b). Temporal system properties that can be considered in tradespace exploration include system vulnerability and resilience (Ross et al., 2014).

10.8 ILLUSTRATIVE EXAMPLE

We use the Multiattribute Tradespace Exploration (MATE) approach to illustrate key concepts associated with concept trade-offs analysis. At a high level, MATE can be implemented using the following seven steps:

1. Determine Key Decision-Makers
2. Scope and Bound the Mission
3. Elicit Attributes and Utilities (preference capture)
4. Define Design Vector Elements (concept generation)
5. Develop Model(s) (evaluation)
6. Generate the Tradespace (computation)
7. Explore the Tradespace (analysis and synthesis)

In this example, MATE is used to identify a high-value design for a maritime security system of systems.

10.8.1 Step 1: Determine Key Decision-Makers

The key decision-maker for this example was a representative of the government agency requesting the study. This representative provided the high-level operational needs statement for the system:
Provide maritime security for a particular littoral Area of Interest (AOI).
Additionally, the decision-maker expressed that stakeholders want a system that **detects**, **identifies**, and **boards** suspicious boats and that is capable of carrying out **search and rescue** missions.

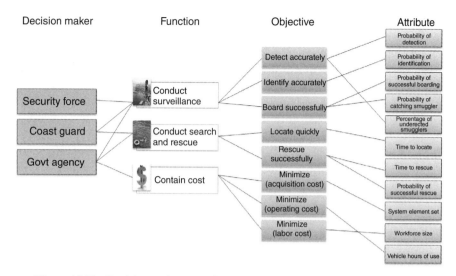

Figure 10.12 Decision-maker to attribute mapping for a maritime security system

10.8.2 Step 2: Scope and Bound the Mission

The goal of the MATE study is to determine what will be the "best" system when considering performance, ilities, and cost in this context. The decision-maker provided bounding constraints that said that the system must leverage inherited ground radar towers and existing manned patrol boats. It was suggested that the study consider UAVs as well. Additionally, the study was to consider the impact of loss of assets during the mission and means for mitigating the impact of such a loss.

10.8.3 Step 3: Elicit Attributes and Utilities (Preference Capture)

Discussions with the decision-maker elicited the function–objective–attribute decomposition shown in Figure 10.12.

For this study, the government agency acted as the actual decision-maker, desiring that both the security force and coast guard stakeholders be treated as separate "missions" to be satisfied. The goal for the government agency was to understand whether and what potential systems could satisfy each mission, as well as the overall cost objectives. Table 10.4 has an example list of attributes considered.

Table 10.4 List of Attributes for Maritime Security Case

Attribute	Comment
Probability of detection	0% to 100%
Probability of identification	0% to 100%
Probability of intercept	0% to 100%, same as successful boarding
Probability of rescue	Contingent on time to locate and time to rescue
Acquisition cost	In $10M
Operations and workforce cost	In $10M/year

ILLUSTRATIVE EXAMPLE

Through (lottery-based) interviews with the stakeholders, the following preferences were elicited:

Security Force

Attribute Name	Weight	Min Acceptable (%)	Max Desired (%)
Probability of detection	0.2	90	100
Probability of identification	0.2	25	100
Probability of intercept	0.6	10	100

Figure 10.13 illustrates the single-attribute utility (SAU) curves for the Security Force mission showing increasing utility with increases in attribute. Of note in these curves is the "$N=$" at the top, which shows how constraining that SAU curve is on the tradespace, effectively showing the resulting fraction of the tradespace still feasible after application of that SAU curve. The asymmetry of the SAU curve is that designs are unacceptable (i.e., infeasible) if they score below $U=0$, while they are not any more valuable, but are still feasible when scoring at or above $U=1$. For this case, it is not possible to score higher than 100% probability, but in general, a design could conceivably score higher than the maximally desirable level of a particular attribute.

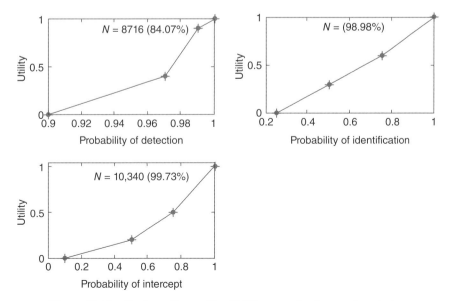

Figure 10.13 Single-attribute utility (SAU) curves for the security mission

Coast Guard

Attribute Name	Weight	Min Acceptable (%)	Max Desired (%)
Probability of rescue	0.4	20	100
Probability of detection	0.3	90	100
Probability of identification	0.2	25	100
Probability of intercept	0.1	10	100

This example will focus on the Security Force mission stakeholder, so Coast Guard SAU curves are not included for brevity.

The weights on the SAUs were elicited and determined to add to one, which coincides with a linear weighted sum multiattribute utility aggregation function:

$$\text{MAU}(X_1 \ldots X_N) = \sum_{i=1}^{N} k_i \text{SAU}_i(X_i) \quad (10.1)$$

where X_i is the value for attribute i, SAU_i is the single-attribute utility curve for attribute i, and k_i is the swing weight for attribute i.

Note: the cost attributes were not rolled up into a utility function in order to help the decision-maker make explicit trades on cost versus benefit. This paradigm enables the decision-maker to explore and update his value model based on discoveries of what is possible and affordable.

10.8.4 Step 4: Define Design Vector Elements (Concept Generation)

Various brainstorming techniques were used to generate seven concepts for the system. These are summarized in Table 10.5.

These concepts were then deconstructed into a number of design variables that parameterize aspects of the system's form and CONOPs, described in Table 10.6.

The expectation is that not only does the form of the system matter (i.e., the number of each type of equipment), but also how it is operated (i.e., the CONOPs, including how tasks are assigned to vehicles).

10.8.5 Step 5: Develop Model(s) (Evaluation)

Once the attributes (outputs) and design variables (inputs) have been defined, the next step is to develop the evaluation model that will determine the attributes from each uniquely defined input concept design specified by enumerated design variable levels. For this study, the performance model was an agent-based discrete event simulator (Figure 10.14), which was developed to allow for analysis of emergent behaviors arising from the interaction of constituent systems and operational choices. Additionally, cost models and utility function evaluation models were used. The models were validated using input–output correlations as well as subject matter experts.

Table 10.5 System Concepts

Concept 1	Use a *heterogeneous* set of unmanned aerial vehicles (UAVs) to perform maritime surveillance (i.e., detection, identification, and interception)
Concept 2	Use *satellite* for communication relay and UAVs for identification and interception
Concept 3	Use *radar tower* for detection and UAVs for identification and interception
Concept 4	Use *mixed manned and unmanned* vehicles for performing surveillance, with manned vehicles (preferably boats) used for boarding and interception
Concept 5	Use UAVs (or satellite or radar tower) for detection, *unmanned boats* for identification, and manned boats for boarding
Concept 6	Use UAVs for detection and identification and *manned boats* for boarding
Concept 7	*All-inclusive* SoS, with UAVs, unmanned boats, radar towers, and satellites for surveillance

Table 10.6 Design Variables Parametrizing System's Form and CONOPs

	Name	Definition Range	Comment
Form	Hermes UAV	2–6	
	Shadow UAV	2–6	
	Manned patrol aircraft	0–4	
	Helicopters	0–3	
	Manned patrol boats	4–12	
	Satellite relay	Yes/No	For now fixed at no due to cost
	Ground radar towers	Yes/No	For now fixed at yes due to cost
CONOPs	Authority	Central/distributed	
	Operators/UAV	2:1/1:4	Many to one, or one to many
	Tasking type	Dedicated/multirole	Each vehicle dedicated or not
	Number of geographic zones	1/2	
	Workforce buffer	0–33%	
	Decoys	0–3	

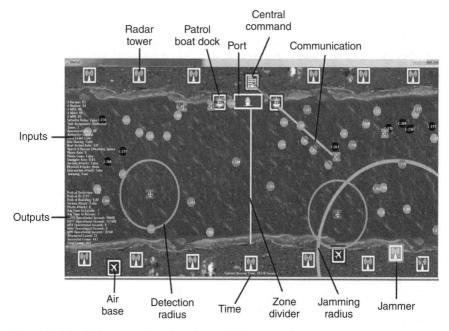

Figure 10.14 GUI screenshot for the maritime security agent-based discrete event simulator

10.8.6 Step 6: Generate the Tradespace (Computation)

In this step, the design space is sampled, with continuous design variables sampled at discrete levels. This resulted in 10,368 unique design concepts to be evaluated in the model. The model then evaluates each point across the mission timeline. The particular mission being evaluated had a number of uncertain elements, including the exact timing and location of boats entering and leaving the AOI. These uncertain elements were treated as stochastic, with values for particular mission timelines determined via sampling of probability distributions. The repeated sampling of the stochastic variables resulted in a large number of different potential mission starting conditions. All design alternatives were exposed to the same set of potential missions. To keep this illustration simple, the results of the simulation are presented as the means of the output scores. This is a simplification for brevity and clarity. In practice, one could consider the full distributions and apply the uncertainty management techniques described earlier in the chapter (e.g., portfolio optimization or real options).

The reason for generating tradespace data rather than just a few point designs is because rich data sets can be explored to reveal complex relationships between design space and value space for generating intuition into a problem. This is especially important for systems that display emergence, have complex/conflicting stakeholder expectations, and are subject to changing contexts where previous experience may bias predictions.

10.8.7 Step 7: Explore the Tradespace (Analysis and Synthesis)

Exploration of the tradespace can take many forms, including using predefined workflows, such as question-driven exploration, or can be more freeform using various interactive visualization tools. For now we will investigate the attractive designs for the Security Force stakeholder, first starting with a point design, moving on to a set of points, and then on to full tradespaces.

As mentioned earlier, when interpreting these figures, the "$N=$" at the top indicates how many designs are considered currently feasible, based on constraints, including preferences (i.e., if a design point does not meet minimum acceptable utility, then it is rejected from the tradespace). The yield, which is the fraction of the originally evaluated tradespace that remains after constraints, is indicated as 82.81% of the original 10,368 evaluated alternatives.

Figure 10.15 illustrates a point design for the maritime security system, with indicated design variable values. In the absence of a tradespace, this design point appears reasonable, with an acquisition cost around $40M and delivering a utility of 0.56. Further investigation reveals that this design has a probability of detection of 100%, probability of identification of 89.5%, and a probability of intercept of 59.5%. It is unclear from looking at the point alone what the trade-offs might be in terms of performance versus cost, for example.

Moving on to identifying an efficient set of points, Figure 10.16 illustrates a set of Pareto frontier points (efficient in terms of Security MAU and Acquisition Cost). All of these designs have in common a central command authority and two operators per UAV. Table 10.7 illustrates the attribute scores, acquisition cost, and MAU for

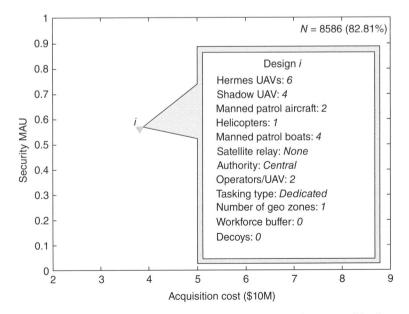

Figure 10.15 Example point design for a maritime security system ($N = 1$)

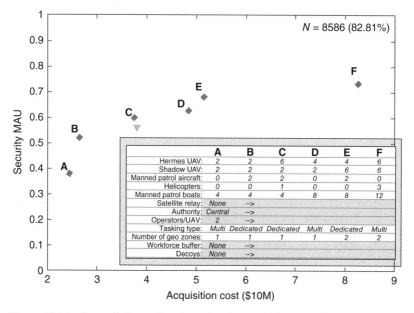

Figure 10.16 Example Pareto frontier points for a maritime security system ($N=6$)

Table 10.7 Attributes, Cost, and MAU for Selected Pareto Points

Metric	A	B	C	D	E	F
Probability of detection	97.5%	99.8%	100.0%	99.9%	100.0%	100.0%
Probability of identification	80.5%	85.2%	89.3%	89.4%	89.8%	90.5%
Probability of intercept	51.7%	56.3%	65.0%	69.1%	75.7%	80.0%
Probability of rescue	40.0%	70.0%	100.0%	50.0%	85.0%	100.0%
Acquisition cost ($10M)	2.5	2.7	3.8	4.9	5.2	8.3
Security MAU	0.38	0.52	0.59	0.62	0.68	0.73

these points. The general trend is that the increased cost moving from A to F gets you more performance in terms of probability of intercept. This coincides with having additional Hermes UAVs, as well as manned patrol boats.

Taking the exploration further results in the tradespace shown in Figure 10.17, with all 10,384 evaluated points considered, of which 82.81% are feasible. Highlighted is the full evaluated Pareto frontier set, expanding the alternatives for consideration in terms of most efficient MAU for Acquisition Cost. Using color-by-design variable, we can investigate design choice impacts on MAU and cost, expanding the scope of insight beyond the handful of Pareto points considered previously.

Figure 10.18 shows the tradespace colored by the number of Hermes and number of Shadows, respectively. The pattern of increases in the number of Hermes in the Pareto points is here again repeated, with three apparent groupings of increases in

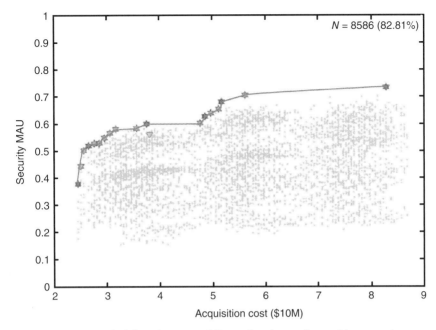

Figure 10.17 Example full tradespace and Pareto frontier set for maritime security system ($N = 8586$).

cost and utility for increases in Hermes. The utility increase from the increase in the number of Shadows is not as clear.

Figure 10.19 shows the tradespace colored by the number of manned patrol boats and type of command authority, respectively. The three groupings in the tradespace are clearly driven by the number of manned patrol boats. This corresponds to their role in boarding suspicious boats, strongly driving MAU (and cost) with increases in number. Central command authority mostly dominates distributed authority in the tradespace.

Figure 10.20 shows the tradespace colored by number of operators per UAV and the number of geographic zones. The apparent general increase in MAU with the two operators over the 0.25 operators per UAV appears counterintuitive (i.e., having each operator control more than one UAV seems as a force multiplier and therefore increases the performance of the system). Upon deeper inspection, we found that cognitive overloading leads to decreased operator performance when each operator must control multiple UAVs. This emerged in the performance model as each operator manages his own queue of tasking during the mission. Task backlogs accumulate as the operators switch between UAVs, operating at a constrained pace. As for number of geographic zones, this reflects the CONOPs trade of whether to allow the vehicles to travel the entire AOI or to constrain them to a fixed zone. In the lower cost region of the tradespace (where designs have fewer vehicles), the highest utility points tend to have one zone, whereas the higher cost region (and larger number of vehicles) the

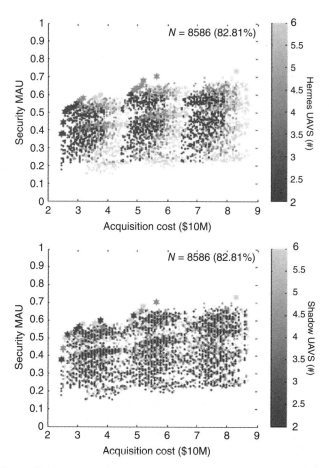

Figure 10.18 Tradespaces colored by number of Hermes and Shadows

highest utility points tend to have two zones. These reflect the general need to have vehicles evenly distributed to increase the chances that they can respond quickly to a target of interest before it moves away.

Investigating a larger tradespace both confirms and expands upon the insights generated through investigation of a handful of design points. At some point, however, additional insights gained can be expected to diminish with increases in the size of the tradespace. The most appropriate size for a particular study should be determined on a case-by-case basis, with the view that tradespaces should scale up, rather than down. As noted previously in this book, the insights gained are only as good as the model developed and the quality of inputs. Through iterative exploration of the tradespace, model goodness can be determined. Tradespace exploration encourages broad sensitivity analysis of both models and preference inputs to better generate insights into

CONCLUSIONS

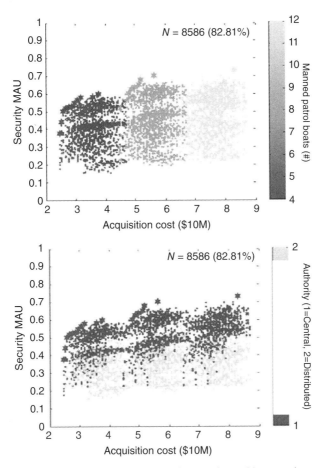

Figure 10.19 Tradespaces colored by number of manned patrol boats and type of command authority

trend relationships between what is desired (value proposition) and what is possible (evaluated design alternatives).

10.9 CONCLUSIONS

Exploration of the concept space is a key aspect of concept trade-offs analysis. Concept trade-offs analysis is a key activity in the conceptual design phase of systems engineering and is part of the overall systems engineering trade-offs analysis process. Similarly, systems engineering trade-off analysis is an integral part of the systems engineering process. The conceptual design phase is characterized by a high degree of

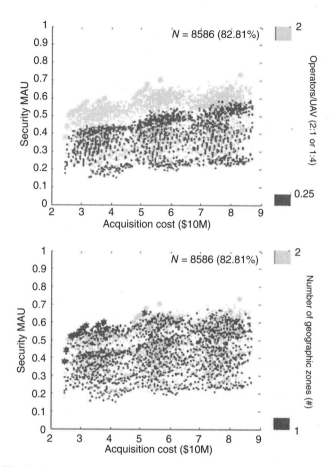

Figure 10.20 Tradespaces colored by operators per UAV and number of geographic zones

uncertainty as customer needs are understood and then incrementally translated into requirements, which are then transformed into system design CONOPS. Trade-offs, performed in this phase, are critical to conceptual design decisions, which determine most of the cost and schedule of the system under development. The quality of conceptual design is determined in large part by the complete specification of the concept design variables and the effective and full exploration of the conceptual design tradespace.

In this chapter, we first defined key terms such as concept trade-offs and concept exploration. Then we discussed key considerations that come into play in exploring the concept space. We also briefly discussed the ongoing transformation in systems engineering as the field is beginning to emphasize the "why," beyond its traditional focus on the "how." Systems engineering decisions and outcomes were discussed. It was emphasized that a good decision does not guarantee a good outcome because of the element of uncertainty and unpredictability that characterizes the behavior of complex systems (Madni and Allen 2011). Then trade-off analysis frameworks were discussed and contextualized within the broader context of systems engineering frameworks. The importance of conceptual trade-offs was discussed from a life cycle perspective. Then, value-based multiattribute tradespace analysis methods were presented in terms of their key features and limitations. Then conceptual tradespace analysis was discussed in terms of tradespace exploration, sensitivity analysis, and system quality attributes that comprise the tradespace. Tradespace exploration was then presented within the context of spiral development and compared to traditional optimization and decision analysis. Finally, the dynamic nature of concept design tradespace was discussed with storytelling-based simulations advanced as an effective way to uncover previously unknown interactions that lead to an expansion of the design variables that contribute to conceptual trade-offs. Storytelling was shown to be a viable and effective approach for dynamic trade-offs analysis (Madni 2015e). We concluded the chapter with an illustrative example using the MATE software tool. This chapter has presented the state-of-the art in conceptual trade-off analysis with the growing convergence among disciplines (Madni, 2015c). Conceptual trade-off analysis can be expected to benefit from new insights offered by these disciplines.

10.10 KEY TERMS

Tradespace: Multivariable mathematical space (trade-off + playspace) used for identifying the optimal boundary space (Pareto frontier).

Storytelling: Conveying of event sequences using words, sounds, images with the objective of informing, educating, and persuading stakeholders about a point of view.

Uncertainty: The state of being unsure about the existence of a state or value about a variable.

Value: Importance, worth, or usefulness of an outcome, feature, or solution alternative.

CONOPS: A user-oriented electronic or paper document that describes system characteristics for a proposed system from the user's viewpoint. Communicated to all stakeholders.

Quality Attributes: Overall factors that affect operational behavior, system design, and user experience.

Optimization: Process of finding greatest/least value of a function for some constraint, which must be true regardless of the solution; find most suitable value for a function within a given domain.

Framework: A layered support structure (potentially comprising concepts, software, hardware, interfaces) or system that allows all elements of a system to be interrelated to each other and be manipulated as needed to support needs of the users.

10.11 EXERCISES

10.1. Give an example that will illustrate the relationship between concept, architecture, design, and system representation. (No credit for the examples discussed in this chapter)

10.2. For an electric vehicle
   (a) Identify eight key system attributes.
   (b) Identify stakeholders associated with each system attribute.
   (c) Identify the uncertainty associated with each system attribute. Do all system attributes include uncertainty?

10.3. What are the key concepts and relationships associated with system CONOPS?

10.4. For the same electric vehicle in question #2, choose 4 key system attributes and discuss how changing one parameter can affect others. Categorize impacts in terms of high, medium, and low. Provide rationale for your categorization.

10.5. For a network of autonomous electric vehicles
   (a) Identify key decision-makers.
   (b) Identify key system and network attributes. Do all vehicles need to have same characteristics (attributes)?
   (c) Map each decision-maker to system attribute(s).
   (d) Identify types of uncertainties that exist in the system and environment, and discuss how they impact tradespace.

10.6. Using the concepts described in this chapter, perform a trade-off analysis for a system concept using the following steps:
   (a) Determine key decision makers.
   (b) Scope and bound the mission.
   (c) Elicit attributes and utilities (preference capture).
   (d) Define design vector elements (concept generation).
   (e) Develop model(s) (evaluation).
   (f) Generate the tradespace (computation).
   (g) Explore the tradespace (analysis and synthesis)

REFERENCES

Bartolomei, J., Hastings, D., de Neufville, R., and Rhodes, D. (2012) Engineering systems matrix: an organizing framework for modeling large scale complex systems. *Systems Engineering*, **15** (1), 41–61.

Chattopadhyay, D., Ross, A.M., and Rhodes, D.H. (2009) *Demonstration of System of Systems Multi-Attribute Tradespace Exploration on a Multi-Concept Surveillance Architecture.* 7th Conference on Systems Engineering Research, Loughborough University, UK.

Derleth, J.E. (2003) *Multi-Attribute Tradespace Exploration and Its Application to Evolutionary Acquisition*, Massachusetts Institute of Technology.

Edwards, W. (1962) Dynamic decision theory and probabilistic information processing. *Human Factors: The Journal of the Human Factors and Ergonomics Society*, **4** (2), 59–74.

Eppinger, S.D. and Tyson, R.B. (2012) *Design Structure Matrix Methods and Applications*, MIT Press.

Fitzgerald, M.E., Ross, A.M., and Rhodes, D.H. (2012) Assessing Uncertain Benefits: A Valuation Approach for Strategic Changeability (VASC). *INCOSE International Symposium*. July 2012, Rome, Italy.

Hazelrigg, G.A. (1996) *Systems Engineering: An Approach to Information-Based Design*, Prentice Hall, Upper Saddle River, NJ.

Howard, R.A. and Abbas, A.E. (2015) *Foundations of Decision Analysis*, Prentice Hall.

Keeney, R.L. and Raiffa, H. (1993) *Decisions with Multiple Objectives--Preferences and Value Tradeoffs*, 2nd edn, Cambridge University Press, Cambridge.

Macharis, C., De Witte, A., and Ampe, J. (2009) The multi-actor, multi-criteria analysis methodology (MAMCA) for the evaluation of transport projects: theory and practice. *Journal of Advanced Transportation*, **43** (2), 183–202.

Madni, A.M. (2014) Generating novel options during systems architecting: psychological principles, systems thinking, and computer-based aiding. *Systems Engineering*, **17** (1), 1–9.

Madni, A.M. (2015a) Expanding stakeholder participation in upfront system engineering through storytelling in virtual worlds. *Systems Engineering*, **18** (1), 16–27.

Madni, A.M. (2015b) Systems Engineering Tradeoffs Analysis: Challenges and Promising Themes. *IIE Annual Conference and Expo*, May 30–June 2, 2015, ISERC (accepted for publication).

Madni, A.M. (2015c) Systems Engineering Transformation: From Process Recipes to Explainable Decisions. *NSF Workshop on Decision Engineering: From Engineering Phenomenon to Value*, October 29–30, 2015, Washington, DC.

Madni, A.M. (2015d) *Systems Architecting (SAE 549) Lecture Viterbi School of Engineering*, University of Southern California.

Madni, A.M. (2015e) Systems Engineering Transformation: From Process Recipes to Explainable Decisions. *Invited Talk, National Science Foundation Workshop on "Decision Engineering: From Engineering Phenomena to Value"*, October 29–30, 2015, Washington DC.

Madni, A.M. and Allen, K. (2011) Options Reasoning for Large-Scale Systems Engineering, Program Management, and Organization Design. *Conference on System Engineering Research (CSER)*, April 15–16, 2011, Redondo Beach, CA.

Madni, A.M., Brenner, M.A., Costea, I., MacGregor, D., and Meshkinpour, F. (1985) Option Generation: Problems, Principles, and Computer-Based Aiding. *Proceedings of 1985 IEEE International Conference on Systems, Man, and Cybernetics*, November, 1985, Tucson, Arizona, pp 757–760.

Madni, A.M. and Madni, C.C. (2004) Context-driven Collaboration During Mobile C2 Operations. *Proceedings of The Society for Modeling and Simulation International Western Simulation Multiconference*, January 18–22, 2004, San Diego, CA.

Madni, A.M., Spraragen, M., and Madni, C.C. (2014a) Exploring and Assessing Complex System Behavior through Model-Driven Storytelling. *IEEE Systems, Man and Cybernetics International Conference*, invited special session "Frontiers of Model Based Systems Engineering", October 5–8, 2014, San Diego, CA.

Madni, A.M. et al. (2014b) Toward an experiential design language: augmenting model-based systems engineering with technical storytelling in virtual worlds. *Procedia Computer Science*, **28**, 848–856.

Madni, A.M. et al (2014c) *Visual Analytix™ SBIR Phase I Report*, Intelligent Systems Technology Inc.

Marais, K. (2005) A new approach to risk analysis with a focus on organizational risk factors. Dissertations. Massachusetts Institute of Technology.

McManus, H., Richards, M., Ross, A., and Hastings, D. (2007) A Framework for Incorporating "ilities" in Tradespace Studies. *AIAA SPACE Conference & Exposition*, September 2007, AIAA Paper 2007-6100.

Mekdeci, B., Ross, A.M., Rhodes, D.H., and Hastings, D.E. (2011) System Architecture Pliability and Trading Operations in Tradespace Exploration. *5th Annual IEEE Systems Conference*, April 2011, Montreal, Canada.

Mikaelian, T., Rhodes, D.H., Nightingale, D.J., and Hastings, D.E. (2008) Managing Uncertainty in Socio-Technical Enterprises using a Real Options Framework. *6th Conference on Systems Engineering Research*, April 2008, Los Angeles, CA.

Myers, S. (1977) Determinants of corporate borrowing. *Journal of Financial Economics*, **5**, 147–175.

Neches, R. and Madni, A.M. (2013) Towards affordably adaptable and effective systems. *Systems Engineering*, **16** (2), 224–234.

Ricci, N., Ross, A.M., and Rhodes, D.H. (2013) A Generalized Options-based Approach to Mitigate Perturbations in a Maritime Security Systems-of-Systems. *11th Conference on Systems Engineering Research*, March 2013, Atlanta, GA.

Roberts, C.J. (2003) *Architecting Strategies Using Spiral Development for Space Based Radar*, Massachusetts Institute of Technology.

Roberts, C.J., Richards, M.G., Ross, A.M., Rhodes, D.H., and Hastings, D.E. (2009) Scenario Planning in Dynamic Multi-Attribute Tradespace Exploration. *3rd Annual IEEE Systems Conference*, March 2009, Vancouver, Canada.

Ross, A.M. (2006) Managing unarticulated value: changeability in multi-attribute tradespace exploration. Doctor of Philosophy Dissertation, Engineering Systems Division, MIT, June 2006, Cambridge, MA.

Ross, A.M. and Hastings, D.E. (2005) The Tradespace Exploration Paradigm. *INCOSE International Symposium*. July 2005, Rochester, NY.

Ross, A.M. and Hastings, D.E. (2006) Assessing Changeability in Aerospace Systems Architecting and Design Using Dynamic Multi-Attribute Tradespace Exploration. *AIAA Space AIAA 2006-7255*, September 2006, San Jose, CA.

Ross, A., Hastings, D., Warmkessel, J., and Diller, N. (2004) Multi-attribute tradespace exploration as front end for effective space system design. *Journal of Spacecraft and Rockets*, **41** (1), 20–28.

Ross, A., McManus, H., Rhodes, D., Hastings, D., and Long, A. (2009a) Responsive Systems Comparison Method: Dynamic Insights into Design a Satellite Radar System. *AIAA SPACE Conference & Exposition*, September 2009, AIAA Paper 2009-6542.

Ross, A.M., McManus, H.L., Rhodes, D.H., Hastings, D.E., and Long, A.M. (2009b) Responsive Systems Comparison Method: Dynamic Insights into Designing a Satellite Radar System. *AIAA Space*, September 2009, Pasadena, CA.

Ross, A.M. and Rhodes, D.H. (2008) Using Natural Value-centric Time Scales for Conceptualizing System Timelines through Epoch-Era Analysis. *INCOSE International Symposium*. June 2008, Utrecht, the Netherlands.

Ross, A.M., Rhodes, D.H., and Hastings, D.E. (2008) Defining changeability: reconciling flexibility, adaptability, scalability, modifiability, and robustness for maintaining lifecycle value. *Systems Engineering*, **11** (3), 246–262.

Ross, A.M., Stein, D.B., and Hastings, D.E. (2014) Multi-attribute tradespace exploration for survivability. *Journal of Spacecraft and Rockets*, **51** (5), 1735–1752. doi: 10.2514/1.A32789.

Shah, N.B. (2004) *Modularity as an Enabler for Evolutionary Acquisition*, SM, Massachusetts Institute of Technology.

Spaulding, T.J. (2003) *Tools for Evolutionary Acquisition: A Study of Multi-Attribute Tradespace Exploration (MATE) Applied to the Space Based Radar (SBR)*, Massachusetts Institute of Technology.

Walton, M. (2002) Managing uncertainty in space systems conceptual design using portfolio theory. PhD. Massachusetts Institute of Technology.

Walton, M. and Hastings, D. (2001) Quantifying Embedded Uncertainty of Space Systems Architectures in Conceptual Design. *AIAA Space*, August 2001, Albuquerque, NM.

Walton, M. and Hastings, D. (2004) Applications of uncertainty analysis to architecture selection of satellite systems. *Journal of Spacecraft and Rockets*, **41** (1), 75–84.

11

ARCHITECTURE EVALUATION FRAMEWORK

JAMES N. MARTIN
The Aerospace Corporation, El Segundo, CA, USA

> True genius resides in the capacity for evaluation of uncertain, hazardous, and conflicting information.
>
> Winston Churchill

11.1 INTRODUCTION

This chapter describes key concepts related to the development and use of an architecture evaluation framework, and a recommended set of steps that could be followed when using these concepts. When we say architecture, we mean architecture as being of *the fundamental concepts or properties of a system in its environment embodied in its elements, relationships, and in the principles of its design and evolution* (ISO/IEC/IEEE 42010, 2010). The evaluation we strive for is to determine if the architecture to be examined is suitable for the expected situation and that it maximizes the satisfaction of stakeholder concerns. A number of these ideas regarding how to deal with stakeholder concerns are expounded upon in Firesmith's "Method Framework for Engineering System Architectures" (Firesmith et al., 2008).

Trade-off Analytics: Creating and Exploring the System Tradespace, First Edition. Edited by Gregory S. Parnell.
© 2017 John Wiley & Sons, Inc. Published 2017 by John Wiley & Sons, Inc.
Companion website: www.wiley.com/go/Parnell/Trade-off_Analytics

378 ARCHITECTURE EVALUATION FRAMEWORK

11.1.1 Architecture in the Decision Space

Architecture is one area within the overall "Decision Space" with regard to doing tradespace analysis as shown in Figure 11.1. The Decision Space represents the range of choices that can be examined during the life of a system. This space is divided into different areas of concern, starting with the operational area of concern, followed by the architectural area of concern. The architecture decision space is the focus of this chapter. Once the architecture is established, then decisions will need to be made regarding the one or more designs that "conform" to the architecture. Even after the design is done, there could be decisions regarding production, testing, deployment, logistics support, decommissioning, and so on.

It is often a good practice to establish the concept of operations before the architecture is created, which is further upstream in the Decision Space. Alternative operational concepts should be examined in the front end of exploring the Decision Space. This could entail a decision, for example, between a manned or an unmanned mission to Mars, or between whether the spacecraft will be autonomous or remotely controlled. These are "concepts" that can be explored in their own tradespace prior to considering how an architecture can best support such an operation.

As we noted earlier, the design of a system must not only honor the tenets of its architecture, but also address many other details that go beyond the scope of the "fundamental concepts and properties" of the system that are expressed in the architecture. So, the design space is further downstream from the architecture space in the

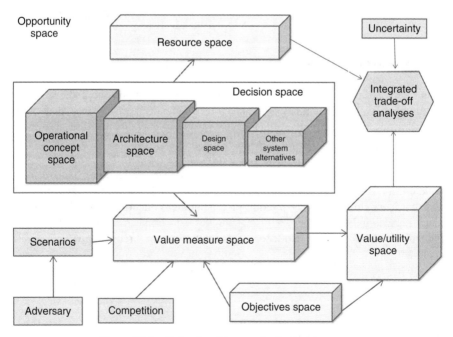

Figure 11.1 Role of architecture in the decision space

overall Decision Space. The rest of this chapter is focused on the establishment of an architecture evaluation framework to aid in exploration of the Architecture Space and describes how to evaluate architecture alternatives within that space using this framework.

11.1.2 Architecture Evaluation

Architecture evaluation is making a judgment or determination about the value, worth, significance, importance, or quality of an architecture. The evaluation effort is aimed at answering one or both of these questions:

- What is the quality of an architecture?
- How well does an architecture address stakeholder concerns?

Quality is a measure of how good or bad something is. An architecture that is being evaluated is usually not "bad" but rather may display varying degrees of goodness. When we evaluate an architecture to determine its value, worth, significance, and so on, we are doing this in terms of who is impacted and how well the architecture addresses the concerns of impacted stakeholders.

We usually examine more than one architecture alternative to ensure that we have more completely explored the potential tradespace. It is a common mistake to not identify and examine alternative architectural approaches (Firesmith et al., 2008). The "tradespace" can be defined as the set of enterprise, program and system parameters, attributes, and characteristics required to satisfy a variety of stakeholder concerns throughout a system's life cycle, while at the same time living within relevant constraints. These various dimensions of the tradespace are often in tension. To increase the performance, one often needs to increase the cost, which is contrary to those to whom cost is a concern. There are many other possibly contrary dimensions in the tradespace: security versus usability, safety versus performance, speed versus accuracy, manufacturability versus maintainability, reliability versus testability, and so on. Our aim during architecture evaluation is to identify the key trade-offs and to understand their implications in terms of delivering the most balanced and robust architectural solution.

Another aim of architecture evaluation is to determine the best way to achieve the greatest amount of goodness given all the constraints and conditions we must contend with. Alexander, who is famous for his concept of architectural patterns, provides a good discussion on how this can be done in his Notes on the Synthesis of Form (Alexander, 1964). Constraints to be dealt with could be financial, technological, sociological, political, and so on. Conditions could entail such things as equitable distribution of work effort, minimal impact on the environment, limited time for doing the evaluation, acceptable risk to the enterprise, and so on.

Architecture evaluation in the context of other architecture-related processes is illustrated in Figure 11.2 (ISO/IEC 42020 2016 draft). Conceptualized architectures are evaluated to determine the suitability of an architecture or to help select among alternatives. More complete models and views of the architecture are elaborated, and

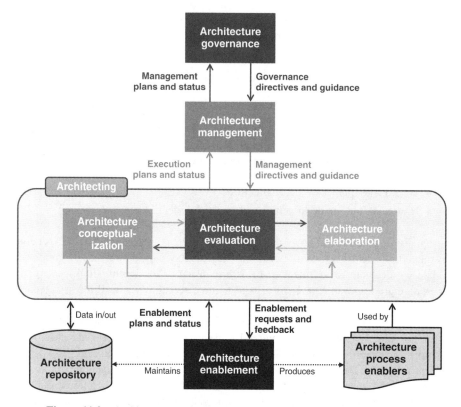

Figure 11.2 Architecture evaluation in context of other architecture processes

these can be evaluated to determine their correctness and completeness with regard to the architecture objectives. Architecture evaluation supports architecture governance in helping determine the possible and most viable and promising ways forward for an enterprise's portfolio of programs, projects, and systems.

11.1.3 Architecture Views and Viewpoints

An architecture is used to characterize the fundamental concepts and properties about a system that address the most important concerns of key stakeholders as illustrated in Figure 11.3. Architecture "viewpoints" are used to help frame the concerns of stakeholders who have an interest in the system, and architecture views and models (specified in the viewpoints) are used to depict key attributes of the architecture under evaluation.

An architecture viewpoint is essentially a "specification" of a particular architecture view that would be found useful in examining the architecture to see how

INTRODUCTION

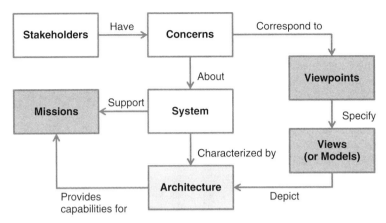

Figure 11.3 Addressing stakeholder concerns through the use of views and models

well it addresses the particular stakeholder concerns represented by that viewpoint. A viewpoint can be thought of as a template to be used when developing architecture views. Example viewpoints are the following:

(a) Operational viewpoint
(b) Maintenance viewpoint
(c) Users viewpoint
(d) Builders viewpoint
(e) Regulators viewpoint
(f) Programmatic viewpoint
(g) Policy viewpoint
(h) Service providers viewpoint
(i) Service consumers viewpoint.

Views are created using these viewpoints, often generated using models of the architecture. For example, an operational view can be composed of one or more of the following models: operational concept diagram, mission reference profile, operational node interaction diagram, operational interchange matrix, operational sequence diagram, conceptual information model, operational activity taxonomy, operational parameter influence diagram, and so on. Similarly, a builder view can be composed of one or more of the following models: system context diagram, functional flow diagram, control flow diagram, state transition diagram, system data model, system block diagram system breakdown structure, system integration sequence diagram, and so on. The key characteristics of architecture views and viewpoints are specified in the international standard on architecture description practices (ISO/IEC/IEEE 42010 2011).

11.1.4 Stakeholders

Stakeholders are individuals or groups that could be impacted – positively or negatively – by the architecture or by the system designs that are derived from the architecture. The impact on a stakeholder can be determined by finding out what "concerns" them. Hence, the concept of stakeholder concerns is crucial in performing an effective architecture evaluation.

User needs and mission needs are often the primary focus of discussion and analysis, but the breadth of stakeholders goes well beyond the needs associated with just users and the mission. It is important to do a thorough examination to determine all those kinds of individuals and groups that could be impacted by the architecture. Examples of stakeholders include users, operators, acquirers, owners, suppliers, developers, builders, and maintainers. It also includes evaluators and authorities engaged in certifying the system for a variety of purposes such as the readiness of system for use, the regulatory compliances of the system, the system's compliance to various levels of security policies, and the system's fulfillment of legal provisions. Furthermore, it is important to pick up stakeholders further upstream and downstream such as company shareholders, supply chain companies, raw material producers, general public for environmental impact, taxpayers for pocketbook impact, and future generations for financial impact. On the other hand, it is sometimes wise to deliberately exclude some stakeholders. This depends on the overall objectives of the evaluation and the intended scope. Some of the stakeholders would be excluded perhaps because the evaluation is purposefully limited in its areas of concern.

11.1.5 Stakeholder Concerns

A concern is something that interests someone because it is important or affects them in some way. This impact could be either positive or negative. It could lead to a benefit for them, or it could be detrimental. Concerns are not specified in any detail but are merely used as dimensions of the problem space that will be characterized during the architecture evaluation process. The detail comes later when these concerns are translated into measurable value assessment objectives and criteria and architecture analysis objectives and criteria. More information on these measurable attributes will be discussed later since these are key components of the architecture evaluation framework that is the main focus of this chapter.

Concerns are not the same as requirements, often because the nature of some concerns is not amenable to the requirements process since requirements must be very precise and unambiguous; requirements must be verifiable in a technical manner and must be usable by engineers in their design process. Firesmith deals with this in the way he addresses "architecturally significant requirements" (2008). Usually, only the acquirer is authorized to specify requirements on the system, and the acquirer (if there is one) is only one of many stakeholders. During architecture evaluation, one should focus on the small number of concerns for each stakeholder. Most often, there is really just one concern for a stakeholder that is most instrumental in getting the architecture right. Sometimes there might be two or three, but rarely are there more than this. Later during the engineering of the system, the design can address more detailed

INTRODUCTION

considerations. But for architecture evaluation, the focus is on those concerns that most impact the architectural features and functions.

Examples of concerns to consider in the assessment include the following items:

- affordability
- agility
- alignment with business goals and strategies
- assurance
- autonomy
- availability
- behavior
- business impact
- capability
- complexity
- compliance to regulation
- concurrency
- control
- cost
- customer experience
- data accessibility
- deadlock
- disposability
- evolvability
- feasibility
- flexibility
- functionality
- information assurance
- interoperability
- interprocess communication
- known limitations
- maintainability
- misuse
- modifiability
- modularity
- openness
- performance
- privacy
- quality of service
- reliability
- resilience
- resource utilization
- schedule
- security
- shortcomings
- state change
- structure
- subsystem integration
- system features
- system properties
- system purposes
- usability
- usage
- viability

Concerns are often expressed as "stories" told by the stakeholders. We can capture these stories in narratives or various other forms such as use-case diagrams, storyboards, influence diagrams, annotated timelines, mission reference profiles, and marketing prospectus. These stories are often essential to giving "voice to the customer," so to speak. It is usually important to capture them in their pure form to help avoid putting our own biases on them, which might lead us to delivering the "right" solution to the wrong problem.

11.1.6 Architecture versus Design

An architecture should be created with stakeholders' concerns in mind. Many of these concerns never make it into the requirements (for the reasons stated earlier). Design, on the other hand, is driven by requirements that have been vetted through the architecture and more detailed analyses of feasibility. Architecture focuses on suitability and desirability, whereas design focuses on compatibility with technologies and other design elements and feasibility of construction and integration.

System architecture deals with high-level principles, concepts, and characteristics represented by general views and models, excluding as much as possible details about implementation technologies and their assimilation such as mechanics, electronics, software, chemistry, human operations, and/or services.

In product line architectures, the architecture is necessarily spanning across several designs. The architecture serves to make the product line cohesive and ensures compatibility and interoperability across the product line. Even for a single product system, the design of the product will likely change over time while the architecture remains constant.

An effective architecture should be as design-agnostic as possible to allow for maximum flexibility in the design tradespace. An effective architecture will also highlight and support trade-offs between conflicting or opposing requirements and concerns. This does not mean that design implementability is ignored during the architecting process. One often learns something about potential designs while evaluating the architecture. Some architectures, although elegant and seemingly addressing the concerns very well, will be discovered to be brittle or possibly unachievable given the available technologies and design approaches. Sometimes design problems are not discovered until well into the design process. In such a case, the architecture evaluation may need to be redone to see what was missed, to possibly come up with alternative architectures that avoid this particular design problem. In some cases, the project might need to be abandoned in favor of more promising ventures.

11.1.7 On the Uses of Architecture

An architecture can be used in other ways than merely as a driver of the system design. An architecture can also be used:

(a) As the basis for operational impact analysis
(b) In support of bid/no-bid and make/buy decisions
(c) To determine where to invest in technologies
(d) In long-term planning to identify follow-on projects
(e) When training operators and end users
(f) To support safety and security analyses
(g) When reviewing features and functions with stakeholders to get their buy-in

11.1.8 Standardizing on an Architecture Evaluation Strategy

This chapter describes the key principles and concepts relevant to the development and use of an architecture evaluation framework. These concepts are intended for inclusion in the international standard that addresses architecture evaluation (ISO/IEC 42030 2016 draft, intended for publication in 2018). These concepts deal with the strategy in which architecture evaluations are organized and recorded for the purpose of:

(a) Evaluating the quality of architectures,
(b) Verifying that an architecture addresses stakeholders' concerns, or
(c) Supporting decision-making informed by the architectures related to the situation of interest.

This strategy addresses the planning, execution, and documentation of architecture evaluations. It prescribes the structure, properties, and work products of architecture evaluations. This approach also specifies provisions that prescribe desired properties of architecture evaluation methods in order to usefully support architecture evaluations. It provides the basis on which to compare, select, or create value assessment and architecture analysis methods.

11.2 KEY CONSIDERATIONS IN EVALUATING ARCHITECTURES

When one evaluates an architecture, it is important to make a clear distinction between the factors that contribute to satisfaction of stakeholder concerns and the factors that affect the key measures of what the system does and how well it does. The former is concerned with how well the fundamental objectives will be met while the latter is focused on how well the means objectives will be met. These two concepts of "value" are often confused. Planning the effort in a systematic manner can help avoid the confusion and help clarify this distinction. As will be shown, the plan drives the overall evaluation effort, which produces a report of what took place and highlights the key findings and recommendations from the effort.

An evaluation framework such as the one shown in Figure 11.4 can help structure the evaluation effort and serve as an easy way to explain to the sponsor what will be accomplished and how this will produce good results. It can help keep the focus on

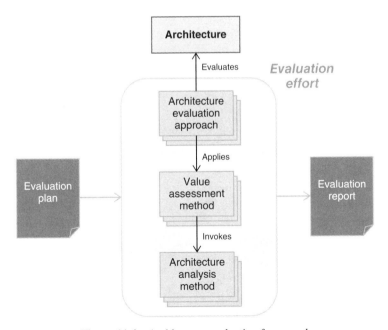

Figure 11.4 Architecture evaluation framework

the objectives of the evaluation rather than getting bogged down in the minutia of the analysis that is to be employed.

11.2.1 Plan-Driven Evaluation Effort

It is important that thoughtful consideration is given to the way in which the evaluation effort will be performed. The discipline of planning can help immensely in providing a structured way of scoping and organizing the work. It is important to develop a plan and have this plan reviewed by the sponsor of the effort to ensure that the evaluation objectives are clear and understood by all parties involved.

The plan should consider which approach is the best one to be taken, sometimes even employing more than one approach for the overall effort. The approach will apply one or more value assessment methods that provide essential information needed to support the evaluation effort. The plan should also specify the value assessment and architecture analysis methods to be used. The typical contents of an architecture evaluation plan are as follows:

(a) Purpose and scope
(b) Evaluation objectives, constraints, criteria, priorities
(c) Schedule and required resources
(d) Evaluation frameworks to be used
(e) Evaluation approaches and methods to be used
(f) Roles and responsibilities of evaluators
(g) Required inputs and reference materials
(h) Expected outputs and deliverables.

The plan should determine how the evaluation results are documented and to whom the report will be shared. The evaluation report should contain the following elements:

(a) Purpose and scope (which might be have changed from the original plan)
(b) Value assessment objectives and criteria used
(c) Architecture analysis objectives and criteria used
(d) Participants involved in the effort (either directly or indirectly)
(e) Inputs used and their sources
(f) Frameworks used
(g) Approaches and methods used
(h) Method results and rationale
(i) Observations and findings generated
(j) Risks and opportunities identified
(k) Recommendations and regrets.

11.2.2 Objectives-Driven Evaluation

As noted in Chapter 2, it is important to understand the distinction between fundamental objectives and means objectives. In the approach described in this chapter, the fundamental objectives are called Value Assessment Objectives, while the means objectives are called Architecture Analysis Objectives.

The Architecture Evaluation Objectives are what the architecture evaluation effort is trying to achieve. For example, the objective for the architecture evaluation effort could be one of the following:

(a) Examine the enterprise portfolio to determine adjustments for better alignment of programs and projects with enterprise goals and objectives.
(b) Examine alternative system designs to establish the basis for a new project.
(c) Examine system designs proposed by competing contractors to determine the winning bid.
(d) Determine the best way to incorporate a new technology coming out of the laboratory.
(e) Find the best way to replace a system in the field that is no longer cost-effective.

The evaluation methods – namely the Value Assessment Methods and the Architecture Analysis Methods – will use the objectives driving them at that level to establish the assessment and analysis criteria to be used, as illustrated in Figure 11.5.

11.2.3 Assessment versus Analysis

Analysis deals with "what" the system does, how well, how often, when, where, under what conditions, but does not deal directly with the value associated with stakeholder concerns (where "value" signifies its worth, importance, significance, or quality). Analysis does not answer the "So What?" question, but instead will focus on the matter-of-fact results of running their models and performing their analytics.

Assessment deals with the "goodness" of the architecture, where goodness is defined in terms of how well stakeholder concerns are addressed. Assessment is primarily focused on answering the "So What?" question. Not all concerns are "technical" in nature. Sometimes there can be "sociopolitical" factors to consider, such as equitable distribution of labor, minimal impact to the environment, securing future funding for the program, minimal impact to jobs in particular governmental districts, and providing "early wins" to the naysayers to help avoid negative pressures.

Architects in general are more likely than systems designers to care about the nontechnical factors. This is one of the key reasons for so much emphasis in this approach on stakeholder "concerns" rather than on stakeholder requirements. The designers and other engineers will take the architecture they are given and optimize such design for the requirements they are given. In fact, this approach can be used as

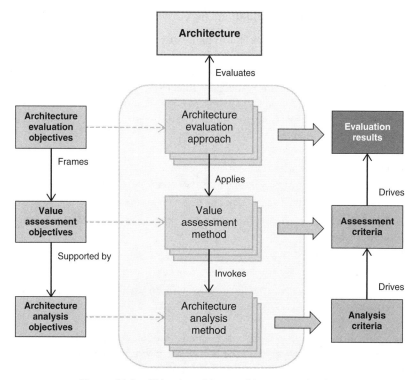

Figure 11.5 Objectives-driven architecture evaluation

the basis for the so-called model-based systems engineering (MBSE) approach where the requirements themselves are (at least at the top level) based on the models of the architecture.

Example stakeholders to consider in the assessment include users, operators, maintainers, owners, sponsors, acquirers, developers, builders, integrators, suppliers, industrial base, labor force, third parties (e.g., environmental impacts), evaluators, policy makers, certification authorities, and auditors.

Value assessment and architecture analysis have rather different goals and results as shown in Table 11.1. The breadth and basis of work are different where assessment is typically a single, unified activity while analysis is usually spread among multiple, separate activities. The overall analysis effort is usually divided into more specific analytical tasks focused on individual concerns or on particular system properties or characteristics. Assessment will integrate the various analysis results and, by taking notice of the competing concerns, must thoroughly examine the tradespace, trying to find the architecture that is most balanced and robust.

Assessment is concerned with utility and value, worth and priorities, ranking and trade-offs (i.e., the fundamental objectives, but sometimes elsewhere called the "ends" objectives), whereas analysis is more concerned with the ways and means of achieving the architectural objectives (i.e., the means objectives). Assessment will

ARCHITECTURE EVALUATION ELEMENTS

Table 11.1 Distinctions between Value Assessment and Architecture Analysis

Characteristics	Value Assessment	Architecture Analysis
Goal Orientation	Fundamental (or ends) objectives (often multilevel)	Means objectives (often multilevel)
Results	Passes "judgment"	Matters of fact
Breadth	Single, unified activity	Multiple, separate activities
Basis of work	Synthesis of analysis results	Technical and other analyses
Scope	Utility, value, worth, priorities, ranking, tTrade-offs	Ways and means
Focus	Effectiveness, efficiencies, equities	Performance determination, limits identification (bounds)
Typical figures of merit	Measures of effectiveness (MOEs), return on investment (ROI), break-even point, key success factors (KSFs)	Measures of performance (MOPs), Key performance parameters (KPPs), technical performance measurements (TPMs), quality metrics
Key items of interest	Competing concerns, performing trade-offs, achieving balance and robustness	Individual concerns, determining system properties and characteristics
Primary questions	So what?, Who cares?, What impacts?, Why?, Why not?	What, where, when, how, how much, how often?

focus on how effective and efficient the architecture will be, leading to its use of things such as measures of effectiveness (MOEs) and key success factors (KSFs). On the other hand, analysis will often focus on things like measures of performance (MOPs), key performance parameters (KPPs), technical performance measurements (TPMs), and other quality metrics.

11.3 ARCHITECTURE EVALUATION ELEMENTS

The key elements of an architecture evaluation are as follows:

(a) Architecture evaluation approach
(b) Value assessment methods
(c) Architecture analysis methods.

11.3.1 Architecture Evaluation Approach

An architecture evaluation approach is a way to deal with the architecture to help determine key characteristics, properties, knowledge or skills of future, current, or past systems related to that architecture. The evaluation approach is a "line of attack"

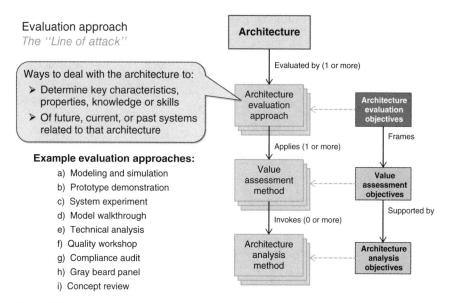

Figure 11.6 Architecture evaluation approaches: choosing one or more "lines of attack"

to be used in evaluating the architecture. Sometimes a multipronged approach is justified, using more than one approach to address different aspects of the situation for that particular evaluation effort. The distinction between approach and method is illustrated in Figure 11.6 and explained in greater detail later. In most cases, a single approach is satisfactory for a particular evaluation effort. However, there are cases where using multiple approaches can be helpful or sometimes even the only practical way to proceed.

Evaluation approaches come in various forms, such as modeling and simulation, prototype demonstration, system experiment, model walkthrough, technical analysis, quality workshop, expert panel, user symposium, concept review, customer focus group, and independent audit.

11.3.2 Architecture Evaluation Objectives

The architecture evaluation approach will be driven by the architecture evaluation objectives and will do this by applying one or more value assessment methods. The approach will integrate the assessment results and will identify key findings and develop recommendations. If there was more than one approach used, then the findings from each separate approach will be examined to identify areas of agreement or conflict and, if necessary, will resolve any discrepancies found. A synthesis of the results will be captured in an evaluation report and communicated to key stakeholders. If applicable, this will be reported to the decision-maker who may either make a decision based on findings and recommendations or, in some cases, ask for some part of the evaluation to be redone.

11.3.3 Evaluation Approach Examples

Here is an example of how this could be applied. Assume the following the evaluation objectives:

(a) Recommend changes to portfolio of systems and technologies
(b) Recommend architecture studies to be conducted

Given these objectives, the following approaches might be chosen:

(a) Modeling and simulation (performance of current systems)
(b) Prototype demonstration (benefits of promising technologies)
(c) System experiment (impact of changing the concept of operations)
(d) Concept review (feedback from key stakeholders after looking at future architecture alternatives and changing conditions and scenarios)
(e) Back-of-the-envelope calculations (to quickly determine which areas to focus on).

An examination of the performance of currently deployed systems could be an important consideration since these systems might be underperforming against their specifications, or not performing well with respect to the competition, or not performing well against the actions of adversaries. It could be helpful to talk with the research department who may have built some advanced prototypes that may have promising features that can be employed in new systems. One could also examine an old system to see if it can be used with different operational methods such that some advantage can be gained; this approach could be useful since it will likely involve a low level of cost and risk. During a concept review, a set of architectural diagrams and mock-ups can be shown to users and operators to get their reactions to the new features and functions. Finally, one should consider doing a "back-of-the-envelope" calculation to determine which of the alternatives can be abandoned without expending a large amount of effort in their examination in detail.

So, it is important to be careful about only picking one approach for evaluating an architecture when multiple approaches might be more cost-effective and timely. The tendency is to only use modeling and simulation when there are often faster, better, and cheaper ways to get the same or better architecture evaluation results.

11.3.4 Value Assessment Methods

The architecture evaluation approach chosen will dictate which value assessment methods are applicable. Whereas an approach is a general, often informal, way in which to do something, a method is a more systematic way of doing something, embodied in an orderly and logical arrangement, often represented as a series of steps. A method if used often will commonly be captured in a specified procedure and posted in a library for later reuse.

A value assessment method is used to determine whether the architecture meets the value assessment objectives, which were specified by the architecture evaluation approach. Each value assessment method specifies the criteria to be used in examination of architecture(s) and applies value assessment criteria to address specific evaluation factors that are derived from one or more value assessment objectives. When necessary, each evaluation factor can be decomposed into lower level factors. This method also will determine if key criteria are met and to what degree stakeholder concerns are addressed.

As illustrated in Figure 11.7, value assessment methods come in various forms, such as portfolio management process, multiattribute utility analysis, mission impact assessment, business case analysis, socioeconomic analysis, strategy-to-task analysis, and user focus group study.

A value assessment method can use manual or tool-based techniques and other suitable enablers. Additional resources such as test environments, discrete event simulations, queuing theory models, and Petri nets, among other architecture model evaluation techniques, can also be used to perform the evaluation. It is not uncommon that these same methods can be applied to operational concepts or to system designs. However, when applied to other things than to an architecture, the value assessment objectives and criteria are often different than those applied to the architecture.

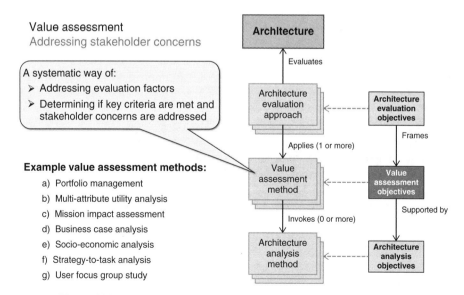

Figure 11.7 Value assessment methods: addressing stakeholder concerns

ARCHITECTURE EVALUATION ELEMENTS

11.3.5 Value Assessment Criteria

Each one of the value assessment criteria applied by the value assessment method is aligned with one or more value assessment objectives. Full coverage of all the objectives in the scope of the architecture evaluation can require the use of multiple value assessment methods.

Benefits should be calculated as net benefits where losses are "subtracted" from gains to ensure a comprehensive look at overall benefit to the wide diversity of stakeholders. Furthermore, some stakeholders might gain while others lose when considering the same architecture criteria. It is important to get a balanced view of the stakeholder concerns' landscape, not just focus on those who benefit.

Of course, determination of net benefit is rarely as easy as doing the simple math. Often, the benefits and costs are evaluated on incommensurate scales. The value assessment method should specify how the measurement scales are used, the protocols for taking the measurements, and how the measurements can be compared. There might need to be a protocol for converting from one scale to another to make comparisons more feasible and accurate.

11.3.5.1 Information Needs The information needed by the value method can be specified in the form of information need statements. Information products will be used by the value assessment method to inform the production of value assessment results.

Some of the information needed could come from architecture analyses. The use of architecture analysis methods is optional since it could be possible to get all the information needed by the value assessment method from sources such as an architecture description, subject matter experts, and system experiments. There is no need to conduct analyses when the information can be readily obtained elsewhere. Analysis can often be time-consuming and expensive and should be used prudently.

11.3.5.2 Value Assessment Example Here is an example of how this could be applied. Assume the following evaluation objectives:

(a) Recommend changes to portfolio of systems and technologies
(b) Recommend architecture studies to be conducted

The relevant value assessment objectives might be the following:

(a) Determine which systems to add, modify, or drop
(b) Determine which technologies to adopt

Given these objectives, the following value assessment methods might be chosen:

(a) Portfolio management
(b) Mission impact assessment
(c) Business case analysis

Portfolio management is a technique for grouping programs and projects into "baskets" and determining the right mix within each basket. The portfolio management process could be fed the results of mission impact assessment and business case analysis since this information can be used to make portfolio adjustment recommendations.

Mission impact is often assessed using some kind of mission models that represent the ways the mission is or will be conducted along with various scenarios and contingencies accounted for. This can be as complicated as a large set of computer-based simulations or could be as simple as a "board game" where the architecture is used as the basis for the game play.

Business case analysis is quite commonly used in the corporate board room to assess business proposals to determine likelihood of success, potential gains and losses, relevant risks and hazards, and alternative pathways to proceed. This method aids decision-making by identifying and comparing alternatives through the examination of mission and business impacts (both financial and nonfinancial), risks, and sensitivities. The strength of the business case is every bit as important as the value inherent in the proposed architecture, which is why it is important that the entire emphasis of architecture evaluation is not on the value metrics alone.

11.3.6 Architecture Analysis Methods

An architecture analysis method examines an architecture in the context of specific evaluation factors. Those factors derive from the architecture analysis objectives specified by the value assessment method, which in turn align with the concerns that define the scope of the architecture evaluation.

As illustrated in Figure 11.8, architecture analysis methods come in various forms, such as functional analysis, object-oriented analysis, performance analysis, behavioral analysis, cost and schedule analysis, risk and opportunity analysis, failure modes, effects and criticality analysis (FMECA), focus group surveys, and Delphi method.

In order to address the architecture analysis objective that support the architecture evaluation objectives, the architecture analysis method needs to identify the architecture attributes of interest and what information about them needs to be analyzed by the architecture analysis using the specified architecture analysis criteria.

11.3.6.1 Architecture Analysis Criteria Architecture analysis criteria can be thought of as conditions on the attributes. The outcomes of applying the attribute criteria are information products that help in determining the degree to which the architecture addresses the stakeholder concerns. Architecture analysis methods use manual or tool-based techniques and other suitable enablers to measure and analyze attributes. Additional resources such as test environments are sometimes required to perform the analysis. Examples of relevant criteria for some analysis objectives are shown in Table 11.2.

Information sources for an architecture analysis method can include such things as architecture description, system description, requirements documents, use case descriptions, design documents, prototypes, system engineering and test plan,

ARCHITECTURE EVALUATION ELEMENTS

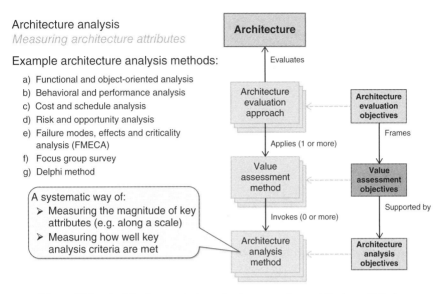

Figure 11.8 Architecture analysis methods: measuring architecture attributes

Table 11.2 Architecture Analysis Objectives and Criteria Examples

Architecture Analysis Objectives	Architecture Analysis Criteria
Determine minimum safe operating condition	Apply the XYZ safety spectrum framework and report the safety level according to the standard levels in the framework
Determine operational throughput	Architecture must enable throughput at least as high as currently deployed systems to be further considered in the analysis
Minimize training time	Determine training hours per year to maintain operator certification levels per standard practice
Determine maximum operating speed	Architecture must enable performance at least as high as currently deployed systems plus an additional 20%

deployed systems, results from related analysis, current system/program risks, issues and management concerns, system analysis results, mission analysis results, business analysis results, stakeholder analysis results, and modeling and simulation results.

11.3.6.2 Measurement Scales and protocols The architecture analysis method could specify a measurement protocol or some other way for gathering information on the attributes and could specify an analysis protocol or some other approach for

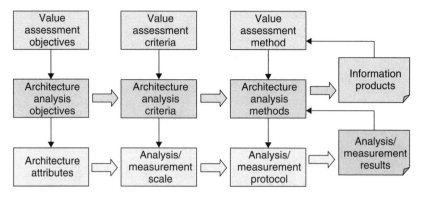

Figure 11.9 Architecture measurement scales and protocols

processing this information to analyze one or more architecture attributes. It does this by applying architecture analysis criteria specified by the method to address the information needs of the higher level value assessment method. This is illustrated in Figure 11.9.

The architecture analysis method, where appropriate, will specify measurement scales and protocols in support of the analysis scales and protocols. The Measurement Process from ISO/IEC 15939 can be used as a basis for using measurement scales and protocols. This International Standard "identifies the activities and tasks that are necessary to successfully identify, define, select, apply and improve measurement within an overall project or organizational measurement structure. It also provides definitions for measurement terms commonly used within the system and software industries" (ISO/IEC 15939 2007). The key elements of the standard's measurement process are illustrated in Figure 11.10.

11.4 STEPS IN AN ARCHITECTURE EVALUATION PROCESS

There is no definitive process for conducting an architecture evaluation. The concepts described in this chapter can be incorporated into a variety of methodologies or corporate procedural guidelines. However, the following steps represent one way of using these ideas in an architecture evaluation effort.

1. Determine purpose and scope of the architecture
2. Identify stakeholders and concerns
3. Establish evaluation objectives
4. Identify or develop architecture alternatives
5. Identify and implement architecture evaluation approaches

STEPS IN AN ARCHITECTURE EVALUATION PROCESS

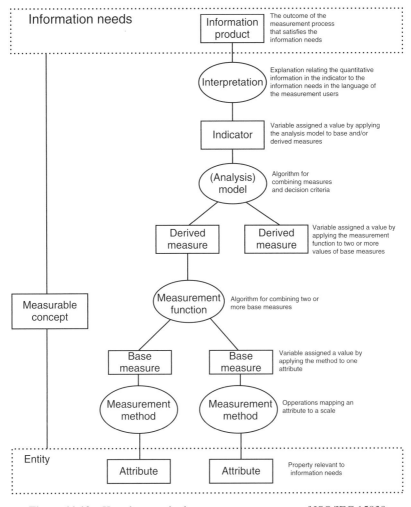

Figure 11.10 Key elements in the measurement process of ISO/IEC 15939

6. Define and implement value assessment methods
7. Define and implement architecture analysis methods
8. Synthesize and present findings and recommendations to decision-maker
9. Document architecture features and functions
10. Communicate results and incorporate feedback

These steps will address all the concepts described in this chapter and provide a reasonable path toward implementing them in an architecture evaluation effort.

11.5 EXAMPLE EVALUATION TAXONOMY

An evaluation framework can be used to help organize the effort using the concepts discussed so far. Part of the framework would be taxonomy of the approaches, assessment methods, and analysis methods. An example is shown as follows. In this example, the following are the evaluation objectives: (i) maximize the mission impact–cost ratio, (ii) maximize the business impact–cost ratio, and (iii) minimize environmental impact. Given these objectives, three evaluation approaches are employed: (i) system experiment, (ii) modeling and simulation, and (iii) prototype demonstration. Three of the value assessment methods for modeling and simulation to be used are, for example, (i) environmental impact assessment, (ii) mission impact assessment, and (iii) business impact assessment.

11.5.1 Business Impact Factors

As illustrated in Figure 11.11, within the business impact assessment, three architecture analysis methods are employed: (i) cost and schedule analysis, (ii) risk and opportunity analysis, and (iii) return on investment.

11.5.2 Mission Impact Factors

Within the mission impact assessment, three architecture analysis methods are employed: (i) performance analysis, (ii) failure analysis, and (iii) user satisfaction (Figure 11.12).

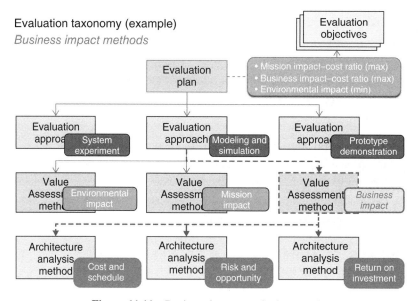

Figure 11.11 Business impact methods example

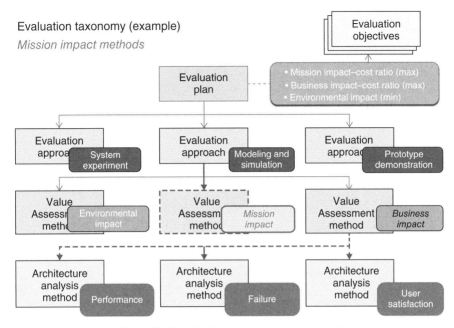

Figure 11.12 Mission impact methods example

11.5.3 Architecture Attributes

Using the same example as earlier, some possibly relevant architecture attributes of interest are shown as follows. As can be seen from this particular example, the large number of things to keep track of and to roll up into the overall evaluation results can be overwhelming. This is why a well-structured evaluation framework can be helpful in getting the work done in an efficient and effective manner (Figure 11.13).

The roll up of results from the architecture analysis level to the value assessment level will be specified in the value assessment method. This roll up is often not merely additive. Sometimes complex mathematics or models are required to do proper integration of the results from one level to another. This is also true as you roll up value assessment results into the evaluation approach level. The evaluation approach should be defined in such a way that you know beforehand how the integration will occur.

This chapter describes how to build an architecture evaluation framework that can be used in the concept exploration activities. See Chapter 10 for details on how to apply mathematical techniques to rolling up the results during concept exploration. Integrating the results at each level in the framework is essential in progressing from the factors examined at the Architecture Analysis level to the next layer during Value Assessment where you need to understand how the architecture supports the "value proposition" as portrayed by the evaluation objectives and criteria.

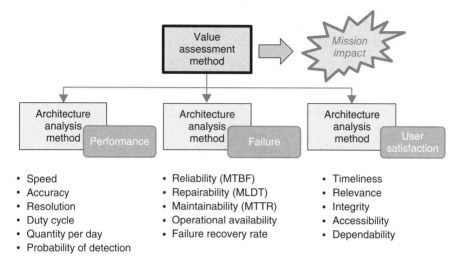

Figure 11.13 Architecture attributes example

11.6 SUMMARY

Evaluating an architecture is an essential step on the way toward exploring the entire Decision Space. The Decision Space represents the range of choices that can be examined during the life of a system. This chapter describes a standard approach for doing architecture evaluation along with guidance on how to use these concepts. Using a structured evaluation framework can help ensure that the full tradespace is considered and that all key stakeholders concerns are properly addressed. It is important to make decisions based on the fundamental objectives that tie to the key stakeholder concerns rather than exclusively focusing on the means objectives that are more closely related to the system requirements. This evaluation can serve to validate proposed requirements for the system to ensure that a more balanced solution is used as the basis for system design.

11.7 KEY TERMS

Architecture: Fundamental concepts or properties of a system in its environment embodied in its elements, relationships, and in the principles of its design and evolution (ISO/IEC/IEEE 42010:2011).

Architecture Analysis Method: A way to examine an architecture in the context of specific analysis criteria. Architecture analysis methods come in various forms, such as functional analysis, object-oriented analysis, performance analysis, behavioral analysis, cost and schedule analysis, risk and opportunity analysis, failure modes, effects and criticality analysis (FMECA), focus group surveys, and Delphi method.

Architecture Attribute: Quality or feature regarded as a characteristic or an inherent part of an architecture as a whole, of the system(s) of interest that conforms to the architecture or of the environment in which the system(s) are situated.

Architecture Evaluation: Making a judgment or determination about the value, worth, significance, importance, or quality of an architecture.

Architecture Evaluation Framework: Conventions, principles, and practices for evaluating architectures that are established specific to a domain of application, a collection of concerns to be examined, or a methodology or set of mechanisms to be applied.

Architecture Evaluation Objective: Something toward which the architecture evaluation work is to be directed, a strategic position to be attained, a purpose to be achieved, or a set of questions to be answered by the architecture evaluation effort (adapted from A Guide to the Project Management Body of Knowledge (PMBOK® Guide) — Fourth Edition).

Concern: Interest in a system relevant to one or more of its stakeholders (ISO/IEC/IEEE 42010:2011).

Criterion: Principle, standard, rule, or test on which a judgment or decision can be based (adapted from A Guide to the Project Management Body of Knowledge (PMBOK(R) Guide) — Fourth Edition).

Evaluation Approach: A general way to deal with the architecture to determine key characteristics, properties, knowledge or skills of future, current, or past systems related to that architecture that are relevant to the evaluation. Approaches come in various forms, such as modeling and simulation, prototype demonstration, system experiment, model walkthrough, technical analysis, quality workshop, expert panel, user symposium, concept review, customer focus group, and independent audit.

Stakeholder: Individual, team, organization, or classes thereof having an interest in a system (ISO/IEC/IEEE 42010:2011). Examples of stakeholders include users, operators, acquirers, owners, suppliers, architects, developers, builders and maintainers, evaluators, and authorities engaged in certifying the system for a variety of purposes such as the readiness of system for use, the regulatory compliances of the system, the system's compliance to various levels of security policies, and the system's fulfillment of legal provisions.

Value Assessment Method: A way to determine whether the architecture meets the assessment criteria. Value assessment methods come in various forms, such as portfolio management process, multiattribute utility analysis, mission impact assessment, business case analysis, socioeconomic analysis, strategy-to-task analysis, and user focus group study.

11.8 EXERCISES

11.1. Identify the stakeholders and their concerns for a family car. Identify the stakeholders and their concerns for a commercial cargo truck. Compare and contrast these lists. Which stakeholders will gain and which ones will lose in each case? What are the key differences between concerns for a car and for a truck?

11.2. Explain in your own words what architecture is. Explain the difference between architecture and design. Why is it important to make this distinction?

11.3. Explain the difference between fundamental objectives and ends objectives. Identify the fundamental objectives and ends objectives for a family car. Identify the fundamental objectives and ends objectives for a commercial cargo truck.

11.4. Explain the distinction between value assessment criteria and architecture analysis criteria. Define the value assessment criteria for a family car. Define the value assessment criteria for a commercial cargo truck. Define the architecture analysis criteria for these two cases.

11.5. Identify some possible architecture evaluation approaches that can be used when evaluating the architecture of a family car. Which approaches would be most appropriate? Map the stakeholder concerns for a family car to the relevant approaches. For each concern, which approach would be most appropriate and why?

11.6. Identify some possible value assessment methods that can be used when evaluating the architecture of a family car. Which value assessment methods would be most appropriate? Map the stakeholder concerns for a family car to the relevant methods. For each concern, which value assessment method would be most appropriate and why?

11.7. Identify some possible architecture analysis methods that can be used when evaluating the architecture of a family car. Which architecture analysis methods would be most appropriate? Map the stakeholder concerns for a family car to the relevant methods. For each concern, which architecture analysis method would be most appropriate and why? Identify the architecture attributes most relevant to each architecture analysis method chosen.

REFERENCES

Alexander, C. (1964) *Notes on the Synthesis of Form*, Harvard University Press.

Firesmith, D., Capell, P., Falkenthal, D., Hammons, C.B., Latimer, D.T., and Merendino, T. (2008) *The Method Framework for Engineering System Architectures*, Auerbach Publications.

ISO/IEC 15939 (2007) *Systems and Software Engineering — Measurement Process*, ISO.

ISO/IEC 42020 (In press, publication expected in 2017) *Systems and software engineering — Architecture Processes*, ISO.

ISO/IEC 42030 (In press, publication expected in 2018) *Systems and software engineering — Architecture Evaluation*, ISO.

ISO/IEC/IEC 42010 (2011) *Systems and software engineering — Architecture Description*, ISO.

12

EXPLORING THE DESIGN SPACE

CLIFFORD WHITCOMB AND PAUL BEERY

Systems Engineering Department, Naval Postgraduate School, Monterey, CA, USA

> Design is not just what it looks like and feels like. Design is how it works.
>
> (Steve Jobs)

12.1 INTRODUCTION

This chapter presents three examples for application of the design of experiments (DOEs) method to systems design. There are two examples from industry: a liftboat example from the offshore oil industry and a cruise ship design. The liftboat uses a fractional factorial design, while the cruise ship uses a Taguchi design. These designs are useful in very early design stages due to the minimal number of points needed. They are useful when information or data is limited or when design synthesis tools are not readily available to develop sets of design variants for complicated systems. Taguchi designs make it possible to explicitly examine "noise" while not including those factors in the design (thereby increasing the number of runs required). Taguchi designs are also resolution III and have interaction effects confounded with main effects, which limit the utility of the results. The third example is for a design of a NATO naval surface combatant ship. This NATO ship example uses a Box–Behnken design. Ship design synthesis tools for naval combatant ships are readily accessible by naval architects, so the design points needed for the DOE method are more readily created. This allows for the use of DOEs that typically require more design points,

---

Trade-off Analytics: Creating and Exploring the System Tradespace, First Edition. Edited by Gregory S. Parnell.
© 2017 John Wiley & Sons, Inc. Published 2017 by John Wiley & Sons, Inc.
Companion website: www.wiley.com/go/Parnell/Trade-off_Analytics

such as Box–Behnken or Central Composite Designs with several variables as factors. These Box–Behnken and Central Composite Designs have resolutions IV and higher, so have fewer confounded effects. This provides for design trade-offs with fewer concerns about interaction effects potentially interfering with design conclusions based on analysis of results.

12.2 EXAMPLE 1: LIFTBOAT

With the current desire to expand offshore drilling for petroleum resources, there has been an expanding market for a capability to maintain offshore rigs. Liftboats provide this capability, especially in the Gulf of Mexico. A liftboat is designed to be a self-elevating, self-propelled vessel with at least one crane and deck space that is open enough to serve multiple purposes. A liftboat is not a conventional jack-up boat or a drilling rig. Liftboats carry the kind of equipment needed to perform platform maintenance, fracking, sand blasting, pipe-laying, and so on. A list of liftboats is available at http://www.shipbuildinghistory.com/history/merchantships/postwwii/liftboats.htm.

Liftboats are classed by leg length. Design considerations allow for leg penetration into the sea bottom, the size of the air gap between the water surface and the vessel hull, and a leg reserve length, which are all part of developing a desired water-depth capability. The operational water depth drives the profits for companies that operate them. In general, the deeper a liftboat can operate, the more profits for the owners. Operational water depths for these vessels approach 300 ft, with leg lengths of over 300 now feasible. As far as the liftboat subsystems are concerned, the legs are the critical subsystem, from both the customer and shipbuilder perspectives. The legs are shown on a liftboat docked in a shipyard in Figure 12.1.

The longer leg lengths needed to provide deeper water capability results in heavier legs, which reduces the lifting capacity of the vessels, as well as reducing the stability both in getting the liftboat to the operational area, and when in operation. In order to look for possible design variations, several leg designs have been attempted with different types of leg internal structures. Figures 12.2 and 12.3 show two typical leg designs of leg structural internals for liftboats. The leg shown in Figure 12.2 uses plate internals, and Figure 12.3 shows a leg with "ladder-" or "lattice-"type internals. The legs are made of high-strength steel, and production is challenging in accomplishing the welding of the internals. Leg construction alternatives that result in a lower displacement are desired for a resulting design, as the lift weight is reduced.

12.2.1 Liftboat Fractional Factorial Design of Experiments

To explore the possibility of constructing liftboats with deeper water capability, a DOE was defined using a two-level five-factor fractional factorial design. This notional example DOE was used to study the design possibilities for reaching deeper operational water depths using various factors of leg diameters, lengths, and thicknesses and responses of displacement and lift weight. The leg internals are not used in this first trade-off, as they would be designed after the overall liftboat

EXAMPLE 1: LIFTBOAT

Figure 12.1 Liftboat docked at the Bollinger Shipyard in Louisiana (Courtesy of Cliff Whitcomb)

Figure 12.2 Liftboat leg internals using a plate construction (Courtesy of Cliff Whitcomb)

Figure 12.3 Liftboat leg internals using a lattice construction (Courtesy of Cliff Whitcomb)

study of displacement and lift weight based on leg diameter, length, and thickness. The results were analyzed using regression analysis using JMP statistical analysis software. The fractional factorial was chosen as the ability to synthesize liftboat designs to cover the design space was limited by the ability of the existing ship design synthesis tools to formulate liftboat designs, from both a technical aspect and the amount of time needed to design ship alternatives. The resulting set of liftboats for the analysis is shown in Table 12.1, which is a condensed version of the full data table limited to three factors and two responses (the full data table is available on the textbook website).

Table 12.1 Fractional Factorial Design for Liftboat

Pattern	Leg Length	Leg Diameter	Leg Thickness	Displacement	Lift Weight
++++++	445	9	3	1000	3840
+−+−+−	445	9	0.757	640	240
++−+−−	445	3	0.757	700	2000
+−++−+	445	3	3	550	500
−+−+−+	130	3	3	300	1200
−++−−+	130	3	3	300	1200
+−−−−+	445	3	3	600	250
−−++−−	130	3	0.757	210	300
++−−++	445	9	3	1000	3840
+−−++−	445	9	0.757	640	240
−−+−++	130	9	3	300	3500
−−−+++	130	9	3	600	5200
+++−−−	445	3	0.757	1050	3000
−+−−+−	130	9	0.757	300	1200

EXAMPLE 1: LIFTBOAT   409

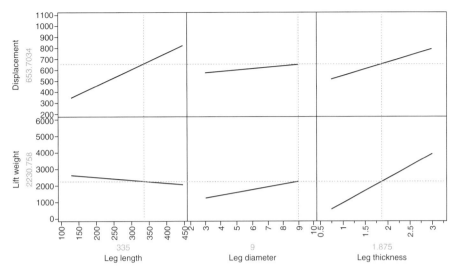

Figure 12.4 Liftboat tradespace for the two responses with respect to the three factors

For displacement, the leg length is found using JMP to be statistically significant, with a *p*-value of 0.0013. For lift weight, both leg diameter, *p*-value of 0.0142, and leg thickness, *p*-value of 0.0209, were found to be statistically significant in the JMP analysis. Figure 12.4 presents an example leg configuration in terms of leg length, leg diameter, and leg thickness. Recall that operating depths of over 300 ft are desirable; accordingly, the leg length is fixed as 335 ft, the leg diameter is set at 9 ft, and the leg thickness is set at 1.875 ft. More detailed examination of this complete trade-off space is required to investigate the appropriateness of the values for leg diameter and leg thickness.

12.2.2 Liftboat Design Trade-Off Space

A potential desired design for a liftboat to achieve 335 ft leg length, indicated in Figure 12.4 by the lines on the *x* and *y* axes, shows a predicted displacement of 654 Ltons and a lift weight of 2,231,000 lb. Figure 12.5 shows a different view of the variable relationships in the form of a "cut" through the trade-off space. The graph shows displacement and lift weight as a function of leg diameter and leg length. Constraint for the designer's desired limit for minimum lift weight is set to 2,200,000 lb and the limit for maximum displacement is set to 675 Ltons. The gray shaded areas are regions that do not meet the constraints where the shaded on the left hand side of the graph corresponds to the leg length and leg diameter combinations that do not satisfy the lift weight constraint and the region shaded in the upper portion of the graph corresponds to the combinations that do not satisfy the displacement constraint. The white, unshaded region is the feasible trade-off space. The crosshairs on the right-hand side of the plot are set to a leg diameter of 9 ft and a leg length of 335 ft based on the

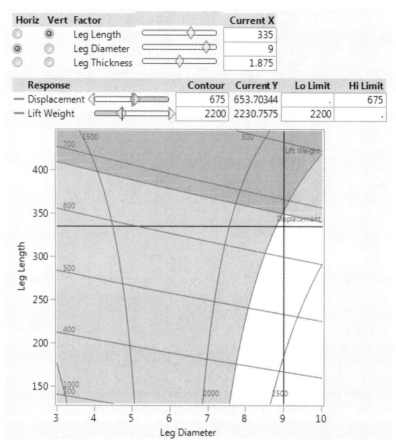

Figure 12.5 Liftboat design tradespace of displacement and lift weight versus leg length and leg diameter, at a leg thickness of 1.875 in

values chosen in Figure 12.4. Further investigation of the trade-off space, with the leg thickness set to 1.875 inches, shows that the projected design appears feasible, as it is within the white space on the graph shown in Figure 12.5.

A quick investigation of existing liftboat designs reveals an existing liftboat, the Robert, that has a leg length of 335 ft, a leg diameter of 9.5 ft, a leg thickness of 1.875 in., with a displacement of 500 Ltons, a maximum water depth of operation at 280 ft, and a lift weight of 1,500,000 lb. Although the Robert's actual lift weight exceeds the constraint associated with this example trade-off, it is a reasonable choice to use for a quick check to verify the output of the liftboat design model as there are a very limited number of existing liftboat designs. The Robert is within the feasible region limit with respect to displacement. A brief study of the predicted design, based on the liftboat model, and an actual design, the Robert, can be accomplished by performing an uncertainty analysis.

EXAMPLE 2: CRUISE SHIP DESIGN

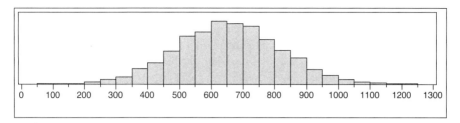

Figure 12.6 Distribution of liftboat model displacement outputs for $N = 5000$

12.2.3 Liftboat Uncertainty Analysis

To investigate the difference between the predicted result and the existing liftboat, a simulation with 5000 data points was performed using JMP to study the distribution of possible displacements in the model. The distribution, Figure 12.6, shows that the model has a mean of 655 Ltons and a standard deviation of 160 Ltons. The model shows a wide variability in the output. This is not surprising as there were few data points for this rough model. If a smaller variation about the mean were desired before using the model for decision-making, more time and effort would have to be put into developing more ship variants or collecting more actual ship data points design to attempt to reduce the variation and improve the prediction capability.

12.2.4 Liftboat Example Summary

This example shows that DOE can be used to define the test points for a model of a liftboat before deciding on the detailed design for construction. The results of model analysis of the test points specified by the DOE can be used as a surrogate for available design tools when data, information, and design tools are limited or not available. The tradespace is then used to investigate the design possibilities, including using uncertainty analysis to further study the resulting liftboat design trade-off space.

12.3 EXAMPLE 2: CRUISE SHIP DESIGN

Traditionally, cruise ship design has been undertaken without consideration of system trade-offs. The two most important characteristics in designing a cruise ship are the economics, such as acquisition cost, profitability, operational cost, and the comfort of the cruise experience (Katsoufis, 2006). The naval architecture technical variable details of the ship characteristics are typically not as important as these business variables to the decision-makers. The trade-off is set up using a set of plots that show how the business variables are related to these technical variables.

12.3.1 Cruise Ship Taguchi Design of Experiments

In order to study the ramifications of early design decisions, a Taguchi screening experiment was performed on a notional cruise ship design. A set of eight factors was chosen to vary the technical aspects of the ship design, and three responses were selected to study their impact on the possible outcomes. The three responses are related to economic aspects are the ones used for decision-making. The fourth, the beam of the ship, is included since the beam can be limiting for transport through various canals, such as the Panama Canal. A Taguchi L18 design was selected for the experiment, Table 12.2.

12.3.2 Cruise Ship Design Trade-Off Space

The cruise ship L18 point designs shown in Table 12.2 were synthesized using a Microsoft Excel spreadsheet model provided on the textbook website. The Microsoft Excel spreadsheet was used to synthesize the ships, while JMP provided the DOE and resulting statistical analysis. The results were analyzed using JMP. The JMP results are shown in Figure 12.7 for the design point with the highest revenue.

The design space can be used to vary the factors to study the possibilities for different ship characteristics, such as number of passengers, propulsion type, power plant, and number of propulsors, and the range of possibilities for costs and revenue, within desired beam restrictions. The technical variable factors on the x-axes of Figure 12.7 can be set interactively in JMP, and the resulting business variable responses on the y-axes. The results show that five factors (operational range, days of stores, propulsion type, number of propulsors, and number of engines) have little or no impact on any of the MOEs that are the primary concern of the stakeholders. These do, however, maintain a link to the variables that the shipbuilders are primarily concerned with, so they are valuable to maintain an ability to discuss the ship characteristics among all stakeholders. The ship characteristics technical variables are set in Figure 12.7 to indicate the maximum revenue, while showing the ship beam in case this becomes an important consideration for passage through canals and other restricted waterways.

The design space can also be explored using the trade-off approach employed for the liftboat example. Figure 12.8 shows revenue, acquisition cost, operating cost, and beam as a function of passenger capacity and brand quality. A $10,000 minimum constraint is set for the Revenue at $10,000, a $1500 maximum constraint is set for the acquisition cost, a $5600 maximum constraint is set for the operating cost, and a 120 ft maximum constraint is set for the beam.

The gray shaded areas correspond to the regions of the design space that do not meet the constraints (where referring to the right hand side of the graph where labels for the respective shaded regions are located the region corresponds to combinations that do not satisfy the revenue constraint, the shaded region just above the white region corresponds to operating cost, the next region up corresponds to the acquisition cost, and top-most shaded region corresponds to the beam). Note that the white, feasible region suggests that as the brand quality Increases, fewer passengers can be included in the ship. Currently, Figure 12.8 identifies a ship configuration with a budget brand quality and 2500 passengers as a feasible ship configuration. Note that the size of the

Table 12.2 Cruise Ship Taguchi L18 Experimental Design

Pattern	Passenger Capacity	Operational Range	Days of Stores	Propulsion Type	Number of Propulsors	Number of Engines	Brand Quality	Power Plant Type	Revenue	Acquisition Cost	Operating Cost	Beam
---- --++	1,000	4,000	10	Pod	2	2	Ultra	Diesel	6,828	914	3,493	92
+-- -++- -0	3,000	4,000	10	Screw	4	2	Budget	Combined	12,049	1,590	5,655	118
--++ --0-	1,000	4,000	20	Screw	2	2	Premium	Gas turbine	5,342	884	3,747	90
++++ +++++	3,000	8,000	20	Screw	4	4	Ultra	Diesel	20,483	2,115	7,816	137
-++- -+--0	1,000	8,000	20	Pod	2	4	Budget	Combined	4,017	730	2,784	83
++-- -+-0+	3,000	4,000	20	Pod	4	2	Premium	Diesel	16,025	1,914	6,837	130
++-+ ---00	3,000	8,000	10	Screw	2	2	Premium	Combined	16,025	1,935	6,976	130
+++- ----	3,000	8,000	20	Pod	2	2	Budget	Gas turbine	12,049	1,577	6,267	118
--+- -+00	1,000	4,000	20	Pod	4	4	Premium	Combined	5,342	857	3,241	90
-+-+ -+0+	1,000	8,000	10	Screw	2	4	Premium	Diesel	5,342	879	3,268	90
++++++ +0	3,000	8,000	20	Screw	4	4	Ultra	Combined	20,483	2,112	7,914	137
---- -+0	1,000	4,000	10	Pod	2	2	Ultra	Combined	6,828	973	3,688	96
++-- +0-	3,000	8,000	10	Pod	4	4	Premium	Gas turbine	16,025	1,899	7,658	130
++++ -++	3,000	4,000	20	Screw	2	4	Budget	Diesel	12,049	1,620	5,669	119
++++ -+-	1,000	8,000	20	Screw	4	2	Ultra	Gas turbine	6,828	1,033	4,366	99
-+-- ++	1,000	8,000	10	Pod	4	2	Budget	Diesel	4,017	803	2,972	88
--++ +--	1,000	4,000	10	Screw	4	4	Budget	Gas turbine	4,017	785	3,351	86
+--- -++-	3,000	4,000	10	Pod	2	4	Ultra	Gas turbine	20,483	2,085	8,615	135

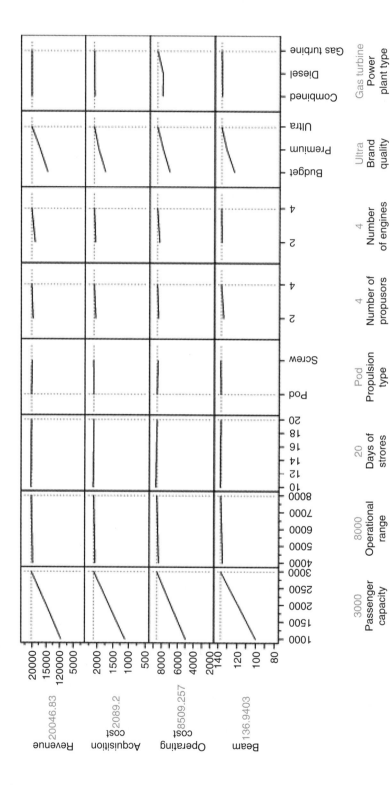

Figure 12.7 Trade-off space for cruise ships with variables set to the point with the highest revenue

EXAMPLE 2: CRUISE SHIP DESIGN

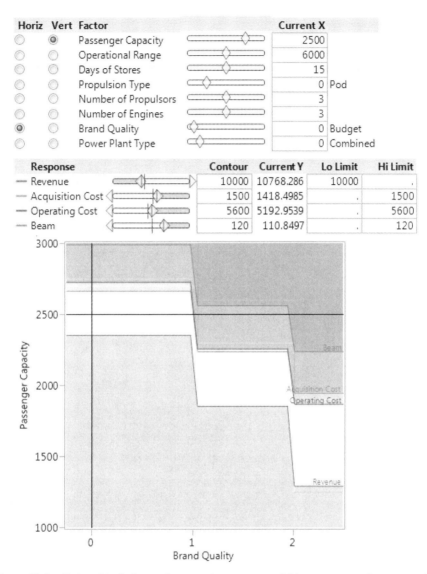

Figure 12.8 Cruise ship design tradespace of revenue, acquisition cost, operating cost, and beam versus passenger capacity and brand quality

feasible region will be reduced if the constraints are altered (Figure 12.9 presents an example).

Figure 12.9 shows the impact of a reduction in the allowable ship beam (note that it has been reduced from 120 ft in Figure 12.8 to 110 ft in Figure 12.9). This has rendered the previous ship configuration of 2500 passengers with a budget brand quality infeasible. Now it is only possible to accommodate 2400 passengers, even with a budget brand quality, due to the reduction in acceptable ship beam.

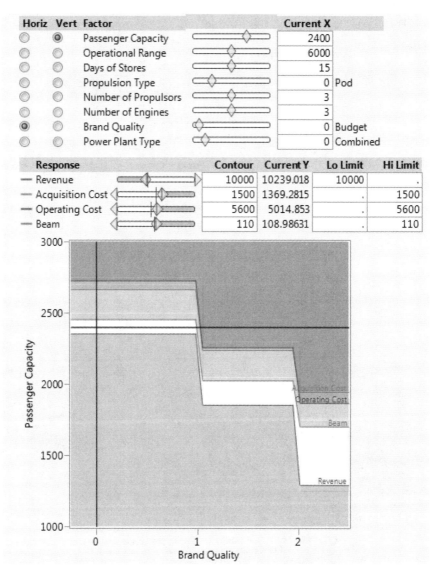

Figure 12.9 Cruise ship design tradespace of revenue, acquisition cost, operating cost, and beam versus passenger capacity and brand quality

12.3.3 Cruise Ship Example Summary

This example shows a more basic trade-off structure where technical variables can be set to indicate the levels of business variable outcomes. The technical variables summarize the various ship variant alternatives that can be designed in order to meet desired business goals, while simultaneously checking a technical variable that might constrain operations in some desired concepts of operation.

12.4 EXAMPLE 3: NATO NAVAL SURFACE COMBATANT SHIP

NATO has planned a surface force concept to enable effective combat in littoral areas (NATO, 2004). The ship is envisioned as an integrated system-of-systems made up of sensors, decision aids, weapons and related support systems as part of a comprehensive maritime-based network. The combined capabilities of this surface strike force (SSF) concept will result in an integrated, multi-dimensional operational capability through which the joint force commander can both project power and protect joint forces from the sea.

In partial support of the overall SSF, a particular operations concept (OpsCon) is envisioned based on guidance for a series of flexible mission combatants, dubbed the flexible surface strike force (FSSF), which is to operate in littoral areas, and provide robust antiair warfare (AAW) self-defense, power projection ashore using gunfire support (GFS) and strike (STK), antisubmarine warfare (ASW), and antisurface warfare (ASUW). The FSSF must operate in conjunction with Aegis ships both as a net-centric asset and in independent operations. It must be capable of performing unobtrusive peacetime presence missions in an area of hostility and immediately respond to escalating crisis and regional conflict. The ship is likely to be forward deployed in peacetime, conducting extended cruises to sensitive littoral regions. Small crew size and limited logistics requirements will facilitate efficient forward deployment.

The FSSF will be among the first naval forces present in a region. It will perform detailed reconnaissance of surface and subsurface topography and the activities of potentially hostile forces. As hostilities intensify, the FSSF may be required to perform preemptive strike, lay mines, and conduct preassault shore bombardment. The FSSF may fall back to escort Amphibious Readiness Groups (ARG), mine counter measure (MCM) groups, or replenishment groups. It will continue to escort these groups or operate independently throughout the course of the conflict, providing flexible and robust ASW, ASUW, and STK capabilities as required. Although AAW emphasis will be on unit self-defense, some local area AAW defense capability will be required, particularly in cooperative engagement. As a conflict proceeds to conclusion, the FSSF will continue to monitor all threats. The FSSF may likely be the first to arrive and last to leave the conflict area. A notional ship design example is shown in Figure 12.10.

The FSSF is envisioned to be part of a NATO developed and deployed force. A reference study, the NATO NAVAL GROUP 6 SPECIALIST TEAM ON SMALL SHIP DESIGN NATO/PfP Working Paper on Small Ship Design, May 2004, is used as a basis for this example of design and acquisition of potential Offshore Patrol Vehicle (OPV) and Small Littoral Combatant (SLC) platforms. These platforms comprise the

Notional FSSF Principal Characteristics			
Length, LBP	93.9 M	Lightship Displacement	1,964 Tonnes
Beam at DWL	13.2 M	Full Load Displacement	2,465 Tonnes
Draft, Full Load, Midships	3.8 M	Arrangeable Deck Area	2,827 MF
Average Hull Depth	9.1 M	Propulsion Plant	Triple Screw CODAG
			Center: 23,150 kW GT, w/ 4.33 M 210 RPM CPP Prop
			Outbd: 6,210 kW diesels, w/ 3.28 M 160 RPM CPP Props
Trial Speed	31.7 knots	Accommodations	110@10.08MF/Accom
Range	4,500 N Mi @ 16 knots	Ships Service Electric Power	Three 880kW SSDG Sets
Endurance	20 Days		

Figure 12.10 Notional FSSF ship design (Source: NATO. Reproduced with permission of NATO)

ships in the FSSF. The primary vessels to be procured as part of the initial effort are the corvette ships, with displacement of approximately 2000 Ltons.

12.4.1 NATO Surface Combatant Ship Stakeholder Need

The first step in this project is to determine the high-level stakeholder needs and requirements. Based on the NATO report, the stakeholder needs for this example are listed in Table 12.3.

Table 12.3 Stakeholder Needs for an FSSF

FSSF Stakeholder Need
Information gathering
Protect sea lines of communications (SLOC)
Protection of high-value units
Embargoes and sanctions
Amphibious operations
Neutralize naval forces
JTF campaign
Support operations
Mobile capability
Independent operation capability
Flexible capability

EXAMPLE 3: NATO NAVAL SURFACE COMBATANT SHIP

Understanding the priorities for these stakeholder needs is useful to accomplish a decision analysis. Using value functions, the stakeholders were found to have priorities as shown in Table 12.4.

Next, the stakeholder value function curves for each of the needs were determined, with the results shown in Figure 12.11.

The individual value functions were rolled up into an alternative value, labeled in this case as Overall Measure of Effectiveness (OMOE), by multiplying the value for each need as indicated in the design outcome by the priority weight for each respective need (see Chapter 2 for further discussion of the additive value model).

Table 12.4 Prioritized Stakeholder Needs

FSSF Stakeholder Need	Priority Weight
Information gathering	0.05
Protect sea lines of communications (SLOC)	0.17
Protection of high-value units	0.17
Embargoes & sanctions	0.02
Amphibious operations	0.17
Neutralize naval forces	0.17
JTF campaign	0.09
Support operations	0.02
Mobile capability	0.03
Independent operation capability	0.02
Flexible capability	0.09

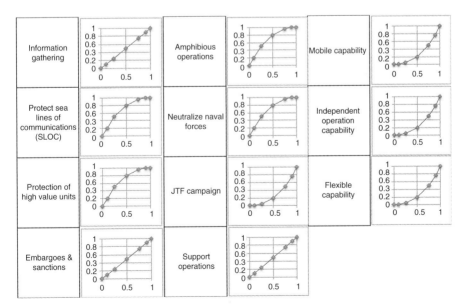

Figure 12.11 Stakeholder value functions for the various high-level needs

12.4.2 NATO Surface Combatant Ship Box–Behnken Design of Experiments

In order to study the possibility for surface combatant ships that might meet the mission requirements, a DOE is conducted. The DOE used is a Box–Behnken design, with four factors at three levels, summarized in Table 12.5.

For this envisioned combatant ship, as is typical for many combatant ship designs, both speed and range have a key interest to both stakeholder and naval architects. The amount of payload that can be carried in this case is used as a surrogate for the weight of the weapons and combat systems equipment that the ship can carry. This is related to the OMOE, as the weapons and combat systems determine the warfighting capability of the ship, but is not modeled explicitly in this example. The margin variable reflects both a design margin, used during ship design to account for design uncertainties, and through-life margin, used to allow for growth in added ship systems that

Table 12.5 DOE Data Table for FSSF Surface Combatant

Pattern	Factors				Responses		
	Speed (knot)	Range (nm)	Payload (Ltons)	Margin (%)	Cost (M$)	OMOE	Displacement (Ltons)
0+−0	30	6000	250	7.5	353.4	0.1776	2299.3
+0−0	34	4500	250	7.5	415.2	0.1355	2476.1
0000	30	4500	300	7.5	360.7	0.5868	2259.6
0−0−	30	3000	300	5.0	352.7	0.5102	2071.8
00−−	30	4500	250	5.0	345.3	0.1010	2107.3
−−00	26	3000	300	7.5	318.1	0.4858	1924.4
+−00	34	3000	300	7.5	422.3	0.5448	2421.2
++00	34	6000	300	7.5	430.7	0.6878	2737.3
+0+0	34	4500	350	7.5	460.8	0.9238	2686.9
−00+	26	4500	300	10	325.1	0.5624	2099.1
0++0	30	6000	350	7.5	394.9	0.9655	2491.4
−0+0	26	4500	350	7.5	356.6	0.8645	2165.0
+00+	34	4500	300	10	436.6	0.6213	2669.7
0+0−	30	6000	300	5.0	356.3	0.6533	2316.6
−00−	26	4500	300	5.0	314.7	0.5523	1974.6
−0−0	26	4500	250	7.5	310.6	0.0766	1948.5
0−0+	30	3000	300	10	365.1	0.5203	2204.1
0−+0	30	3000	350	7.5	391.1	0.8224	2229.7
00−+	30	4500	250	10	357.7	0.1111	2240.3
+00−	34	4500	300	5.0	420.6	0.6112	2512.7
0−−0	30	3000	250	7.5	349.7	0.0345	2052.6
00+−	30	4500	350	5.0	386.6	0.8889	2287.5
0+0+	30	6000	300	10	369.0	0.6634	2460.2
−+00	26	6000	300	7.5	321.6	0.6288	2153.4
00++	30	4500	350	10	399.4	0.8990	2492.2

EXAMPLE 3: NATO NAVAL SURFACE COMBATANT SHIP

can be accommodated without negatively impacting the level of the design waterline, in the ship during the operation phase of the ship life cycle. The larger the margin, the more flexible the ship is in its ability to allow for engineering changes or upgrades to weapons and other ship systems, while still being able to meet ship operational requirements such as stability and maximum speed through the maintenance of the ship at the design waterline through the life cycle. Margin then has an upside value to stakeholders as it provides for a hedge against uncertainty during design and for a more robust ship over the operational life cycle in that it can adapt to technology advancements in ship systems. The downside to margin is that it adds to displacement at the expense of payload in the end result.

The ship variants for this design study are synthesized using a Microsoft Excel spreadsheet computer program with the factors as inputs. Each ship variant is created individually and balanced from a naval architecture perspective to achieve a feasible ship for each variant.

12.4.3 NATO Surface Combatant Ship Cost-Effectiveness Trade-Off

Displacement and length are model responses of interest to the naval architects and marine engineers, and for comparison to other existing ships during a decision-making process, so they are kept in the model results. The key responses of interest for the decision-makers are the cost and the OMOE, so these are the focus of this example. These results are summarized in Table 12.6, with each variant given a unique alternative design number, labeled as A1 through A25.

A plot of the cost versus OMOE is shown in Figure 12.12.

The variants that fall on the Pareto frontier are the ones with the highest OMOE at the lowest cost. These are clustered along a line that traces a path along the upper left edge of the plot. This region is magnified in Figure 12.13 to show the four variants that are nondominated. The four variants range in cost from $314.7 to $394.9 M, and OMOE from 0.629 to 0.966. The selection of the best variant is up to the decision-makers in trading off cost versus effectiveness, as all four solutions are equally optimal in a multiple criteria sense. Affordability considerations (Chapter 4) will be important in this decision. All other variants are dominated by these four, and unless there is significant uncertainty in the variants, these do not need to be considered further in the decision-making process.

12.4.4 NATO Surface Combatant Ship Design Tradespace

In order to more thoroughly explore the design space, the factors and response outputs from the DOE are studied using interaction profilers as well as contour profilers. The variables for the design variants are explored by studying their relationship to the outputs used for overall decision-making; in this case, cost and effectiveness, as indicated by OMOE, are used as the basis for decision-making. The resulting Effects plot is shown in Figure 12.14.

Table 12.6 Cost and OMOE for FSSF Surface Combatant Variants

Ship Variant	Cost	OMOE
A1	$353.4	0.1776
A2	$415.2	0.1355
A3	$360.7	0.5868
A4	$352.7	0.5102
A5	$345.3	0.101
A6	$318.1	0.4858
A7	$422.3	0.5448
A8	$430.7	0.6878
A9	$460.8	0.9238
A10	$325.1	0.5624
A11	$394.9	0.9655
A12	$356.6	0.8645
A13	$436.6	0.6213
A14	$356.3	0.6533
A15	$314.7	0.5523
A16	$310.6	0.0766
A17	$365.1	0.5203
A18	$391.1	0.8224
A19	$357.7	0.1111
A20	$420.6	0.6112
A21	$349.7	0.0345
A22	$386.6	0.8889
A23	$369.0	0.6634
A24	$321.6	0.6288
A25	$399.4	0.899

By visually inspecting the curves, lines with steep slopes correspond to variables that have substantial impacts on each MOE. The factor that impacts cost the most is speed, while the factor that impacts the OMOE the most is payload. This is confirmed by examining the JMP Effects Pareto Plots, Figures 12.15 and 12.16, with the x-axis as percent effect.

Since payload and speed have the most impact on the design decisions, as indicated by the longest bars in Figures 12.15 and 12.16, respectively, these factors are the primary ones varied to explore the design space. This exploration provides a more detailed look at design possibilities beyond the four Pareto solutions by including variations in the design variables that fall between the data points used for creation of the bounds of the design space.

12.4.5 NATO Surface Combatant Ship Design Trade-Off

To begin with, a series of design space portions have to be viewed on trade-off space plots generated by JMP. The plots can be studied interactively in JMP, but static

EXAMPLE 3: NATO NAVAL SURFACE COMBATANT SHIP

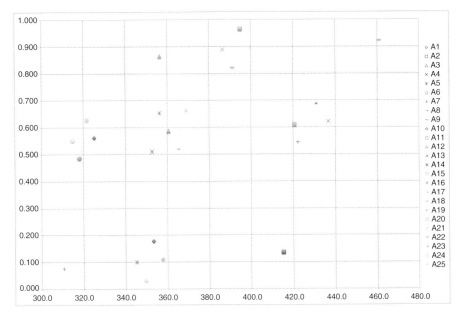

Figure 12.12 Cost versus OMOE for FSSF surface combatant variants

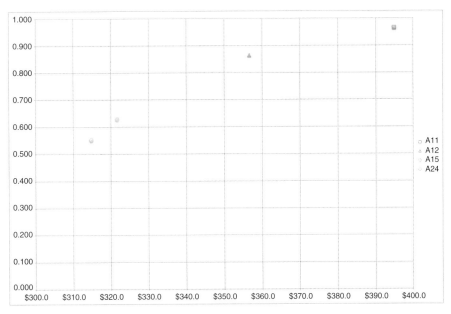

Figure 12.13 Magnified plot region showing only Pareto frontier of nondominated variants for cost versus OMOE for FSSF surface combatants

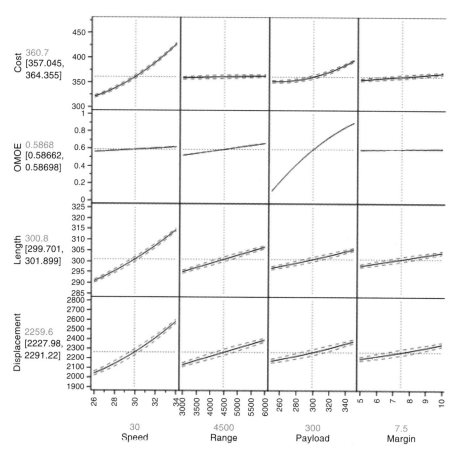

Figure 12.14 Effects plot for NATO FSSF surface combatant ship

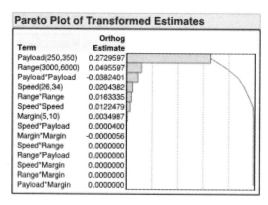

Figure 12.15 OMOE effects Pareto

EXAMPLE 3: NATO NAVAL SURFACE COMBATANT SHIP

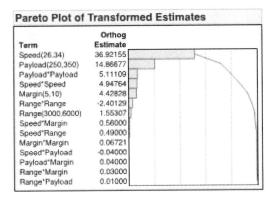

Figure 12.16 Cost effects Pareto

images are used here to demonstrate a process for investigating the tradespace. The two factors of payload versus speed are plotted with several responses overlaid. For now, the high cost limit is set to $400 M and the low OMOE value is set to 0.83. The resulting feasible design decision space is now shown with a white background in Figure 12.17. The trade-off example is not exhaustive and is used only to show how the process of studying a trade-off space can be accomplished.

The design variant defined by Figure 12.17 is detailed in Table 12.7. Note that Ship A11 is one of the original variants from the DOE table.

Ship variant A11 is the best performing alternative with respect to the OMOE that also satisfies the cost constraint of $400 M. However, it may be useful to examine alternative scenarios. As designs mature from concept to preliminary stages, information generated about the design details may make a design impossible to achieve. Unforeseen design changes may also make it impossible to actually create Ship A11. Examination of Figure 12.17 suggests that it may be interesting to examine an alternative ship (termed Ship A11a) that is unable to achieve a total payload of 350 Ltons and can only achieve a payload of 320 Ltons. A visualization of this ship is shown in Figure 12.18; note that it is incapable of satisfying the OMOE constraint of 0.83 and is therefore currently an infeasible system alternative.

The utility of the interactive design space exploration approach is easily demonstrated through presentation of additional design trade-off spaces. For instance, while Figure 12.18 suggests that Ship A11a, which can only achieve a payload of 320 Ltons, is incapable of satisfying the OMOE constraint, examination of an alternative visualization of the trade-off space suggests a potential solution. Figure 12.19 presents a visualization of the same trade-off space but presents the speed and the range (rather than speed and payload, as shown in the previous visualizations).

Note that Ship A11a is still shown as infeasible, but the alternative visualization suggests that by increasing the range to 6600 nm, the ship (now termed Ship A11b in Figure 12.20) becomes feasible.

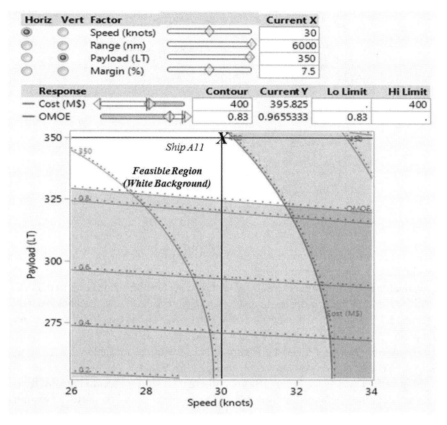

Figure 12.17 Trade-off space of speed versus payload (Ship A11)

Table 12.7 Design Variant

Ship Variant	Cost	OMOE	Length (ft)	Displacement (Ltons)	Margin (%)	Payload (Ltons)	Speed (knot)	Range (nm)
A11	$395.8	0.9655	311	2491	7.5	350	30	6000

The impact of this change can be visualized more clearly by a reexamination of the original trade-off space of speed and payload. Figure 12.21 presents an example of Ship A11b from this perspective. Note that Ship A11b is feasible with respect to both the OMOE and cost despite only having a payload of 320 Ltons (compared to 350 Ltons for the original Ship A11). This is a result of the increase in range from 6000 nm for Ship A11 to 6600 nm for Ship A11b.

Further study of the feasibility of Ship B with respect to payload versus speed at 5% margin is shown in Figure 12.22. This ship alternative is within the feasible

EXAMPLE 3: NATO NAVAL SURFACE COMBATANT SHIP

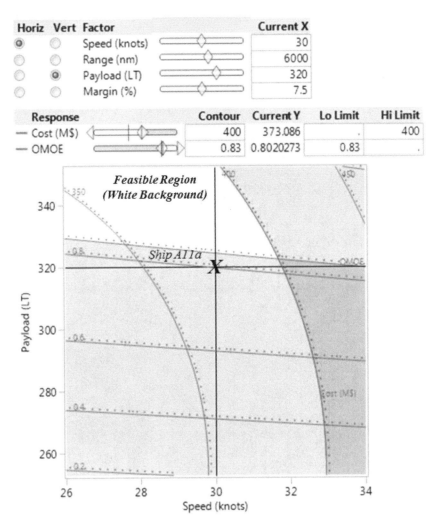

Figure 12.18 Trade-off space of speed versus payload where payload is restricted and the system becomes infeasible (Ship A11a)

region, indicated with a white background. This example highlighted the utility of the design trade-off space exploration approach for scenarios where constraints are imposed that cause the system to become infeasible. The design trade-off space exploration approach is also useful when requirements are imposed by a stakeholder that may conflict with analysis results. For example, a stakeholder may desire a modification to Ship A11 that increases the speed from 30 to 32 knot while also reducing the payload to 325 Ltons. A visualization of such a scenario is shown in Figure 12.22 (the ship is now termed Ship A11c).

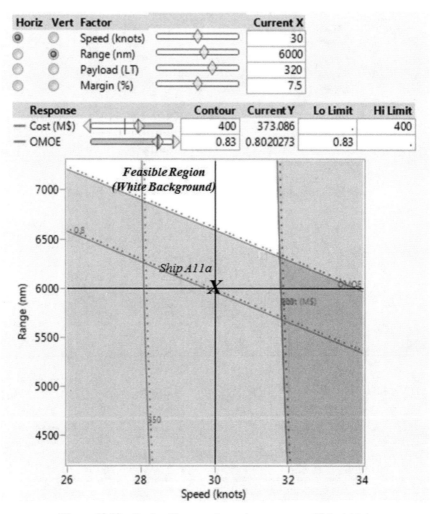

Figure 12.19 Trade-off space of speed versus range (Ship A11a)

The stakeholder requirement for increased speed, coupled with the decrease in the acceptable payload, has caused the ship to become infeasible with respect to cost. Note that the system can actually become feasible by decreasing the speed to the original value of 30 knot. However, if the stakeholder is unwilling to accept a restricted speed, the design trade-off approach can explore alternative views to define an acceptable system. Figure 12.23 presents a visualization of the trade-off space between the margin and the payload. Note that there is a small feasible region associated with ship configurations with a payload of 325 Ltons and a margin of

EXAMPLE 3: NATO NAVAL SURFACE COMBATANT SHIP

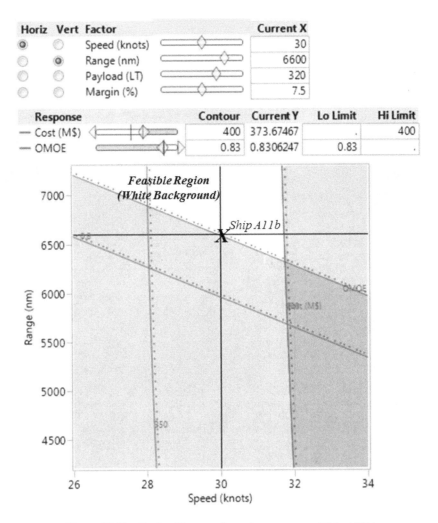

Figure 12.20 Trade-off space of speed versus range (Ship A11b)

5% (reduced from the original value of 7.5%). This ship configuration is plotted as Ship A11d.

Ship A11d is feasible within the cost and OMOE constraints, even though the ship speed is increased to 32 knot and the payload is restricted to 325 Ltons. This highlights the alternative usage of the design trade-off space exploration approach, which allows a stakeholder to examine the impact that system requirements may have on the size of the total number of feasible system configurations as well as the interactions between input factors that must be investigated as a result of each new system requirement.

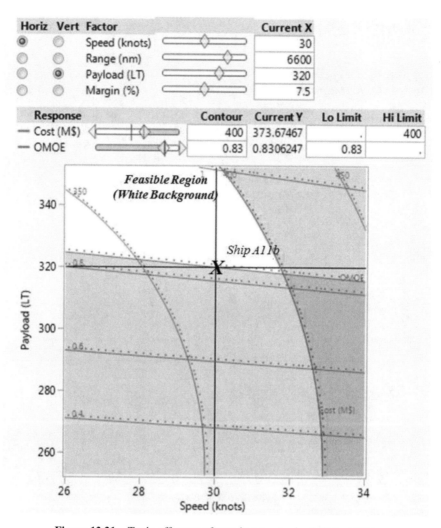

Figure 12.21 Trade-off space of speed versus payload (Ship A11b)

12.4.6 NATO Surface Combatant Ship Trade-Off Summary

This example shows a design trade-off beginning with consideration of stakeholder needs and requirements and continuing to the development of a DOE trade-off space. Several cost versus Effectiveness optimal solutions are identified from the DOE results. Next, a possible design trade-off is then studied by varying design variable factors and responses to investigate the space for solutions that are interpolated between the DOE points. The resulting designs are summarized in a form that decision-makers can review as a series of variant alternatives that are both feasible and reach the limits of the trade-off since the variables cannot be set to achieve all of the available values simultaneously.

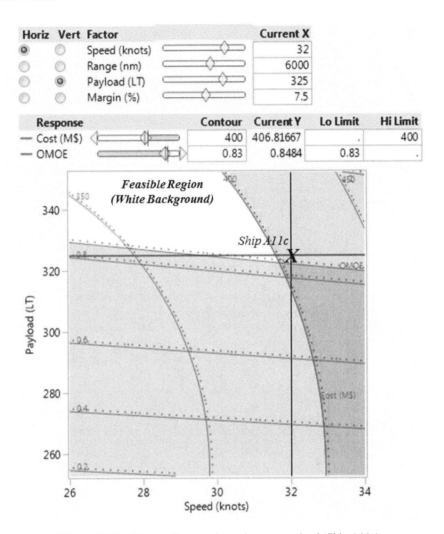

Figure 12.22 Trade-off space of speed versus payload (Ship A11c)

12.5 KEY TERMS

Box–Behnken Design: A Box–Behnken design is a three-level experimental design. It has no design points at the vertices of the cube defined by the ranges of the factors. This is sometimes useful when it is desirable to avoid these points due to engineering considerations. The price of this characteristic is the higher uncertainty of prediction near the vertices compared to the central composite design (SAS Institute, 2015a,b).

Central Composite Designs: A central composite design combines a two-level fractional factorial and two other kinds of points, center points and axial points.

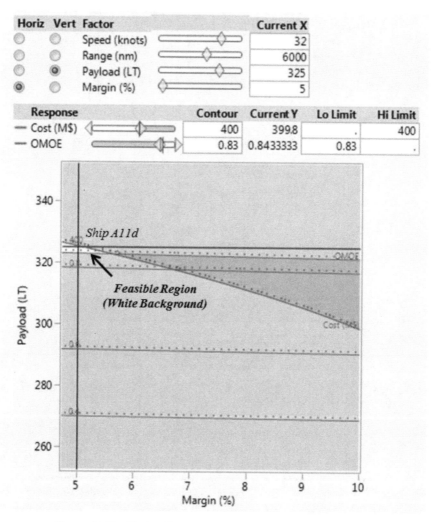

Figure 12.23 Trade-off space of margin and payload (Ship A11d)

The center points are defined such that all the factor values are at the zero (or midrange) value. The axial points are defined such that all but one factor are set at zero (midrange) and that one factor is set at outer (axial) values (SAS Institute, 2015a,b).

Design of Experiments: A designed experiment is a controlled set of tests designed to model and explore the relationship between factors and one or more responses (SAS Institute, 2015a,b).

Design Space: A region of interest for a design described by the experimental range of the factor settings chosen for study (SAS Institute, 2015a,b).

Effects Plots: Graphs that indicate the relationships among the effects in a design.

Factor Fractional Factorial Design: A design that uses a portion, or fraction, of a full factorial design set of points for the design factors.

Feasible Region: The area of a design space that contains the points that satisfy constraints and provide solutions.

***p*-Value:** The calculated probability of finding the observed result when testing a statistical hypothesis.

12.6 EXERCISES

The following exercises are based on a notional early-stage ship design effort for a small surface combatant, as previously discussed in Section 12.4. This example requires the utilization of JMP to conduct trade-off analysis. Note that the previous example in Section 12.4 utilized a different experimental design and a different dataset, these exercises are focused on the same system but are not related to the data used in Section 12.4.

12.1. Upload the following dataset into JMP. Note that the design is a central composite design generated using JMP. The three responses (cost, OMOE, and displacement) are functions of the four factors (speed, range, payload, and margin). Use the Analyze → Fit Model functionality in JMP to develop prediction formulas for each of the responses. When prompted to Construct Model Effects, include all two-way interactions (using the Factorial to Degree functionality) as well as all quadratic effects (using the Polynomial to Degree functionality). Conduct Least Squares Regression and save the prediction formula (Save Columns → Save Prediction Formula) for each of the responses.

12.2. Open a Contour Profiler (Graph → Contour Profiler) using each of the prediction formulas developed in Problem 1. Impose constraints for each of the responses. Impose a high limit on cost of $1750 M, a low limit on OMOE of 0.50, and a high limit on displacement of 4750 Ltons. Describe the feasible space.

12.3. Based on your response to Problem 2, explore alternate two-dimensional projections (projections other than speed and range) and suggest at least two potential feasible ship configurations without altering any of the constraints imposed on the responses.

Pattern	Factors				Responses		
	Speed (knot)	Range (nm)	Payload (Ltons)	Margin (%)	Cost (M$)	OMOE	Displacement (Ltons)
----	26	3000	250	5	398.81185	0.0104	3484.287
---+	26	3000	250	10	675.443203	0.3020	3848.332
--+-	26	3000	350	5	656.090758	0.5811	3707.986
--++	26	3000	350	10	911.019546	0.8458	4028.422
-+--	26	6000	250	5	1828.8852	0.0338	3976.728
-+-+	26	6000	250	10	2295.84548	0.3125	4330.991
-++-	26	6000	350	5	2169.71101	0.6055	4206.921
-+++	26	6000	350	10	2642.01684	0.9904	4541.402
+---	34	3000	250	5	950.393424	0.0245	5444.087
+--+	34	3000	250	10	1210.58148	0.3369	5741.125
+-+-	34	3000	350	5	1268.01422	0.6970	5505.467
+-++	34	3000	350	10	1532.41456	0.9007	5961.524
++--	34	6000	250	5	2474.35812	0.0498	5998.325
++-+	34	6000	250	10	2920.50349	0.3519	6204.49
+++-	34	6000	350	5	2725.02865	0.7573	6132.663
++++	34	6000	350	10	3185.3991	0.9697	6477.513
a000	26	4500	300	7.5	1379.07059	0.4224	4063.652
A000	34	4500	300	7.5	1978.00227	0.4873	6052.191
0a00	30	3000	300	7.5	969.966418	0.4399	4727.896
0A00	30	6000	300	7.5	2466.02425	0.4951	5212.494
00a0	30	4500	250	7.5	1517.93628	0.1128	4759.32
00A0	30	4500	350	7.5	1750.40849	0.7623	5032.142
000a	30	4500	300	5	1490.98832	0.2887	4870.674
000A	30	4500	300	10	1797.51409	0.6535	5151.25
0	30	4500	300	7.5	1629.584	0.4304	4882.302
0	30	4500	300	7.5	1586.39436	0.5131	4934.937

12.4. After selecting a feasible ship configuration (from Problem 3), reset the axes such that the speed is shown on the x-axis and the range is shown on the y-axis. Compare the feasible and infeasible spaces shown to the original feasible space identified in Problem 2. The constraints have not changed and the two-dimensional projection has not changed, why has the shape of the feasible region changed?

12.5. Provided that there is a feasible region shown in the answer to Problem 4, select a design point in the feasible region (if there is not, alter the payload and/or margin to create a feasible region). Next, change the y-axis from range to payload. Is the selected design point still feasible? Change the y-axis to Margin, is the selected design point still feasible? Discuss why changing the axes does not alter the feasibility of the selected design point.

12.6. Reset each of the factor values to their initial levels (30 knot speed, 4500 nm range, 300 Ltons payload, and 7.5% margin). Increase the OMOE constraint to 0.80. Identify the changes necessary to define a feasible set of ship configurations. Is it possible to select a ship configuration with a speed of 30 knot? If not, what constraint must be relaxed to enable the selection of such a ship? Is it possible to select a ship with a maximum range of 5000 nm? If not, what constraint must be relaxed to enable the selection of such a ship?

12.7. The previous question suggested that it may be desirable to relax constraints to enable the selection of ships with higher speed or a higher range. However, this example utilizes three responses that address the cost, effectiveness, and physics (through the displacement of each ship). Discuss how such a tool can be used to highlight potentially unnecessary increases to factors such as speed and range.

REFERENCES

Katsoufis, G. (2006) A decision making framework for cruise ship design, Master thesis, MIT.

NATO (2004) NATO/PfP Working Paper on Small Ship Design, NATO Naval GROUP 6 SPECIALIST TEAM ON SMALL SHIP DESIGN.

SAS Institute Inc (2015a) *JMP® 12 Design of Experiments Guide*, SAS Institute Inc., Cary, NC, ISBN 978-1-62959-443-9.

SAS Institute Inc (2015b) *Using JMP® 12*, SAS Institute Inc., Cary, NC, ISBN 978-1-62959-479-8.

13

SUSTAINMENT RELATED MODELS AND TRADE STUDIES

JOHN E. MACCARTHY
Systems Engineering Education Program, Institute for Systems Research, University of Maryland, College Park, MD, USA

ANDRES VARGAS
Department of Industrial Engineering, University of Arkansas, Fayetteville, AR, USA

> The sustainment key performance parameter (KPP) (Availability) is as critical to a program's success as cost, schedule, and performance.
>
> (DoDI 5000.02 (2015))

13.1 INTRODUCTION

This chapter develops a number of "first-order" sustainment-related models[1] for a relatively simple fictional remotely piloted air vehicle (i.e., drone) system that illustrates modeling techniques that are currently used to support reliability, availability, and maintainability (RAM) analysis, total life cycle cost analyses, and cost–RAM performance trade-off analyses. While the models developed in this chapter are based

---

[1] In this chapter, the term "first-order model" is taken to mean a model that provides values for one or more output metrics based on input values for the most important input factors and based on a set of simplifying assumptions that permit the use of relatively simple equations.

Trade-off Analytics: Creating and Exploring the System Tradespace, First Edition. Edited by Gregory S. Parnell.
© 2017 John Wiley & Sons, Inc. Published 2017 by John Wiley & Sons, Inc.
Companion website: www.wiley.com/go/Parnell/Trade-off_Analytics

on simplifying assumptions and fictional data, they can provide a modeling framework for more complex RAM and life cycle cost analysis of real systems that use real data and fewer simplifying assumptions.

Sustainment trade studies are important from two perspectives – life cycle cost and system performance. Generally, 50% (or more) of a typical system's life cycle cost is associated with operations and maintenance (INCOSE, 2015). A large portion of these costs are typically associated with maintenance (which depends on system reliability). In addition, system availability (which depends on reliability and maintainability) is a system performance metric of particular importance to the operator.

Given the clear importance of system reliability, maintainability, and availability, the DoD has established a "Sustainment Key Performance Parameter (KPP)" that requires programs to include technical performance requirements for reliability and availability and cost estimates for maintenance (CJCSI 3170.01I, 2012). In addition, the Department of Defense Office of Acquisition, Technology, and Logistics (AT&L) has established guidance on sustainment that emphasizes the importance of including quantitative requirements for system reliability, maintainability, and availability (DoDI 5000.02, 2015).

For the purposes of this chapter, **reliability** is defined as the probability that a system or product [element/item/service] "will perform in a satisfactory manner for a given period of time, when used under specified operating conditions." (Blanchard and Fabrycky, 2010). The principal metric used to specify a system's reliability is the mean time between failures (MTBF). Maintainability may be defined in two ways depending on whether one is most interested in the time required to repair the system or the cost of repairing the system. In the first case, **maintainability** is defined as "the probability that an item will be retained in or restored to a specified condition within a given period of time, when maintenance is performed in accordance with prescribed procedures and resources." (Blanchard and Fabrycky, 2010). One common metric used to describe a system's maintainability is the mean downtime (MDT). In the second case, **maintainability** is defined as "the probability that the maintenance cost for the system will not exceed y dollars per designated period, when the system is operated in accordance with prescribed procedures." (Blanchard and Fabrycky, 2010). One may use the system's mean life cycle maintenance cost as a metric for this aspect of maintainability. Finally, a system's operational **availability** is defined as "the probability that a system or equipment, when used under stated conditions in an actual operational environment, will operate satisfactorily when called upon." (Blanchard and Fabrycky, 2010). The relationship between a system's availability, reliability, and maintainability is discussed in Section 13.2.1.1.

This chapter develops the RAM and life cycle cost models and demonstrates how a number of analytical techniques may be applied to find the most cost-effective design and to determine the performance uncertainty associated with different design solutions. Section 13.2 provides an example of how to build a model to calculate the reliability and availability of an unmanned air vehicle (UAV) system based on the system's logical system architecture. It also shows how such a model may be used to perform trade-off analyses between different system architectures (and reliability allocations to different elements), different operational concepts, different maintainability requirements, and different resulting system availabilities. Section 13.3 provides an example of how to build a model for the system's sustainment costs and how to use this model in tandem with the previously developed availability model to perform cost–performance (total life cycle cost vs A_o) trade-off analyses. Section 13.4 uses the modeling framework developed in Section 13.3 to develop an Excel model that employs an evolutionary optimization technique to find the lowest cost design solution that meets the system's availability requirement. The Excel model is also used to provide a "cost-effectiveness" tradespace curve and to perform a deterministic parameter sensitivity study to guide the development of the Monte Carlo (MC) model developed in Section 13.5. Finally, Section 13.5 provides an example of how to develop a Monte Carlo extension to the availability model developed in Section 13.2 and how such a model can be used to determine the confidence that one may have that a given design will be able to meet its availability requirement.

13.2 AVAILABILITY MODELING AND TRADE STUDIES

This section develops an analytic model for the operational availability of a simple Forest Monitoring Drone System (FMDS) as a function of a variety of system and system element parameters. It then demonstrates how the model may be used to perform availability trade studies to identify architecture modifications that will enable the system to achieve a required system availability. The section also demonstrates how such a model may be used to perform sensitivity analyses. We begin by describing the FMDS mission, system architecture, operational concept, and maintenance concept. We then develop a state model for each element and use the existing body of research on how to model complex, standby systems. The result is a reliability block diagram (RBD) model that may be used to calculate the system-level mean time between critical failures (MTBCFs) and to develop the associated availability model.

13.2.1 FMDS Background

This section provides a quick overview of the system's mission, availability requirement, conceptual design/physical architecture, and concept of operation.

In our illustrative example, the USDA Forest Service is considering using drones to monitor a forest for fires. They want to be able to provide 24/7 coverage of the forest. The drone will generally fly an established repeated flight path ("orbit") over the forest. It will be launched from an air field (base) that is about a 10 min flight from the forest. This geometry is outlined in Figures 13.1 and 13.2.

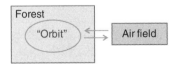

Figure 13.1 System operational concept

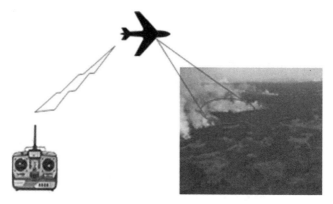

Figure 13.2 System operational concept (cont.)

We further suppose that the Forest Service has specified that the system must provide "an operational availability of coverage" of 80% (i.e., there must be a drone on orbit, under control and successfully reporting sensor data, 80% of the time).

The "System of Interest" (or system) consists of the mission elements, the maintenance element, and the personnel required to operate and maintain the mission elements. The mission elements consist of a number of drone/air vehicle elements (N_d) and a number of "control elements" (N_c), where N_d and N_c are to be determined through this analysis.

The drone will be launched from the air field and travel to its orbit. The drone carries enough fuel for up to 4 h of flight (maximum flight time T_{mf}). The drone is controlled by the "control element" that is located at the air field. The control element exercises control of the drone via a dedicated communications link supported by communications components of the drone and control elements. The control element also receives and displays the continuous, real-time sensor data provided by the drone via a dedicated communications downlink that is supported by communications components of the drone and control elements.

The typical mission timeline for a drone consists of preparing it for flight, launching it, flying it to station, monitoring the area (time on station), returning to base, landing, and postflight maintenance. Table 13.1 provides a summary of the nominal times associated with each mission activity. The "Time on Station Margin" reflects an expectation that one will generally want to return the drone to the airbase with some fuel to spare.

We can see from the aforementioned timeline that the Time on Station (3.1 h) will be less than the maximum flight time (4 h). It should also be noted that there will be uncertainties associated with each of the nominal times indicated earlier.

In order for the Forest Service to maintain 24/7 monitoring of the forest, a second drone will need to be prepared for flight, launched, and reach its orbit to the first drone having to begin its return to base. Figure 13.3 provides and illustration of this timeline (not to scale). While each drone requires a dedicated control element to control its flight, any control element may be used with any drone.

AVAILABILITY MODELING AND TRADE STUDIES

Table 13.1 Mission Activities and Nominal Times

Activity/Time	Symbol	Nominal Time (h)
Maximum flight time	T_{mf}	4.0
Preflight preparation time	T_{pfp}	0.5
Time to launch	T_{lch}	0.1
Time to station	T_{ts}	0.2
Time to return to base	T_{rtb}	0.2
Time to land	T_{lnd}	0.1
Time on station margin	T_m	0.3
Time on station	T_{os}	$T_{mf} - (T_{lch} + T_{ts} + T_{rtb} + T_{lnd} + T_m) = 3.1$
Postflight maintenance time (drone)	T_{pfm}	1.0

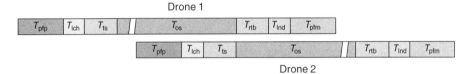

Figure 13.3 Mission timeline

The drone is expected to have a mean time between critical failures ($MTBCF_d$) of 50 h. The mean time between critical failures for the control element ($MTBCF_c$) is expected to be 75 h. For the purposes of this chapter, a critical failure is a failure that results in an element being unable to perform its intended purpose, could lead to loss of the element, or injury to person or property. An element that experiences a critical failure may not be used until the failure is corrected.[2] If a drone element experiences a (critical) failure, it must return to base and undergo unscheduled (corrective) maintenance. The downtime associated with unscheduled maintenance (T_{dum}) will be discussed in more detail in the following section. If a second drone is not already in the process of preparing for flight, launching, or flying to station, it must begin preparation for flight and take the place of the failed drone as quickly as possible (see timeline). A drone that is undergoing maintenance (scheduled or unscheduled) may not begin flight preparation until its maintenance is completed.

If a control element experiences a failure and a second control element is available, it may be used. The downtime associated with "swapping in" the backup control element (T_{cs}) is nominally expected to be 0.1 h. The failed control element will then undergo unscheduled maintenance. The unscheduled maintenance downtime associated will be addressed in the following section.

---

[2] For simplicity, we will assume that all critical failures permit a return to base (i.e., there are no "catastrophic failures."

13.2.1.1 The FMDS Analytic Availability Model This subsection develops the analytic availability model for the FMDS that will be used to perform a series of availability trade studies. A system's (steady-state) operational availability is defined as its uptime divided by the sum of its uptime and downtime. Given this definition, and ignoring the effects of scheduled maintenance, if MTBCF_S is the system's MTBCF and MDT_S is the MDT associated with a critical system failure, it follows that:

$$A_o = \frac{\text{MTBCF}_S}{\text{MTBCF}_S + \text{MDT}_S} \quad (13.1)$$

Now consider a system that has some number of independent failure modes (N_f), each characterized by a constant failure rate λ_i and MDT_i.[3] The system failure rate will be $\lambda_s = \sum \lambda_i$ and the mean system downtime (MDT_S) will be a frequency-weighted average of the individual wait times, $\text{MDT}_S = \sum (\text{MDT}_i * \lambda_i / \lambda_s)$. Using the fact that $\text{MTBCF}_S = 1/\lambda_s$ we have:

$$A_o = \frac{\frac{1}{\lambda_s}}{\frac{1}{\lambda_s} + \sum_{i=1}^{N_f} \frac{\text{MDT}_i * \lambda_i}{\lambda_s}} = \frac{1}{1 + \sum_{i=1}^{N_f} \left(\frac{\text{MDT}_i}{\text{MTBCF}_i}\right)} \quad (13.2)$$

The $\sum (\text{MDT}_i/\text{MTBCF}_i)$ may be thought of a normalized "failure rate-weighted downtime."

Section 13.2.1.2 develops the reliability models required to identify the failure modes and to calculate the MTBCF for each identified failure mode. Section 13.2.1.3 provides a maintenance concept for the system and develops expressions for the MDT associated with each failure mode. Furthermore, Section 13.2.1.4 provides an influence diagram for the FMDS System. Finally, Section 13.2.1.5 discusses an integrated Excel implementation of equation 13.2, and the MTBF_i and MDT_i expressions are developed Sections 13.2.1.2 and 13.2.1.3.

13.2.1.2 Reliability Models and Failure Modes This section develops an RBD for our system (the FMDS), which is then used to identify the principal system-level failure modes. We will see that the FMDS is a relatively simple example of what is termed a "complex structure".[4] Some of the techniques developed in the current literature on complex structures[5] will be used to develop expressions that will permit us to calculate values for system-level MTBFs, based on the element-level MTBFs and associated downtimes.

---

[3] For simplicity it is assumed that both times between failures and down times are exponentially distributed.
[4] In this chapter, the term "complex structure" is taken to mean a system composed of multiple instantiations of multiple types of elements that interact with one another, which differ significantly in their form and function. The elements are characterized by multiple failure modes, a mixture of series and parallel reliability architectures, and load sharing (see Birolini, 2007).
[5] See Amari et al. (2008), Amari et al. (2008), Birolini (2007), Boddu & Xing (2012), Jacob & Amari (2005), Kuo & Zuo (2003), Lad et al. (2008), Morrison & Munshi (1981), Sandler (1963), van Gemund & Reijns (2012), Wang & Loman (2012), Zuo et al. (2003).

AVAILABILITY MODELING AND TRADE STUDIES

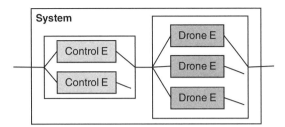

Figure 13.4 System reliability block diagram

Figure 13.4. provides the RBD for our system of interest. We see that the FMDS consists of two "substructures" (a control substructure and a drone substructure) that operate in series. Each substructure is in turn composed of two or more elements organized in a parallel manner. The "disconnections" in the structures indicate that the "standby elements" do not "kick in" until the active element fails and that there is a "downtime" associated with the switch. Such elements are referred to as "cold standby" elements. The control and drone "substructures" are examples of "1-out-of-N cold standby" structures.[6]

We define a "system failure" to be an event that results in a potential interruption to time on station. There are two types of system failures: (i) an element failure where a standby element is available (Type 1 failure) and (ii) an element failure where a standby element is unavailable[7] (Type 2 failure). Generally, the mean system downtime associated with a Type 1 failure will be much shorter than the MDT associated with a Type 2 failure.

An "element failure" is defined to be an event that requires one element to be replaced by an identical backup element. Given this definition, for simplicity, we will assume that the only event that would lead a control element to have to be replaced by a backup would be a control element critical failure. In the case of Type 1 control element critical failures, the mean time between critical control element failures will be denoted by "$MTBCF_{1c}$" (which was earlier assumed to have a nominal value of 75 h). The mean system downtime associated with this kind of failure will be denoted "MDT_{1c}." This is just the time it takes to power up the backup control element (hardware and software) and to establish the required communications links. We assume a nominal value of 0.1 h. In the case of Type 2 control element failures, the mean time between Type 2 control element failures will be denoted by "$MTBCF_{2c}$" and the associated downtime is denoted by "MDT_{2c}." The value of MDT_{2c} depends on the maintenance concept and is a topic of discussion in the next subsection. Birolini (2007) (and Kuo & Zuo, 2003) developed a recursion formula that may be used to calculate the $MTBCF_{2c}$. A "structure" (or substructure) consisting of n parallel elements in which *k* of the elements must be operating for the system to operate and the

---

[6]This is special case of a more general theoretical framework that address "k-out-of-N" hot, cold, or warm standby structures.
[7]Because all standby elements are in some stage of maintenance.

remaining $n - k$ elements are not operating (cold), but serve as "standbys" in the case of a failure of one of the operating units is referred to as a "k-out-of-n cold standby" structure. When an element fails, it goes into repair. The number of failed elements than may go be simultaneously repaired is limited by the number of "repair crews." In our analysis, we will consider the case of "1-out-of-n cold standby" substructures where there is only one repair crew. In this case, the recursion formula takes the simpler form:

$$\text{MTBCF}_S(0) = \text{MTBCF}_e + \text{MTBCF}_S(1)$$
$$\text{MTBCF}_S(j) = \frac{1 + \lambda_e \text{MTBCF}_S(j+1) + \mu * \text{MTBCF}_S(j-1)}{\lambda_e + \mu_e}$$
$$\text{MTBCF}_S(n-1) = \frac{1 + \mu \text{MTBCF}_S(n-2)}{\lambda_e + \mu_e} \quad (13.3)$$

where the subscript "e" indicates either a control or drone element, $\mu_e = 1/\text{MDT}_e$, $\lambda_e = 1/\text{MTBCF}_e$, and $\text{MTBCF}_S(0)$ is the MTBCF for the substructure.

Applying this recursion formula to the case of a substructure consisting of one active unit and one standby unit (a "1-out-of-2 cold standby" structure), with a single repair crew[8], the control substructure's MTBF may be shown to be:

$$\text{MTBCF}_{2c}(1 \text{ of } 2) = \text{MTBCF}_{1c} * \left(2 + \frac{\text{MTBCF}_{1c}}{\text{MDT}_{2c}}\right) \quad (13.4)$$

The case of the drone is a bit more complicated. Here we assume that there are two events that could lead to a Drone Element having to be replaced. First, a drone runs low on fuel and must return to base (a scheduled return to base "failure"). We denote the mean time between scheduled returns to base (SRTB) events by "MTBSRTB_{1d}," which may be determined using:

$$\text{MTBSRTB}_{1d} = T_{lch} + T_{ts} + T_{os} = T_{mf} - (T_{rtb} + T_{lnd} + T_m) = 3.4 \, \text{h} \quad (13.5)$$

Second, a drone may experience a critical failure and have to return to base earlier than planned. This is just the MTBCF_S for the drone, denoted as "MTBCF_{1d}" (which was earlier assumed to have a nominal value of 50 h).

Each of these failure modes results in a different mean system downtimes. We denote the mean system downtime associated with a SRTB by "MDT_{1srtb}." If a standby drone is prepared and launched in time to reach orbit prior to the on-station drone having to return (which is assumed to be the case), there is no associated system downtime (i.e., $\text{MDT}_{1srtb} = 0$). We denote the mean system downtime associated with a critical failure is denoted by "MDT_{1d}." Since this is unexpected,

---

[8]Theoretical frameworks also exist for addressing systems with m repair crews.

the system is down until an available standby drone may be prepared, launched and transit to station. As such, a nominal value for MDT_{1d} may be calculated using:

$$MDT_{1d} = T_{pfp} + T_{lch} + T_{ts} = 0.8\,h \quad (13.6)$$

In the case of Type 2 drone element failures, if one assumes that the mean system downtime associated with bringing a replacement drone that is currently in maintenance back to active on-station status (denoted by "MDT_{2d}") is the same regardless of whether the returning drone is returning due to a SRTB or a critical failure, one may define an "effective" $MTBCF_S$ for the drone $MTBCF_{1de}$, that includes both critical failures and SRTBs, as

$$MTBCF_{1de} = \frac{1}{\left(\frac{1}{MTBCF_{1d}} + \frac{1}{MTBSRTB_{1d}}\right)} = 3.18\,h \quad (13.7)$$

and use the recursion formula for a "1-out-of-3 cold standby" structure (assuming one active and two standby drones) to calculate the drone substructure's $MTBCF_{2d}$.

$$MTBCF_{2d}(1\text{ of }3) = MTBCF_{1de} * \left[3 + \frac{2 * MTBCF_{1de}}{MDT_{2d}} + \left(\frac{MTBCF_{1de}}{MDT_{2d}}\right)^2\right] \quad (13.8)$$

The value of MDT_{2d} depends on the maintenance concept and will be a topic of discussion in the next subsection.

Given these definitions, Table 13.2 summarizes the five failure modes described above.

Again, for simplicity we assume that the time between failures may be approximated by an exponential distribution (including SRTBs).[9]

Table 13.2 Summary of Failure Modes

Failure Mode	$MTBCF_i = 1/\lambda_i$ (h)	MDT_i (h)
Control element critical failure with standby	$MTBCF_{1c} = 75$	$MDT_{1c} = 0.1$
Control element critical failure without standby (2 elements)	$MTBCF_{2c} = 263$	$MDT_{2c} = 50$ (from next section)
Drone element SRTB w. standby	$MTBSRTB_{1d} = 3.4$	$MDT_{1srtb} = 0.0$
Drone element critical failure with standby	$MTBCF_{1d} = 50$	$MDT_{1d} = 0.8$
Drone element failure (SRTB or critical failure) without standby (3 elements)	$MTBCF_{2d} = 13.2$	$MDT_{2d} = 6.8$ (from next section)

[9] While this may be unrealistic for the case of SRTB "failures," it does enable us to develop an analytical availability model. Monte Carlo studies may be used to determine the degree to which the results are sensitive to this assumption.

13.2.1.3 The Maintenance Concept and Substructure Downtimes

The drone element is maintained by the maintenance element, which is collocated with the control element at the air field. The maintenance element consists of the following: (i) a repair facility; (ii) the tools required to maintain, diagnose, and repair the drones; (iii) a limited supply of spare parts; and (iv) a drone storage area.

The maintenance element and its associated personnel are responsible for performing preflight preparation and postflight maintenance of the drone element. We assume that there is only one drone element "repair crew" and that the control operators are able to perform routine maintenance on the control elements.

The control element is assumed to require very little (or no) scheduled maintenance. If a control element experiences a critical failure, a control operator will remove it from service and either send it out for repair or purchase a new element. In either case, the nominal MDT (associated with the repair or replacement of the failed unit) is assumed to be $\text{MDT}_{2c} = 50\,\text{h}$.

Drone element preflight preparation consists of fueling the aircraft, performing preflight diagnostics and corrective actions, and moving the drone from storage to fueling, diagnostics, and launch areas. As indicated in Table 13.1, the nominal MDT associated with this is assumed to be $T_{\text{pfp}} = 0.5\,\text{h}$.

Drone element postflight maintenance consists of performing "standard" postflight diagnostics and preventative maintenance as well as any corrective maintenance required as the result of either a critical failure or any failures discovered as the result of standard postflight diagnostics and maintenance. It also includes moving the drone from the landing area to the maintenance area and from the maintenance area to storage. As indicated earlier, the nominal MDT associated with this is assumed to be $T_{\text{pfm}} = 1.0\,\text{h}$.

If a drone replacement part is not immediately available, there will be a logistics delay associated with obtaining the part. We will assume a nominal MDT associated with logistics delay of $\text{MDT}_{\log} = 50\,\text{h}$. We will also assume a nominal probability that a replacement part is available for Pap = 0.90.

Given this, one may calculate the total mean maintenance downtime for a drone element to be:

$$\text{MDT}_{\text{dm}} = T_{\text{pfp}} + T_{\text{pfm}} + (1 - P_{\text{ap}}) * \text{MDT}_{\log} = 6.5\,\text{h} \quad (13.9)$$

Since we are interested in Time on Station availability, a drone would also be considered unavailable (or "effectively down") during its launch, transit to station, return to base, and landing. As such, the total EFFECTIVE MDT for a drone element is:

$$\text{MDT}_{\text{de}} = \text{MDT}_{\text{dm}} + T_{\text{lch}} + T_{\text{ts}} + T_{\text{rtb}} + T_{\text{lnd}}$$
$$\text{MDT}_{\text{de}} = 7.1\,\text{h} \quad (13.10)$$

The MDT_{2d} referred to in the previous section is just

$$\text{MDT}_{2d} = \text{MDT}_{\text{de}} - (T_{\text{rtb}} + T_{\text{lnd}}) = 6.8\,\text{h}. \quad (13.11)$$

AVAILABILITY MODELING AND TRADE STUDIES

As we can see, there are many factors that contribute to the effective downtime of the drone and that this effective downtime may be significantly larger than the expected "repair time." For computational simplicity, we will assume that the total effective downtime exhibits an exponential distribution.[10]

13.2.1.4 An Influence Diagram for the FMDS Availability Model Influence Diagrams are a particularly useful way to summarize information about the relationships between parameters that make up a model. Figure 13.5 provides such a diagram for the FMDS availability model using the conventions described in Chapter 6, with the addition of calculated uncertainties represented by double ellipses. The arrows indicate calculation influences.

We can see from the diagram that even this relatively simple model for the FMDS yields a nontrivial network of inputs and calculation dependencies. The segmented line captures all the parameters associated with drone operational availability, which

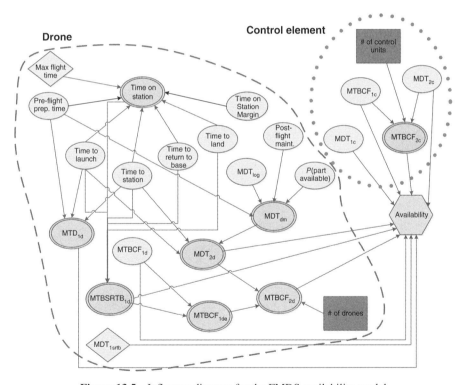

Figure 13.5 Influence diagram for the FMDS availability model

---

[10]While this is admittedly unrealistic (for a variety of reasons), it does enable us to develop and exercise an analytical availability model. Again, initial Monte Carlo studies have indicated that the results are relatively insensitive to how this is modeled (provided the distributions exhibit "tails").

is influenced by the mission activity times and other parameters regarding drone maintenance. Analogously, the dotted line encloses the parameters that influence control element availability. The MTBF and MDT values for drones and control elements are used to calculate the system operational availability using equation 13.2.

13.2.1.5 An Excel-Based System-Level Analytic Availability Model for the FMDS[11] Table 13.3 provides an example of an Excel instantiation of the FMDS analytic availability model (and nominal input parameter values) developed in the previous sections. The yellow cells indicate inputs to the model, the green cells indicate cells with calculated values, and the blue cell indicates the principal output of the model (the system's operational availability). (*The reader is referred to the online version of this book for color indication.*) The first column indicates the system failure mode. The second column indicates the MTBF associated with each failure mode. The third column indicates the number of each element type that makes up the system of interest. The fourth column indicates the mean system downtime associated with each failure mode. The fifth column indicates the contribution of that failure mode to the ratio sum that is used to calculate A_o (see equation 13.2), and the sixth column indicates the availability the system would exhibit if the indicated failure mode was the only failure mode present. The "Net Eff Failures" $MTBF_i$ entry simply represents the inverse of the sum of the SRTB and critical failure rates (i.e., equation 13.7) that is used to calculate the MTBCF for "no standby drone" failure mode.

We can immediately see that the model indicates that the system, as currently designed, has an $A_o = 0.58$ that falls far short of its 80% availability requirement.

Table 13.3 Excel Instantiation of the FMDS Analytic Availability Model

Element	$MTBF_i$ (h)	N_i	MDT_i (h)	$MDT_i / MTBF_i$	A_o
Control					
- Crit Failures (with SB)	75.00		0.10	0.001	0.999
- Crit Failures (no SB)	262.50	2.0	50.00	0.190	0.840
Drone					
- SRTB (with SB)	3.40		0.00	0.000	1.000
- Crit Failures (with SB)	50.00		0.80	0.016	0.984
- Net Eff Failures	4.63				
- Eff Failures (no SB)	13.23	3.0	6.80	0.514	0.660
System				0.722	*0.581*

[11]The "Ch 13.2 Avail Model.xlsx" Excel file contains the availability models used to support the analyses performed in the following sections.

AVAILABILITY MODELING AND TRADE STUDIES 449

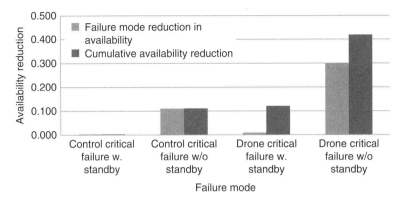

Figure 13.6 Reduction in availability due to each failure mode

We can also see that this spreadsheet model can be very useful in guiding trade studies aimed at improving system availability. Specifically, the ratio of the $MDT_i/MTBF_i$ to the sum of $MDT_i/MTBF_i$ indicates the degree to which each failure mode contributes to the reduction in the value of A_o. Figure 13.6 indicates the "reduction in availability" due to each failure mode (RA_i). RA_i may be calculated using the following expression:

$$RA_i = (1 - A_o) \frac{MDT_i/MTBF_i}{\sum MDT_i/MTBF_i} \qquad (13.12)$$

Given this, we see that the principal drivers in the 0.286 reduction in A_o (from 1.0 to 0.714) are the "no standby" failure modes. We can further see that the drone "no standby" failure mode is responsible for ~60% of this reduction in A_o. As such, it makes sense to begin our set of trade studies by looking for ways to reduce this.

13.2.2 FMDS Availability Trade Studies

Table 13.4 summarizes the system architecture/design parametric input factors that affect the system's availability metric.

While in this section, for the sake of brevity, we will consider the effect of varying the value of the nine indicated model input factors, it should be noted that some of these factors are themselves functions of lower-level factors (e.g., MDT_{2d} is itself a function of six lower-level factors). A more detailed examination would break these out.

Given the model's indication that the "no standby" drone failure mode is principally responsible for the low system A_o, we will begin our trade study by looking for ways to reduce its impact. An examination of the model indicates there are three effective ways to reduce the value of $MDT_i/MTBF_i$: (i) increase the $MTBSRTB_{1d}$

Table 13.4 Summary of Factors Affecting System Availability

Factor Category	Input Factor Variable	Description	Dependence
Number of elements	N_{ce}	Number of control elements in system	Determines equation that will be used to calculate $MTBCF_{2c}$
	N_{de}	Number of drone elements in system	Determines equation that will be used to calculate $MTBCF_{2d}$
MTBFs	$MTBSRTB_{1d}$	Mean time between scheduled return to bases	$= T_{lch} + T_{ts} + T_{os}$
	$MTBCF_{1c}$	Mean time control element critical failures	
	$MTBCF_{1d}$	Mean time between drone element critical failures	
MDTs	MDT_{1c}	Mean downtime for the system given a control element failure (w. standby available)	
	MDT_{2c}	Mean downtime for the system given a control element failure (w. no standby available)	
	MDT_{1d}	Mean downtime for the system given a drone element critical failure (w. standby available)	$= T_{pfp} + T_{lch} + T_{ts}$
	MDT_{2d}	Mean downtime for the system given a drone element SRTB or critical failure (w. no standby available)	$= T_{pfp} + T_{pfm} + (1 - P_{ap}) * MDT_{log} + T_{lch} + T_{ts}$

(by using drones with greater maximum flight times)[12]; (ii) increase the probability of having a spare part (by keeping more in inventory)[13]; and (iii) increase the number of spare drones (so one is less likely to experience no standby event). The impact of each of these is summarized in Table 13.5. Recall that the reference value for $MDT_i/MTBF_i$ for this failure mode was ~0.51 and the reference value for A_o was 0.58.

We see that doubling the maximum flight time of the drone (increasing the $MTBSRTB_{1d}$) appears to be more effective in increasing the availability of the system, but that no single change is able to get the system to the required availability of 0.80. It should also be noted that increasing the value of N_{de} from three to four changes the form of the equation for $MTBF_{2d}$ to the following (using Birolini's

---

[12] Increasing $MTBF_{1c}$ will have very little effect.
[13] One could also look for ways to decrease the logistics delay time.

AVAILABILITY MODELING AND TRADE STUDIES

Table 13.5 Availability Sensitivity/Trade Study ("No Standby" Drone Failure Mode)

Factor	Reference Factor Value	Revised Factor Value	Revised Value of $MDT_i/MTBF_i$	Revised A_o	Comment
$MTBSRTB_{1d}$	3.4	7.4	0.182	0.72	Corresponds to increasing T_{mf} from 4 to 8 h
MDT_{2d}	6.8	4.3	0.269	0.68	Corresponds to increasing P_{sp} from 0.9 to 0.95
N_{de}	3	4	0.359	0.64	

recursion formula for a "1-out-of-n cold standby" structure):

$$MTBCF_{2d}(1 \text{ of } 4) = MTBCF_{1de}\left[4 + \frac{3 * MTBCF_{1de}}{MDT_{2d}} + 2\left(\frac{MTBCF_{1e}}{MDT_{2d}}\right)^2 + \left(\frac{MTBCF_{1e}}{MDT_{2d}}\right)^3\right]$$

(13.13)

Table 13.6 provides the output of the Excel availability model for the case where $MTBSRTB_{1d}$ was increased from 3.4 to 7.4 h. We see that once this is done the "no standby" control element critical failure mode now becomes dominant and limiting (the model indicates that if this were the ONLY failure mode, the system would still fall short of meeting the 0.80 availability requirement).

Given this result, we should consider the ways in which we might decrease value of the $MDT_i/MTBF_i$ associated with the "no standby" control element failure mode. We see that there are essentially three ways to do this: (i) increase the $MTBF_{1c}$, (ii) decrease the logistics delay time; and (iii) increase the number of spare control elements (so one is less likely to experience no standby event). The impact of each of

Table 13.6 Effect of Doubling the Maximum Flight Time

Element	$MTBF_i$ (h)	N_i	MDT_i (h)	$MDT_i/MTBF_i$	A_o
Control					
- Crit Fail (with SB)	75.00		0.10	0.001	0.999
- Crit Fail (no SB)	262.50	2.0	50.00	0.125	0.889
Drone					
- SRTB (with SB)	7.40		0.00	0.000	1.000
- Crit Fail (with SB)	50.00		0.80	0.016	0.984
- Net Eff Failures	6.45				
- Eff Fail (no SB)	37.35	3.0	6.80	0.182	0.846
System				0.390	0.719

Table 13.7 Availability Sensitivity/Trade Study ("No Standby" Control Element Failure Mode)

Factor	Reference Factor Value	Revised Factor Value	Revised Value of $MDT_i/MTBF_i$	Revised A_o	Comment
$MTBCF_{1c}$	75	150	0.067	0.79	
MDT_{2c}	50	25	0.067	0.79	
N_{ce}[a]	2	3	0.081	0.78	

[a]Note that increasing the N_{ce} to three requires us to use the 1-out-of-3 MTBF equation 13.8 in place of the 1-out-of-2 equation 13.4 for the no standby control element failure mode.

these is summarized in Table 13.7. Recall from Table 13.6 that the reference value for $MDT_i/MTBF_i$ for this failure mode is 0.190 and the (new) reference value for the system A_o is 0.72.

The model indicates that increasing the control element MTBCF by a factor of 2 has the same effect as reducing its downtime by 50% and that either of these brings us closer to meeting the $A_o \geq 0.80$ requirement (but still short). Table 13.8 provides the output of the Excel availability model for the case where $MTBCF_{1c}$ was increased from 75 to 100 h. We can see that by doing this the "no standby" control element failure mode is no longer limiting. The drone (no standby) failure mode is once again the limiting factor.

Table 13.9 provides the model results for the same case as for Table 13.8, but where the probability of having needed drone spare parts is increased to 0.95 (yielding an $MDT_{2d} = 4.3$ h). We see that in this case we are able to meet the A_o requirement.

Table 13.8 Effect of Increasing the Reliability of the Control Element

Element	$MTBF_i$ (h)	N_i	MDT_i (h)	$MDT_i/MTBF_i$	A_o
Control					
- Crit Fail (with SB)	150.00		0.10	0.001	0.999
- Crit Fail (no SB)	750.00	2.0	50.00	0.067	0.938
Drone					
- SRTB (with SB)	7.40		0.00	0.000	1.000
- Crit Fail (with SB)	50.00		0.80	0.016	0.984
- Net Eff Failures	6.45				
- Eff Fail (no SB)	37.35	3.0	6.80	0.182	0.846
System				0.265	0.790

AVAILABILITY MODELING AND TRADE STUDIES 453

Table 13.9 Effect of Increasing the Maximum Flying Time of the Drone, the Reliability of the Control Element, and the Probability of Having Drone Spare Parts

Element	$MTBF_i$ (h)	N_i	MDT_i (h)	$MDT_i/MTBF_i$	A_o
Control					
- Crit Fail (with SB)	150.00		0.10	0.10	0.999
- Crit Fail (no SB)	*750.00*	2.0	50.00	0.067	0.938
Drone					
- SRTB (with SB)	7.40		0.00	0.000	1.000
- Crit Fail (with SB)	50.00		0.80	0.016	0.984
- Net Eff Failures	6.45				
- Eff Fail (no SB)	*53.15*	3.0	4.30	0.081	0.925
System				0.164	*0.859*

13.2.3 Section Synopsis

This section provided an example of how a build an analytical availability model for a system that has a relatively simple "complex structure" from an associated set of reliability and maintainability models and simplifying assumptions. We found that even this simple availability model required input values for ~20 parameters.

The section then examined how the model could be used to guide and perform sustainment-related sensitivity studies and trade studies on how changes in architecture, design, operations, and maintenance affect system availability. Specifically, the model was used to perform sensitivity studies to determine what parameters are likely to have the largest effect on improving (or maintaining) availability and in identifying the range of values for those parameters that are worth considering. We found that there were limits to the degree to which changing individual parameter values can improve system availability. By trading-off improvements in performance for a number of parameters, we were able to find at least one solution that met the availability requirement. We saw other solutions were also possible. To find an optimal solution (from a cost-effectiveness perspective) we need to estimate the life cycle cost associated with implementing potential changes to each parameter's value. This will be addressed in Section 13.3.

While this section focused on the development of an analytical model, we could also have developed a Monte Carlo simulation to estimate the system's availability (see Exercise 2 associated with this section). We chose to use an analytic implementation for the following reasons: (i) it is easier to implement; (ii) it is more transparent; (iii) one does not have to wait for the simulation to reach a steady state (so it takes less time to run); and (iv) one does not have to worry about calculating a standard error for a given result. While these are all nice properties for the analytic model, there are a number of nice properties associated with the implementation of a Monte Carlo simulation: (i) one is not confined to exponential distributions; (ii) one is not required to make as many simplifying assumptions and/or approximations; and (iii) one can observe stochastic variation in the availability of the system over time (which gives a better sense of what one is likely to observe on a day-to-day basis in a real system).

When performing trade studies it is often useful to use analytic models and Monte Carlo simulations together. Analytical models are suited to providing a quick, "coarse grain" understanding of the trade space (since they run more quickly and are often easier to develop), while Monte Carlo simulation is best suited to providing a more "fine grain" understanding of the more promising regions of the trade space.

13.3 SUSTAINMENT LIFE CYCLE COST MODELING AND TRADE STUDIES[14]

In Section 13.2 we saw that the reference design for the FMDS system resulted in an availability that was significantly less than requirement (0.71 vs 0.90). We performed a series of trade studies that examined a number of system design options for improving the system's availability to the point where it would be able to meet its availability requirement.

In order to make an appropriate design decision, we need to know the total life cycle cost impact of each option. We would also like be able to estimate the most cost-effective design for providing an availability that exceeds the system's current requirement. For further information on life cycle cost modeling see Chapter 4.

In this section, we develop a life cycle cost model for the FMDS system that may be used to calculate the total system life cycle cost (TSLCC) associated with a given system design option. We then use this model to perform a series of trade studies to determine the least costly design option that enables us to meet the availability requirement. Finally, we examine how the framework developed here may be used to develop a cost-effectiveness curve.

Again, the focus of this chapter is on illustrating cost modeling and cost-effectiveness trade-off techniques and establishing a general framework for cost-effectiveness trade-off analyses (and not on reproducing specific analyses that were performed for real systems). As such, it uses fictional data for a fictional system and makes liberal use of simplifying assumptions. The resulting framework may be used to develop more complex models for real systems through the use of more realistic data and fewer simplifying assumptions.

13.3.1 The Total System Life Cycle Model

Generally, the TSLCC (C_{tlc}) of a system may be expressed as the sum of the following terms:

---

[14] The "Ch 13.3 LCC Model.xlsx" Excel file contains the cost model used to support the analyses performed in this section.

(A) Total development costs (C_{td})
(B) Total procurement costs (C_{tp})
(C) Total operations and support (O&S) costs (C_{tos})
(D) Total retirement/disposal costs (C_{trd})

Given these categories, we will develop our total system life cycle model based on the following framing assumptions:

1. $C_{td} = \$20.0\,M$.
2. $C_{tp} = N_u * (N_{cpu} * C_{cp} + N_{dpu} * C_{dp})$, where
   (a) N_u is the number of units that make up the system
   (b) N_{cpu} and N_{dpu} are, respectively, the number of control and drone elements per unit
   (c) C_{cp} and C_{dp} are, respectively, the per element procurement costs of each control element and drone.
3. C_{tos} is a complex function of the system architecture, operational concept, and maintenance concept. The next subsection is devoted to the development of this model element.
4. $C_{trd} = N_u * (N_{cpu} * C_{cd} + N_{dpu} * C_{dd})$, where C_{cd} and C_{dd} are, respectively, the per element disposal costs of each control element and drone.

Table 13.10 provides a TSLCC model for the reference system of interest. The input parameters are highlighted in yellow, intermediate calculations are highlighted in green, and the TSLCC is highlighted in blue. (*The reader is referred to the online version of this book for color indication.*) The values indicated for the O&S life cycle costs were obtained from O&S cost model developed in the subsection that follows.

Table 13.10 Model Input Parameters and LCC Calculations

Value Function/Output Parameter	Variable	Value	%
Total Life Cycle Cost	C_{tlc}	$107,173,484	
- System Development Cost	C_{td}	$10,000,000	9.3
- Total Procurement Cost	C_{tp}	$3,200,000	3.0
- Total Operations and Support Cost	C_{tos}	$93,671,484	87.4
- Total Retirement/Disposal Cost	C_{trd}	$302,000	0.3
Procurement Cost per Control Element	C_{cp}	$10,000	
Procurement Cost per Drone	C_{dp}	$100,000	
Number of Units	N_u	10	
Number of Control Elements/Unit	N_{cpu}	2	
Number of Drones/Unit	N_{dpu}	3	
Disposal Cost per Control Element	C_{cd}	$100	
Disposal Cost per Drone	C_{dd}	$10,000	

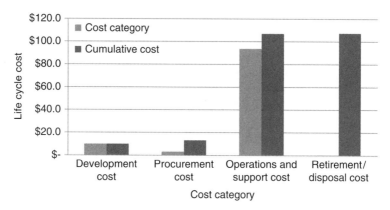

Figure 13.7 Cost category contributions to the TSLCC

Figure 13.7 provides a plot of the contribution of each major TSLCC cost element to the TSLCC, as well as a cumulative percentage as one proceeds through the system's life cycle. We can see that in this particular case, the O&S cost accounts for more than 85% of the system's total life cycle cost. We now turn to the task of developing the O&S cost model.

13.3.2 The O&S Cost Model

In order to develop an activity-based cost model (Chapter 4), one must first establish an appropriate work breakdown structure (WBS). Different WBSs are appropriate for different stages of a system's life cycle. In this section, we use the WBS structure developed by the Office of the Secretary of Defense (OSD) Director of Cost Assessment and Program Evaluation (CAPE) to develop estimates for operating and support (O&S) costs.[15]

The six top-level CAPE WBS O&S cost elements are defined as follows:

1.0 Unit-Level Manpower. Cost of operators, maintainers, and other support manpower assigned to operating units. May include military, civilian, and/or contractor manpower.

2.0 Unit Operations. Cost of unit operating material (e.g., fuel and training material), unit support services, and unit travel. Excludes material for maintenance and repair.

3.0 Maintenance. Cost of all system maintenance other than maintenance manpower assigned to operating units. Consists of organic and contractor maintenance.

---

[15] Office of the Secretary of Defense Cost Assessment and Program Evaluation, "Operating and Support Cost-Estimating Guide," Chapter 6, March 2014.

4.0 Sustaining Support. Cost of system support activities that are provided by organizations other than the system's operating units.

5.0 Continuing System Improvements. Cost of system hardware and software modifications.

6.0 Indirect Support. Cost of support activities that provide general services that lack the visibility of actual support to specific force units or systems. Indirect support is generally provided by centrally managed activities that provide a wide range of support to multiple systems and associated manpower.

In developing our O&S cost model, we will make the following simplifying assumptions (that lead to a *de facto* mathematical model):

1. The system will operate for a given system life time (T_l).
2. The cost estimates are in constant (now year) dollars.[16]
3. For the vast majority of its operational life, it will operate in a steady state. As such, we may approximate the total expected life cycle O&S cost (C_{tos}) as the product of T_l and the mean annual O&S Cost (C_{maos}):

$$C_{tos} = T_l * C_{maos} \qquad (13.14)$$

4. A "Unit" consists of given number control elements (N_{cpu}) and drone elements (N_{dpu}).
5. Once the system has achieved its steady state, no new elements will be produced and no units are "lost" (all units are repairable).
6. A separate control element must be used for each drone in flight.[17]
7. Each "Base" is the home for a single unit.[18]
8. The **mean annual Cost of Unit-Level Manpower (C_{amp})** [19] is roughly proportional to: the number of units in operations (N_u); the number of operators (N_{opu}), maintainers (N_{mpu}), and support personnel (N_{spu}) assigned to each unit; and the average annual cost per person (C_{app}), that is,

$$C_{amp} = N_u * (N_{opu} + N_{mpu} + N_{spu}) * C_{app} \qquad (13.15)$$

9. The **mean annual Cost of Unit Operations (C_{ao})** [20] is roughly proportional to: the total mean annual number of hours of operations (N_{aoh}) and the mean

---

[16] This assumption may be modified with relative ease to address inflation (and then-year cost estimates) and the discount rate (and the net present value of money). See problems at the end of this section.

[17] This ensures that control elements and drones have the same number of operational hours.

[18] This simplifying assumption is associated with the assumption that follows it. If a base contains multiple units, one may have a single maintenance crew that is expected to serve multiple units (which would be contrary to the assumption that follows).

[19] Generally, each active unit has a given number of staff that does not vary significantly with its use.

[20] Generally, fuel is one of the larger components of unit operations costs. The use of fuel is proportional to the number hours that a system is used.

operational cost per operational hour (C_{opoh}), that is,

$$C_{ao} = N_{aoh} * C_{opoh} \tag{13.16}$$

10. N_{aoh} depends on the number of units (N_u), the total expected (required) annual on station hours per unit (T_{aospu}), the number of on-station operational hours per flight (T_{ospf}), and the drone's mean mission flight time (i.e., $T_{mmf} = T_{mf} - T_m$), that is,

$$N_{aoh} = N_u * T_{aospu} * \frac{T_{mmf}}{T_{os}} \tag{13.17}$$

Note that assuming 24/365 coverage, $T_{aospu} = 365*24$ os h/flt $= 8760$ os h/flt.[21]

11. The mean annual number of flights per unit (N_{afpu}) follows as:

$$N_{afpu} = \frac{T_{aospu}}{T_{os}} \tag{13.18}$$

12. The **mean annual Cost of Maintenance (C_{am})** is roughly proportional to the following: the total mean annual number of hours of operations (N_{aoh}); the mean number of failures requiring a part replacement (or external repair) per hour for each element (or failure rate, R_{frri}[22]); and the mean cost to replace a failed part (C_{rfpi}), that is,

$$C_{am} = N_{aoh} * (R_{frrc} * C_{rfpc} + R_{frrd} * C_{rfpd}) \tag{13.19}$$

13. The ratio of failures requiring replacement to critical failures is r_c for control elements and r_d for drones, that is,[23]

$$R_{frrc} = \frac{r_c}{MTBCF_c} \tag{13.20}$$

$$R_{frrd} = \frac{r_d}{MTBF_d} \tag{13.21}$$

14. The mean annual Cost of Sustaining Support (C_{ass}) is assumed to be small and may be neglected to first order.

---

[21] An "OPTEMPO" factor may be defined and used in the cases where the operational tempo is less than one (i.e., the number of operational hours per unit is less than 24/365).

[22] Where $i = $ "c" (for control elements) and $= $ "d" (for drone elements).

[23] There are generally many failures caught as a part of postflight maintenance that have not yet resulted in a critical failure. Some of these failures will require the replacement (or external repair) of a part (that will result in a nonlabor cost). Since the preventative maintenance that is part of postflight maintenance is designed to detect and correct potential failures before they occur, the number of failures requiring replacement will be much larger than the number of critical failures.

SUSTAINMENT LIFE CYCLE COST MODELING AND TRADE STUDIES   459

15. The mean annual Cost of Continuing System Improvement (C_{ao}) is assumed to be small and may be neglected to first order.
16. The mean annual Cost of Indirect Support (C_{ais}) is assumed to be small and may be neglected to first order.

It should be noted that in the case where one or more of the aforementioned assumptions do not hold, one may make alternative assumptions and extend (and complicate) the model in a rather straightforward manner to account for these changes.

In addition to the assumptions regarding the system's design, operations concept, and maintenance concept, the following assumptions are made regarding the availability of data:

1. Estimates for many of the parameters listed earlier should be available from the documentation supporting the system's life cycle cost estimate.
2. Historical data on similar systems may also be used to develop estimates for the ratios of failures requiring replacement to critical failures (r_i), the mean cost to fix a failure (C_{ffi}), as well as for some of the other parameters.
3. Estimates for some factors may also be obtained from the system design and from operational and maintenance concepts and analyses.

Table 13.11 provides a screenshot of an Excel implementation of this O&S cost model that uses reference values as inputs (highlighted in yellow), which was used to generate the value of C_{tos} that was used in Table 13.10. (*The reader is referred to the online version of this book for color indication.*)

13.3.3 Life Cycle Cost Trade Study

From the availability model developed in Section 13.2, we see that system availability is constrained primarily by the drone effective failures for which no standby drone is available. We see that there are four ways in which we can reduce the value of $MDT_i/MTBF_i$ due to this failure source: (i) decrease MDT_i by reducing the logistics delay time; (ii) decrease MDT_i by decreasing the probability of a part not being available; (iii) increase the $MTBF_i$ by increasing the maximum flight time of the drone (which increases the MTBSRTB); and (iv) increase the $MTBF_i$ by increasing the number of drones per unit. For simplicity, we will only consider cases 3 and 4 (i.e., we will assume limited storage for parts and that the logistics downtime cannot be reduced further).

Figure 13.8 summarizes the values for A_o that are obtained using the availability model from Section 13.2 as one varies the drone's maximum flight time (T_{mf}), the number of drones per unit, and the number of control elements per unit (and reference system input values are used for the remaining parameters). Reference values were used for all other input parameters. The green highlight indicates the parameter space that just meets the requirement. The yellow highlight indicates a design option that almost meets the requirement. (*The reader is referred to the online version of this book for color indication.*)

Table 13.11 Total O&S Life Cycle Cost Model (Reference Values)

Value Function/Output Parameter	Variable	Value	Units
Total Life Cycle O&S Cost	C_{tos}	$93.67	$M
Mean Annual O&S Cost	C_{maos}	$9.37	$M/yr
Annual Cost Elements			
1 Annual Manpower Cost	C_{amp}	$4.00	$M/yr
2 Annual Unit Ops Cost	C_{ao}	$3.14	$M/yr
3 Annual Maintenance Cost	C_{am}	$2.23	$M/yr
4 Annual Sustaining Support Cost	C_{ass}	$0.00	$M/yr
5 Annual Cost of Continuing System Improvement	C_{acsi}	$0.00	$M/yr
6 Annual Indirect Support Cost	C_{ais}	$0.00	$M/yr
Factor/Input Parameter	**Variable**	**Value**	**Units**
System Life	T_i	10	yrs
Number of Units	Nu_u	10	unit
Number of Controls/Unit	N_{cpu}	2	c/u
Number of Drones/Unit	N_{dpu}	3	d/u
Number of Operators/Unit	N_{opu}	4	op/u
Number of Maintainers/Unit	N_{mpu}	2	mp/u
Number of Support Personnel/Unit	N_{spu}	2	sp/u
Required Annual On-Station Hours per unit (Mean Annual)	T_{aospu}	8,760	OS hrs/yr
Maximum Flight Time (per flight)	T_{mf}	4	Hrs/Flt
Non-OS Ops Time/Flt (Mean) (per flight)	T_{nospf}	0.6	Hrs/Flt
Time on Station Margin (per flight)	T_m	0.3	Hrs/Flt
Time on Station (per flight)	T_{os}	3.1	op hrs/Flt
Number of Op Hrs (Total Mean Annual)	N_{aoh}	104,555	op hrs/yr
Cost Per Personnel (Mean Annual)	C_{app}	$50,000	$/per yr
Operations Cost/op hr (Mean Annual)	C_{opoh}	$30	$/op hr
Mean time between control element critical failures	$MTBF_c$	75	op hrs
Mean time between drone critical failures	$MTBF_d$	50	op hrs
Ratio of control element failures requiring replacement to critical failures	r_c	1	
Ratio of drone failures requiring replacement to critical failures	r_d	5	
Cost to replace a failed control part (mean)	C_{rcp}	$100	$/failure
Cost to replace a failed drone part (mean)	C_{rdp}	$200	$/failure

The reference design is shown in bold redline. We see that if we are going to achieve the A_o requirement of 0.80, we must increase the maximum flight time (T_{mf}) of the drone, the number of drones per unit, and the number of control elements per unit. In order to explore the life cycle cost implications of the indicated design options, we must expand the model developed in Section 13.2. Specifically, increasing the maximum flight time capability of an aircraft generally requires a larger aircraft,

Two variable trade/sensitivity study ($N_c = 2$)				
N_d	Max flight time			
	4	5	6	7
3	0.58	0.64	0.67	0.70
4	0.64	0.69	0.72	0.74
5	0.68	0.72	0.75	0.77
6	0.70	0.74	0.76	0.78

Two variable trade/sensitivity study ($N_c = 3$)				
N_d	Max flight time			
	4	5	6	7
3	0.62	0.68	0.73	0.76
4	0.69	0.74	0.78	0.81
5	0.73	0.78	0.81	0.84
6	0.76	0.80	0.83	0.85

Figure 13.8 A_o as a function of the maximum flight time (T_{mf}), number of drones per unit (N_{dpu}), and number of control elements (N_{cpu})

which in turn generally results in: (i) increased procurement cost; (ii) use of more fuel; and (iii) more expensive replacement parts.[24] We will assume the following power law functions for drone production cost, annual operational cost per operational hour, and cost to replace drone part.[25]

$$C_{dp} = \$0.10\,\text{M} * \left(\frac{T_{mf}}{4.0\,\text{h}}\right)^2 \tag{13.22}$$

$$C_{opoh} = \$30 * \left(\frac{T_{mf}}{4.0\,\text{h}}\right)^{0.5} \tag{13.23}$$

$$C_{rfpd} = \$200 * \left(\frac{T_{mf}}{4.0\,\text{h}}\right)^{0.5} \tag{13.24}$$

These functions may be used to calculate input values for these parameters for use in the cost model developed above. Table 13.12 provides a screenshot of an integrated implementation of the Excel TLCC models developed in Sections 13.3.1 and 13.3.2 for the design option in bolded blue (with an $A_o = 0.80$), i.e., $N_{dpu} = 6$ and $T_{mf} = 5.0\,\text{h}$. The bold red items in the model indicate the values that changed from the reference case described in Tables 13.10 and 13.11 (i.e., N_{dpu}, N_{cpu}, T_{mf}, C_{ppd}, C_{opoh}, and C_{rdp}). (*The reader is referred to the online version of this book for color indication.*)

Table 13.13 summarizes the TSLCC associated with each N_{dpu}, T_{mf} pair (for $N_{cpu} = 3$). We see that the lowest TSLCC design solution that meets the requirement ($A_o = 0.80$) is $N_{dpu} = 6$, $T_{mf} = 5\,\text{h}$. It has a cost of \$118 M (vs. our reference case TSLCC of \$107 M with an $A_o = 0.58$).

---

[24]For simplicity, over the range of maximum flight times indicated, we assume that the drones have the same reliability characteristics and do not require additional manpower.
[25]While the selection of these equations is somewhat arbitrary, in general, one may obtain these kinds of cost estimating relationships (CERs) by performing regression analyses on historical procurement and operational data for a variety of systems that are similar to the one of interest but differ in their maximum flight times. The resulting CERs are often found to exhibit power law behavior.

Table 13.12 Integrated Life Cycle Cost Model for $N_{dpu} = 6$, $N_{cpu} = 3$, and $T_{mf} = 5$ h

Value Function/Output Parameter	Variable	Value	Units
Total Life Cycle Cost	C_{tcl}	$117.75	$M
System Development Cost	C_{tsd}	$10.00	$M
Total Procurement Cost	C_{tp}	$9.68	$M
Total Operations and Support Cost	C_{tos}	$97.48	$M
Total Retirement/Disposal Cost	C_{tp}	$0.05	$M
Procurement Cost per Control Element	C_{ppc}	$10,000	$/ce
Procurement Cost per Drone	C_{ppd}	*$156,250*	$/de
Disposal Cost per Control Element	C_{dpc}	$100	$/ce
Disposal Cost per Drone	C_{dpd}	$10,000	$/de
Value Function/Output Parameter	**Variable**	**Value**	**Units**
Total Life Cycle O&S Cost	C_{tos}	$97.48	$M
Mean Annual O&S Cost	C_{maos}	$9.75	$M/yr
Annual Cost Elements			
1 Annual Manpower Cost	C_{amp}	$4.00	$M/yr
2 Annual Unit Ops Cost	C_{ao}	$3.37	$M/yr
3 Annual Maintenance Cost	C_{am}	$2.38	$M/yr
4 Annual Sustaining Support Cost	C_{ass}	$0.00	$M/yr
5 Annual Cost of Continuing System Improvement	C_{acsi}	$0.00	$M/yr
6 Annual Indirect Support Cost	C_{ais}	$0.00	$M/yr
Factor/Input Parameter	**Variable**	**Value**	**Units**
System Life	T_i	10	yrs
Number of Units	N_u	10	unit
Number of Controls/Unit	N_{cpu}	*3*	c/u
Number of Drones/Unit	N_{dpu}	*6*	d/u
Numbers of Operators/Unit	N_{opu}	4	op/u
Number of Maintainers/Unit	N_{mpu}	2	mp/u
Number of Support Personnel/Unit		2	sp/u
Required Annual On-Station Hours per unit (Mean Annual)	T_{aospu}	8,760	OS hrs/yr
Maximum Flight Time (per flight)	T_{mf}	5.0	Hrs/Flt
Non-OS Ops Time/Flt (Mean) (per flight)	T_{nospf}	0.6	Hrs/Flt
Time on Station Margin (per flight)	T_m	0.3	Hrs/Flt
Time on Station (per flight)	T_{os}	4.1	op hrs/Flt
Number of Op Hrs (Total Mean Annual)	N_{aoh}	100,420	ops hrs/yr
Cost Per Personnel (Mean Annual)	C_{app}	$50,000	$/per yr
Operations Cost/op hr (Mean Annual)	C_{opoh}	$34	$/op hr

SUSTAINMENT LIFE CYCLE COST MODELING AND TRADE STUDIES

Table 13.12 (*Continued*)

Factor/Input Parameter	Variable	Value	Units
Mean time between control element critical failures	$MTBF_c$	75	op hrs
Mean time between drone critical failures	$MTBF_d$	50	op hrs
Ratio of control element failures requiring replacement to critical failures	r_c	1	
Ratio of drone failure requiring replacement to critical failures	r_d	5	
Cost to replace a failed control part (mean)	C_{rcp}	$100	$/failure
Cost to replace a failed drone part (mean)	C_{rdp}	**$223.6**	$/failure

Table 13.13 Total Life Cycle Cost (C_{tlc}) as a Function of the Maximum Flight Time (T_{mf}) and Number of Drones per Unit (N_{dpu})

Max Flt Time	N_{dpu}			
	3	4	5	6
4	$107	$108	$109	$111
5	$113	$114	$116	$118
6	$119	$121	$123	$126
7	$125	$128	$131	$134

Figure 13.9 provides the trade space associated with the A_o and TSLCC (from Table 13.13) for different design options (from Figure 13.8) for $N_{cpu} = 3$. The box in the lower left indicates the reference design $N_{dpu} = 3$, $N_{cpu} = 2$, $T_{mf} = 4.0$ h). Depending on affordability considerations, the customer may use this plot to trade increases

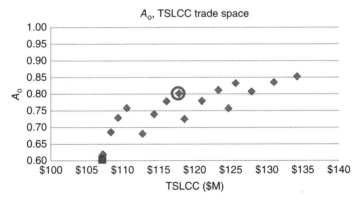

Figure 13.9 A_o as a function of total system life cycle cost (TSLCC) for the designs provide in Figure 13.8 ($N_{cpu} = 3$)

in A_o for increases in TSLCC. The figure may be used to find the least expensive design option that provides the required A_o (which is circled).

13.4 OPTIMIZATION IN AVAILABILITY TRADE STUDIES

While the previous section illustrated a manual approach to finding an optimal design solution, this section illustrates how optimization techniques can be applied to determine the minimum cost design option that meets the $A_o \geq 0.90$ availability requirement. This section is structured as follows. Section 13.4.1 identifies the value/objective function, the principal decision variables, and constraint equations. It then expresses the optimization problem in canonical form and identifies the optimization technique that will be used to find an optimal design solution. Section 13.4.2 describes the Excel instantiation of the optimization problem. Section 13.4.3 discusses the results obtained from this instantiation and Section 13.4.4 provides a deterministic sensitivity study of associated with the examining the impact of the uncertainties associated with the values assigned to the model input parameters.

13.4.1 Setting Up the Optimization Problem

In order to specify an optimization problem in canonical form one must identify the objective function to be optimized, the nature of the optimization, and the decision variables upon which it depends, and the constraint functions. In our case, the objective function is the TSLCC which must be minimized. This cost objective function is specified by the TSLCC model developed in the previous section. As we saw, this TSLCC model was driven principally by the value of three decision variables, the number of drones (N_d), the number of control elements (N_c), and the maximum time of flight of the drone (T_{mf}). For the sake of simplicity, the optimization problem fixes T_{mf} to a value of 4 h and will only consider varying N_d and N_c. The minimum and maximum values for N_d and N_c are modeled to be 2 and 8 respectively.

Given this, the optimization problem may be stated in canonical form as:

- Minimize: $C_{tlcc}(N_d, N_c)$
- Subject to:
  - $A_o(N_d, N_c) \geq 0.80$
  - $N_d \geq 2$
  - $N_d \leq 8$
  - $N_c \geq 2$
  - $N_c \leq 8$

In order to select an optimization method, we need to examine the properties of the functions and decision variables. Since the TSLCC (C_{tlcc}) and A_o are non-linear functions of the decision variables, N_d and N_c can only take on integer values, and T_{mf} is constrained (somewhat artificially) to four values we must use an optimization technique that is appropriate. To this end we have selected the "Evolutionary" method implemented in Excel.

13.4.2 Instantiating the Optimization Model

The optimization problem was instantiated in Excel for two reasons. First, both the cost model and the availability models are complex non-linear functions that were developed in Excel and were thus easy to import. Second, Excel Solver provides the Evolutionary optimization solver that is appropriate for solving non-linear, non-smooth optimization problems.

The Optimization Excel model (CH13 FMDS Deterministic and Optimization.xlsx; *excel available online as supplementary material*) contains three spreadsheets or tabs, the Control Panel tab, the Calculation tab, and the Life Cycle Cost tab. The principal elements of each tab used to implement the optimization algorithm are described below. Other tab elements are used to construct useful graphs and to support sensitivity analysis (see Section 13.4.4). Additional information regarding each tab may be found in the Excel file.

Figure 13.10 provides an example of the "A_o Input Parameters" portion of the *Control Panel* tab. It contains all the input parameter values associated with the calculation of A_o. For the purposes of the optimization analysis, only the "Base" values for each parameter are used. The "Worst" and "Best" values for these parameters (as well as the "Index") are used to perform the deterministic sensitivity study in Section 13.4.4.

Figure 13.11 provides an example of the "Decision Variables, Constraints and Results" portion of the *Control Panel* tab. It indicates the constraints on the decision variables N_c (cells D44 and D45), N_d (cells D46 and D47), and A_o (D43). One must also put in "initialization" values for N_c and N_d (cells B39 and B40). These values yield initial values for A_o (B50) and the TSLCC (B51). Once the optimization algorithm is run, the initial values for N_c and N_d (in cells B39 and B40) are replaced

Operational Availability Model for a Forest Monitoring Drone System

NOTE: PRESS SHIFT + F9 TO UPDATE RESULTS AFTER ANY CHANGES

Parameters

Legend: Input; Calculation; Objective; Decisions

Index: 2

Drone Cycle	Notation	Units	In Use	Index Base	Worst	Base	Best
Pre-Flight Preparation Time	T_{pfp}	hours	0.5	2	1.0	0.5	0.3
Time to Launch	T_{lch}	hours	0.1	2	0.20	0.1	0.05
Time to Station	T_{ts}	hours	0.2	2	0.3	0.2	0.0
Time on Station	T_{os}	hours	3.1				
Time to Return to Base	T_{rtb}	hours	0.2	2	0.3	0.2	0.0
Time to Land	T_{lnd}	hours	0.1	2	0.20	0.1	0.05
Time on Station Margin	T_m	hours	0.3	2	0.5	0.3	0.20
Post-Flight Maintenance Time	T_{pfm}	hours	1.0	2	2.0	1.0	0.5

	Availability	Notation	Units	In Use	Index Base	Worst	Base	Best
Control	MTB Critical Failure w. Standby	$MTBCF_{1c}$	hours	75	2	50	75	125
Control	MDT due to Critical Failure w. Standby	MDT_{1c}	hours	0.1	2	0.20	0.1	0.05
Control	MDT due to Critical Failure w/o Standby	MDT_{2c}	hours	50	2	75	50	25
Drone	MTB Critical Failure w. Standby	$MTBCF_{1d}$	hours	50	2	30	50	80
Drone	MDT due to Scheduled Return to Base w. Standby	MDT_{1prtb}	hours	0				

Other	Notation	Units	In Use	Index Base	Worst	Base	Best
MDT due to Logistics Delay to Obtain Part	MDT_{log}	hours	50	2	75	50	25
Probability Replacement Part is Available	P_{sp}	Probability	0.90	2	0.80	0.90	0.95

Figure 13.10 The "A_o Input Parameters" portion of the Control Panel tab

	A	B	C	D
34	**OPTIMIZATION**			
35	Minimize:			
36	Total System Life Cycle Cost ($M)	$ 107.17		
37				
38	**Decision Variables**			
39	Number of Controls per Unit	2		
40	Number of Drones per Unit	3		
41				
42	Subject to:			
43	80% Availability Requirement	58.08%	>=	80%
44	Min. Number of Controls	2	>=	2
45	Max. Number of Controls	2	<=	8
46	Min. Number of Drones	3	>=	2
47	Max. Number of Drones	3	<=	8
48				
49	**Results:**			
50	Availability	58.08%		
51	Total System Life Cycle Cost ($M)	$ 107.17		

Figure 13.11 The "Decision Variables, Constraints, and Results" portion of the Control Panel tab

	A	B	C	D	E	F	G	H
1								
2					$MTBF_i$		MDT_i	
3	**Control Element Failure Modes**			Notation	Value	Notation	Value	$MDT_i/MTBF_i$
4	1. Critical Failure w. Standby			$MTBCF_{1c}$	75	MDT_{1c}	0.1	0.00133333
5	3. Critical Failure w/o Standby*			$MTBCF_{2c}$	262.5	MDT_{2c}	50	0.190
6	**Drone Failure Modes**							
7	4. Scheduled Return to Base w. Standby			$MTBSRTB_{1d}$	3.4	MDT_{1srtb}	0	0
8	5. Critical Failure w. Standby			$MTBCF_{1d}$	50	MDT_{1d}	0.8	0.016
9	9. Critical Failure w/o Standby*			$MTBCF_{2d}$	13.23	MDT_{2d}	6.8	0.514
10							Sum:	0.722
11	10.System Availability:	58.08%						

Figure 13.12 The Calculations tab

by the optimum values and the initial values for A_o and TSLCC (in cells B50 and B51) are replaced by the resulting A_o and minimum TSLCC.

Figure 13.12 provides an example of the *Calculations* tab. It is used to calculate the value of A_o (Cell B11) that results from the values of the input and decision variables provided in the Control Tab.

Figure 13.13 provides an example of the summary-level portion of the *Life Cycle Cost* tab. The Life Cycle Cost tab contains values for all the life cycle model inputs (that were addressed in the Control Panel), intermediate cost calculations (in green), and the calculation of the total life cycle cost (Cell D7). Recall it is this value that is

OPTIMIZATION IN AVAILABILITY TRADE STUDIES

Figure 13.13 The life cycle cost tab

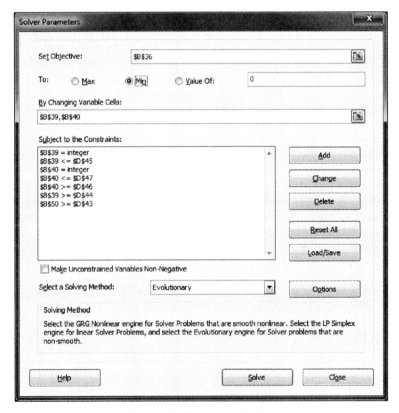

Figure 13.14 Solver window

to be minimized. (*The reader is referred to the online version of this book for color indication.*)

Once all input parameters and decision variable values are specified, optimization can be performed to the model using Solver's Evolutionary Method. Figure 13.14 demonstrates how the Solver window is used to minimize the objective (the TSLCC

	A	B	C	D
34	**OPTIMIZATION**			
35	**Minimize:**			
36	Total System Life Cycle Cost ($M)	$ 110.98		
37				
38	**Decision Variables**			
39	Number of Controls per Unit	7		
40	Number of Drones per Unit	6		
41				
42	**Subject to:**			
43	80% Availability Requirement	80.20%	>=	80%
44	Min. Number of Controls	7	>=	2
45	Max. Number of Controls	7	<=	8
46	Min. Number of Drones	6	>=	2
47	Max. Number of Drones	6	<=	8
48				
49	**Results:**			
50	Availability	80.20%		
51	**Total System Life Cycle Cost ($M)**	$ 110.98		

Figure 13.15 Optimization results

in cell B36), over the range of decision variables (provided in cells B39 and B40), subject to the indicated constraints on A_o, N_c, and N_d in the Control Panel tab. Given these values one selects "Solve."

Figure 13.15 provides an example of how the "Decision Variables, Constraints and Results" portion of the *Control Panel* tab changes as a result of the optimization. We can see that the values in N_c and N_d (cells B39 and B40) are now populated with the decision variable solution ($N_c = 7$, $N_d = 6$), the resulting constraint-satisfying value for A_o (80.20%) populating cell B50, and minimum TSLCC that results ($111 M) in cell B51.

13.4.3 Discussion of the Optimization Model Results

The solution obtained in the previous section has an estimated TSLCC of $111 M which is ~$7 M less than the solution obtained by hand in Section 13.3.3. We see that automated implementation of an optimization algorithm allows us to find more optimum solutions with much less effort than can be found using a manual search.

The automated search showed that it was far less expensive to adopt a design solution with additional control and drone elements, than it was to adopt the design with a greater maximum flight time. It should be noted that the solution might well change if one increases the number of units that are to be procured or other changes are made to either the availability model or the TSLCC model.

OPTIMIZATION IN AVAILABILITY TRADE STUDIES

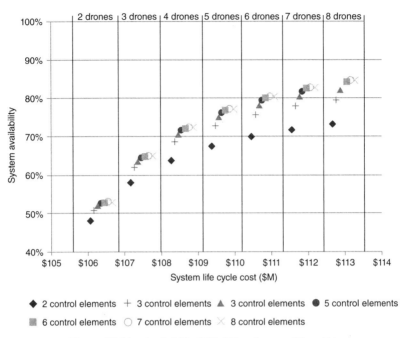

Figure 13.16 Availability/TSLCC tradespace ($T_{tf} = 4$ h).

Figure 13.16 was obtained using the data tables constructed in the Control Panel tab of the Excel model described in the previous section. It indicates the availability/cost trade space associated with keeping $T_{mf} = 4.0$ h. It shows the cost associated with each of the 49 design solutions that resulted from taking values for N_c and N_d that increased incrementally from 2 to 8.[26]

We can see from this graph that very little improvement in A_o is achieved by increasing N_d beyond about 6 or 7 or N_c beyond about 4.

13.4.4 Deterministic Sensitivity Analysis

Prior to performing a Monte Carlo analysis of a system, one should determine which uncertainties are likely to have the greatest effect on the metrics of interest. This is typically done by performing a single factor sensitivity analysis. In such analysis one first determines the (deterministic) "base" (expected) value for the metric of interest (in this case A_o) based on assigning "base" (expected) values to each of the input

---

[26]Note that this figure differs from Figure 13.9 in that the tradespace for Figure 13.9 was N_d and T_{mf} (N_c constant) versus N_c and N_d (T_{mf} constant).

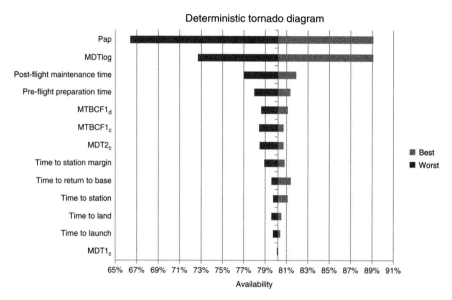

Figure 13.17 Tornado diagram for $N_c = 7$ and $N_d = 6$

factors. One then systematically varies the value of each of the input factor from its "worst" value (i.e., the value that results in a lower A_o) to its "best" value (yielding a higher A_o), while holding all other inputs factors at their base values.

The Optimization Excel model developed in Section 13.4.2 may be used to perform such an analysis. Columns J, K, and L of the "A_o Input Parameters" portion of the *Control Panel* tab are used to specify the worst, base, and best values for each input factor. These values are then used to determine the resulting "swing" in the value of A_o that would result from such changes in input values (this is done in cells A55–D105). The results of such an analysis may be presented as a "Tornado Diagram." Figure 13.17 provides an example of such a diagram that was obtained using the "optimum system design" of $N_c = 7$ and $N_d = 6$.

This diagram provides a great deal of useful information. It tells us that uncertainties in Pap and postflight maintenance time are the greatest sources of uncertainty in the expected value of A_o (they can swing it by ~9%–14% in either direction). It also shows that uncertainties MDT_{log}, Preflight Prep. Time, $MTBCF_{1d}$, $MTBCF_{1c}$, and $MDT2_c$ can result in A_o swings of about 1–8% in either direction. This implies that in developing a Monte Carlo model, one should certainly model the first two variables as random and possibly the next five as well. Since uncertainty in the remaining six variables has a relatively small impact on the value of A_o, they may be treated as constants (equal to their base values). Finally, the asymmetric nature of many of these uncertainties (there is greater "downside" impact than "upside" impact) may be expected to give rise to an "expected" Monte Carlo value of A_o that is lower than the "base" deterministic A_o. We will see that this is the case in the following section.

13.5 MONTE CARLO MODELING

There are at least two different ways in which Monte Carlo modeling may be done to support the kind of sustainment analyses described in this chapter. The first approach is to develop a "Scenario/Mission-based" Monte Carlo simulation for system availability that models the takeoff, flight, landing, failure, and maintenance of each drone and the failing and replacement/repair of each control element over some time period of interest. Such a model could be used to validate the analytic model developed in Section 13.2, explore the implications of more realistic distributions for key parameters, explore transient (as opposed to steady-state) behavior, and get better feel for the day-to-day variability in system availability that one could expect to see. Such models are generally time-consuming to develop and are left as an exercise for the reader (see Exercise 2 at the end of this chapter).

The second approach is to use Monte Carlo simulation to develop a sense of the degree to which uncertainties in input factor values can yield uncertainties in model output metric values. It is this later approach that is considered in this section.

The models developed in the previous sections (except Section 13.4.4) did not address uncertainties in the ability to achieve designs or the values of various cost parameters. While such deterministic modeling is useful for establishing a modeling framework and for obtaining crude point solution "expected values" for important system metrics, it does not give one a sene of the uncertainty and risk associated with achieving those expected values. This section provides an example of how to develop Monte Carlo extensions to the deterministic availability optimization and cost models developed in Section 13.4 and illustrate how such extensions may be used to determine performance and cost risk. The model can be found in the file CH13 FMDS Monte Carlo Analysis.xlsx available online as supplementary material.

13.5.1 Input Probability Distributions for the Monte Carlo Model

Uncertainty can be incorporated into FMDS model by adding probabilistic distributions to lower level parameters. These are represented in the influence diagram in Section 13.1.1 by the parameters circled by an ellipse. By adding uncertainty to these parameters, the resulting availability value will differ from the one obtained through deterministic analysis. The degree of such variation is dependent on the probabilistic distributions assigned to each parameter. As discussed in Section 13.4.4, probabilistic distributions should be assigned only to those input parameters that have determined to have the greatest impact on the output metric of interest (i.e., the availability) through deterministic sensitivity analysis. For this reason, triangular distributions were added to the top seven uncertainties in Figure 13.17. Figure 13.18 illustrates the probability density function for the triangular distribution embedded to drone postflight maintenance time as an example of the added distributions using Palisade's @Risk package for Excel.

The minimum, peak, and maximum values for postflight maintenance time (T_{pfm}) are 0.5, 1, 2 h respectively (from Figure 13.18). For this factor, the minimum value

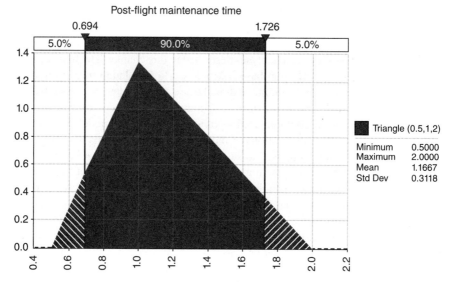

Figure 13.18 Postflight preparation time triangular distribution

corresponds to the "best case," that is, it results in a larger value for A_o.[27] It should be noted that distribution is skewed toward higher values of T_{pfm}. As a result, the mean of the distribution is higher than the "peak" values. As such, one would expect that the resulting (Monte Carlo) mean value for A_o would be lower than the one predicted using the deterministic model.

13.5.2 Monte Carlo Simulation Results

Once all triangular distributions have been incorporated to low-level parameters, Monte Carlo simulation can be performed to obtain the expected system availability when uncertainty is present in the model. Figure 13.19 shows the cumulative density functions for 7 controls/6 drones, 8 controls/7 drones, and 5 controls/8 drones. Besides 7 controls/6 drones, these combinations were considered since they were the ones that approached the 80% requirement at the lowest TSLCC.

The leftmost cumulative density function corresponds to the 7 controls/6 drones combination that resulted in the least expensive design that was able to meet the $A_o \geq 0.80$ requirement. We see that in this case, the MC model provided a mean $A_{\text{omc}} = 0.7716$, which is lower than the $A_{\text{od}} = 0.8020$ obtained from the deterministic model in Section 14.4.2. In order to determine whether this is significant, we need to calculate the uncertainty in A_{omc}. The "Standard Error"(SE) provides a measure of this uncertainty. It is calculated from the standard deviation (SD) and number of runs (N_r) using:

$$\text{SE} = \frac{\text{SD}}{\sqrt{N_r}} = \frac{0.059}{31.62} = 0.0019 \qquad (13.25)$$

---

[27] It should be noted that the minimum value of a parameter is not always associated with the "best case" (e.g., the minimum for value MTBF_d corresponds to the "worst case," that is, a lower value for A_o).

MONTE CARLO MODELING   473

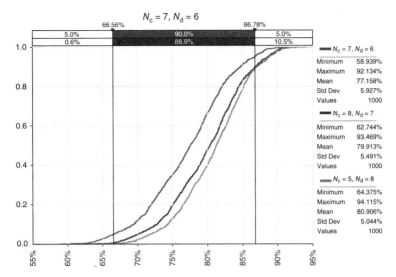

Figure 13.19 Cumulative density functions for control/drone combinations

Given this, one should technically report the value of the MC mean as:

$$A_{\text{omc}} = 0.7716 \pm 0.0019$$

Since the difference between A_{omc} and A_{od} is more than two standard errors, we can conclude that the difference is statistically significant. Generally, there are two potential sources for such differences. The first is skewness in the input distributions. The second is nonlinearity of the functions used to determine the value of the output value.

The cumulative probability distribution generated for A_o in Figure 13.19 may be used to determine the "confidence" that a given design will be able to meet its requirement. This permits us to perform the following confidence/design trade studies. As an example, in the $N_c = 7$, $N_d = 6$ case, we see that about a 68% of runs resulted in values of A_o less than 0.80, corresponding to a 32% confidence that the design will meet the requirement. If we increase N_c to 8 and N_d to 7, only about 48% of runs fall below $A_o = 0.80$, corresponding to a 52% confidence that this design will meet the requirement. Alternatively, if we decrease N_c to 5 and increase N_d to 8, only about 40% of runs fall below $A_o = 0.80$, corresponding to a 60% confidence that this design will meet the requirement.

13.5.3 Stochastic Sensitivity Analysis

The Monte Carlo simulation performed in the previous section also serves as a tool to conduct a stochastic sensitivity analysis for the optimal solution found in Section 13.4.3. Specifically, the most sensitive uncertainties determined from the deterministic sensitivity analysis in Section 13.4.4 can be analyzed to determine their respective contribution to the variability in system availability when probabilistic distributions

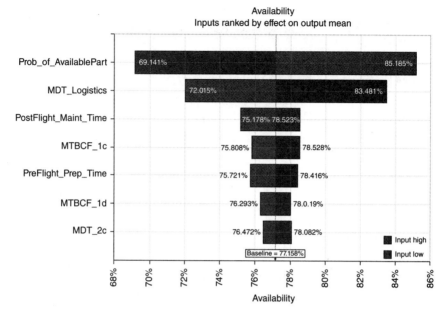

Figure 13.20 Tornado diagram for the FMDS system

are added. This can be done through @RISK's Change in output mean tornado diagram functionality as shown in Figure 13.20.

This plot differs from the deterministic tornado diagram in that the lower (upper) values of A_o for each parameter are the mean of the 10% of Monte Carlo runs that had the worst (best) random values for that parameter. [28] The results of the stochastic tornado diagram display some important differences relative to its deterministic equivalent. First, we see that the expected (base) value has changed. Second, we see that there is a decrease in the maximum availability that could be possibly met when varying the most sensitive parameter (Prob. of Available Part). The deterministic sensitivity analysis indicated that the value of A_o could exceed 89% as shown in Figure 13.17, whereas A_o only reaches 84.94%. Third, the sensitivity bars associated with the stochastic diagram are more symmetric than those associated with the deterministic diagram. Finally, the stochastic analysis suggests there is a change in the order of most sensitive uncertainties. When stochastic analysis is performed, $MTBCF_{1d}$ moves from the fifth position in the order to the least sensitive parameter. Similarly, $MDT2_c$ moves from seventh position to the sixth most sensitive parameter. Post-flight maintenance time and $MTBCF_{1c}$ also change positions. One of the reasons for the changes in the magnitude and symmetry of the effects, and their order of importance, is that the stochastic analysis reflects the mathematical coupling between parameters, while the deterministic analysis does not.

---

[28] The probabilistic tornado diagram is performed by @Risk by sorting the input variable into 10 ascending bins. The plot shows the mean of the output measure of the 1st and 10th bins. (Other options are available in @Risk).

13.6 CHAPTER SUMMARY

The availability of a system is an important operational performance parameter. The associated reliability and maintainability requirements are major drivers of system's TSLCC, especially those associated with operations and support (which generally account for the majority of total life cycle cost). As such, it is important to have models that provide decision-makers information regarding the cost-effectiveness of different designs and different operational and maintenance concepts.

To this end, Section 13.2 developed a first-order performance model for the availability of the FMDS as a function of a variety of design, operational, and maintenance factors and demonstrated how such a model could be used to perform a variety of sensitivity and trade-off analyses related to system design and to associated operational and maintenance concepts.

Section 13.3 developed a first-order total life cycle cost model for the FMDS (with a special focus on life cycle O&S costs) within the context of standard DoD cost WBSs. It then demonstrated how to integrate the cost model with the performance model developed in Section 13.2 and how to use such an integrated cost–performance model to perform a cost-effectiveness trade-off analysis.

Section 13.4 demonstrated how one could develop an Excel model that employs an evolutionary optimization technique to find the lowest cost design solution that meets the system's availability and automatically generate a "cost-effectiveness" tradespace curve. It also showed how one could use a tornado diagram to determine which input factors have the greatest effect on a given output metric. We saw that this helped us identify the most important parameters to model as random variables in a Monte Carlo model and provided information that could be used to determine the shape of the associated random number generators.

Section 13.5 provided an example of how to develop a Monte Carlo extension to the deterministic availability model developed in Section 13.2 and how such a model can be used to determine the confidence that one may have that a given design will be able to meet its requirement. It also showed how one could perform a stochastic sensitivity analysis to determine the degree to which each input parameter affects the output metric, as the other parameters are varied stochastically.

The models developed in this chapter illustrate important modeling and trade-off analysis techniques and lessons. One of the most important lessons in modeling is that if one attempts to model everything, one will successfully model nothing. As such, this chapter demonstrated how to develop first-order models based on simplifying assumptions that may be used to provide a framework for initial studies and for elaboration, in spiral fashion, to develop more complicated models that incorporate fewer simplifying assumptions.

Other important "takeaways" from this chapter include the following:

- Cost-effectiveness trade studies provide information that is essential for many design decisions.

- Generally, a cost-effectiveness trade study requires the development of two types of models: (i) one or more system performance (system effectiveness) models and (ii) one or more cost models.
- One should develop a cost model that reflects how different design options will affect the TSLCC, not just the cost associated with one portion of the cycle (e.g., development or production).
- Performance models and LCC models permit one to structure the problem and guide the analysis. Even first-order (performance and life cycle cost) models can be complicated and require the use of many input parameters.
- As such, one should initially focus on identifying and modeling only the most important value functions and associated factors, relationships, and effects that affect them.
- Performance models and LCC models should be extensible so that they may be modified to address changes in simplifying assumptions and/or additional information that is uncovered during the course of the study.
- It is important to develop integrated performance and cost models so that one may observe how a change in a parameter value can simultaneously affect both the performance model (e.g., availability) and the life cycle cost model.

The purpose of this chapter is to illustrate techniques for developing models that can be used to perform RAM-related cost-effective trade studies and for performing such trade studies (not to provide a detailed trade study of a specific, real system). As such, the models are based on a variety of illustrative simplifying assumptions and make use of fictional data. The resulting **modeling framework** can then be extended to develop more detailed and accurate models for real systems, based on real data and fewer simplifying assumptions. The exercises at the end of this chapter provide the reader an opportunity to explore some of these extensions and a wider range of sensitivity and trade-off analyses than were covered in the chapter.

13.7 KEY TERMS

Availability: the probability that a system or equipment, when used under stated conditions, will operate properly at any point in time.

Complex Structure: a system composed of multiple instantiations of multiple types of elements that interact with one another, which differ significantly in their form and function. In addition, each element is characterized by multiple failure modes, a mixture of series and parallel reliability architectures, and load sharing.

Cost Estimating Relationship: a mathematical function that indicates how a set of variables are related to one another. It is often obtained through regression analysis.

Cost Model: a mathematical model that calculates the cost associated with some aspect of a system based on the input values of some set of factors.

KEY TERMS

Critical Failure: a failure that requires an element to abort its mission and seek immediate repair.

Cold Standby: the case in which a "standby elements" does not become active until the active element fails. There is generally a short system "downtime" associated with this replacement.

Development Cost: the total cost associated with designing, production, and testing of system prototypes and/or engineering development models, as well as that associated with requirements development, technology development, system analysis, and system trade studies.

Deterministic Model: a model in which none of the variables are random.

Downtime: the time a system or element is inoperable following a failure or associated with a scheduled maintenance. Generally, it includes the time to obtain the parts required to repair the item, the time required to repair it, and any additional time required to return it to service. In the case of a standby element, it consists of the time between when the active element fails and the standby element is able to take its place and the system is able to resume operation.

Influence Diagram: a diagram that consists of decision nodes, uncertainty nodes, and value nodes connected by directional arcs that indicate that the behavior/value of the source node influences the behavior/value of the end node.

Integrated Performance/Cost Model: a mathematical model that calculates the cost of some aspect of a system and the performance of the system (with respect to one or more response variables of interest) based input values for some set of cost and performance factors.

***k*-Out-of-*n* Cold Standby System:** a parallel structure composed of n elements in which k elements must operate simultaneously for the system to be operational and the remaining $n-k$ elements are nonoperating (cold) "standby elements." If one of the k operating elements experiences a failure, it is switched off and one of the standby elements takes its place (after a short replacement downtime).

Maintainability: the probability that an item will be retained in or restored to a specified condition in a given period of time, when maintenance is performed in accordance with prescribed procedures and resources.

Markov Process: a system that is characterized by a set of states and a set of transition probabilities that depend only on the current state of the system.

Mean Downtime: the average downtime experienced by a system or element.

Mean Time Between Critical Failures (MTBCF): the average time that a system or elements operates without experiencing a critical failure (i.e., the average operating time between critical failures).

Monte Carlo Simulation: a simulation that makes use of repeated random sampling of one or more random variables in order to obtain numerical results. Within the context of this chapter, the term is used to refer to "Scenario/Mission-based" Monte Carlo simulations and "Uncertainty Analysis" Monte Carlo simulations.

Operational Availability: the probability that a system or equipment, when used under stated conditions in an actual operational environment, will operate satisfactorily when called upon.

Operations and Support (O&S) Cost: the total cost incurred from system deployment through end of system operations. It includes the costs of operating, maintaining, and supporting a fielded system.

Performance Model: a mathematical model that estimates the performance of a system with respect to one or more response variables based on the input values of some set of factors.

Procurement Cost: the total cost of producing and deploying all of the units that make up the system over its operating life.

Reliability: the probability that a system or product [element/item/service] will perform in a satisfactory manner for a given period of time, when used under specified operating conditions.

Reliability Block Diagram: a diagram that identifies how the reliability of different elements contributes to the reliability of a system. Components are drawn as being in series or parallel configurations.

Retirement/Disposal Cost: the cost associated with retirement and disposal of the system, including the cost of disposing of hazardous materials.

Sensitivity Analysis: a type of trade-off analysis in which one evaluates the degree to which a change in one or more input factors affects the value of an output metric.

Standby Element: a redundant element that is able to replace an active element that is required for system operation, if the active element fails.

Stochastic Model: a model in which one or more variables are random.

Structure: a collection of elements connected to one another in a series and/or parallel architecture.

Total Life Cycle Cost: the total cost of a system over its entire life cycle, from conception, through development, production, operations and maintenance, and retirement/disposal.

Tornado Diagram: a diagram that indicates the degree to which an output metric varies as one changes the value of input parameters from their lowest to their highest values. The input variables that lead to the greatest change in output metrics are placed at the top, and those that lead to the smallest variation are placed on the bottom.

13.8 EXERCISES

13.1. Construct the analytical availability model provided in Figure 13.5 using Excel.

(a) Reproduce the results.

EXERCISES

(b) Develop a graph showing how availability of the system varies the reliability of the control element ($MTBCF_{1c}$) from 50 to 200 h in 25 h steps.

(c) Develop a graph showing how the availability of the system varies as the reliability of drone element ($MTBCF_{1d}$) from 20 to 100 h in 20 h steps.

(d) How would one determine whether it is better to focus development effort on improving control element or drone element reliability (a design decision)?

(e) Develop a graph showing how the availability of the system varies as the logistics downtime (MDT_{log}) for the drone element increases from 10 to 80 h in 10 h steps. Note that this represents decrease in maintainability.

(f) Develop a graph showing how the availability of the system varies as the preflight preparation time (T_{pfp}) for the drone element increases from 0.2 to 1.0 h in 0.2 h steps. Note that this also represents decrease in maintainability.

(g) Does it make more sense to focus on decreasing the logistics downtime or the preflight preparation time (a maintenance concept decision)? Explain.

(h) How would one determine whether it is better to focus improving design (MTBCF) or the maintenance concept (MDT_{log} or T_{pfp})? Note this is a decision regarding a trade-off between design and maintenance concept.

(i) Develop a graph showing how the availability of the system changes as one increases the number of drone elements in a system from 2 to 5.

(j) Under what conditions would one want to increase the number of drones in a unit (i.e., system architecture), as opposed to increasing element reliability (a change in design)?

(k) Develop a graph showing how the availability of the system changes as one increases the maximum flight time of the drone element from 3 to 7 h in 1 h steps (a change in design and possibly operational concept).

(l) Under what conditions would it make sense to trade an increase in maximum flight time for a decrease in number of drone elements per unit?

(m) Develop a graph showing how the availability of the system changes as one increases the time to station of the drone element from 0.1 to 0.5 h in 0.1 h steps (a change in operational concept).

13.2. Develop a "Scenario/Mission-based" Monte Carlo simulation for the FMDS using the parameter values provided in Table 13.3. In this case, one should model the takeoff, flight, landing, failure, and maintenance of each drone and the failings and replacement/repair of each control element over some time period of interest. This will include the generation of random failure times and maintenance times and the control logic that will cause one drone to take off to replace a drone returning to base (scheduled or unscheduled).

(a) How long did it take you to develop the analytical model used in Problem 1?

(b) How long did it take you to develop this working Monte Carlo model?

(c) Run the MC simulation until the mean value for the availability "settles down" (does not change by more than about 10%).

i. For what value of system time did this occur?

ii. What value of A_o was obtained?

(d) How does this result compare to the result obtained using the analytical model? If there are differences, what are some likely explanations of the differences?

(e) How long did you have to wait to obtain the Monte Carlo result? How long did it take to get an answer using the analytical model?

(f) Perform 100 repetitions of this Monte Carlo simulation and determine the Mean Availability and the standard error.

(g) How do these results differ from those obtained in c? Which is more likely the true mean? Explain.

(h) How do the results in f. differ from the analytical model? Are the differences significant? Why or why not?

(i) If the differences found in h. are significant:

i. Identify the likely sources of these differences.

ii. Indicate which result is more likely to be observed in a real operating system. Explain your answer.

(j) Under what conditions would you want to use the analytical model?

(k) Under what conditions would you want to use the Monte Carlo simulation?

(l) How might one use the two types of models in a synergetic manner?

13.3. Explore the cost-effectiveness of reducing the MDT_i associated with Effective Drone Failures with no standby available.

13.4. Explore the cost-effectiveness of increasing the MTBF for the control element (see Section 13.2). What costs would be affected by doing this? Develop a model for determining these costs.

13.5. Assuming a constant inflation rate of 2% per year, calculate the total O&S cost of the system described in Table 13.11 in "then-year" dollars.

13.6. Assuming a discount rate of 3%, calculate the net present value of the O&S cost of the system described in Table 13.11. (Hint: see net present value analysis is Chapter 4.)

13.7. What is the difference between the O&S cost calculated in "current-year" dollars and "then-year" dollars and net present value?

13.8. Suppose you are uncertain about the value of the following parameters, but expect them to be in the ranges indicated as follows:

EXERCISES

Parameter	Min	Max
System life	15 yr	30 yr
Operators per unit	4	6
Annual cost per personnel	$80,000/per	$130,000/per
Operations cost/hour	$30/yr	$60/hr
Mean time between critical failures (drone)	30 h	80 h
Ratio of drone failures requiring part replacements to critical failures	1	10
Mean cost to replace a drone part	$250/part	$500/part

(a) Develop a Pareto Chart that shows the degree to which uncertainty in these values can affect the Total O&S Cost.

(b) Which of these would it make the most sense to use in Monte Carlo cost simulation? Why?

13.9. Develop a Monte Carlo simulation for the cost model provided in Table 13.11. Assume a triangular distribution with the following properties, for the following parameters:

Parameter	Min	Most Likely	Max
Operations cost/hour	$30/yr	$40/h	$60/h
Mean time between critical failures (drone)	30 h	50 h	80 h
Ratio of drone failures requiring part replacements to critical failures	1	3.0	10

(a) Find the mean Total O&S Cost.

(b) Find the standard deviation associated with this mean.

(c) How many runs of the Monte Carlo simulation were required to get an appropriate mean and standard deviation? How did you determine this?

(d) What is the 80% confidence Total O&S Cost for this system?

13.10. Develop a TSLCC model for a system of interest to you.

(a) Determine the TSLCC in "current-year" dollars.

(b) Determine the TSLCC in "then-year" dollars.

(c) Perform a sensitivity study with respect to one or more input parameters.

(d) Perform a trade-off study with respect to one or more input parameters.

(e) Based on your trade-off study results, provide a recommendation as to what the program should do with respect to those values and the rationale for your recommendation.

REFERENCES

Amari, S. (2012) Reliability of k-out-of-n standby systems with gamma distributions. *IEEE Transactions on Reliability*.

Amari, S., Zuo, M.J., and Dill, G. (2008) O(kn) Algorithm for analyzing repairable and non-repairable k-out-of-n: G systems, in *Handbook of Performability Engineering* (ed. K.B. Misra), Springer, pp. 309–320.

Birolini, A. (2007) *Reliability Engineering: Theory and Practice*, 5th edn, Springer.

Blanchard, B. and Fabrycky, W. (2010) *Systems Engineering and Analysis*, 5th edn, Pearson.

Boddu, P. and Xing, L. (2012) Redundancy Allocation for k-out-of-n:G systems with mixed spare types. *IEEE Transactions on Reliability*.

CJCSI 3170.01I (2012) *Joint Capabilities Integration and Development System (JCIDS)*.

Defense, D. o (2014) Operating and Support Cost-Estimating Guide. Chapter 6, Office of the Secretary of Defense Cost Assesment and Program Evaluation.

DoD Instruction 5000.02 (2015) Operation of the Defense Acquisition System.

INCOSE (2015) *Systems Engineering Handbook: A Guide for System Life Cycle Processes and Activities*, 4th edn, Wiley.

Jacob, D. and Amari, S. (2005) Analysis of complex repairable systems. *IEEE Transactions on Reliability*.

Kuo, W. and Zuo, M. (2003) *Optimal Reliability Modeling: Principles and Applications*, Wiley.

Lad, B., Kulkarni, M., and Misra, K. (2008) Optimal reliability design of a system, in *Handbook of Performability Engineering*, Springer.

Morrison, M. and Munshi, S. (1981) Availability of a v-out-of-m+r:G system. *IEEE Systems Transactions on Reliability*, **R-30** (2), 200–201.

Sandler, G. (1963) *System Reliability Engineering*, Prentice Hall.

van Gemund, A. and Reijns, G. (2012) Reliability of k-out-of-n systems with single cold standby using Pearson distributions. *IEEE Transactions on Reliability*.

Wang, W. and Loman, J. (2012) Reliability/availability of k-out-of-n systems with m cold standby units. *Proceedings of Annual Reliability and Maintainability* XE "Maintainability" *Symposium*.

Zuo, M., Huang, J., and Kuo, W. (2003) Multi-state k-out-of-n systems, in *Handbook of Reliability Engineering* (ed. H. Pham), Springer.

14

PERFORMING PROGRAMMATIC TRADE-OFF ANALYSES

GINA GUILLAUME-JOSEPH
MITRE Corporation, McLean, VA, USA

JOHN E. MACCARTHY
Systems Engineering Education Program, Institute for Systems Research, University of Maryland, College Park, MD, USA

> It is not the beauty of a building you should look at; it's the construction of the foundation that will stand the test of time.
>
> David Allan Coe
>
> Great services are not canceled by one actor or by one single error.
>
> Benjamin Disraeli

14.1 INTRODUCTION

Systems engineering is an interdisciplinary field of engineering to design and manage complex engineering systems over their life cycle. Systems engineering defines the processes and methods from design, requirements engineering, performance, testing and evaluation, and many other aspects necessary for successful system development, implementation, and ultimate decommission.

System engineering is a robust approach to the design, creation, and operation of systems. In simple terms, the approach consists of identification and quantification of

---

Trade-off Analytics: Creating and Exploring the System Tradespace, First Edition. Edited by Gregory S. Parnell.
© 2017 John Wiley & Sons, Inc. Published 2017 by John Wiley & Sons, Inc.
Companion website: www.wiley.com/go/Parnell/Trade-off_Analytics

system goals, creation of alternative system design concepts, performance of design trades, selection and implementation of the best design, verification that the design is properly built and integrated, and postimplementation assessment of how well the system meets (or met) the goals.

NASA (1995)

As systems grow ever more complex, tools, methods, and processes are required to ensure their success throughout the life cycle. This chapter introduces methods that allow early detection of possible failures to allow systems engineers the ability to explore the issues early on the development process to make critical decisions. One very important component of systems engineering is modeling and simulation. Models can be defined as mathematical, conceptual, and physical tools to assist decision-makers in the systems engineering process. Models are used in trade studies to provide estimates of system effectiveness, performance, and cost based on a set of known or estimated quantities. Models help to identify a set of meaningful quantitative relationships among its inputs and outputs to express correlation and causality.

Section 14.2 develops a modeling framework to support analyses associated with the acceptance decision. It begins by providing a relatively simple decision tree model with three decision options (accept, reject, or fix and retest) and a number of uncertain events. Given input values for costs and probabilities, the model may be used determine the decision option with the lowest expected life cycle cost and to perform associated sensitivity and trade studies. The section then develops models that may be used to calculate values for the probabilities associated with each uncertain event and the costs associated with each branch of the decision tree from data that is typically available for projects following an acceptance test. The section also includes a discussion of hypothesis testing and its use in the design of an acceptance test. The section provides examples of use of these models to perform associated sensitivity and trade studies.

Section 14.3 highlights the system's mindset in addressing failure to introduce a software-specific predictive analytics model that accurately predicts software project outcomes of failure or success and identifies opportunities for incorporation in the federal and commercial space. The section describes how to systematically learn from historical software project failures to train a confidence level model to make project failure predictions. The section demonstrates how to train a generalized model with a range of failure factors combined to efficiently and accurately predict software project failures. The results of the model would be used during acquisition, prior to project initiation, and throughout the software development life cycle. It is a decision analysis tool to assist decision-makers in making the crucial decisions early in the life cycle to cancel a project predicted to failure or to identify and implement mitigation strategies to improve project outcome.

Section 14.4 describes the methods to effectively decommission legacy systems. An analysis was conducted on several Human Resources (HR) applications within the Department of Treasury portfolio of legacy applications to determine whether or not retiring several of the legacy systems written in COBOL and/or Assembly Language was the best course of action. A cost–benefit analysis coupled with net present value (NPV), return on investment (ROI), and break-even analysis was performed to come to a decision to retire the systems. The Ship and Submarine Recycling Program (SRP) is the US NAVY retirement and decommission program for nuclear-powered vessels. Since 1954, nearly 200 submarines have be deployed into production, and many have new reached their service life as well as their safety window where their time in operation does not justify their high maintenance operating cost. Before SRP can begin, the NAVY has instituted critical steps to be carried out at designated facilities. These steps are required because improper handling of the nuclear reactors of these vessels could pose a major environmental catastrophe. A decommissioning concept called rigs-to-reefs for retiring Offshore Oil and Gas Platforms in California allows leaving the decommissioned platforms in place to become a habitat for sea life. These massive structures span the size of a football field with carbon footprints levels so large that air quality regulatory agencies would not issue the permits to completely remove them from the seabed. The PLATFORM tool was developed to determine the best decommissioning alternative for 27 oil and gas platforms nearing their service life. The tool analyzed the impact of each decommission decision and strategy based on costs, fishery production, and air emissions.

Decision-makers will benefit from an end-to-end collaborative decision management process that engages all stakeholders from the design to the decommissioning and retirement of systems, and this chapter identifies some of the tools and methods to support sound decisions throughout the system's life cycle.

14.2 SYSTEM ACCEPTANCE DECISIONS AND TRADE STUDIES

The acceptance decision is one of the most important decisions made by a customer. Typically, 80% of a system's life cycle cost is incurred by the postdevelopment costs that follow this decision (INCOSE, 2015). In this section, we develop a relatively simple modeling framework for the acceptance decision that may be used to support trades studies on the expected costs associated with accepting a system, delaying acceptance of a system, or rejecting the system based on the results of an acceptance test and estimates of the cost to correct and retest a system that does not pass an acceptance test. The same modeling framework may be used to support the setting of acceptance criteria, the design of an acceptance test, and performing related sensitivity studies.

Section 14.2.1 develops an "acceptance decision model" that may be used to guide trade studies related to the acceptance decision. The remainder of the section provides more complex models that permit one to derive values for the inputs to this acceptance decision model. Specifically, Section 14.2.2 develops the theoretical framework to determine the "Confidence" that a system's actual performance equals or exceeds some specified threshold value, given a mean value for that metric obtained from an acceptance (or other) test. The results of this section may be used to provide values for two of the independent uncertain factors that are inputs to the decision model developed in Section 14.2.1. Section 14.2.3 develops the theoretical framework that may be used to "design an acceptance test." More specifically, it may be used to determine how long one must test a system (T_d) of actual MTBF (M_a) to obtain a desired Power and Confidence value associated with demonstrating that a system has achieved a given required MTBF (M_r). This theoretical framework may be used to perform trade studies in which T_d, M_a, and M_r may be traded to achieve different values of Power and Confidence. The results of this section, when combined with the modeling framework developed in Section 14.2.4, may be used to establish estimates for the remaining parameters associated with the acceptance decision model. Section 14.2.4 develops a cost model that may be used to estimate the cost implications of different acceptance test design approaches. Section 14.2.5 provides an example of a trade study that integrates the models developed in Sections 14.2.1–14.2.4 and shows how this modeling framework may be used to carry out a wide range of important system trade studies for testing and acceptance decisions. Finally, Section 14.2.6 provides some conclusions.

The models developed in these sections are based on simplifying assumptions and fictional data. They are meant to illustrate useful modeling and analysis techniques and to show how they may be combined to provide an integrated acceptance decision modeling framework that may be extended to real systems based on real data and fewer simplifying assumptions.

14.2.1 Acceptance Decision Framework[1]

One of the most important customer decisions associated with a system is whether to (i) move forward with production and deployment of a system ("Accept" the system), (ii) cancel continued development/procurement of a system ("Reject" the system), or (iii) delay the acceptance decision until flaws in system design or manufacturing have been corrected and a second acceptance test is performed. In this section, we develop a simple parametric model that may be used to structure and perform trade studies acceptance decisions.

While we want to "Accept" a "Good" system, there are generally significant costs associated with rejecting a system or accepting a "Bad" system. In this section, we consider these costs in a simple and systematic way. We begin by identifying and defining the metric of interest and the factors that influence this metric (i.e., we will identify and define the value function and associated trade space).

For the purpose of this model, the value function is the expected value for the sum of the "future costs" (C_f) associated with production of the system (C_p), delay in the acceptance of the system (C_d), and additional acceptance-related design and testing of the system (C_{dt}), that is,

$$C_f = C_p + C_d + C_{dt}.$$

---

[1]The "Ch 14.2 System Acc Dec Tree.xlsx" Excel file contains the models used to support the analyses performed in this section.

SYSTEM ACCEPTANCE DECISIONS AND TRADE STUDIES

For this model, we make the following simplifying assumptions:

1. All calculations are performed in "base year dollars."
2. The performance of the system may be characterized as "Good" or "Bad."
3. If the system is "Bad," it will not be used; if it is "Good," it will be used.
4. An acceptance test has been performed.
5. Based on the acceptance test data, we can calculate the probability that the system is "Good" ($P(G)$).

There are three decision options available: Accept the system, Fix and Retest the system, or Reject the system (and cancel the program).

1. We only execute one Fix and Retest cycle (i.e., the only decision options following Fix and Retest are Accept or Reject).
2. There is a fixed annual cost associated with not having/using a "Good" new system in place (which may be associated with the additional cost of maintaining an existing old system or with "lost opportunity"). This is the "annual delay cost" (C_{da}).
3. Execution of the Fix and Test option will result in a delay in the deployment date for T_d years (if the fixed system is accepted).
4. There are two costs associated with the Fix and Test option: a fix and test cost (C_{ft}) and a delay cost ($C_d = T_d * C_{da}$).
5. If the system passes the acceptance test associated with the Fix and Test option, the customer will accept the system.
6. A "Good" system will not require any product improvements over the course of its life.
7. If a system is "Accepted," it will begin full production (regardless of whether it is good or bad).
8. If a system is "Accepted," there will be no product improvements over its lifetime (regardless of whether it is good or bad).

Note that if one or more of these assumptions are invalid, we can derive a more complex model through modifications of the simple model.

Given the aforementioned set of assumptions, there are a total of six possible outcomes. Table 14.1 identifies these outcomes (or end states) and indicates the cost associated with each outcome. In a later section, we will provide a framework for estimating these costs.

Given an understanding of these costs and the potential outcomes, we may express this decision problem as the decision tree provided in Figure 14.1.

This decision tree provides a framework (or pattern) that may be used to evaluate a wide range of acceptance decisions. It is characterized by one decision node, three uncertain events (each with two associated probabilities), and six potential outcomes (or end states).

Table 14.1 Outcome Future Costs

	Outcome	Future Cost
1	Accept a Good System (no F&T)	$C_{tf} = C_{tp}$
2	Accept a Bad System (no F&T)	$C_{tf} = C_{tp} + T_1 * C_{da}$
3	Reject a System (G or B, no F&T)	$C_{tf} = T_1 * C_{da}$
4	Accept a Good System (F&T cycle)	$C_{tf} = C_{tp} + C_{ft} + T_{ft} * C_{da}$
5	Accept a Bad System (F&T cycle)	$C_{tf} = C_{tp} + C_{ft} + T_1 * C_{da}$
6	Reject a System (G or B, F&T cycle)	$C_{tf} = C_{ft} + T_1 * C_{da}$

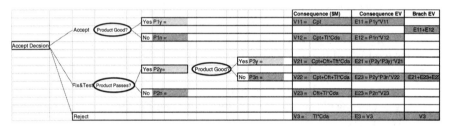

Figure 14.1 Generic decision tree for acceptance decision

To solve this decision tree (to find the decision sequence that will result in the lowest expected cost), we need to know the cost associated with each end state (provided in Table 14.1) and the values for the six probabilities. Table 14.2 defines these probabilities and the relationships between them. In a later section, we provide a framework for estimating $P(G)$, $P(P)$, and $P(G|P)$.

Given this framework, the Expected (Future) Cost ($E(C_{tf})$) associated with each path may be determined. Table 14.3 provides the equations that may be used to calculate these expected costs. Upon calculating these expected costs, we select the decision option with the smallest Expected Cost.

We can perform these calculations by hand, implement them in a spreadsheet, or use a decision tool such as PrecisionTree™.

Now we consider an example of how this decision framework may be used. Table 14.4 provides example values for the parameters of interest (yellow background indicates a model input, green background indicates a calculation (*the reader is referred to the online version of this book for color indication*)). Figure 14.2 provides an Excel decision model that was used to solve the problem.

Figure 14.2 indicates that for this set of input parameters, the expected cost for accepting the system without a fix/test cycle is ~$160 M, the expected cost of the Fix/Test option is ~$183 M, and the expected cost of rejecting the system (without a fix/test cycle) is ~$200 M. Based on these inputs, we would accept the system (without a fix/test cycle).

We can also use the information provided in the aforementioned decision tree to examine the risk associated with each decision option. This is done in Figure 14.3.

SYSTEM ACCEPTANCE DECISIONS AND TRADE STUDIES

Table 14.2 Decision Tree Probabilities

	Probability	Notation
1	Probability that the system is **Good** (given it is accepted based on the results of the first acceptance test)	$P_1(G)$
2	Probability that the system is **Bad** (given it is rejected based on the results of the first acceptance test)	$P_1(B) = 1 - P(G)$
3	Probability that the system will **Pass** the second acceptance test (given improvements in the design/manufacturing of the system and the design of the second acceptance test)	$P_2(P)$
4	Probability that the system will Fail the second acceptance test	$P_2(F) = 1 - P(P)$
5	Probability that the system will be **Good**, given it **Passed** the second acceptance test	$P_3(G|P)$
6	Probability that the system will be **Bad**, given it **Passed** the second acceptance test	$P_3(B|P) = 1 - P(G|P)$

Table 14.3 Decision Tree Expected Costs

	Decision Option	Expected Cost
1	Accept	$E_1(C_{tf}(A_{cc})) = P_1(G)\, C_{tp} + P_1(B)\, (C_{tp} + T_1 * C_{da})$
2	Reject	$E_3(C_{tf}(R_{ej})) = T_1 * C_{da}$
3	Fix and Test	$E_2(C_{tf}(F\&T) = P_2(P)\, [P_3(G|P)*(C_{pt}+C_{ft}+T_{ft}*C_{da})$ $+ P_3(B|P)*(C_{pt}+C_{ft}+T_1*C_{da})] + P_2(F)\,[C_{ft} + T_1 * C_{da}]$

Table 14.4 Example Decision Tree Inputs

Times	
Decision Horizon (yrs):	20
Fix/Test Delay Time (yrs):	2
Costs	
Cost to Produce System ($M):	100
Cost of Fix/Test Cycle ($M):	20
Annual Delay Cost ($M)	10
Probabilities	
Prob Syst Good, Given Passed 1st Test:	0.70
Prob Syst Bad, Given Passed 1st Test:	0.30
Prob Syst Passes 2nd Test (assumes me	0.85
Prob Syst Fail 2nd Test:	0.15
Prob Syst Good, Given Passed 2nd Test	0.80
Prob Syst Bad, Given Passed 2nd Test:	0.20
Prob Syst Bad, Given Passed 2nd Test:	0.80

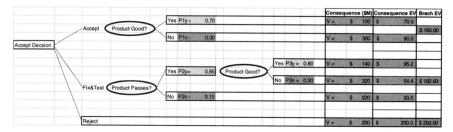

Figure 14.2 Excel model used to solve the example problem

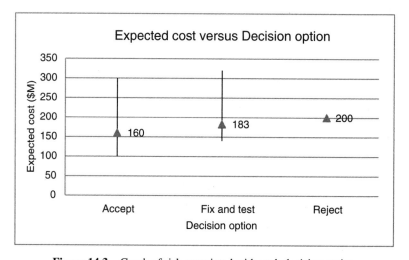

Figure 14.3 Graph of risk associated with each decision option

This graph shows the high, low, and mean cost associated with each decision option. The graph indicates that while the "Accept" decision has the lowest expected cost, there is a possibility that its cost could exceed cost associated with the "Reject" option.

The acceptance decision model developed in this section may be used to: (i) clarify what parameters are important to the acceptance decision; (ii) identify what studies need to be performed to determine the costs and probabilities relevant to the decision (i.e., the model inputs); (iii) determine how sensitive the result is to uncertainties in input costs and probabilities; and (iv) perform trade studies with respect to the degree to which system performance should be improved and the design of the second acceptance test. We will examine this last use in more detail in a following section.

One can also examine the risk associated with the three alternatives and look for deterministic and/or stochastic dominance by comparing the cumulative cost profiles of each.

SYSTEM ACCEPTANCE DECISIONS AND TRADE STUDIES   491

14.2.2 Calculating the Confidence[2] That a System Is "Good"[3]

This section develops the theoretical framework required to determine a confidence that the actual mean (M_a) of a metric (e.g., the Mean Time between Failures (MTBF)) is greater than some specified threshold (required) value (M_r), given a measured (observed) mean (M_o) obtained from a set of N_o measurements, that is, it indicates how confident we may be that the system is "Good." While not strictly a probability, this confidence value may be used to provide an estimate for $P(G)$ in the decision model developed in Section 14.2.1.[4]

In general, we obtain an observed value for a system's MTBF (M_o) by measuring the number of observed failures (N_{fo}) that occur over the course of a test that is performed over a time period of T_d operational hours ($M_o = T_d/N_{fo}$). In this case, we use the right-tailed Chi-squared function to establish a one-sided Confidence Interval in which we may be $C\%$ confident that the system's actual MTBF (M_a) will be larger than some lower bound on the MTBF (M_l). The following equation[5] expresses the relationships between the key parameters for "Time-Terminated" tests[6] (Ebeling, 2010; O'Connor & Kleyner 2012):

$$M_l = \frac{2T_d}{\chi^2_{\alpha, 2(N_{fo}+1)}} = \frac{2T_d}{\text{CHIINV}(\alpha, 2(N_{fo}+1))} \qquad (14.1)$$

where $\alpha^7 = 1 - C$ and CHIINV is the Excel right-tailed inverse Chi-squared function.

---

[2]The term "Confidence" here has a different meaning in this section than it does in Section 14.2.3. Here it refers to the confidence that the system is "good," given an observed measure of the mean of the parameter of interest (in this case the system's MTBF). In Section 14.2.3 the term refers to the probability that the system will fail an acceptance test, given the system is "bad."
[3]The "Ch 14.2 Test Analysis.xlsx" Excel file contains the models used to perform the analyses provided in this section.
[4]Technically, the term probability only refers to random events [Navidi]. The actual value for M (M_a) is what it is (it is not random). As such, Confidence (C) may be more accurately viewed as the probability that the method used to calculate the one-sided confidence interval has probability C of including M_a.
[5]Different equations are used to calculate confidence intervals depending on the nature of the parameter being measured. This expression is appropriate for measuring Poisson processes. The t distribution and binomial distribution are used for parameters that have Normal and Binomial properties. See O'Connor & Kleyner (2012) and MIL-HDBK-781A (1996).
[6]A time-terminated (Type I) test is one that ends after a fixed number of operational hours (Ebeling, 2010; O'Connor & Kleyner 2012; MIL-HDBK-781A, 1996). A "failure-terminated" (Type II) test is one that is terminated following the kth failure. For a failure terminated test, $M_l = \frac{2T_d}{\chi^2_{\alpha, 2 \text{nfo}}} = \frac{2T_d}{\text{CHIINV}(\alpha, 2N_{fo})}$. For a Type I test, the number of degrees of freedom is $v = 2(N_{fo}+1)$. For a Type II test, $v = 2N_{fo}$.
[7]Technically, α is the "p-value" associated with the null hypothesis that $M_a < M_l$ and the measurement N_{fo} over time T_d. α is the probability that one will observe N_{fo} over time T_d, given $M_a < M_l$, that is, it is a measure of the degree to which the evidence supports the null hypothesis (that the system is "bad"). As such, $C = 1 - \alpha$ is a measure of the degree to which the evidence supports the "alternative hypothesis" that the system is "good" ($M_a \geq M_l$) (Navidi, 2011).

To find the confidence that the $M_a > M_r$, given a measured value of M_o, we use the aforementioned equation to construct a table or graph of M_1 versus C. One then finds the value of C that corresponds to $M_1 = M_r$.

As an example, consider the case where $M_r = 100\,\text{h}$, $T_d = 4000\,\text{h}$, and we measure $N_{of} = 34$ failures (so $M_o = 117.6\,\text{h}$). We use these values and the aforementioned equations to construct Table 14.5.

This table indicates the confidence one may have that the actual MTBF is in a one-sided confidence interval whose lower bound is M_1, given $N_{of} = 34$ and $T_d = 4000$ (i.e., $\mathbf{M_o = 117.6\,h}$). It also indicates the values for α and CHIINV() that were used to calculate M_1. The values in **red bold** indicate that we may be 80% confident that given what was measured, the actual value for the system's MTBF (M_a) is greater than 100.4 h (M_r), that is, that the system meets its requirement.

The parameters associated with the aforementioned equation are summarized in Table 14.6.

Table 14.5 Table of MI versus C (Time-Terminated Test, $T_d = 4000\,\text{h}$ and $N_{fo} = 34$)

$C = 1 - \alpha$	α	CHIINV (Rt Tailed)	M_1
0.1	0.9	55.3	144.6
0.2	0.8	59.9	133.6
0.3	0.7	63.3	126.3
0.4	0.6	66.4	120.5
0.5	0.5	69.3	115.4
0.6	0.4	72.4	110.6
0.7	0.3	75.7	105.7
0.8	**0.2**	**79.7**	**100.4**
0.9	0.1	85.5	93.5

Table 14.6 Parameters Used to Calculate C

Type of Parameter	Symbol	Parameter	Value
Probability	Confidence that $M_a > M_r$ (i.e., the system is good)	C	80%
Design Factor	Test Duration	T_d	4000 h
Measurement	Observed Number of Failures	N_{fo}	34 f
Requirement	Required MTBF	M_r	100 h

14.2.3 Acceptance Test Design and Trade Studies[9]

In this section, we develop the theoretical framework required to design an acceptance test that has a desired Confidence and Power, based on M_r, T_d, the expected MTBF (based on the system design), and the test acceptance threshold MTBF value (M_{at}).

Before we discuss the theory behind this aspect of test design, we need to define a number of related concepts and terms. **Consumer risk** is the probability of accepting a bad system ($P(A|B)$). **Producer risk** is the probability of rejecting a good system ($P(R|G)$). Generally, we want to design a test that minimizes both consumer risk and producer risk.

This aspect of test design is related to the theory of Hypothesis Testing. Generally, in hypothesis testing, we define two hypotheses – the "null hypothesis" (H_0) and the "alternative hypothesis" (H_1). By convention, the "no effect" or "reference" hypothesis is taken as the null hypothesis. Since from the consumer's perspective, one has to prove the system is "good," the reference hypothesis (H_0) is that $M_a < M_r$ (i.e., the system is "bad").

Given this selection of the null hypothesis, we define four potential outcomes for a test. Each outcome may be expressed in a number of ways, and each outcome has an associated conditional probability. Table 14.7 provides a summary of each potential outcome and the terms and probabilities that are used to describe it.

Note the conditional probability $P(P|G)$ should be read as "the probability that the system will Pass the test, **given** the system is Good."

Table 14.7 Expressions of Possible Outcomes of Tests and Associated Probabilities

Practical Outcome	Hypothesis Outcome	Term	Associated Conditional Probability
A good system passes the acceptance test	The null hypothesis is correctly rejected	True negative (H_0 is correctly rejected)	$P(P\|G) = 1-P(F\|B)$ = Power = $P = 1-\beta$
A bad system passes the acceptance test	The null hypothesis is incorrectly rejected	False positive Type I error	$P(P\|B) = \alpha$
A good system fails the acceptance test	The null hypothesis is incorrectly accepted	False negative Type II error	$P(F\|G) = \beta$
A bad system fails the acceptance test	The null hypothesis is correctly accepted	True positive (H_0 is correctly accepted)	$P(F\|B) = 1-P(P\|B)$ = Confidence = $C = 1-\beta$

---

[8]The "Ch 14.2 Test Analysis.xlsx" Excel file contains the models used to perform the analyses provided in this section.
[9]Recall that the term "Confidence" has a different meaning in this section that it does in Section 14.2.2. See footnote in Section 14.2.2.

Table 14.8 Matrix of Possible Test Outcomes of Tests and Associated Probabilities

	H_o is True (System Is Bad)	H_o is False (H_1 Is True, System Is Good)
System passes test (P) (rejecting H_o)	False positive (Type I error) $P(P|B) = \alpha$	True negative $P(P|G) =$ Power
System fails test (F) (accepting H_o)	True positive $P(F|B) =$ Confidence	False negative (Type II error) $P(F|G) = \beta$

Table 14.9 Test Design Trade Space

Type of Parameter	Symbol	Parameter
Probability	Confidence ($P(F|B)$)	C
Probability	Power ($P(P|G)$)	P
Design factor	Test duration	T_d
Design factor	Acceptance threshold number of failures (MTBF threshold $M_{at} = T_d/N_{fat}$)	N_{fat}
Design factor	Anticipated (design) MTBF	M_d
Calculation factor	Actual MTBF	M_a

Table 14.8 provides another view of the relationships between hypothesis, test results, terms, and the associated conditional probabilities and symbols.

Generally, we want to design a test that has both a Confidence and a Power ≥ 0.8. The Confidence and Power associated with a given test design may be determined by examining a "Receiver Operating Characteristic" (ROC) curve or table for that design. Table 14.9 provides a summary of the parameters associated with generating the ROC curve/table (that serve as the test design's "trade space"). The "Type of Parameter" column indicates the parameters that define the aspects of the test's design relevant to this discussion.

The ROC curve/table may generated as follows:

1. Specify T_d and N_{fat}.
2. Using the cumulative Poisson Distribution to develop a table/curve that indicates the probability of accepting the system ($P(A)$) versus different values for the actual MTBF (M_a), given the specified T_d and N_{fat}[10] (i.e., calculate $P(A|T_d, N_{fat}; M_a)$)

$$P(A) = \sum_{i=0}^{N_{fat}} e^{-T_d/M_a} \frac{\left(\frac{T_d}{M_a}\right)^i}{i!} \quad (14.2)$$

(or $P(A) = \text{POISSON}((N_{fat}),(T_d/M_a),1)$ using Excel)

[10] Note that $M_{at} = N_{fat}/T_d$ is the "acceptance test threshold MTBF."

SYSTEM ACCEPTANCE DECISIONS AND TRADE STUDIES 495

3. The values for C, P, α, and β may be determined from the table/graph as follows:
   - α is determined by finding the P(A) where $M_a = M_r$ and $C = 1 - \alpha$
   - P is determined by finding the P(A) associated with $M_a = M_d$ and $\beta = 1 - P$.

As an example, consider the case where $T_d = 4000$ h, $N_{fat} = 34$ ($M_{at} = 117.6$), and $M_r = 100$ h (the design basis for the test considered in the previous section). Table 14.10 is generated using equation 14.2 and varying the N_{fat} over a range of potential "actual" values of M_a (which would result in an expected $N_{fe} = T_d/M_a$). The value of P(A) is the probability that number of observed failures (N_{fob}), will be less than the N_{fat} value, given the actual MTBF is $M_a(=T_d/N_{fe})$). We see for $M_a = M_r = 100$ h, $P(A) = 0.194$. This indicates that this design is characterized by $\alpha = 0.194 = P(P|B)$ and a **Confidence** of **C = 0.806**, that is, the probability of passing a bad system is $P(P|B) = 0.194$ and the probability of failing a bad system is $P(F|B) = 0.806$. We also see that if we expect our current design to provide an actual MTBF of $M_{d1} = M_a = 143$ h, $P(A) = 0.888$. This indicates that the **Power** of the test design is $P = 0.89$ and $\beta = 0.11$, that is, the probability of passing a good system is $P(P|G) = 0.89$ and the probability of failing a good system is $P(F|G) = 0.11$.

Table 14.10 Receiver Operating Characteristic Table ($T_d = 4000$ h, $N_{fat} = 34$)

$M_a = T_d/N_{fe}$	N_{fe}	$P(A) = P(N_{fob} \leq N_{fat}|N_{fe})$ $= P(M_{ob} \geq M_{at}|M_a)$
200.0	20	0.9985
181.8	22	0.9936
166.7	24	0.9794
153.8	26	0.9472
142.9	**28**	**0.8879**
133.3	30	0.7973
125.0	32	0.6792
117.6	34	0.5454
111.1	36	0.4115
105.3	38	0.2914
100.0	**40**	**0.1939**
95.2	42	0.1214
90.9	44	0.0718
87.0	46	0.0401

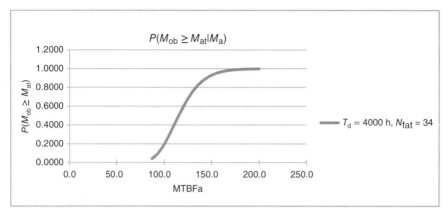

Figure 14.4 Receiver operating characteristic curve (for $T_d = 4000$ h, $N_{fat} = 34$)

Figure 14.4 provides the curve generated by Table 14.10. It may be used to come to the same conclusions. Note that $P(A) = 0.5$ when the actual MTBF is equal to the threshold MTBF ($M_a = M_{at}$).

Let us now consider three specific examples of the many types of test design trade studies that may be performed using equation 14.2.

Our first trade study involves understanding the trade between the Power of a test design and the expected (design) MTBF (M_d). Suppose that as we get closer to the date for the acceptance test designed earlier, developmental test data indicates that the expected actual MTBF of the system (M_d) is expected to be $M_{d2} \sim 125$ h (vs $M_{d1} = 143$ h), which is still greater that M_r and therefore "good." By examining Table 14.10 (or Figure 14.4), we note that this reduction in the expected M_a reduces the probability that the system will pass the acceptance test as it is currently designed (even though the system is "good"). Specifically, we see that the Power of the test drops from $P \sim 0.89$ to $P \sim 0.68$.

As a second trade study, let us consider how we might change the design of the test to increase the probability of passing the "good" system (i.e., increasing the test's Power). One way to do this would be to reduce the acceptance MTBF threshold. Again we consider the case where $T_d = 4000$ h, and $M_r = 100$ h. Let us suppose that we are willing to accept a value of $P = 0.80$. We want to use equation 14.2 to find a value for M_{at} that will result in a value of $P \geq 0.80$, given we now expect $M_a = M_{d2} = 125$ h (vs 143 h). While there are a variety of ways to do this, we will select a method that allows us to compare the ROC tables/curves that result from alternative test designs. To do this, we generate ROC tables (like 14.2.10) for decreasing values of N_{fat} until we find one for which $M_a = M_{d2} = 125$ h yield a $P(A) \sim 0.8$. Table 14.11 provides the result of such a search and Figure 14.5 compares the ROC curve obtained from the new test design (where $N_{fat2} = 36$ or $M_{at2} = 111$ h) with that obtained using the previous (reference) test design (where $N_{fat1} = 34$ or $M_{at1} = 143$ h).

We can see that the second test design (with the lower acceptance threshold of $M_{at2} = 111$ h) provides a Power of ~ 0.80. However, we also see that this new test

SYSTEM ACCEPTANCE DECISIONS AND TRADE STUDIES 497

Table 14.11 Receiver Operating Characteristic Table ($T_d = 4000$ h, $N_{fat1} = 34$, $N_{fat2} = 36$)

$M_{a12} = T_{d12}/N_{fe12}$	N_{fe12}	$P(A) = P(N_{fob} \leq N_{fat1}\|N_{fe})$ $= P(M_{ob} \geq M_{at1}\|M_a)$	$P(A) = P(N_{fob} \leq N_{fat2}\|N_{fe})$ $= P(M_{ob} \geq M_{at2}\|M_a)$
200.0	20	0.9985	0.9996
181.8	22	0.9936	0.9978
166.7	24	0.9794	0.9918
153.8	26	0.9472	0.9756
142.9	28	**0.8879**	0.9411
133.3	30	0.7973	0.8804
125.0	**32**	0.6792	**0.7901**
117.6	34	0.5454	0.6744
111.1	36	0.4115	0.5442
105.3	38	0.2914	0.4138
100.0	**40**	**0.1939**	**0.2963**
95.2	42	0.1214	0.1999
90.9	44	0.0718	0.1273
87.0	46	0.0401	0.0766

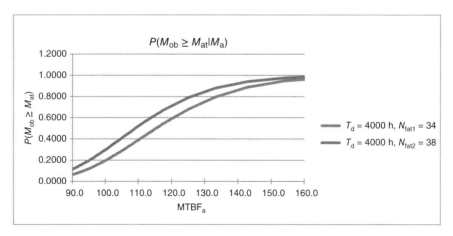

Figure 14.5 Receiver operating characteristic curves (for $T_d = 4000$ h, $N_{fat1} = 34$, $N_{fat2} = 38$)

Table 14.12 Finding T_d and N_{fat} that Yield Desired Confidence and Power ($M_{d3} = 125$ h)

T_d	N_{fat}	$M_{at} = T_d/N_{fat}$	$P(A|M_a \leq M_r)$	$P(A|M_a \geq M_d)$
4000	34	117.6	0.1939	0.6792
4500	38	118.4	0.1665	0.6699
5000	43	116.3	0.1798	0.7162
5500	48	114.6	0.1917	0.7556
6000	52	115.4	0.1666	0.7465
6500	57	114.0	0.1767	0.7801
7000	**62**	**112.9**	**0.1860**	**0.8091**
7500	67	111.9	0.1945	0.8340
8000	71	112.7	0.1713	0.8265

design has decreased the test's Confidence ($P(R|B)$) from 0.80 to ~0.70, that is, by changing the design we traded Confidence for Power.

As a third trade study, consider how we might change the design of the test to increase the probability of rejecting a bad system (the test's Confidence), without reducing its Power. This may be done by extending the duration of the test (T_d). To see this, consider the case where we expect M_a to be $M_{d3} = M_{d2} = 125$ h and $M_r = 100$ h. We want to keep $P \sim 0.8$, but want to increase C to ~0.8. One approach to finding such a design is to develop a table of $P(A|M_a = M_r)$ and $P(A|M_a = M_d)$ versus T_d and N_{fat} (and M_{at}). In this table, we look for the lowest value of T_d for which a value of N_{fat} (and M_{at}) can be found such that $1 - C < 0.20$ and $P > 0.80$. Table 14.12 illustrates this approach and was developed using the following algorithm:

- The table begins with the initial test duration ($T_d = 4000$ h) and is increased by fixed increments (in this case $dT_d = 500$ h)
- For each value of T_d:
  o The N_{fat} is increased until one finds the lowest value that will provide a confidence > 0.8 (or $P(A|M_a < M_r) = 1 - C < 0.2$)
  o The value the Power ($P(A|M_a = M_d)$) is checked
  o If P is greater than the desired Power, one has found the combination of lowest value of T_d (and the associated M_{at}) that achieves the C and P objectives.
  o If P is less than the desired Power, one moves to then next value for the test duration.

Table 14.12 indicates that a test duration of $T_{d3} = 7000$ h and an acceptance threshold of 62 failures ($M_{at} = 113$ h) is required to achieve a $C \sim 0.81$ and a $P \sim 0.81$ (given

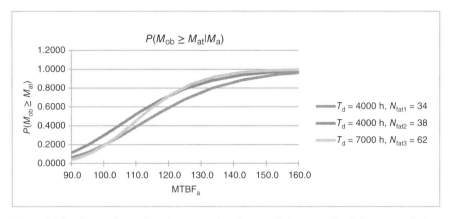

Figure 14.6 Comparison of receiver operating characteristic curves for different test designs

the expected MTBF is $M_d = 125$ h). Such a table may be used to perform a variety of trades related to T_d, M_d, and desired C and P.

To complete our analysis of this case, we compare the ROC curve that results from this third test design ($T_{d3} = 7000$ h, $N_{ft3} = 62$ ($M_{at} = 113$ h)) to the ROC curves associated with the previous two designs. This comparison is provided in Figure 14.6.

We can see that the third ROC curve permits us to achieve both goals of a high C and a high P (given a reduced expected $M_d = 125$ h), but it does so at the cost of increased test duration. In the next section, we develop a cost model that will enable us to assess the costs of an increase in the test duration.

14.2.4 A "Delay, Fix, and Test" Cost Model[12]

This section develops a theoretical framework that may be used to estimate the costs associated decision model described in Section 14.2.1 (with a special focus on the three outcomes associated with the "Fix and Test" decision option). The aim of the models developed in this section is to provide the simplest set of models that are relatively complete with respect to the variables that affect the acceptance decision and to illustrate common cost estimating relationship techniques.

We begin by noting that the total "Delta Cost" (C_{dt}) associated with each branch of the "Fix and Test" option is a sum of three delta cost components. The first component is the "Direct Delta Cost" (C_{dd}) associated with performing the fix and test activities. The second component is the "Production Delta Cost" (C_{dp}), which reflects any increase in the "per unit cost" that results from implementing the fixes. The third component is the "Schedule Delay Cost" (C_{sd}) associated with any additional costs of extending the duration of the development contract and the "lost opportunity" cost

---

[11] The "Ch 14.2 Acc Dec Cost Model.xlsx" Excel file contains the models used to perform the analyses provided in this section.

associated with a delay in the fielding of the system. The total delta cost is then

$$C_{dt} = C_{dd} + C_{dp} + C_{sd} \qquad (14.3)$$

The "Direct Delta Cost" (C_{dd}) is itself the sum of four components – the delta development cost to fix the design (C_{ddes}), the delta production cost associated with fixing the test units (C_{dtu}), the delta cost associated with development testing (DT) of the fixes (C_{ddt2}), and the cost of performing an additional acceptance/operational test (OT) (C_{ot2}), that is,

$$C_{dd} = C_{ddes} + C_{dtu} + C_{ddt2} + C_{ot2} \qquad (14.4)$$

While there are many ways in we may model the components of C_{dd}, Table 14.13 provides a summary of the mathematical models and assumptions used in this chapter. The table is followed by a discussion of the rationale for each equation.

Equation 14.5 (for C_{ddes}) assumes that majority of the direct cost associated with modifying the design of the system will be "design team" labor costs. While the average cost/staff month (R_{mdf}) is relatively easy to determine, it is somewhat more

Table 14.13 Simple Mathematical Models for the C_{dd} Components

Equation #	Equation	Parameter Definitions
(14.5)	$C_{ddes} = R_{smfd} * S_{mfd}$	R_{smdf} = Cost per staff month to fix design S_{mfd} = Staff months to fix design
(14.6)	$S_{mfd} = S_{mfdo} + A_{df}((M_{dg}/M_c)^{B_{df}} - 1)$	S_{mfdo} = Fixed develop time in staff months A_{df} = Development time coefficient M_{dg} = MTBF design goal M_c = Current MTBF (from most recent test) B_{df} = Development time exponent
(14.7)	$C_{dtu} = N_{tu} * C_{dmfuc}$	N_{tu} = number of test units C_{dmfuc} = cost to fix test units
(14.8)	$C_{dmfuc} = A_{mf}((M_{dg}/M_c)^{B_{df}} - 1)$	A_{mf} = production coefficient B_{mf} = production exponent
(14.9)	$C_{ddt2} = C_{ddt1}$	C_{dtt1} = cost of the last development test
(14.10)	$C_{ot2} = C_{otf2} + R_{ot2} * T_{dot2}$	C_{otf2} = fixed costs of the operational test (OT) R_{ot2} = Cost per operational hour of the OT T_{dot2} = duration of second OT in operational hours

challenging to estimate the staff months (sm) that will likely be required to "fix" the design (S_{mfd}).

Equation 14.6 (for S_{mdf}) assumes that there are two components. One is a fixed number of sm that will be incurred regardless of the magnitude of the change required (S_{mfdo}) and a second component that depends on the magnitude of the desired change. While there are many functions could be selected for the second component, in this chapter, we assume the rather simple expression "$A_{df} ((M_{dg}/M_c)^\wedge B_{df} -1)$." While in principle, values for A_{df} and B_{df} may be determined from a regression analysis (were such data available), in this chapter, we assume (somewhat arbitrarily) that $A_{df} = 0.1*C_{des1}$ (the Design Costs to date) and $B_{df} = 1$.

Equation 14.7 (for C_{dtu}) assumes that there is a per unit cost associated with modifying each test unit to reflect the redesign (C_{dmfuc}). While there are many functions that we could select to calculate C_{dmfuc}, equation 14.8 assumes that the greater the required design change, the greater will be cost of the modification. Equation 14.8 follows equation 14.6 by using rather simple expression "$A_{mf} ((M_{dg}/M_c)^\wedge B_{mf} -1)$." While in principle, values for A_{mf} and B_{mf} may be determined from a regression analysis (if the data is available), we assume (again somewhat arbitrarily) that $A_{mf} = 0.1*C_{uc1}$ (the expected unit cost prior to making the design changes) and $B_{mf} = 1$.

Equation 14.9 (for C_{ddt2}) assumes that the DT costs associated with verifying that the implemented design changes meet their requirements would be about the same as the cost of the DT that was performed to verify the implementation of the previous design.

Finally, equation 14.10 (for C_{ot2}) assumes that an operational/acceptance test consists of two components. The first component is a fixed cost associated with planning, setting up, and analyzing the data collected from the OT (C_{otf2}) and a second component that depends on the "duration"[12] of the test ($R_{ot2}*T_{dot2}$). In this chapter, for simplicity, we assume that $C_{otf2} = C_{otf1}$ (the fixed cost portion of the previous OT event) and that $R_{ot2} = R_{ot1}$ (the cost per operational hour of the previous OT event).

We note that mathematical models for the cost components developed earlier (and those that follow) make simplifying (if questionable) assumptions. They are meant to illustrate how we might incorporate many of the important factors that drive the "Fix and Test" option costs. The purpose of the models provided here is to provide a simple cost framework that may be modified and extended through the use of more complex (and accurate) models that reflect the data that is available and relevant to a specific program of interest.

The "Production Delta Cost" (C_{dp}) model is relatively simple:

$$C_{dp} = N_p * (C_{u2} - C_{u1}) \qquad (14.11)$$

---

[12] Here, the term duration refers to the number of operational hours required and not the "schedule time" for the test. As we shall see, the "schedule time" may be reduced by increasing the number test units and will have to be increased to reflect the fact that not all units can be expected to operate 24/7 for the duration of the test.

Here, N_p is the number of units to be produced, C_{u2} is the expected unit cost of the fixed units, and C_{uc1} is the reference unit cost (without fixes). For simplicity, we assume that $C_{u2} - C_{u1} = 0.5 C_{dmfuc}$ (i.e., the product unit delta cost will be 50% that of the "per unit cost" to fix the test units).

The "Schedule Delay Cost" (C_{sd}) may be approximated as the product of the total "fix and test" schedule delay (in months) and the sum of the per month cost of extending the program (R_{ped}) and the per month lost opportunity cost (R_{lo}), that is,

$$C_{sd} = S_{dpt} * (R_{ped} + R_{lo}) \qquad (14.12)$$

One simple way to obtain a rough estimate for R_{ped} would be to take the current program burn rate (R_p) and subtract the burn rate associated with design team (R_{des}), which may be approximated as $R_{des} = N_{des} * R_{smdf}$, where N_{dt} is the size of the design team (in full-time equivalents).

Obtaining a value for R_{lo} is one of the more challenging tasks. What constitutes "lost opportunity" depends on the system. It could be the higher monthly O&S costs associated with continued use of the system that is being replaced. It could be the monthly profit lost by delaying the release of a product. It could be some other estimate of the monetary monthly value of having the system in place.[13] In this chapter, we take R_{lo} to be the expected difference in monthly O&S costs between the new system and the system it is replacing.

It is assumed that the total schedule delay (S_{dpt}) may be approximated as the sum of the schedule delay due to development (S_{dpd}) and the schedule delay due to the additional OT (S_{dot}).[14]

$$S_{dpt} = S_{dpdev} + S_{dot} \qquad (14.13)$$

Table 14.14 provides a summary of the mathematical models and assumptions used in this chapter to estimate values for S_{dpdev} and S_{dot}. The table is followed by a discussion of the rationale for each equation.

Equation 14.14 (for S_{dpd}) indicates that the developed schedule delay is simply the sum of the design delay (S_{dpdes}), the test unit repair delay (S_{dpdm}), and the delay associated with a DT that verifies the implemented design (S_{dpdt}). Equation 14.15 (for S_{dpdes}) assumes that the design team consists of N_{dt} designers working full time (as such, it is likely a lower bound).

Equation 14.16 (for S_{dpot}) assumes that the schedule for the OT consists of two parts (fixed and variable). The fixed schedule time (S_{otf2}) consists of the time it takes to set up the test (once test units are available) and the time required to perform analysis of the results following completion of the test. The variable schedule time (S_{otv2}) consists of the time it takes to perform the test. Equation 14.17 (for S_{otv2}) reflects that the time required to perform the test will depend on: (i) the number of operational

---

[13] As an example, if the new system is safer, R_{lo} may be a reduction in costs associated with reducing injury or death.

[14] That is, it is assumed that there are no other sources of schedule delay (e.g., funding delays and delays associated with setting up dates for OT or with scheduling a second decision milestone).

SYSTEM ACCEPTANCE DECISIONS AND TRADE STUDIES

Table 14.14 Simple Mathematical Models for the S_{dpt} Components

Equation #	Equation	Parameter Definitions
(14.14)	$S_{dpdev} = S_{dpdes} + S_{dpdm} + S_{dpdt}$	S_{dpdes} = Schedule delay due to design S_{dpdm} = Schedule delay associated with realizing the design (fixing the test units) S_{dpdt} = Schedule delay associated with DT.
(14.15)	$S_{dpdes} = S_{mfd}/N_{dt}$	S_{mfd} = Design team staff months (Table 14.13) N_{dt} = Size of the design team (in full-time equivalents)
(14.16)	$S_{dpot} = S_{otf2} + S_{otv2}$	S_{otf2} = "Fixed" schedule time associated with setting up the second OT and analyzing the results S_{otv} = The "Variable" schedule time associated with T_{dot2}
(14.17)	$S_{otv2} = T_{dot2}/(N_{ot2} * O_{hpdpu2} * N_{tdpm2})$	T_{dot2} = Duration of the second OT in operation hours N_{ot2} = Number of test units O_{hpdpu2} = Average hours per test day that a unit operates N_{tdpm2} = Number of test days per month

hours required (T_{dot2}); the number of test units (N_{ot2}); and the number of test days per month (N_{tdpm2}). In this chapter, we assume that values for S_{otf2}, N_{otw}, O_{hpdpu2}, and N_{tdpm2} will be roughly the same as they were for the previous OT.

Finally, before we work on an example, we consider how to determine two other costs that play a role in the Acceptance Decision model developed in Section 14.2.1, the Production Costs associated with the original and the "fixed" units (C_{p1} and C_{p2}), and the Program Cancelation Costs associated with the second and third decision options (C_{c1} and C_{c2}). Equations 14.18, 14.19 provide models that may be used to estimate C_{p1} and C_{p2}:

$$C_{p1} = N_{p1} * C_{pu1} \qquad (14.18)$$

$$C_{p2} = N_{p2} * (C_{pu1} + C_{pdu}) \qquad (14.19)$$

where N_{p1} and N_{p2} are the are the number of field units to be produce and C_{pdu} is the delta cost associated with implementing the design change. For simplicity, in this chapter, we assume $N_{p2} = N_{p1}$ and $C_{pdu} = 0.01 C_{pu1}$.

Equations 14.20 and 14.21 provide models that may be used to estimate C_{c1} and C_{c2}:

$$C_{c1} = N_m * R_{lo} \qquad (14.20)$$

$$C_{c2} = (N_m - S_{dpt}) * R_{lo} \qquad (14.21)$$

where N_{m1} is the planned life of the system in months.

Now that we have developed a cost model that may be used to provide estimates for inputs to the acceptance decision model; we consider an example of how the model might be used. Table 14.15 provides an Excel implementation of this model. Inputs to the model are highlighted in yellow, and calculations are highlighted in green. The table is divided into three parts. The first part summarizes the reference total production cost (C_{tp1}), the ratio of the desired to the current MTBF, and T_{dot2} (highlighted in green). A number of these parametric values are used as inputs to the following two parts. The second part provides the model that is used to calculate the total direct delta costs (C_{dd}), production delta costs (C_{dp}), and schedule delay costs (C_{sd}). These three values are highlighted in blue. The third part provides the model used to calculate the lost opportunity cost associated with canceling the program in the absence of a fix and test cycle (C_{c1}) and with canceling the program following a fix and test cycle (C_{c2}). These values are highlighted in blue. (*the reader is referred to the online version of this book for color indication*)

We see that from this model that the total delta Cost associated with the Fix and Test option is $C_{dt} = C_{dd} + C_{dp} + C_{sd} = \$9.4\,M + \$1.1\,M + \$25.6\,M = \$36.1\,M$ and the drivers are the schedule delay costs ($C_{sd} = \$25.6\,M$) and the cost of performing a second OT event ($C_{dot2} = \$8.0\,M$).

In the next section, we see how the cost model developed may be integrated with the acceptance test design model developed in Section 14.2.3 and the confidence model developed in Section 14.2.2 to estimate values for the inputs to the Acceptance Decision model developed in Section 14.2.1.

14.2.5 The Integrated Decision Model[16]

This section demonstrates how the cost model developed in Section 14.2.4 can be integrated with the acceptance test design model developed in Section 14.2.3 and the confidence model developed in Section 14.2.2 to estimate values for the inputs to the Acceptance Decision model developed in Section 14.2.1.

Suppose the system has an MTBF requirement of $M_r = 100\,h$ and that the first acceptance test had a test duration of $T_{d1} = 4000$ operational hours and the system MTBF was $M_{ob} = 95\,h$. We see that Section 14.2.2 enables us to estimate a confidence that the system is good at $C(Good) = 0.33$.

Assuming that we believe that we can improve the MTBF from $M_c = 95\,h$ to $M_{dg} = 125\,h$, Section 14.2.3 indicated that in order for us to have an acceptance/OT with a Confidence of $C = 0.80$ (80% probability of failing a bad system) and a Power of $P = 0.80$ (80% probability of passing a good system), we need a test duration of $T_{d2} = 7000\,h$ and an acceptance threshold of $M_t = 113\,h$.

We may use Section 14.2.2 to determine the confidence that the system is good (given it passes this second acceptance test) by setting $T_{d2} = 7000\,h$ and setting $M_{ob} = 113\,h$. This yields a confidence of $C(Good) = 0.81$.

---

[15] The "Integrated Decision Acceptance Model" consists of using the "Ch 14.2 System Acc Dec Tree.xlsx" Excel file with cost inputs from "Ch 14.2 Acc Dec Cost Model.xlsx" and probability inputs obtained using "Ch 14.2 Test Analysis.xlsx."

Table 14.15 Example Excel Implementation of the Cost Model

Ref Prod Costs	Ctp 1 =	100	M$		
		Np 1 =	100	units	
		Cuc1 =	1	M$	
Design Inputs	Mdg/Mc =	1.11			
		Mc =	113	hr	Current MTBF
		Mdg =	125	hr	Expected MTBF from Test Design
OT Parameters	Tdot2 =	7000	hr		
		Cg =	0.8		Goal Confidence
		Pg =	0.8		Goal Power
		Mr =	100	hr	Requirements
Direct Delta Costs	Cdd =	9.4	M$		
	Design	Cddes =	0.35	M$	
		Rmfd =	0.03	M$	
		Smfd =	11.6	sm	
		Smfdo =	1	sm	
		Adf =	100		
		Bdf =	1	sm	
	TU Prod	Cdtu =	0.08	M$	
		Ntu =	4		
		Cdmfuc =	0.02	M$	
		Amf	0.2		
		Bmf	1	M$	
	DT	Cddt2 =	1	M$	
	OT	Cdot2 =	8.0	M$	
		Cotf2 =	1	M$	
		Rot2 =	0.001	M$/hr	

(*continued*)

Table 14.15 (Continued)

Prod Delta Cost	Cdp =	1.1	M$			
		Cu2 − Cu1 =	0.010619469	$M		
Prog Sched Delay Cost	Csd =	25.6	M$			
		Rped	1			M$/mo
		Rlo	1			M$/mo
	Prog Sched Delay					
		Sdpt =	12.8	mo		
		Des	Sdpdev =	4.3		
			Sdpdes =	2.3	mo	
				Ndt =	5	fte
			Sdpm =	1	mo	
			Spdt =	1	mo	
		OT Delay	Sot2 =	8.5	mo	
			Sptf2 =	2	mo	
			Sotv2 =	6.5	mo	
				Ntdpm2 =	30	d/mo
				Not2 =	3	units
				Ohpdpu2 =	12	hr
Prog Cancel (LO) Cost	Cc1 =	240.0	M$			
		Nyop =	20	yrs		
	Cc2 =	227.2	M$			

SYSTEM ACCEPTANCE DECISIONS AND TRADE STUDIES

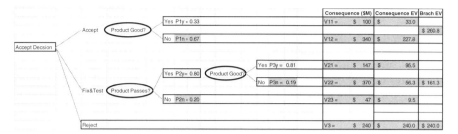

Figure 14.7 Acceptance decision model

This provides us the values of the input probabilities for the Acceptance Decision model.

In order to calculate the costs associated with a fix and test option with the indicated design goals and acceptance test design, we use the cost model developed in Section 14.2.4. The output of this model provides the cost inputs to the Acceptance Decision model. Table 14.16 provides the results that were obtained. With the exception of M_c, this implementation of the cost model uses the same inputs that were used in Section 14.2.4.

Table 14.17 indicates how to translate the values in Table 14.16 into the consequence values that go into the acceptance decision model.

Figure 14.7 provides the Acceptance Decision Model that results from these inputs. We can see that given these inputs, the accept solution has the highest expected cost ($E_a = \$260.8$ M), the cancel program option has lower expected cost ($E_c = \$240.0$ M), while the "Fix and Test" option has the lowest expected cost ($E_{ft} = \$161.3$ M).

Figure 14.8 shows the risk associated with each decision option. The graph indicates that while the "Fix and Test" decision has the lowest expected cost, it also has an outcome that spans those associated with the other option.

We can use this collection of models to perform sensitivity studies to determine the degree to which uncertainties in the input parameters affect which decision option is optimal or perform trade-off analyses between cost, system design, and the design of the acceptance test.

As a specific example of a sensitivity study, we can use the cost model to determine the change in cost and schedule that would be expected if the observed value for the MTBF in the first acceptance test were increased from $M_o = M_c = 95$ h (in Table 14.16) $M_o = M_c = 113$ h (in Table 14.15). In this case, we see that there would be an expected reduction in the sum of C_{dd}, C_{pd}, and C_{sd} from \$47.4 M to \$36.1 M and a reduction in the schedule delay from 17.0 to 12.8 months. There are a variety of approaches that one may take to performing more extensive sensitivity studies.[16]

---

[16] One approach is to assume a normal distribution for each input parameters and calculate the resulting variance for the high-level cost parameters. Tools such as Crystal Ball™ may also be used to determine the uncertainty in higher-level cost parameters based on the distributions assumed for the input parameters. One may also consider the impact of uncertainties in the probabilities used in the acceptance model.

Table 14.16 Cost Model Results

Ref Prod Costs	Ctp =	100	M$			
		Np 1 =	100	units		
		Cuc1 =	1	M$		
Design Inputs	Mdg/Mc =	1.32				
		Mc =	95	hr			Current MTBF
		Mdg =	125	hr			Expected MTBF from Test Design
OT Parameters	Tdot2 =	7000	hr			
		Cg =	0.8				Goal Confidence
		Pg =	0.8				Goal Power
		Mr =	100	hr			Requirements
Direct Delta Costs	Cdd =	10.2	M$			
Design		Cddes =	0.98	M$		
		Rmfd =	0.03	M$		
		Smfd =	32.6	sm		
			Smfdo =	1	sm	
			Adf =	100		
			Bdf =	1	sm	
TU Prod		Cdtu =	0.25	M$		
			Ntu =	4		
			Cdmfuc =	0.06	M$	
				Amf	0.2	M$
				Bmf	1	
DT		Cddt2 =	1	M$		
OT		Cdot2 =	8.0	M$		
			Cotf2 =	1	M$	
			Rot2 =	0.001	M$/hr	

Prod Delta Cost	Cdp =	3.2	M$			
		Cu2 − Cu1 =	0.031578947	$M		
Prog Sched Delay Cost	Csd =	34.0	M$			
		Rped	1	M$/mo		
		Rlo	1	M$/mo		
Prog Sched Delay						
	Sdpt =	17.0	mo			
	Des	Sdpdev =	8.5	Sdpdes =	6.5	mo
				Ndt =	5	fte
			Sdpm =	1	mo	
			Spdt =	1	mo	
	OT Delay	Sot2 =	8.5	Sptf2 =	2	mo
				Sotv2 =	6.5	mo
				Ntdpm2 =	30	d/mo
				Not2 =	3	units
				Ohpdpu2 =	12	hr
Prog Cancel (LO) Cost	Cc1 =	240.0	M$			
		Nyop =	20	yrs		
	Cc2 =	223.0	M$			

Table 14.17 Translation for Acceptance Decision Model

Consequence	Equation	Value ($M)
V11 =	Ctp =	100.0
V12 =	Ctp + Cc1 =	340.0
V21 =	Ctp + (Cdd + Cdp + Csd) =	147.4
V22 =	V21 + Cc2 =	370.4
V23 =	V21 − Ctp =	47.4
V3 =	Cc1 =	240.0

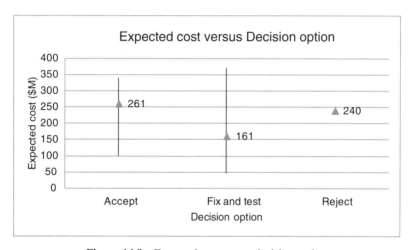

Figure 14.8 Expected cost versus decision option

We see that the principal trades associated with this decision model are between the total "fix and test" delta cost ($C_{dt} = C_{dd} + C_{dp} + C_{sd}$), the confidence ($C$) and power ($P$) one requires for the second acceptance test, and the value one establishes as the system requirement (M_r). The example provided in this section illustrated a the process that may be used to perform such trade studies:

- The model developed in Section 14.2.3 was used to calculate values for T_d, M_t, and M_d from given values for C, P, and M_r.
- The model developed in Section 14.2.4 was used to calculate the C_{dt}-related and S_{pdt} inputs to the Decision model from the values for these values of T_d and M_d.
- The model developed in Section 14.2.2 was used to calculate the confidence that the system is good ($C_2(G)$) following the second acceptance test (based on M_r and the value of M_t obtained in the first step).

- The value for P and the values for C_{pt}, S_{pdt}, and $C_2(G)$ that resulted from the given values of C, P, and M_r were then used as inputs to the acceptance decision model.

Using this approach, we may generate curves of expected costs for each decision option as one varies C, P, and/or M_r.

14.2.6 Conclusions

This section provided an integrated modeling framework for making a system acceptance decision. The focus was on illustrating a number of important modeling and analysis techniques and show how they could be combined to provide an integrated acceptance decision modeling framework. Fictional data and a host of simplifying assumptions were employed to keep the models as simple as possible. The resulting framework can be extended to real systems through the use of real data and a relaxation of some of the simplifying assumptions.

Section 14.2.1 developed a relatively simple decision tree model for the product acceptance decision. Three options were provided: accept, fix and retest, reject. Important inputs to the decision tree model were the costs associated with each decision branch, the confidence one may have that a system is good given it passed its initial acceptance test, the probability that the system will pass a second acceptance test (given product improvement), and the confidence one may have that the system is good, given it passed the second acceptance test. We saw how this model could be used to perform quick sensitivity studies and trade studies. The remainder of Section 14.2 dealt with somewhat more complicated task of how to calculate these inputs from data that is typically available to a program.

Section 14.2.2 made the simplifying assumption that the single criterion for acceptance was that the system meet its system reliability requirement.[17] Given this assumption, we learned how to calculate the confidence that a system is "Good" given the observed mean value and standard deviation of the metric, the requirement threshold for the metric, and the test duration. We found that one could use the equations developed here to perform important sensitivity and trade studies related to the duration of the test and the confidence that one could have that the system was good. An example was provided that illustrated the application of the techniques developed in the section.

Section 14.2.3 made use of the theory associated with hypothesis testing to design a second acceptance test that would provide (i) a probability that a system would pass (i.e., the measure reliability exceeds some test threshold), given the actual reliability of the system is greater than some specified value, and (ii) a probability that the system would fail, given the actual reliability of the system is less than some specified value. We saw that these probabilities were referred to, respectively, as the "Power" and "Confidence" provided by the test. We also note that this use of the term "Confidence"

---

[17] In the case of multiple acceptance criteria, one may develop a Multi-Attribute Value/Utility function that may be used.

is different from the way the term was used in Section 14.2.2. Generally, one wants to design a test with both a high Power and a high Confidence. We examined how the Power and Confidence were functions of the test duration, the expected (design) value for the metric of interest (e.g., reliability), the test pass/fail threshold for the metric of interest, and the required value for the metric of interest. We found that these expressions could be used to perform a variety of important sensitivity and trade studies related to test design. An example was provided that illustrated the application of the techniques developed in the section.

Section 14.2.4 developed a cost modeling framework to provide the cost-related inputs required by the decision model developed in Section 14.2.1. It focused on developing the simplest set of models that were relatively complete with respect to the variables that affect the acceptance decision. We saw that even a "simple" cost model can be rather complicated. We also saw how such a cost model may be used to perform important sensitivity and trade studies with respect to system redesign and testing. An example was provided that illustrated the application of the techniques developed in the section.

Finally, in Section 14.2.5, we provided an example of how the theoretical framework developed in Sections 14.2.1–14.2.4 could be integrated into a complete model that could be used to perform a wide variety of sensitivity and trade studies related to the acceptance decision.

While the modeling framework developed in this section was extensively deterministic, it may also be extended in a rather straightforward manner to address modeling uncertainties using Monte Carlo (MC) techniques. Such extensions are explored in some of the exercises that follow.

14.3 PRODUCT CANCELATION DECISION TRADE STUDY

14.3.1 Introduction

The research shows that software projects often fall short of quality and successful implementation goals. According to a 2011 Government Accountability Office (GAO) report on software project failures, planned federal information technology (IT) projects with investment costs totaling approximately $81 billion frequently incurred cost overruns and schedule slippages and contributed little to mission-related outcomes (GAO, 2011).

When improperly executed, the systems engineering process can result in complex, incomplete, and inconsistent systems where variables of design, reliability, quality, user satisfaction, budgets, and schedules are often poorly considered and estimated throughout the Systems Engineering Life Cycle (SELC) (Eisner, 2005). Project failure often stems from the fact that systems are becoming increasingly complex as systems engineers strive to integrate multiple, disparate subsystems through the use of software (Ryan et al., 2014) (Madni & Sievers, 2014) (ISO/IEC/IEEE, 12207, 2008a).

Software project failure was a concern in 1968 when the North Atlantic Treaty Organization (NATO) held its first conference on software engineering to address the problem. Experts from 11 countries convened at the conference to discuss many of the same failure factors that systems engineers are still facing today: changing or unclear requirements, insufficient technical knowledge, problematic technology, project cost overruns, and project schedule delays (Stevens, 2011) (Ewusi-Mensah, 2003). The conference was organized to shed light on these very issues in software engineering and to discuss possible techniques to solve them in order to improve project outcomes. An attendee at the conference recommended that projects develop a feedback loop to monitor the system, through which performance data could be collected for use in future improvements (Naur & Randell, 1969).

The authors set out to understand the underlying causes of software project failure in order to develop a predictive analytics model to accurately predict future outcomes for software projects using data from past software project performance. The predictive model will point to some of the root causes of past software project failures to assist in decision-making that can steer projects to successful outcomes in the future. The decision to cancel projects occurs too late in the life cycle when costs and schedules have escalated out of control. The most cost-effective point in the life cycle to cancel a failing project is at project initiation and prior to development. In this section, we will develop a predictive model to determine project success or failure to make an informed decision to cancel the project before large monetary, resources, and time investments are made.

Failed development projects represent lost investments and revenue. They often lead to poor project team moral due to wasted effort. In truth, there are too many unknowns to accurately predict the likelihood of project success. However, finding and uncovering these unknowns quickly and coming to a decision to cancel a program in the early stages is the ideal situation in order to save money and time and free up resources to focus on potentially successful projects (SPs).

14.3.2 Significance

Despite the ambiguous, uncertain, and intangible nature of software, investment in software represents one of the largest proportions of IT spending. Forrester Research, Inc. (2013) forecasted global IT spending of a staggering $2.06 trillion across IT services by enterprises and government for hardware and software. According to Forrester, software spending is the largest investment category within global IT spending, with $542 billion in spending in 2013. The biggest changes and greatest innovations in technology are software-intensive and include cloud-based Software-as-a-Service (SaaS) implementations, mobile applications, and smart computing such as Analytics and Business Intelligence (Lunden, 2013).

Gartner's top technology priority list of IT spending in 2013 is noted as Analytics and Business Intelligence. The Gartner "survey highlighted the need for CIOs to set aside old rules and adopt new tools" (Pettey & van der Muelen, 2013). Analytics and business intelligence were named as top priority because CIOs understood the importance of not repeating the past by leveraging these IT innovations to harvest data and information to extend the past into the future. A McKinsey & Company survey observed that IT leaders all agree that the need for better data and analytics is a challenge that has grown in importance in recent years as support of managerial decision-making has become top priority (Arandjelovic et al., 2015).

In 1991, Boehm observed that no tools were available for assessing the probability of the risk factors associated with project outcomes. Eisner (2005) observed that complex systems present technical and management challenges that "require stronger and more insightful thinking patterns to be successful" (Ryan et al., 2014). Additionally, a panel of experts on Systems Engineering Science at the International Council on Systems Engineering (INCOSE) International Symposium (IS) held in Henderson, Nevada, in 2014, discussed leveraging systems thinking to improve project outcomes. A member of the panel indicated that the "systems engineering toolbox needs a mechanism that predicts software project outcomes. Understanding why natural systems fail should be in the toolbox for systems engineers. Unfortunately it is not … it is a research problem that should have effort put into it" (Singer, 2014).

The model is software-specific and was developed using software-intensive project data. The data contained a combination of failed and successful software projects to ensure that its predictive capabilities were substantiated by both sets of outcomes. The predictive model developed in this research allows organizations to quantify a software product's probability of success or failure before, during, and after its development. The model predicts whether or not the software functionality will be successfully implemented prior to getting to the end state and before large financial investments have been made. It helps decision-makers in making early and informed decisions about which projects in the portfolio to proceed with and which to cancel. The predictive model is a tool for planning, strategic thinking, and decision-making.

14.3.3 Defining Failure

Researching and analyzing the outcomes of software projects, as well as data from the failed software projects and related GAO reports, enabled the identification of one or more of the following attributes or characteristics of failure (GAO, 2006; GAO, 2009; GAO, 2011; Tan, 2011; The Standish Group, 2013; GAO, 2014; DoD, 2015):

- Catastrophic or deadly outcomes
- Completely abandoned projects or projects never released into production
- Project cancelation
- User dissatisfaction leading to abandoning the system postdeployment
- Quality goals not achieved
- Significant cost overruns that exceed 30% or more of the estimated budget
- Significant schedule slippage that exceeds 30% or more of the scheduled duration.

In defining software project failure, the authors identified key failure factors that directly correlate to project failure in the failed software project data analysis and predictive model development (Tan, 2011; Ewusi-Mensah, 2003).

14.3.3.1 Factors Present in Software Failure The succeeding sections will define and outline the failure factors and discuss their impact on software project failure and success. A set of 202 software projects were identified, analyzed, and used in the development of the model. The research determined that software project failure factors spanned the project life cycle and were interrelated through their causal relationships. Software project failure resulted from many failure factors working together. (Lehtinen et al., 2014).

14.3.3.1.1 Unrealistic Project Goals and Expectations The major goal and objective of systems engineering, software engineering and project management is to deliver high-quality software on time and within budget. However, unrealistic project goals and expectations contribute to software project failures both in the literature and in the 202 projects studied to develop the predictive model. This factor stems from the inability of the project team to specify project goals, expectations, and outcomes or from lack of agreement and consensus on the project goals, leading to confusion about what the project is expected to achieve. This factor is one of the most significant.

14.3.3.1.2 Changing or Unclear Requirements Delivery of high-quality software on time and within budget begins with excellent requirements. However, changing and unclear requirements posed several challenges leading to failed software projects. Poor requirements management was among the major challenges encountered in this research of failed software projects. Vague and ambiguous requirements were identified as a cause of failure because developers often had to guess the true intent of the requirement. The data from the failed projects (FPs) studied indicated that scope and feature creep plagued the project because a process to manage the requirements was lacking. Scope creep indicates that requirements were not identified during elicitation or that key user stakeholder groups were overlooked or not identified. As with Unrealistic Goals and Expectations, this factor was also very significant.

14.3.3.1.3 Insufficient Technical Knowledge Insufficient technical knowledge is a leading cause of software project failure. This can manifest as a lack of technical experience on the part of the project team. Inexperience with the technology required to develop the software system resulted in additional risks because the project team required significantly more time to learn the technology and ramp up. Lack of experience in the technology leads to a lack of clear understanding of the requirements, design, development, and test planning for the software project. Insufficient technical knowledge leads to failure because the project team lacks the technical competency to solve the software design problem in a manner that can lead to successful implementation.

14.3.3.1.4 Problematic Technology Cutting-edge technology is often the platform for innovative and successful software projects. However, problematic, immature, or changing technology was identified in the research as a cause of failure. During the analysis of FPs, the inability of the technology infrastructure to support the software project was often identified as a factor in project failure. Technology that is outpacing the rate at which the project team can develop the software leads to substantial rework to catch up. The opposite is true where the technology is so new that the project team has a difficult time learning and applying it to the design, development, and test planning.

14.3.3.1.5 Lack of Executive Leadership Support Executive leadership support is crucial to software project success. Unclear or inadequate support from executive leadership leads to poor project governance arrangements that precipitate unclear lines of responsibility. Failure to establish effective leadership in the business, technical, and organizational domains often results in a lack of project monitoring and control that results in a project running out of control. Lack of support also leads to failure to establish clear decision-making ownership, resulting in indecision and project confusion. Management leadership must be aware that the project will encounter setbacks throughout the life cycle and that, therefore, they must be prepared to support and rally behind the project despite the setbacks. A lack of consistent leadership support and direction often contributed to high staff turnover, which in turn caused project delays and in turn contributed to project failure.

14.3.3.1.6 Insufficient User Commitment Insufficient user commitment – lack of identification, engagement, and feedback from the required users and other stakeholders of the software system – is also a factor in the failed software projects identified in this research and in the literature. User engagement is necessary throughout the software system life cycle, as their needs, desires, and expectations are captured as requirements from which the system is developed and tested. Lack of commitment results in failed systems because the crucial requirements for how the system will operate and the intended interactions with the users are not effectively captured and thus cannot be accurately validated.

14.3.3.1.7 Project Cost Overruns Project cost overruns are a contributing factor to project failure and abandonment. Cost overruns are defined as software project cost estimates, identified during the acquisition and planning stages, being exceeded by more than 30% during the software development life cycle. Several policies are in place on government acquisition programs to notify the appropriate leadership and decision-makers when costs exceed guidelines and thresholds, including the Nunn–McCurdy Act. Programs in Nunn–McCurdy breach status that have exceeded original cost baselines by 30% are reported to Congress and those breaching by 50% or more face possible termination (AcqNotes, n.d.).

14.3.3.1.8 Project Schedule Delays The impact of schedule delays is detrimental to project success. Throughout this research on software project failures, schedule delays were identified as a leading cause of termination and abandonment. Project schedule delays breaches are defined in the DoDI 5000.02 Directive (2015) as a critical change in schedule that causes a delay of either 1 year or 25% or more of the baseline schedule. Adding more resources to an already late software project causes resource strain, with the adverse effect of still lower team performance and continued delays (Brooks, 1995).

14.3.3.1.9 Insufficient Project Management and Control Project management and control implement success measures and criteria throughout the software project life cycle. However, as the research uncovered, lack of sufficient project management and control stemmed from a failure to plan the project and its outcomes. The research indicated that a lack of project management and control leads to poor decision-making because there are no checks and balances to steer the software project in the right direction.

14.3.3.1.10 Project Failure in Testing Phase Testing is a critical activity in the development of successful software projects because, as noted earlier, software is intangible and cannot be quantified by a physical product. Undercutting or eliminating adequate testing can prove fatal to a software project. Without clear testing procedures, it is impossible to determine whether the project has met systems, business, and user requirements of the software project before moving into the production environment. Software project failure in testing signifies that defects have not been uncovered and reported immediately and that tests are not conducted against proper requirements. A buggy system deployed into production can lead to user abandonment, cancelation, or worse, lethal failure.

14.3.3.2 Software Project Failure Cases Systems and software products can be found in most every major industry, including the defense, aerospace, automotive, medical, telecommunications, industrial, and semiconductor fabrication industries (Walls, 2006) (ISO/IEC/IEEE, 12207, 2008a). The data were derived from failed and successful software projects found throughout those same industries, with selected examples depicted in the snapshot shown in Table 14.18. The project names were converted to project identification numbers when structuring the data within the database in preparation for building the software predictive models. FP in the table stands for FP and the numbers identify each specific project. These software project failure examples spanned a historical timeline dating from the 1990s until 2013.

In addition to the snapshot in Table 14.18, two specific examples of software project failure should be described. One example is the $327.6 million Mars Climate Orbiter (MCO) developed by NASA's Jet Propulsion Laboratory in 1998. Communication with the Orbiter was lost minutes after it approached Mars as a result of a software defect that caused a navigation error. The failure occurred because different groups within the engineering team were using different units of measurement. One team working on the thrusters measured in English units of pounds-force-seconds; the others used metric Newton-seconds. Researchers from the navigation team identified the defect; however, their concerns were dismissed and the result was the thrusters operated at a level 4.45 times more powerful than expected by the operators. According to the report developed by the NASA MCO Mishap Investigation Board (MIB), if this root-cause software defect had been addressed, it could have been corrected prior to the launch. Unfortunately, trajectory correction maneuver number 5 was not

Table 14.18 Snapshot of Failed Software Projects ID for Coding

Project ID	Result of Failure	Estimated Cost	Estimated Start Date
FP1	Not Delivered	$185,000	
FP2	Abandoned	1 billion	2005
FP11	Abandoned	3 billion	2004
FP13	Abandoned	260 million	2001
FP29	Canceled	$54.4 million	2004
FP30	Abandoned after deployment	$400 million	2004
FP31	Abandoned	$527 million	2004
FP34	Canceled	$170 million	2002
FP35	Canceled	$33 million	2002
FP38	Canceled	$130 million	2001
FP43	Canceled	$4 billion	1997
FP44	Canceled	$40 million	1997
FP46	Rocket explodes	$350 million	1996
FP47	Canceled	$25.5 million	1995
FP48	Canceled	$2.6 billion	1994
FP49	Canceled	$44 million	1994
FP51	Canceled	$600 million	1993
FP52	Abandoned	$130 million	1993
FP53	Canceled	$11.25 million	1990
FP54	Abandoned	$15 million	1993
FP56	Canceled	$165 million	1992
FP105	Canceled	$560 million	2008

performed and the result of that software error is now a spacecraft that is lost forever (NASA, 1999).

Another example of a single catastrophic software failure is the Patriot Missile System Bug of February 21, 1991, which occurred during Operation Desert Shield. The Patriot Missile System was deployed as a defense against enemy aircrafts and missiles. The tracking software for the Patriot Missile System predicts the enemy target in a time series by using the velocity of its target and the present time. The development team was aware of an existing defect in the targeting software that caused the internal clock to slowly drift away from accurate time and developed "the workaround" to reboot the system periodically. The reason is that the longer the system was left running, the greater the clock's time drift. However, the system was left running for 100 h without "the workaround" reboot, causing a time drift and lag of 0.34 of a second. On that day, an Iraqi missile was launched toward the US airfield in Dhahran, Saudi Arabia. The Iraqi missile was detected by the Patriot Missile System; however, when the system tried to calculate the next location of the Iraqi missile, the system was erroneously looking at an area over half a kilometer away from the missile's true location. Not seeing the target in the location, the Patriot Missile System assumed that there was no enemy missile and canceled the interception. The enemy missile carried on to its destination over the Dhahran airfield where it claimed the lives of 28 soldiers and injured 98 (GAO-IMTEC-92-26, 1992; Naur & Randell, 1969).

Why are so many software projects failing? This research identifies factors that contribute to software project abandonment, cancelation, nondelivery, and ultimate failure to develop a predictive model to predict the outcomes of future software development projects.

14.3.4 Developing the Predictive Model

Predictive analytics encompasses a range of methods to anticipate outcomes. There are four main reasons for a business to use analytics: to predict, to identify opportunity, to calculate demand, and to identify and prevent fraud and risk. Roy (2013) developed a predictive model for disease prevention to predict patient readmission to the hospital and patient death after discharge. Accurate prediction of those outcomes would allow health care professionals to develop strategies and measures to reduce these serious and costly risks. The patient outcome tool calculates the probability of the outcome for a set of patients and derives a risk score for each patient.

The authors identified several pitfalls to avoid in using predictive analytics, one of which is that organizations cannot simply build a model once and apply it to everything (Fitzgerald, 2014). A new model may be required for every question asked. Why are software projects failing at such staggering rates and how can outcomes be improved? Why are software projects continuing to fail despite the vast amounts of research on the topic? Why are managers and decision-makers continuing to fund failing software project efforts? In order for the authors to understand this phenomenon, a new model was developed that included data from both sets of projects outcomes: success and failure. The authors also identified and included factors present in the four software engineering life cycle stages of Requirements, Design, Development, and Test to determine if the factors identified in the failed and successful software project data and used to develop the model contributed to software project failure.

Fitzgerald (2014) pointed out another trap to avoid in developing predictive models: creating models that do not scale and are too complex and expensive to be reused easily. The model for software predictions in this research was developed with the open source, free R software tool so that it can be easily transferred to and used broadly within any size software development organization. Larger organizations as well as small startups that are more vulnerable to software project failures because they have limited IT budgets and resources can use the model presented in this research to make better informed, evidence-based decisions.

14.3.4.1 Data Collection In developing the model, project failure and success data were collected for 202 projects using publicly available case studies, news articles, surveys, industry websites, and congressional reports. The authors also gathered data from reports generated by the GAO (www.gao.gov), court litigation cases, and other government artifacts. Through content analysis (Tsao et al., 2012), they defined key factors of project failure and success and developed a database to capture information pertinent to each individual project. They reviewed multiple sources for each software project to ensure identification of the most informative and accurate information. The researchers performed content coding to extract the most consistent information and to create groups of software failure factors for each project. To identify common software failure factors, they compared each project to the others with cross-case analysis of each failure factor's impact on project failure. This analysis was crucial to developing the predictive model, as it helped to understand and overcome the challenge of structuring the model and transforming the data into a format readable by the R software.

The goal was to develop a methodology to support and enable sound decision-making about any software project, its product, and its associated risks and to use the predictive model to create a "manage-by-fact environment" in which decision-makers listen to their organization's observed and empirical data (Baldridge, 2012).

14.3.4.2 Building the Model The authors began developing the predictive model by structuring the software project outcome data. Because the outcomes could be only one of two options – Failure or Success – the authors chose logistic regression to build the model that would explain the relationship between the independent variables (Failure Factors) and the outcome or dependent variable (Failure). Because regression analysis is one of the most commonly used methods of prediction, the authors chose to use it to predict individual project binomial outcomes where the dependent variable (Failure) is a dichotomous variable that returns either Failure = 1 or Not Failure/(Success) = 0. It models a linear relationship between the independent variables and a function of the dependent variable. This allows for nonlinear effects on the dependent variable; however, the model itself is linear in its predictors for failure or success. (Lattin et al., 2003; Everitt & Hothorn, 2006).

The underlying statistical concept using logistic regression is the generalized linear model (glm) function (R-Development, 2011; Kuhn, 2008). The glm is a flexible generalization of linear regression that allows the linear model to be "linked" to the response variable via the logit function to enable a wide range of disparate problems, such as the 10 Failure Factors, to come together in a powerful yet flexible framework. The logit of a probability is the log of the odds of the response with a value of 1. The authors want to know the odds of failure, and equation 14.22 is the logistic function used in the development of the models (Everitt & Hothorn, 2006):

$$\text{Logit}(\pi) = \log\left(\frac{\pi}{1-\pi}\right) = \beta_0 + \beta_1 x_1 + \beta_2 x_2 + \ldots + \beta_q x_q \quad (14.22)$$

Therefore, the logit of a probability is the log of the odds of response or outcome with the value of 1 as written in equation 14.23. This will be demonstrated with the software project probability outcomes of the predictive model.

$$\pi(x_1, x_2, \ldots, x_q) = \frac{\exp(\beta_0 + \beta_1 x_1 + \ldots + \beta_q x_q)}{1 + \exp(\beta_0 + \beta_1 + x_1 + \ldots + \beta_q x_q)} \quad (14.23)$$

The researcher used equations 14.24 and 14.25 to fit the software project predictive model. Equation 14.25, derived from Equation 14.24, is a variation of the logit and probability functions in R. Because project failure stems from multiple factors, they are all deemed significant in building the model, as the research will demonstrate.

$$\text{Logit}(\pi) = \log\left(\frac{\pi}{1-\pi}\right) = \beta_0 + \beta_1 \text{FFA} + \beta_2 \text{FFB} + \ldots + \beta_q \text{TestF} \quad (14.24)$$

$$\text{Model} = \text{glm(Failure} \sim \text{FFA} + \text{FFB} + \text{FFC} + \text{FFD} + \text{FFE} + \text{FFF} + \text{FFG}$$
$$+ \text{FFH} + \text{FFI} + \text{TestF, traindata, family} = \text{"binomial"}) \quad (14.25)$$

where π = probability of failure

In developing the model, the researcher randomly split the original $n = 202$ project success and failure data set into three separate data sets. The first training set was used to functionalize or fit the model. This set comprised $n = 182$ projects containing both successes and failures. The second test set comprised $n = 10$ projects randomly selected and containing both successes and failures. The third hold-out validation set comprised $n = 5$ projects also containing both successes and failures. The validation set was developed to assist in performing exploratory analysis to identify the best model fit. The authors realized that the sum of the three sets is only 197. This was to ensure that the last validation set did not contain any data points that the model had "seen" prior to the validation, as well as to distinguish, for the purposes of this research, between test and validation. Additionally, a smaller validation set was selected because the authors wanted to ensure that the model would scale for future deployment. The research included all the failure factors identified in Table 14.19 in

Table 14.19 Failure Factor ID Coding

Failure Factor ID	Failure Factors
FFA	Unrealistic Project Goals and Expectations
FFB	Changing or Unclear Requirements
FFC	Insufficient Technical knowledge
FFD	Problematic Technology
FFE	Lack of Executive Leadership Support
FFF	Insufficient User Commitment
FFG	Project Cost Overruns
FFH	Project Schedule Delays
FFI	Insufficient Project Management and Control
TestF	Project Failure in Testing Phase

```
 91
 92  #make Prediction
 93  pr=FALSE
 94  if(bestType==0 | bestType==2)pr=TRUE
 95
 96  p=predict(modelLib, scale(xTest, attr(s, "scaled:center"), attr(s, "scaled:scale")), proba=pr, decisionvalues=TRUE)
 97
 98  #display confusion matrix
 99  res=table(p$predictions, yTest)
100  print(res)
101
102  #Compute Balanced Classification Rate
103  #BCR=mean(c(res[1,1]/sum(res[,1]), res[3,3]/sum(res[,3]), res[4,4]/sum(res[,4])))
104  #print(BCR)
105
106
107  #fit the model with formula glm(y~x1+x2) where y=failure and x1,x2,...=failure factors including TestFailures
108  model=glm(Failure~FFA+FFB+FFC+FFD+FFE+FFF+FFG+FFH+FFI+TestF, traindata, family="binomial")
109  model2=glm(Failure~FFA+FFB+FFC+FFD+FFE+FFF+FFH+TestF, traindata, family="binomial")#remove non-significant factors FFC, FFD, FFG and FFI
110  modelReg=glm(Failure~FFA+FFB+FFC+FFD+FFE+FFF+FFG+FFH+FFI+TestF, traindata, family="binomial")#remove non-significant factors identified by LibLinear
111
112  #summary of model
113  summary(model)
114  summary(model2)
115  summary(modelReg)
116
117  #get odds of a failure as it relates to each factor
118  exp(model$coef)
119  exp(model2$coef)
120  exp(modelReg$coef)
121
122  #create a prediction objects for training and test data
123  predtrain = predict(model, type='response', newdata=traindata)
124  predtest = predict(model, type='response', newdata=testdata)
125  pred2 = predict(model2, type='response', newdata=testdata)
126  predReg =predict(modelReg, type='response', newdata=testdataReg)
127
128  #Look at the results of the predictions for the Test data and training data
129  table(predtrain)
130  table(predtest)
131  table(pred2)
132  table(predReg)
```

Figure 14.9 R coding for the predictive model

the model. FFA through FFI and TestF denote the independent variables or Failure Factors. Equation 14.25 is used to calculate the probability of a failure for each line item within the original data set. The probability score is a value between 0 and 1. The authors set a cutoff value of 0.5, with increased likelihood of failure for any results above 0.5 based on analysis of the original data set.

The model is built using the R open-source statistical tool. Figure 14.9 provides a sample of the R code used to develop the predictive model to determine project success or failure. The model could have been developed in any number of statistical tools; however, the free open-source R tool makes it readily available for anyone to use.

14.3.5 Research Results

The authors developed two models to determine the most effective in its predictive abilities. The results of each model and the most optimized one are described.

14.3.5.1 Analysis of Predictive Model A Predictive Model A was run using all the failure factors defined in Equation 14.24 of the original data set. The purpose of this model was to determine whether or not the data from the original data set could in fact accurately predict software project failure. The figures are the resulting outputs when the authors ran and tested the validity and fit of the model throughout the build process. The resulting *p*-values from Figure 14.10 indicate that FFA, FFB, FFC, FFE, FFF, FFH, and TestF are significant predictors of a software failure in the presence of the other predictors. All the failure factors with the exception of FFC, FFD, and FFG have significance level *p*-values less than the industry standard of 0.05. The low *p*-values indicate that the project failure attributing to these factors was unlikely to occur simply by chance (Higgins & Green, 2011). Though FFC, FFD, and FFG are not statistically significant in Model A, further modeling approaches, such as

PRODUCT CANCELATION DECISION TRADE STUDY

```
glm(formula = Failure ~ FFA + FFB + FFC + FFD + FFE + FFF + FFG +
    FFH + FFI + TestF, family = "binomial", data = traindata)

Deviance Residuals:
    Min       1Q    Median       3Q      Max
-2.77443  -0.06671   0.03867   0.11212  2.85520

Coefficients:
            Estimate Std. Error z value Pr(>|z|)
(Intercept)  -5.4133     1.7115  -3.163  0.00156 **
FFA           2.8028     1.0260   2.732  0.00630 **
FFB           3.7279     1.1439   3.259  0.00112 **
FFC           0.6196     1.0059   0.616  0.53791
FFD           1.5375     1.0374   1.482  0.13834
FFE           2.6292     1.1466   2.293  0.02184 *
FFF          -3.8406     1.5269  -2.515  0.01189 *
FFG           0.5103     1.0123   0.504  0.61418
FFH           4.9002     1.1569   4.236  2.28e-05 ***
FFI          -0.6935     1.0813  -0.641  0.52130
TestF         2.9599     1.3222   2.239  0.02518 *
---
Signif. codes:  0 '***' 0.001 '**' 0.01 '*' 0.05 '.' 0.1 ' ' 1

(Dispersion parameter for binomial family taken to be 1)

    Null deviance: 213.996  on 181  degrees of freedom
Residual deviance:  41.579  on 171  degrees of freedom
AIC: 63.579
```

Figure 14.10 *p*-Value significance and AIC scores for Model A

regularization and cross-validation, find these particular factors to be significant later in the analysis of Predictive Model B.

Figure 14.11 identifies the Failure Factors resulting in actual project outcomes for each of the projects identified in the $n = 10$ Test Set. Project ID SP7 in row 22 depicts the actual Failure = 0 for a SP and the corresponding predicted value 0.04829964 probability of failure. SP7 probability of failure is less than the probability score cut-off of 0.5 and thus accurately predicts a SP. Project ID SP51 in row 111 also accurately depicts success with a probability score of 0.25483709 and actual Failure = 0. Alternatively, Project ID FP23, FP52, FP59, FP81, FP104, FP119, and FP129 accurately

row.names	Project.ID	FFA	FFB	FFC	FFD	FFE	FFF	FFG	FFH	FFI	TestF	Failure
22	SP7	1	0	0	0	0	1	1	0	0	1	0
37	FP23	1	0	1	1	0	0	0	0	0	0	1
94	FP52	0	1	0	0	1	0	1	1	1	0	1
101	FP59	1	0	0	1	0	0	0	1	1	0	1
111	SP51	1	0	0	1	0	0	0	0	0	0	0
125	FP69	1	0	1	1	0	0	1	0	1	1	0
141	FP81	1	0	0	0	0	0	1	1	1	1	1
164	FP104	1	0	0	0	0	0	1	1	1	1	1
179	FP119	0	1	0	1	1	0	1	1	1	0	1
189	FP129	0	1	0	1	1	0	1	1	1	0	1

```
HTML> predtest
         22         37         94        101        111        125        141        164        179        189
 0.04829964 0.38855638 0.99653232 0.95826183 0.25483709 0.91079014 0.99373460 0.99373460 0.99925268 0.99925268
```

Figure 14.11 Actual project outcome versus prediction for Model A test data

depict Failure = 1 for an FP based on their predicted probability scores greater than 0.5. Model A incorrectly predicted row 125 Project ID FP69 as success when in fact the actual outcome was a project that failed in industry. FP69 actual outcome is Failure = 1 with probability of 0.91079014 of failure. This was a concern that the Model performance may not be adequate in correctly classifying project success or failure. Further analysis would prompt the authors to develop another model.

The ROC Curve in Figure 14.12 measures the ability of Predictive Model A, run with the training set and the test set, to correctly classify those projects that failed versus those that were successful. It plots the true positive rate (sensitivity) against the false positive rate (1-specificity) of the model against the given data set. The closer the test follows the left-hand border and the top portion of the ROC curve, the more accurate the test Figure 14.12 demonstrates that the training set produces a more accurate test than the test set. The test set has never been seen before by the model, so it does not perform as well. This is confirmed by Figures 14.13–14.15.

The Confusion Matrices in Figure 14.13 demonstrate that the model correctly classifies project success or failure in the training set approximately 96% of the time as opposed to only 80% for the test set as identified in the proportions correct field of

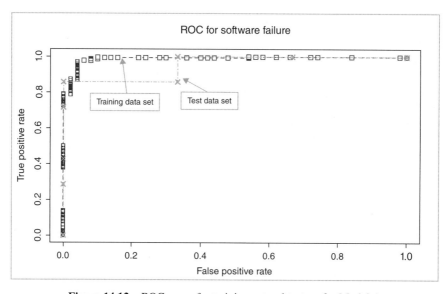

Figure 14.12 ROC curve for training set and test set for Model A

```
HTML> mattrain              HTML> mattest
     obs                         obs
pred  0   1                 pred 0  1
   0 46   2                    0 2  1
   1  4 130                    1 1  6
attr(,"class")              attr(,"class")
[1] "confusion.matrix"      [1] "confusion.matrix"
```

Figure 14.13 Confusion matrix for train and test of Model A

threshold	AUC	omission.rate	sensitivity	specificity	prop.correct	Kappa
0.5	0.9524242	0.01515152	0.9848485	0.92	0.967033	0.916232

Figure 14.14 Accuracy measures for training data set Model A

threshold	AUC	omission.rate	sensitivity	specificity	prop.correct	Kappa
0.5	0.7619048	0.1428571	0.8571429	0.6666667	0.8	0.5238095

Figure 14.15 Accuracy measures for test data set Model A

Figures 14.14 and 14.15, respectively. Additionally, the training set performs better for the Area under Curve (AUC) calculations, the omission rate, sensitivity, specificity, and Kappa compared to those same values for the test set. Due to the degradation in performance of Predictive Model A when run with the test set, the researcher created a second Predictive Model B to compare their performances and to identify the best model.

14.3.5.2 Analysis of Predictive Model B Based on the low accuracy results of the test set for Model A, the researcher created Predictive Model B and used the hold-out validation set $n = 5$ not used in the training of Model A to fine-tune the performance of the model. The researcher used the LiblineaR function in building Predictive Model B. LiblineaR allows estimation of predictive linear models for Regularized logistic regression to select the best factors leading to failure (Fan et al., 2008). LiblineaR produces seven types of glms by combining several types of regularization schemes and loss functions:

- 0 – L2-regularized logistic regression
- 1 – L2-regularized L2-loss support vector classification (dual)
- 2 – L2-regularized L2-loss support vector classification (primal)
- 3 – L2-regularized L1-loss support vector classification (dual)
- 4 – multi-class support vector classification by Crammer and Singer
- 5 – L1-regularized L2-loss support vector classification
- 6 – L1-regularized logistic regression
- 7 – L2-regularized logistic regression (dual)

Regularizations prevent overfitting by eliminating the random error or noise and preserving instead the underlying relationship between the dependent and independent variables, especially because this research contained a small number of training examples. Overfitting occurs when a model has too many parameters relative to the number of observations. A model that has been overfit may often exhibit poor predictive performance due to exaggeration of minor nuances in the data. Using regularization will allow Model B to scale to larger sample data sets in the future, with significant computational savings. Companies such as Facebook, Google, Yahoo, and

Microsoft use regularization – one can only imagine the vast amounts of data analyzed within each organization.

A drawback to regularization is that it may introduce a constraint cost penalty proportional to the size of every coefficient. The cost constraint is the trade-off between correct data classification and regularization. If this cost constraint is large, the model fit will tend to keep all failure factor parameters used. If a smaller cost constraint penalty is used, the end result may be the elimination of too many failure factors and, as a consequence, bias or poor fit in the model will result. The authors sought a solution to minimize the total cost from constraint violations. To account for the cost constraint violation risk in arbitrarily selecting one of the L1 or L2 regularization schemes, the authors went one step further to prevent bias by using the 10-fold cross-validation technique to identify the best regularized model with the best cost parameter and the best accuracy prediction. In developing the model and running the program through different regularization cost parameters, best Type = [6], which corresponds to L1-regularized logistic regression, best Cost = [100] for the cost constraint, and best Accuracy = [0.956044] for the highest accuracy of all the model options were identified.

Predictive Model B was therefore developed to determine how well the results of the analysis will generalize to the independent validation set. Though it is very important to remove failure factors in the model that were not significant because they unnecessarily take up degrees of freedom without adding to predictive ability, Predictive Model B with L1 regularized logistic regression and cross-validation combined proved that all the failure factors were contributors to project software failure and were integral in accurately predicting software project outcomes.

The authors tested Model B performance using the $n = 5$ validation data set and based on the results of the accuracy measures, confusion matrix, the ROC curve, and AUC, Predictive Model B performed better than Model A, as demonstrated by Figures 14.16–14.19.

```
glm(formula = Failure ~ FFA + FFB + FFC + FFD + FFE + FFF + FFG +
    FFH + FFI + TestF, family = "binomial", data = traindata)

Deviance Residuals:
    Min       1Q   Median       3Q      Max
-2.77443  -0.06671  0.03867  0.11212  2.85520

Coefficients:
            Estimate Std. Error z value Pr(>|z|)
(Intercept)  -5.4133     1.7115  -3.163  0.00156 **
FFA           2.8028     1.0260   2.732  0.00630 **
FFB           3.7279     1.1439   3.259  0.00112 **
FFC           0.6196     1.0059   0.616  0.53791
FFD           1.5375     1.0374   1.482  0.13834
FFE           2.6292     1.1466   2.293  0.02184 *
FFF          -3.8406     1.5269  -2.515  0.01189 *
FFG           0.5103     1.0123   0.504  0.61418
FFH           4.9002     1.1569   4.236 2.28e-05 ***
FFI          -0.6935     1.0813  -0.641  0.52130
TestF         2.9599     1.3222   2.239  0.02518 *
---
Signif. codes:  0 '***' 0.001 '**' 0.01 '*' 0.05 '.' 0.1 ' ' 1

(Dispersion parameter for binomial family taken to be 1)

    Null deviance: 213.996  on 181  degrees of freedom
Residual deviance:  41.579  on 171  degrees of freedom
AIC: 63.579

Number of Fisher Scoring iterations: 8
```

Figure 14.16 Significance p-values and AIC for Model B

```
HTML> mattest2
      obs
pred 0 1
   0 1 0
   1 0 4
attr(,"class")
[1] "confusion.matrix"
```

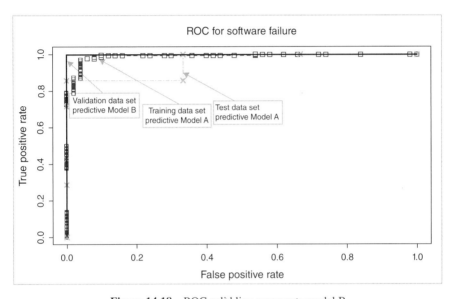

Figure 14.17 Confusion matrix for Model B

Figure 14.18 ROC solid line represents model B

	threshold	AUC	omission.rate	sensitivity	specificity	prop.correct	Kappa
1	0.5	1	0	1	1	1	1

Figure 14.19 Accuracy measures for Model B

Actual Outcomes versus the Predictions for Predictive Model B are displayed in Figure 14.20 to confirm the model's good performance. As demonstrated by Figures 14.16–14.19, Predictive Model B correctly predicts each project outcome. Figure 14.20 displays the results of project data entered into the predictive model to determine whether each project would result in success of failure. Project ID FP1, FP18, FP1, and FP10 with rows in the excel spreadsheet, respectively, 8, 32, 59, and 68 indicated that these projects would fail if they were continued. Project FP18 on line 32 specifically indicated that there would be a 99.98% failure rate due to the failure factors it exhibited throughout the development life cycle. Using a quantifiable mechanism such as this predictive model, the project manager would conduct evidence-based project management to inform decisions to cancel or continue the program.

```
row.names  ProjectID  FFA  FFB  FFC  FFD  FFE  FFF  FFG  FFH  FFI  TestF  Failure
8          FP8        1    0    1    1    0    0    0    1    0    0      1
32         FP18       0    1    0    0    1    0    1    1    1    1      1
51         SP21       1    0    1    1    0    1    1    0    0    0      0
59         FP1        1    1    0    0    0    0    1    0    1    0      1
68         FP10       0    1    0    1    1    0    1    1    1    0      1
HTML> predReg
          8           32          51          59          68
     0.98841950  0.99981969  0.02223404  0.71793497  0.99925268
```

Figure 14.20 Actual project outcomes versus predictions for Predictive Model B

Neither regularization nor 10-fold cross-validation is a novel concept in predictive analytics and logistic regression; however, the synergistic combination of the two methods identified a sound predictive model that can be used to predict software project failure in industry. This software product developed with a free open-source tool is based on data of failed and successful software-intensive projects found in the federal government and commercial industries. The data was collected from all domains of industry, all sizes of organizations, and varying software project costs. As such, the predictive model can be leveraged as an analytic tool in any of these domains to improve software project outcomes by making sound evidence-based decisions on the current development health of the software system and overall products developed within the organization in the past.

14.3.6 Model Implementation In Industry

14.3.6.1 Model B Use on New Projects Not Yet Started The authors propose that Predictive Model B be implemented in the systems engineering and software engineering processes at project inception and planning when decisions are being made about which projects to embark upon in the project portfolio. Managers would begin by collecting data of past projects developed within the project portfolio for the organization to determine how well past projects performed. If an organization is newly formed and developing their very first software product, the authors recommend collecting publicly available project performance data from competitors in the same market, similarly to how the data was collected for this research.

Once data has been collected, it would be entered into the Model B to provide a probability of failure score for the particular software project. The failure probability break-out scores are described as follows:

- **Scores of 0–40%** indicate a healthy and mature organization with processes in place to mitigate risk and ensure that goals, objectives, and requirements are clearly defined during planning and inception and are monitored throughout the software development stages. This organization develops robust and successful software systems that are deployed on time and on budget and satisfy user requirements. There are no perfect software projects; therefore, probability scores within individual factors tracking near failure should be documented as risks and monitored and controlled throughout the project life cycle.

- **Scores 40.01–50%** indicate a medium probability of failure. The potential risks to the program as identified by the model should be documented and risk remediation strategies must be developed and utilized throughout the life cycle if the decision-makers decided to proceed with the project. Process improvement efforts would be implemented within the phase or phases that posed the most risk to the health of the software system.
- **Scores 50.01–70%** indicate a high probability of failure. The risks to the program as identified by the model should be documented and risk remediation strategies must be developed and utilized throughout the life cycle. Strict monitoring and control as well as regular audits of the program should be performed throughout the life cycle. Process improvement efforts must be implemented within the phase or phases that posed the most risk to the health of the software system and tracked throughout the life cycle.
- **Scores 70.01–100%** indicate a severe probability of failure of projects within that portfolio. Decision-makers should proceed with great caution as the stakes are high and future failure is imminent. The risks to the program as identified by the model should be documented and risk remediation strategies must be developed and utilized throughout the life cycle. Strict monitoring and control as well as regular audits of the program should be performed throughout the life cycle. Process improvement efforts must be implemented within the phase or phases that posed the most risk to the health of the software system and tracked throughout the life cycle. These are programs the research would recommend not proceeding with and replace with others that would be more beneficial to the overall health of the organization and its software project portfolio.

14.3.6.2 Model B Use on Software Projects in Flight Though it is recommended that the model be used before project inception, the model is an excellent tool to measure the health of a software project during any stage of the software development life cycle to determine the present and future health of the software project. The model would be a tool to identify the current state of the project and to determine the improvements to undertake to ensure success. A single project data point at the current stage in the development life cycle would provide the input to the model. The results would be interpreted as follows:

- **Scores of 0–40%** indicate a healthy software project on track to successful completion with high quality, on time and on budget. However, there are no perfect software projects; therefore, probability scores within individual factors tracking near failure should be documented as risks and monitored and controlled throughout the project life cycle.
- **Scores 40.01–50%** indicate a medium probability of failure. The potential risks to the program as identified by the model should be documented and

risk remediation strategies must be developed and utilized throughout the remaining life cycle. Process improvement efforts would be implemented within the phase or phases that posed the most risk to the health of the software system with continuous monitoring and audits through continued use of the model.
- **Scores 50.01–70%** indicate a high probability of failure. The risks to the program as identified by the model should be documented and risk remediation strategies must be developed and utilized throughout the remaining life cycle stages. Strict monitoring and control as well as regular audits of the program should be performed throughout the life cycle by continued use of the model. Process improvement efforts must be implemented within the phase or phases that posed the most risk to the health of the software system and tracked throughout the life cycle. If the project is identified with a high probability of failure in the development or testing phases, decision-makers should weigh the options to potentially cancel the project and reallocate the resources to other more viable software projects identified by the model.
- **Scores 70.01–100%** indicate a severely troubled software project that is doomed for failure if risk mitigation strategies are not applied as soon as possible. Decision-makers should proceed with great caution as the stakes are high and future failure is imminent. Strict monitoring and control as well as regular audits of the program should be performed throughout the remaining life cycle stages by continued use of the model. Process improvement efforts must be implemented within the phase or phases that posed the most risk to the health of the software system and tracked throughout the remaining life cycle stages. These are software projects the research would recommend not proceeding with and implement others that would be more beneficial to the overall health of the organization and its software project portfolio.

14.3.7 Predictive Model Deployment in Industry

Ineffective executive governance coupled with ineffective project management allows software projects to continue development at staggering rates of failure. Insight and valuable information of troubled projects is not being communicated upstream in a way that decision-makers can understand and take action. The literature states that organizations do not have adequate processes in place for dealing with troubled software projects and a majority wait until the project has missed time and budget targets before taking action despite many executives' knowledge that early intervention on troubled projects would facilitate better management of limited resources.

In developing the model, the research did not identify concrete examples to make predictive tools and their use a mainstream requirement to improve software project outcomes.

14.3.7.1 The Model as a Legislative Tool As a legislative tool, the authors recommend that federal government projects conduct this analysis during project planning and throughout the life cycle and submit results to the Government

Accountability Board or other government authority for review prior to moving to the next stage of the software life cycle. DoDI Directive 5000.02 (2015) states, "The Defense Acquisition System exists to manage the Nation's investments in technologies, programs, and product support necessary to achieve the National Security Strategy and support the United States Armed Forces. In that context, our objective is to acquire quality products that satisfy user needs with measurable improvements to mission capability at a fair and reasonable price." The software intensive aspects of Major Defense Acquisition programs (MDAP) and Major Automated Information Systems (MAIS) programs carry the greatest risk consequences in terms of management level, reporting requirements, and documentation and analysis to support program decisions. The model will assist the Major Decision Authority (MDA) and supporting staff organizations at program decision reviews using evidence-based data derived from the model to facilitate the examination of the system to allow the MDA to decide whether a program is ready to proceed to the next milestone.

The authors identified several segments of the directive in which the model and its subsequent probability results can be recommended for use. In depiction of the generic acquisition milestone and decision points in Figure 14.21, the directive states

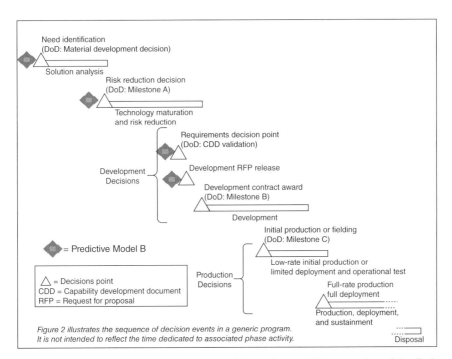

Figure 14.21 Generic acquisition phases, decision points, and incorporation of Predictive Model B use in DoDI directive 5000.02 operation of the Defense Acquisition System in determining if the program is viable and whether contractors are fully capable of delivering a successful system within scope, cost, quality, and schedule

that in practice all decisions must be made prior to RFP release; however, for DoD, the Development RFT Release Decision Point is where plans for the programs must be carefully scrutinized to ensure that all risks are identified, understood, and are under control. It is the critical decision point in acquisitions and determines if "the program will either successfully lead to a fielded capability or fail, based on the soundness of the capability requirements, the affordability of the program, and the executability of the acquisitions strategy" (DoD, 2015). The strategic incorporation of the model results will be beneficial in meeting the following objectives within the directive as depicted in Figures 14.21 and 14.22:

1. To ensure that the program plan is sound, its objectives are clearly understand and the program will be successfully executable and remain within budget prior to releasing the RFP in section 6 of DoDI 5000.02 Directive.
2. To ensure that contractors have a track record of delivering quality software systems that meet the user needs, within cost and schedule parameters by including language and justification for them to provide the results of the model to show past performance in their responses as a qualification criteria.

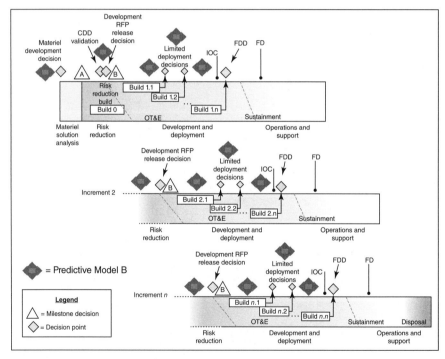

Figure 14.22 Incrementally deployed software intensive program milestone decisions, decision points, and incorporation of Predictive Model B in DoDI directive 5000.02 operation of the Defense Acquisition System to inform these critical software project decisions in determining project cancelation or continuation

3. To perform periodic reviews of the program in progress and identify opportunities for adjustment, improvement or redirection by the Configuration Steering Boards as in paragraph 5d(5)(b). Performance data will be identified by the model to inform the RFP Release Decision point and throughout the project life cycle.
4. To assess the performance of incrementally deployed Software Intensive Programs. The development process in Figure 14.22 describes systems that are deployed in increments of 1–2 year cycles. The use of the model will determine whether or not each increment is a fully functional new capability. The model will assess each past deployed increment to predict the likelihood of success for the succeeding increment and to identify and mitigate any risks that might impede success.
5. To regularly check the health of high-risk projects and report their health to Congress and designated decision-makers. A GAO Report (2006) describing High-Risk IT projects provided the recommendation that the Director of the Office of Management and Budget (OMB) establish a process for government agencies to update high-risk projects on a regular basis. The model would be an invaluable tool in that health assessment.

Another example of the model's legislative integration is within the Clinger-Cohen Act (CCA), enacted by Congress in 1996, to reform and improve the way Federal agencies acquire and manage IT resource. The act was mandated by senior government officials in the implementation of strategies to control development risks, to better manage IT spending, and succeed in achieving measureable improvements in software project performance. The use of the model as a predictive tool may be considered a required aspect of CCA compliance at the selected Decision gates in the software engineering life cycle for federal government agencies. CCA regulatory compliance is an aspect of the DoD 5000.02 Directive where noncompliance is reported to Congress. The use of the Model and the generated probability of failure score for a program would better define the areas of noncompliance and the extent to which the software project has deviated to inform an evidence-based decision to cancel or to proceed with strict risk mitigation strategies.

14.3.7.2 The Model as a Procurement Decision Tool The authors recommend the Model be used as procurement decision tool requiring this analysis and its results in the Request for Proposal for all contractors bidding on software procurement contracts. The tool can be used as a forcing function requiring contractors to list out all work performed and their outcome. The tool would facilitate transparency and provide a probability of success score for each contractor. This will allow decision-makers the ability to compare past performance with quantitative results to make evidence-based decisions on contractor selection.

The Federal Acquisition Regulation (FAR) governs the acquisition process by which the Federal Government acquires services and goods to regulate government personnel activities in carrying out that process. This tool can be embedded in the FAR (2014) Part 9 Contractor Qualifications. Subpart 9.2(a)(1) Qualification

Requirements states the policy in carrying this out, "The head of the agency or designee shall, before establishing a qualification requirement, prepare a written justification." The justification would be that software projects are failing at staggering rates. Despite rigid controls, project management and systems engineering rigor, software project failures still persist. The qualification requirement would consist of using the Predictive Model B to assess the overall health of the contractors' software development programs to maximize competition and to ensure that the federal government achieves the highest possible value for software products and services.

14.3.7.3 The Model as a Systems Engineering, Software Engineering, and Project Management Enabling Tool Systems engineering is an interdisciplinary approach enabling the realization of successful systems by defining customer needs and required functionality early in the development cycle. Systems engineering considers both the business and technical needs of all stakeholders with the goal of providing a quality product that meets user needs, within cost, schedule, scope, and quality parameters, all the while minimizing undesirable consequences. This can be accomplished through the inclusion of and contributions from experts across relevant disciplines and coordinated by the systems engineer. Systems engineers are commissioned to explore issues and undesirable consequences to make critical decisions expeditiously (INCOSE, 2010).

According to IEEE, the definition of software engineering is the application of a systematic, disciplined, quantifiable approach to the development, operation, and maintenance of software. Software engineering can also be defined as the study of these approaches or the application of engineering to software. Software engineering's primary objective is the production of programs that meet specifications, are produced on time, and within budget.

Project Management is the application of a collection of techniques, tools, and processes to manage and direct the use of resources in the successful execution of a complex, unique, one-time task with constraints of scope, quality, cost, and time. Each task calls for a unique set of tools and techniques specific to the task environment and its life cycle (PMI, 2015). The underlying theme of the three disciplines is the development of successful projects on time, on budget, within scope and quality parameters. According to the Systems Engineering Handbook, schedule and cost overruns are lessened with increased systems engineering rigor. The handbook went on to note that cost and schedule overrun predictions are very difficult to gauge with low systems engineering effort. However, the staggering rate of software project failure indicates that more needs to be done. The authors propose the incorporation of the model in key decision processes and areas not just of systems engineering, but within software engineering and project management as well.

Why should an organization care about analytics, systems engineering, software engineering, and project management? To borrow a phrase from the SE Handbook, "to better understand, evaluate, control, learn, communicate, improve, predict, and certify the work performed." Figure 14.23 identifies the overlaps in the three disciplines and the areas where the predictive tool would be integrated to predict issues

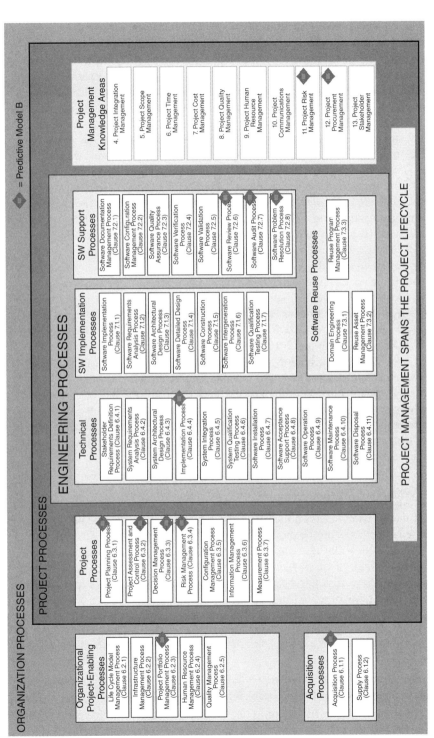

Figure 14.23 Project management, ISO/IEC/IEEE 15288 and 12207 systems and software life cycle processes overlay and predictive model incorporation

and facilitate decision-making. For the purposes of this research, the authors propose incorporating the predictive model within the IEC/ISO/IEE 15288 and 12207 guidance as well as the PMI process and knowledge areas. The model as a tool would assist stakeholders within each discipline to collaborate one with the other in achieving the goal of delivering a successful software system. The model would allow full transparency among those responsible for a project's success. Project outcome probability scores would be made available during project acquisition and procurement in making source selections where the past performance would drive the identification of key processes and mitigation strategies to implement to ensure success. The model would be used in project planning to identify areas of concern and as an assessment and control process tool to ensure that the project is on track. The model would be incorporated into the software implementation process to monitor its development and to make critical decisions at designated milestone gates throughout the life cycle. The model would be used within the software audit and review process to ensure that the user needs are being met within budget, schedule, and quality constraints. The model can be leveraged in the problem identification and risk management where key failure factors are documented and mitigation strategies are developed and tracked to ensure a successful outcome.

"[The] growth of complexity is accelerating and we need tools to influence SPs. How do we use modeling to take on a systems engineering challenge that is quantitatively different from the past?" (Wade, 2014) One answer, and the focus of this research, is the emerging technology of predictive analytics that uses data and information to predict future software project outcomes. Effective implementation using a select group of synergistic system engineering tools, practices, and processes requires vision. This research can help to turn the tide on project failure to push software development projects toward success by utilizing good systems engineering practices. The output of the model can be used to guide the project team to ensure that all stakeholders have the same vision and drive toward SP outcome.

14.3.8 When the Decision Has Been Made to Cancel the System

Now that the decision has been made to cancel the project, overcommunicate this with the team members on the project ensuring that EVERYONE on the team has an opportunity to provide feedback. This input can be incorporated into the lessons learned and the project reviews. Many working on a project that needs to be canceled almost always knows it should be canceled. When the decision has been made to cancel, there is a sense of relief from the team that the right decision has been made, leading to improved morale. This is the time to establish cancelation best practices:

- Communicate, communicate, and communicate the cancelation decision to everyone on the project.
- Collect feedback from everyone on the project team to gain their perspective and to remove the emotional effects of a perceived failure.
- Develop lessons learned document and archive all project documentation and work products. In the archiving include hardware, software, licenses, and

services that have been paid for. There may be modules that could be of value on another project.
- Develop a final accounting of the project actual cost and benefit. Ensure that there are no outstanding invoices, software license agreements are up to date, and so on.
- Develop a resource reassignment plan to identify other projects on which the talents of the team can be used to make those projects successful.
- Conduct a final wrap-up meeting to close out the project and ensure that the team stays clear away from finger-pointing and blame.

14.3.9 Conclusion

This research will assist in carrying out quantifiable Systems Engineering that aids in making informed decisions. During the Introduction of the SE Vision 2025 at the INCOSE IS in 2014, a panel member asked the key questions that all decision-makers and project stakeholders must ask during project planning: "How do you measure value? What are the differences between success and failure? (Wade, 2014)". This same panel member immediately followed those questions with a crucial response: "Organizations that have data do analysis to make decisions in a decisive way" (Wade, 2014). The results of the model and ultimately this research are to assist decision-makers and stakeholders in developing mitigation strategies in their systems and software engineering programs to prevent project failure by implementing risk reduction strategies that decrease the occurrence of defects in the system. Complete defect prevention is not realistic, given the nature of software and the increasing complexity of systems. However, a case study evaluating project decisions sums up this research by stating, "Effective decisions are crucial to the success of any software project, but to make better decisions you need a better decision-making process to systematically evaluate portfolio decisions and avoid the bad choices that lead to project failure" (Hoover et al., 2010).

This research demonstrates that there is no one silver bullet that will solve the problem of software project failure (Brooks, 1995). In fact, a combination of tools, techniques, and processes is required to improve software project outcomes. Systems engineers, software engineers, and project managers need information for planning, estimating, and tracking project work. The predictive model can be a support mechanism within software development. Information is needed to project the future, to educate, and to communicate, for identifying and resolving problems and, finally, for making evidence-based decisions in software programs.

The research for this paper did have some limitations, including the relatively small data set and selection bias in the research data. However, the information sources were very reliable in that the bulk of the research consisted of failure and success data extracted from GAO reports, court litigation cases, and Government Investigation Board Reports. The authors would like to expand the predictive model to measure its performance on large project databases such as the Standish Chaos database, which consists of 50,000 projects (The Standish Group, 2013).

The authors have demonstrated a predictive tool that can be deployed across many domains with the one clear objective of delivering successful software projects on time, on budget, and of the highest quality. It is a tool to help stir legislation to adopt policies in software engineering to facilitate and mandate disclosure of past and present software project performance. It is a tool to assist in effective acquisition and procurement to acquire knowledgeable, capable, and success-driven resources with the track record of achieving software project success. It is a tool that is easily assimilated into the systems engineering, software engineering, and project managers' current toolbox to bind the three disciplines under one common goal, agenda, and objective: to deliver successful software.

Ineffective executive governance coupled with ineffective project management allows projects to continue development at staggering rates of failure. Insight and valuable information of troubled projects are not being communicated upstream in a way that decision-makers can understand and take action. Organizations do not have adequate processes in place for dealing with troubled software projects and a majority wait until the project has missed time and budget targets before taking action despite many executives' knowledge that early intervention on troubled projects would facilitate better management of limited resources. Canceling a project is never easy; however, it may be the only right way to proceed.

14.4 PRODUCT RETIREMENT DECISION TRADE STUDY

14.4.1 Introduction

Decisions to retire, decommission, or sunset legacy systems are not often made in a timely and cost-effective manner. All too often projects languish in limbo where money and resources are spent on the upkeep of a system that has exceeded its service life. At one organization, it was identified that many systems were still kept running despite not been used in several years or they were used by fewer and fewer users mostly for data extraction. In this organization, the maintenance cost of the growing number of legacy systems in production presented a challenge to the overall performance of these systems. Users often complained about the antiquated technology and the many steps to complete transactions in addition to the slow performance of the program. More often than not, these systems are laden with defects and workarounds making them burdensome for the users.

An analysis of the open defect ticket reports for one legacy system in the organization's portfolio was performed. Over 1000 open and active defects were identified with some dating back over a decade. The original intent of the defect analysis was to categorize and prioritize them to generate an integrated prioritized master defect list to update and upgrade the existing system. The analysis team suggested additional analysis to identify the capability needs of the organization and the continued reliance on the system as well as to perform an architecture analysis. The architecture analysis would identify the application environment and determine if the system was in alignment with the organization's present Enterprise Architecture (EA) strategy.

However, our suggestions were not followed through and the legacy system is still in production with the mounting list of active defects.

Our suspicion was that the legacy system should have been retired and any functionality required for business continuity could have been developed into a new system or included into an existing system as an upgrade. Another significant challenge noted with that legacy system was the incompatibility with new systems developed within the organizational program portfolio as these were some of the defects documented in the defect report. Incompatibility of the legacy data to integrate with the new fully integrated systems results in continued use of the legacy system as a workaround until the data challenges are fixed. However, our experience noted that the data challenges do not usually get resolved because it is easier and more convenient to continue using the legacy system data with the increased risk to data integrity and vulnerability.

Why do organizations do such a bad job at retiring aging systems? Our experience determines that many organizations do not have a process to decommission legacy systems. It is often something that just happens, often when the system simply dies of old age. This section of the book outlines the importance of proper decommissioning and decision analysis mechanisms that are employed in this important but often neglected final step in the systems life cycle.

14.4.2 Legacy HR Systems

14.4.2.1 Collecting the Data An analysis was conducted on several HR applications within the Department of Treasury portfolio of legacy applications to determine whether or not retiring several of the legacy systems written in COBOL and/or Assembly Language was the best course of action. A cost–benefit analysis coupled with NPV, ROI, and break-even analysis was performed to come to a decision to retire the systems. Additionally, recommendations were made to transfer of the legacy systems' capabilities to new web-enabled technologies with database storage capabilities to provide easy access to a consolidated user environment (Istiaque et al., 2014). The tangible and intangible benefits achieved in retiring the legacy systems and developing new technologies were identified and included:

- Developing new cost-effective systems reallocates those dollars from maintenance of antiquated systems to development of new state-of-the-art web-enabled systems that do not require the high cost of COBOL and Assembly Language specialized development maintenance support.
- Targeted enhancements address current and relevant business needs that the legacy systems could not address.
- The new system provides a modernized technical architecture that is flexible and scalable for future growth and changes in technology platforms.

Table 14.20 outlines the one-time costs associated with developing the new web-enabled HR system and migrating the data from the legacy system over to the new. These costs represent the tangible or quantitative costs for systems engineering

Table 14.20 One-Time Cost of Decommissioning the Legacy HR System and Migrating its Functionality to the New Web-Enabled Platform

One-Time Cost	Technical Skills Needed	Tangibility (T/I)	Resource Type	FTE	Cost per Hour	Total Cost
Software Analysis and Design	System business Analysis	T	Internal	1	36	$1,440
Software Development/Training	Software developers with re-engineering legacy systems experience	T	External	3	65	$7,800
Website design	Web software developer	T	Internal	1.5	38	$2,280
Oracle developer/DBA	DB setup/data migration	T	Internal	1.25	47	$2,350
Software testing	QA/testers	T	Internal	1.5	40	$2,400
Software most free (i.e., open source codes)	Developers/Systems Admins	T	N/A	N/A		$0
Hardware (HP DL ProLiant with Raid disks)	Procurement Manager	T	N/A	N/A		$6,500
Total FTE, Hours, and Cost				8.25	226	$22,770

Recurring Annual Cost	Tangibility (T/I)	Total Cost
Software support (optional)	T	$0
Hardware maintenance	T	$850

Source: Data from Istiaque et al. 2014.

PRODUCT RETIREMENT DECISION TRADE STUDY

Table 14.21 Cost Gains from Decommissioning the Legacy HR System and Migrating Its Functionality to Faster Web-Enabled Technology Platform

Type of Economic Benefit Gained by Re-engineering HR Legacy Systems	Tangibility (T/I)	Cost	Total Saving
Availability of secured centralized, which allows data sharing and reduces data redundancy	I	N/A	N/A
Added additional systems functionality that improves productivity and efficiency	I	N/A	$3000
Incorporated a report tool for faster decision-making	I	N/A	$1800
Modernized web-based architecture gives flexibility for future growth and changes in technology.	I	N/A	N/A
Saved on the cost of yearly maintenance supported legacy software (i.e., COBOL etc.)	I	N/A	$2500
Total			$7300

Source: Data from Istiaque et al. 2014.

of the new system. In this example, the costs and the task of decommissioning the system consisted solely of the data migration activity. The total one-time cost to develop and decommission was $22,770 with a recurring yearly hardware maintenance cost of $850.

Table 14.21 outlines the total cost savings and gain in decommissioning the legacy HR system and migrating the data and functionality to the new system. The Intangible benefits of the decommissioning are highlighted with several containing cost savings, including the elimination of the yearly maintenance cost related to a COBOL and Assembly Language developer for the legacy system. In this example, the recurring software support cost is $0 as the new system operates on free open-source software.

In calculating the new development costs and the decommissioning economic benefits gained, the NPV calculations could be derived to determine whether or not it was economically sound to decommission the systems. Often, cost benefits alone may not be sufficient to make an informed decision to retire the system.

14.4.2.2 Using Net Present Value for Driving Retirement and Decommission Decisions

NPV helps to determine whether a project will result in net profit or loss. A positive NPV results in profit and a negative NPV results in loss. NPV is a useful method for the time value of money to appraise long-term projects. In this context, NPV will be used in the decision to retire a system by determining the amount of value the project adds to the organization in $t = 5$ years. Using the results for the cost benefit and economic gain calculations in the previous section, NPV is calculated with the following formula:

$$\text{NPV} = \frac{R_t}{(1+i)^t} \tag{14.26}$$

where

t: number of time periods of the cash flow;

i: discount rate (the rate of return that could be earned on a project with similar risk); the opportunity cost of capital;

R_t: Net cash flow at time t

If R_t is negative, then the project is in discounted cash outflow status in time t and thus becomes a candidate for a retirement decision without developing new systems to replace it. Therefore, projects with a positive NPV with an acceptable level of risk may be chosen to be decommissioned and the capability migrated to more cost effective platforms. Table 14.22 describes the retirement decision matrix as it relates to NPV outcomes.

The results of the NPV calculations where $i = 12\%$, $t = 5$ years, and $R_t = \$22,770$ are displayed in Table 14.23. Notice that the Net Economic benefit value of \$7300 and the Recurring Cost of \$850 were incorporated into the NPV calculation. This was to ensure that the NPV of all the benefits and all the costs weighed into the final decision to decommission only or to decommission and develop the new system. The overall NPV is a positive \$481, and based on the matrix in Table 14.22, the project investment will add value to the organization and supports a decision to decommission the legacy system and replace it with the new web-enabled technology identified in the initial analysis.

To ensure that the decision made from the NPV calculation was sound, ROI and break-even analysis were calculated, and both confirmed the decision to retire the legacy system, develop the new system, and migrate the data. ROI measures the bottomline return on an investment.

$$\text{ROI} = \frac{\text{Gains} - \text{Cost}}{\text{Cost}} = \frac{26{,}315 - 25{,}834}{25{,}834} = 0.01862 \qquad (14.27)$$

Calculating ROI for this example provides the project team additional insight into making a sound decision to decommission the legacy system and develop the new

Table 14.22 The Retirement Decision Matrix When Translating the Resulting Net Present Value of a System

If	Definition	Recommended Decision
NPV > 0	The project investment will add value to the organization	Decommission the legacy and develop the new system to replace it
NPV < 0	The project investment will detract value from the organization	Retire the system but do not develop the new capability
NPV = 0	The project investment is neutral, with neither gain nor lose to the organization	A retirement decision should be made based on other criteria

Table 14.23 Calculations for the NPV, ROI, and Break-Even Analysis for Decommissioning the Legacy System and Developing the New Web-Enabled System

	Net Present Value (NPV), ROI and Break-Even Analysis						
		Year of Project					
	Year 0	Year 1	Year 2	Year 3	Year 4	Year 5	Totals
Net economic benefit	$0	$7,300	$7,300	$7,300	$7,300	$7,300	
Discount rate (12%)	1	0.8929	0.7972	0.7118	0.6355	0.5674	
NPV of benefits	$0	$6,518	$5,820	$5,196	$4,639	$4,142	
NPV of all benefits	$0	$6,518	$12,338	$17,534	$22,173	$26,315	$26,315
One-time costs	($22,770)						
Recurring costs	$0	($850)	($850)	($850)	($850)	($850)	
Discount rate (%12)	1	0.8929	0.7972	0.7118	0.6355	0.5674	
NPV of recurring costs	$0	($759)	($678)	($605)	($540)	($482)	
NPV of all costs	($22,770)	($23,529)	($24,207)	($24,812)	($25,352)	($25,834)	($25,834)
Overall NPV							$481
Overall ROI							0.01862
Break-even analysis							
Yearly NPV cash in flow	($22,770)	$5,759	$5,142	$4,099	$4,099	$3,660	
Overall NPV cash out flow	($22,770)	$17,011	$11,869	$7,278	$3,179	$481	
Project Break-even occurs between year 4 and year 5							
Break-even ratio = (4099 − 3179)/4099							0.22
Actual break-even occurred at 4.3 years							

Source: Data from Istiaque et al. 2014.

web-based technology. Over the 5-year time period, the project will yield 1.8% return on the initial investment of $22,770 or $423.

Break-even measures the time required for the cash inflows to equal the original cash outflow. It measures the risk to the organization in retiring the system and migrating the data to the new system and helps to determine if the idea is worth pursuing. Looking at the spreadsheet, the break-even falls somewhere between years 4 and 5. Calculating the breakeven ratio yields = 4099 − 3179/4099 = 0.22; therefore, break-even occurs at 4 years and 3 months where 0.22 year is rounded up as 0.25 year or 3 months.

In the analysis of the legacy systems and to determine the best decision to move forward with systems decommissioning, several decision analysis methods were used to ensure that the risks were quantified, project costs and benefit gains were evaluated, and NPV was determined. These methods helped to ensure a sound data-driven and evidence-based decision was made to decommission and retire the legacy HR system.

14.4.3 The US NAVY Retirement and Decommission Program for Nuclear-Powered Vessels

"Carrier strike groups are expensive to buy and to operate. Factoring the total lifecycle costs of an associated carrier air wing, five surface combatants and one fast-attack submarine, plus the nearly 6700 men and women to crew them, it costs about $6.5 million per day to operate each strike group (Hendrix, 2013)." For example, the cost of the USS George H W Bush comes in at $7 billion dollars. The USS Gerald R Ford costs $13.5 billion and has a service life of 50 years (Hendrix, 2013).

The SRP is the decommissioning and retirement process used by the US Navy to retire nuclear vessels. The first nuclear submarine built in the United States was Nautilus in 1954. The Navy has been operating nuclear-powered submarines beginning with USS Nautilus deployment into operation in 1955. Since Nautilus, nearly 200 US submarines have been delivered into production. Many of these vessels have reached their useful life where their time in military operability does not justify their continued operating cost. Also, arms control treaties have forced the Navy to remove a significant number of vessels from operation due to their personnel safety, environment safety, and national security concerns (US Department of State, 2015). The primary power circuit in the vessel consists of the reactor, the pumps, the steam generators, and the connecting pipes located in the reactor compartment. The reactor cores may house 200–300 fuel assemblies each containing tons of fuel rods. The thermal power of these vessel reactors can range from 10 MW to 200 MW (Kopte, 1997). An aging nuclear vessel is an extremely dangerous collection of radioactively contaminated components; therefore, timely decommissioning is critical to personnel and environmental safety.

SRP is carried out at the Puget Sound Naval Shipyard (PSNS) in Bremerton, Washington. Before SRP can begin, however, several critical steps must take place on site or at other designated sites.

> Step 1. The nuclear fuel is removed in a process called defueling. The vessel is referred to as "USS Name" prior to the fuel being removed. After defueling,

the "SSS" is dropped and the vessel is simply referred as "ex-Name." Defueling of the vessels is carried out at five ship repair facilities on the West Coast. The spent nuclear fuel removed from the decommissioned ships and submarines is shipped to the Naval Reactor Facility at the Idaho National Laboratory (INL) in Idaho where it is stored in special fuel containment canisters. The nuclear fuel is not reprocessed.

Step 2. Reusable equipment is removed and compartmentalized accordingly for recycling or for sale to other organizations.

Step 3. The hulls are towed to PSNS and thus begins SRP where the submarine is cut into four segments. The first segment is the reactor compartment that is removed from the vessel, sealed at both ends, and shipped to the Department of Energy's (DOE) Hanford Nuclear Reservation in Washington State where they will be buried in reactor compartment trenches as seen in Figure 14.24. The expectation is that the nuclear compartments will be able to contain reactors for more than 600 years without losing their integrity (Knot & Allen, 2012).

Step 4. The Missile compartment comprises the second segment and is dismantled according to the strategic Arms Reduction Treaty (Department of State, 2015).

Step 5. The last two segments the aft and the forward of the vessel are joined together. Materials considered hazardous by the Environmental Protection

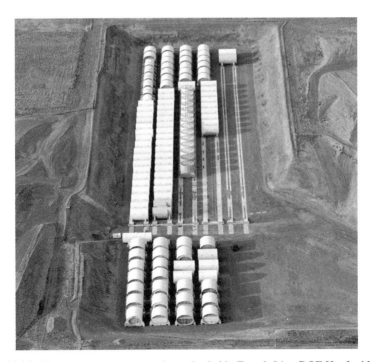

Figure 14.24 Reactor compartment packages buried in Trench 94 at DOE Hanford Nuclear Reservation in Washington state as of November 2009 (Source: Knot 2012)

Agency and United States Coast Guard are removed from these sections. In order to reduce the costs, these remaining submarine sections are recycled, all hazardous and toxic wastes are identified and removed, and reusable equipment is removed and put into inventory. Scrap metals and all other valuable materials are sold to private industry. The overall process is not profitable, but does provide some cost relief.

The retiring of submarines by the SRP costs the Navy approximately $25–$50 million per submarine. The high cost of decommissioning is due to the fact that one cannot simply turn off the ignition switch on the submarine and park it at the Navy Yard. The nuclear reactors of vessels that have not been defueled must be cooled permanently by circulating coolant through the primary circuit by running electricity to the vessel. Failure to "keep the lights on" would result in the reactor overheating and a nuclear meltdown would ensue. In addition, maintenance crew members must remain with the vessel to ensure its safe operation at all times. The annual cost to keep one nuclear submarine idling at port is $32 million and approximately 630,000 man hours (Kopte, 1997).

In 1959, the NSS Seawolf was decommissioned and its reactor compartment was dumped into the Atlantic Ocean at a depth of 2700 meters and 200 km east of Delaware. However, planned decommissioning did not commence until nearly 25 years later in the 1970s. The first two vessels removed from operation were the USN SSN Thresher and Scorpion that sank due to accidents. The USS Thresher imploded and sank 100 miles east of Cape Code, Massachusetts, during its test voyage killing all 129 naval members and 17 civilian observers onboard. The USS Scorpion sank 400 miles southeast of the Azores killing 99 crewmen aboard (Navy, Submarine Chronology, 2000) (Kopte, 1997). Though these vessels were decommissioned, a strategy to safely decommissioning first began with SRP. With adequate funding, SRP was deployed in the early 1990s as an integrated program with special processes, production facilities and infrastructure to retire, dismantle, and dispose of nuclear-powered vessels. The program was allocated $2.7 billion to decommission more than 100 nuclear submarines in the naval fleet.

Nuclear-powered submarines are complex weapons carrier systems that are quite difficult to dismantle. The biggest threat is the disposal of the nuclear and conventional torpedoes and missiles built into the system. After about 25–30 years in operation, the nuclear reactor and the parts surrounding the reactor will become radioactively contaminated. The uranium fuel used to power the vessel is far more highly enriched than that of civilian reactors in order that the vessels may go longer periods between refueling the reactor: up to 10 years between refueling. In fact, each refuel and overhaul cycle costs minimum $250 million per vessel. As such, the Navy has decommissioned some newer class LOS ANGELES submarines as they were coming due for this midlife refueling. In the face of current political, economic, and military environments, Navy officials could not justify the $250 million price tag to refuel each vessel (Toppan, 2000).

To date, submarines have been the only vessels decommissioned; however, the first aircraft carrier due for decommissioning and entering the SRP is the USS Enterprise

Figure 14.25 USS Enterprise is the first nuclear-powered aircraft carrier commissioned in 1961 and decommissioned in 2011 (Source: US Deparment of the Navy 2015)

(CVN 65). Commissioned in 1961 and withdrawn from service in 2012, ENTERPRISE was the oldest active duty ship in US Naval Fleet. The mammoth vessel shown in Figure 14.25 was powered by eight nuclear reactors; two reactors were harnessed together to power each of the ships' four propeller shafts. Dubbed the world's first nuclear-powered aircraft carrier, ENTERPRISE was constructed in 3 years and 9 months, and in October of 1962, it was dispatched to its first international crisis to set up quarantine and intercept all military equipment shipments into Cuba to help end the Cuban Missile Crisis (Navy U., 2015).

An environmental assessment was conducted to decommission the USS Enterprise with several alternatives considered. The first alternative was the No Action alternative, which would place ENTERPRISE at a long-term storage facility for nuclear-powered ships located at Bremerton, Washington. The estimated cost for this alternative would require $11 million dollars to construct and install the necessary structures to moor the ship and the additional annual cost for support and maintenance services. These structures and additional services included fire and flooding alarm systems, dehumidification system, cathodic protection and lighting, and the associated electrical power distribution upgrade because the current systems are not adequate to meet the demands of the ENTERPRISE. The periodic maintenance and inspection costs for long-term storage include the extensive 8-year hull inspection and the 15-year dry dock inspection and repair. It was noted that this alternative was not preferred as it simply delays ultimate permanent reactor compartment disposal and produces an increased risk of a nuclear reactor accident to the maintenance personnel and the environment. Maintenance costs would further increase as the ENTERPRISE hull deteriorates with age, requiring the necessary repairs to maintain watertight integrity (Knot & Allen, 2012). The USS LONG BEACH, for instance, has been in waterborne long-term storage for over 15 years with an extension to the 15-year dry dock inspection recently approved. LONG BEACH selection of the No Action Alternative may not have been a cost-effective decision as the alternatives comparison in Table 14.24 shows that progression to decommissioning in the SRP program is the best alternative based on many factors identified in the Environmental Assessment by Knot & Allen, 2012. This is an indication that decommissioning and retirement planning for these vessels was not laid out during the early design phase in the systems engineering life cycle. In fact, when nuclear-powered vessels were built, disposing of the reactors at the end of their service life was simply an afterthought. The engineers assumed that the reactors could be dumped in the ocean; the United States disposed of USS Seawolf at sea (Toppan, 2000).

The preferred alternative is decommissioning of ENTERPRISE expected to arrive at PSNS and IMF at Bremerton, Washington, defueled and under tow in 2017. There the reactor compartments would be removed and prepared for disposal as reactor compartment packages, the remnant hull sections would be recycled, and the reactor packages would be transported to Trench 94 for disposal at the Department of Energy Hanford Site. This alternative under the SRP program is estimated to cost $500 million. The environmental concern included the exposure of PSNS and IMF employees at a rate of not more 0.5 rem per year of radiation exposure. The reactor compartment disposal of ENTERPRISE will require approximately 6 to 8 years to complete. Figure 14.26 provides a pictorial rendering of the ENTERPRISE reactor compartment packaging and loading concept for the preferred alternative to decommission under the SRP program.

Nuclear vessels and their respective reactors do not last forever. As these vessels age, the need for maintenance and repair increases and thus becomes more expensive. Additionally, their technology platform may become obsolete, preventing them from living up to the present-day requirements for combat (Kopte, 1997). Table 14.24 was produced using the data outlined in the 2012 Environmental Assessment report by Knot & Allen, 2012 and aids the decision-makers in coming to an informed decision to decommission the ENTERPRISE. By the end of 1994, the United States had built 180 submarines. Nearly 80 have been decommissioned with 43 nuclear reactor compartments at the burial site in Hanford. The Navy plans to decommission the Sturgeon and Benjamin Franklin classes and has begun decommission of the Los Angeles and the Ohio class of vessels (Toppan, 2000).

In 2004, the Navy began engineering the Virginia (SSN 774) class of submarines with the most recent commission of the USS Minnesota in 2013. Minnesota's nuclear reactor plant is designed to last the ships' 30-year service life to help reduce life

Table 14.24 Several Relevant Factors in Determining the Decision to Retire/Decommission the ENTERPRISE

Decision	Cost	Years	Environmental Effect	Socioeconomic Effect	Carcinogenetic Effects	Radiological Effects
No-Action Alternative	$11 million + $32 million per year indefinitely	Indefinite	Nuclear breach could release radioactive waste into the environment	630,000 man hours and $32 million per year indefinitely	Nuclear breach could lead to the death of crewmembers	Nuclear breach could release radioactive waste in large doses
Decommission under SRP Alternative	$500 million total expenditure	8 years	0.06 curies of radioactive waste will be released into the environment	850,000 man-days of work with total cost of $300 million to $500 million assuming $400 to $500 per day	0.1 latent cancer fatalities based on one latent cancer fatality per 2500 rem of worker exposure	<25 millirem per year

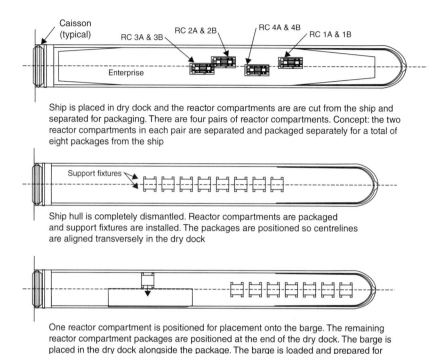

Figure 14.26 Enterprise reactor compartment package barge loading concept for preferred alternative (Source: Knot 2012)

cycle costs due to the refueling (Hawkins, 2013). Six additional Virginia class vessels have begun construction with two more under contract within the fleet (Navy, United States Navy Fact File, 2014). However, it is not certain if the engineers included the total cost of decommissioning through the SRP as a segment of the total life cycle cost of the Virginia Fleet of nuclear-powered vessels. "The United States, always an innovating nation, must break out of its ossified force structure and not only get ahead of the strategic curve, but plan a graceful transition that stops building carriers, plans a path for those already built to see them through their service life and creates new means of operational effectiveness in the future (Hendrix, 2013)." Embedded in the strategy of building systems, a retirement program must be outlined during the early planning stages of the systems life cycle to retire and decommission systems in a cost-effective and environmentally conscious manner. Cost, schedule, and resources must be preallocated to effectively carry out the retirement strategy for a system to prevent its costly, slow, and languishing demise. As commission decisions are made for the USS Minnesota and the ninth ENTERPRISE (CVN 80), it is recommended that a decommission strategy is laid out in parallel to prevent the costly storage of these giant vessels at astronomical costs to the federal government and taxpayers.

14.4.4 Decision Analysis for Decommissioning Offshore Oil and Gas Platforms in California

There are over 6500 oil and gas platforms worldwide that the industry is looking to decommission by 2025. A concept currently exists in some regions called the rigs-to-reefs that allows leaving decommissioned platforms where they stand to become a habitat for sea life; however, most regions of the United States and the world require complete removal. Complete removal of these massive structures will cost the industry billions of dollars and will leave a massive carbon footprint (Macreadie et al., 2011). The analysis of the air emissions from complete removal of one of these structures contained 29,400 tons of carbon dioxide, 600 tons of NO_x, and 21 tons of fine particulates (PM10); levels at such magnitudes would not allow air quality regulatory agencies to provide the permits to decommission the platform (Henrion et al., 2015).

The basic structure of the oil and gas offshore platform is depicted in Figure 14.27 and comprises the topsides, which include the gas and oil piping and processing equipment; the well conductors are pipes for drilling mud down and oil and gas back up for production, the jacket is the steel lattice structure that supports the deck and anchors it to the seafloor, and finally, the shell mound and drill cutting debris on the seafloor around the platform (Henrion et al., 2015). Most of these fixed platforms stand the size of the Eiffel Tower and have a footprint the size of a football field (Macreadie et al., 2011).

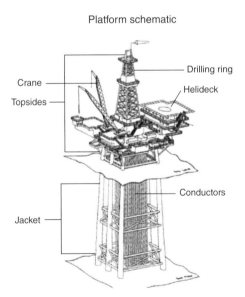

Figure 14.27 The major components of a generic offshore oil and gas platform (Source: http://www.lumina.com/case-studies/a-win-win-solution-for-californias-offshore-oil-rigs. Reproduced with permission of Max Henrion, Lumina Decision Systems, Inc)

Henrion et al. (2015) developed the PLATFORM tool for decommissioning offshore oil and gas platforms as part of a policy analysis conducted for the State of California. In determining the best decommissioning alternatives for the 27 oil and gas platforms nearing their oil and gas production service life between 2015 and 2030, many alternatives were evaluated. The key alternatives that helped to clarify and assess the decision strategies included complete removal of the oil and gas platform or partial removal to 85 ft below the water line. The PLATFORM software tool structured and analyzed the impact of each option based on costs, fishery production, and air emissions. The results of the project was instrumental in the decision to pass legislation to expand California's "rigs-to-reefs" program to include a cost sharing between the state and the operators of the oil and gas rigs (Henrion et al., 2015).

A decision tree was created to depict the decommission alternatives considered in the study. Figure 14.28 lists the option alternatives used in the development of the PLATFORM software model developed in Analytica. One of PLATFORM's key objects was to "provide stakeholders a tool for interactive exploration of decision strategies from varying perspectives (Henrion et al., 2015)." The green boxes denote two alternative options of complete removal and partial removal (*the reader is referred to the online version of this book for color indication*). The different branches indicate that different decisions may be appropriate for each platform as each differed in age, size, ocean depth, and location. These differing characteristics affect the platforms' decommission costs, environmental effects, and their ability to support reefing. Another important observation was that the decision analysis would consider entire decision options for some or all the platforms as a combined rather

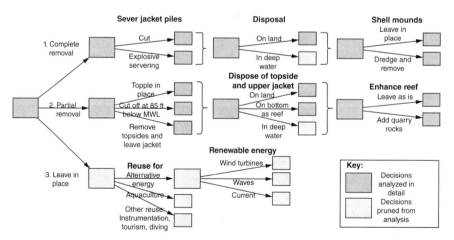

Figure 14.28 Decision tree showing the decommissioning alternatives considered in the study. Options with green boxes were analyzed in greater detail and gray boxes were omitted from quantitative analysis (Source: http://www.lumina.com/case-studies/a-win-win-solution-for-californias-offshore-oil-rigs. Reproduced with permission of Max Henrion, Lumina Decision Systems, Inc) (*The reader is referred to the online version of this book for color indication*)

than each platform individually. One of the largest costs to the decommissioning effort was the procurement of a heavy lift vessel (HLV) with a 4000 tons lifting capacity to hoist platform sections from the ocean into transport barges. Cost sharing of the HLV rental and transport with multiple platforms in a combined decommissioning effort greatly reduced the cost.

Analytica is a software environment for building decision models, and Figure 14.29 displays the graphical user interface (GUI) of PLATFORM with separate views to define decision options, perform cost analysis, and conduct multiattribute analysis of all the attributes of each platform. PLATFORM included all the options in the decision tree.

Influence diagrams depicting the decommissioning variables and influences used to calculate the program costs for decommissioning. The diagrams identify that data sources, uncertainty, decisions and results, and arrows of influence between each of them as shown in Figure 14.30.

Decommissioning stakeholder objectives were researched, analyzed, and organized as attributes in the influence diagram in Figure 14.31. Several attributes are easy to quantify; however, others may pose some difficulty due to lack of quality data and lack of understanding of the causal factors and their processes. Some attributes such as cost are readily quantified; however, attributes such as impact on marine mammals, benthic or seafloor impacts are difficult to quantify due in part to inadequate data

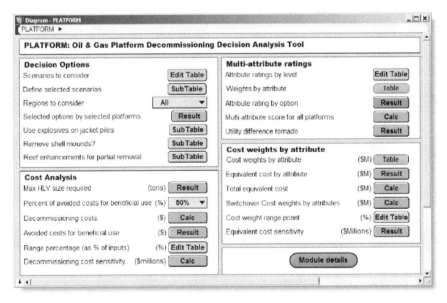

Figure 14.29 Graphical user interface (GUI) for PLATFORM with separated components to define decision options, perform quantitative cost analysis of the scenarios, and conduct multi-attribute analysis including all attributes (Source: http://www.lumina.com/case-studies/a-win-win-solution-for-californias-offshore-oil-rigs. Reproduced with permission of Max Henrion, Lumina Decision Systems, Inc)

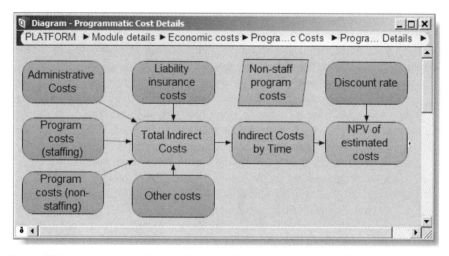

Figure 14.30 Analytica influence diagram showing selected variables and influences involved in calculating the programmatic costs for decommissioning (Source: http://www.lumina.com/case-studies/a-win-win-solution-for-californias-offshore-oil-rigs. Reproduced with permission of Max Henrion, Lumina Decision Systems, Inc)

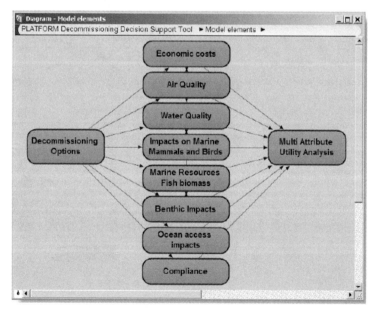

Figure 14.31 Influence diagram showing how the multiattribute analysis is based on the results of analysis of the eight key attributes used to evaluate the costs and benefits of alternative decommissioning options (Source: Henrion et al. 2015)

and/or the lack of a clear understanding of causal processes (Henrion et al., 2015). The PLATFORM tool takes into account all the attributes identifed through analysis to evaluate the cost benefits of alternative decommissioning options. Henrion et al. (2015) used multiattribute analysis to incorporate all the identified attributes in their analysis where both quantitative and qualitative attributes were treated as important aspects to any stakeholder. Table 14.25 summarized each attribute, describes their treatment and the characteristics ascribed to each one.

Both qualitative and quantitative attributes were treated differently in the PLATFORM tool. For the qualitative attributes, a five-point scale ordered from worst to best plausible outcome was developed. For example, Impact of Marine Mammals was modeled, and Figure 14.32 describes the levels for "Disturbance, disorientation and possible mortality, identifies the corresponding decision option possibly producing the outcome and the score for each level falling between 0% and 100%." The qualitative attributes of decommissioning costs, fish production, and changes in ocean access represented a small portion of California's and oil and gas companies' annual spending and so was handled as a linear utility default function in PLATFORM.

As the analysis progressed, PLATFORM combined attributes and swing weights. Figure 14.33 is the user interface screen to assist stakeholders in specifying swing weights for the attributes. The user selects an attribute they view as important, set the level to the highest swing weight of 100, and then they order the weights from second most important to least important. Finally, a swing weight between 0 and 100 is specified for each attribute relative to the most important.

The analysis into the decommissioning of the 27 oil and gas rigs helped to identify the differing stakeholder views on the relative importance of the attributes. Some are concerned with the potential environmental impact and others consider the huge cost of complete removal as critical and so on. The PLATFORM tool is capable of identifying the sensitivity of the decommission conclusion to such differences in value. The tornado chart in Figure 14.34 indicates that only two attributes favor complete removal Compliance and Ocean access with the reaming variables favoring partial removal.

Figure 14.35 shows the recommended decommission decision for partial removal and complete removal for each of the 27 platforms as function of the swing weight assigned to Strict Compliance. The bottom row shows the number of all the platforms recommended for complete removal from 0 to all 27 as swing weight for Strict Compliance increases from 0 to 100. The table orders the platforms from the shallowest to the deepest, and at intermediate swing weights, it tends to recommend complete removal of shallower platforms because the decommissioning costs and environmental impacts are higher for the deeper platforms.

Decommissioning these 27 California oil and gas platforms is a large undertaking with complex trade-offs of cost, public controversy, environmental factors, and differing stakeholder values and perspectives. When it comes to cost, it is estimated that complete removal of the 27 structures will exceed $1 billion. Henrion et al. (2015)

Table 14.25 Summary of Finding and Characteristics of the Eight Attributes Included in the Multiattribute Analysis. The Analysis Focused on Identifying the Difference Between the Complete and Partial Removal Alternatives Across all Eight Attributes

Attribute Description	Characteristics
Costs: The direct costs of decommissioning, including acquiring requiring permits, obtaining equipment such as heavy lift vessels (HLVs), cutting up the platform, removing some or all parts, transporting them to a disposal or recycling site, and processing removed equipment. Programmatic costs included for reefing option	• Quantified in US dollars (2009) • Actions identical in both options (e.g., deck removal) did not affect choice of option
Air quality: Much of the equipment used to dismantle, lift, and transport the elements of the platform runs on fossil fuel, usually diesel, emitting carbon dioxide and criteria pollutants. Only on-site emissions are considered, excluding emissions from transit of heavy lift vessels (HLVs) from the North Sea or east Asia	• Quantified for worst case, the largest platform (Harmony) • Quantified for other platforms based on size comparison with Harmony
Water quality: Removal of platforms, oil and gas processing equipment, and dredging of shell mounds and debris below the platform may have some impact on water quality due to dispersal of contaminants	• Qualitative based on relative risk of spills, dispersal, past experience
Marine mammals: Seals, sea lions, and other marine mammals often visit platforms due to the local concentration of fish. Complete removal will remove this food source. Removal of platforms, especially if explosives are used to sever steel supports, may disturb or injure marine mammals in vicinity	• Qualitative based on use of explosives, relative amount of vessels traffic, behavior and migration patterns, past experience
Marine birds: Marine birds use platforms for roosting, enabling them to feed with shorter flights than from onshore roosting. At the same time, there are some fatalities from flight collisions with platforms. Both options will remove surface structures, having the same impact on birds	• Qualitative based on past studies • No difference between options, therefore did not affect choice

Benthic impacts: The benthic zone is the ecological region on the seafloor, including surface and subsurface sediments. Complete removal of platforms will have some impact from anchoring the HLV, extracting the jacket piles, piping, and cabling, and, dredging or covering the shell-mounds. Partial removal will have much smaller impacts on the benthos

- Qualitative based on relative amount of size of platform and shell mound, relative degree of disturbance, past studies

Fish productivity: Biological productivity around the platforms provides sustenance for fish, including rock fish of value to commercial fishermen, and is an attraction for recreational divers. Complete removal will remove all such habitat and reduce productivity

- Quantified as Kg/year by platform
- Model included amount of habitat per platform, data from monitoring surveys, population dynamics (i.e., reproduction, settlement, growth, survival/mortality rates)

Ocean access: Partial removal option increases ocean area accessible for shipping and some fishing vessels, but reduces or leaves unchanged access to other user groups. Value of each option depends on the specific user group

- Quantified changes to access in square nautical miles
- Qualitative for other aspects
- User group preferences classified as pro, con, or neutral for each option
- Most socioeconomic impacts not considered because of data gaps, large uncertainties, and small size relative to local economy

Source: Data from Istiaque et al. 2014.

Attribute: Impacts on Marine Mammals			
Level	Description	Decisions	Score
Best	Status quo, no effect	No action	100
Good			
Medium	Slight effect son movement or migration of marine mammals	Partial removal	70%
Poor	Some disturbance or disorientation	Complete removal without explosive severing	50%
Worst	Disturbance, disorientation, and possible mortality	Complete removal with explosive severing	0

Figure 14.32 Definition of levels for impact on marine mammals is a qualitative attribute and contains a description and conditions that would give rise to that level. Scores of 70% and 50% are example scores to illustrate user input (Source: Henrion et al. 2015)

Assessing swing weights by attribute				
Attributes	Type	Best outcome	Worst outcome	Swing weight
Costs	Quantitative	Status quo: $0	Complete removal: $250 million	100
Air quality	Qualitative	Status quo: Zero emissions.	Complete removal: Emissions from 4400 ton HLV onsite for 113 service days for complete removal.	40
Water quality	Qualitative	Status quo: No impact	Complete removal: Accidental discharge of contaminated material at surface, or shell mound removal with toxic sediment contaminates water column.	15
Marine mammals	Qualitative	Status quo: No impact	Complete removal: Explosive severing for complete removal causes disturbance, disorientation, and some mortality to marine mammals.	20
Birds	Qualitative	Deck removal: Reduced mortality from flight collisions. Loss of offshore roosting replaced by new	Deck removal: Loss of offshore roosting reduces fitness and survival, which outweighs reduced flight collisions.	10
Benthic impacts	Qualitative	Status quo: No impact	Complete removal: Anchoring or shell mound removal leads to widespread impact and spreading contaminants.	10
Fish production	Quantitative	Status quo: 10,000 Kg/y	Complete removal: Zero fish production	25
Ocean access	Quantitative	Removal: Adds 2 Sq N MI	Status quo: Limits access	20
Strict compliance	Qualitative	Complete removal complies with lease	Partial or no removal violates lease.	50

Figure 14.33 User interface screen to assist users in assessing swing weights for each attribute in estimating the value to a stakeholder by changing each attribute from its Worst to its Best outcome, relative to most important attribute. Cost are identified as most important attribute and assigned swing weights of 100 (Source: http://www.lumina.com/case-studies/a-win-win-solution-for-californias-offshore-oil-rigs. Reproduced with permission of Max Henrion, Lumina Decision Systems, Inc)

adopted decision analysis methods such as decision trees to identify policy strategies, influence diagrams to structure the analysis, probability distributions to express uncertainties, multiattribute utility model representing stakeholder objectives, and sensitivity analysis to explore the effects on varying assumptions, including the importance stakeholders ascribed to their objectives to inform a decommission decision. PLATFORM and the final decommissioning project report are available on the Ocean Science Trust (OST) website http://www.oceansciencetrust.org/project/oil-and-gas-platform-decommissioning-study/ for use with the free Analytica Player.

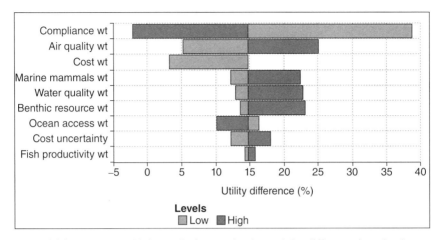

Figure 14.34 Range sensitivity analysis tornado chart of the difference in value between complete removal and partial removal for platform Harmony, changing the swing weights for each attribute from 0 low to 100 high and cost uncertainty from 10th to 90th percentile while keeping the other variables as their base values (Source: http://www.lumina.com/case-studies/a-win-win-solution-for-californias-offshore-oil-rigs. Reproduced with permission of Max Henrion, Lumina Decision Systems, Inc)

14.4.5 System Retirement and Decommissioning Strategy

As demonstrated by the Navy's SRP program, a decommissioning strategy is a project in and of itself that requires, funding, and resources of people, processes, and technologies. The decommissioning strategy must adhere to strict quality control mechanisms to prevent negative impacts to the project and the organization. In the case of the nuclear vessel decommissioning, quality control mechanisms were in place to prevent catastrophic nuclear fallout (Kopte, 1997).

A decommission and retirement plan should be included in the artifacts required during project planning and inception, before full-on development commences as depicted in Figure 14.36. The Decommission Plan would describe the importance of properly decommissioning obsolete systems and the risks incurred if not carried out in an efficient manner. The Decommission Plan should determine the total cost for the decommissioning effort and identify the stakeholder parties involved in the decommission effort and the initiator. It should describe the phases of the decommissioning, identifying the resources that must be decommissioned and instructions on how to identify these. The Project Manager will leverage the decommissioning plan to estimate the required effort and total duration of the decommissioning effort. The Procurement team would leverage the plan to determine all the components used by the decommissioned system to determine whether licenses and support contracts can be canceled or reused for other projects. As in the Navy SRP program, they identified components that could be sold or reused and which components required special disposal such as the nuclear reactors and the spent fuel.

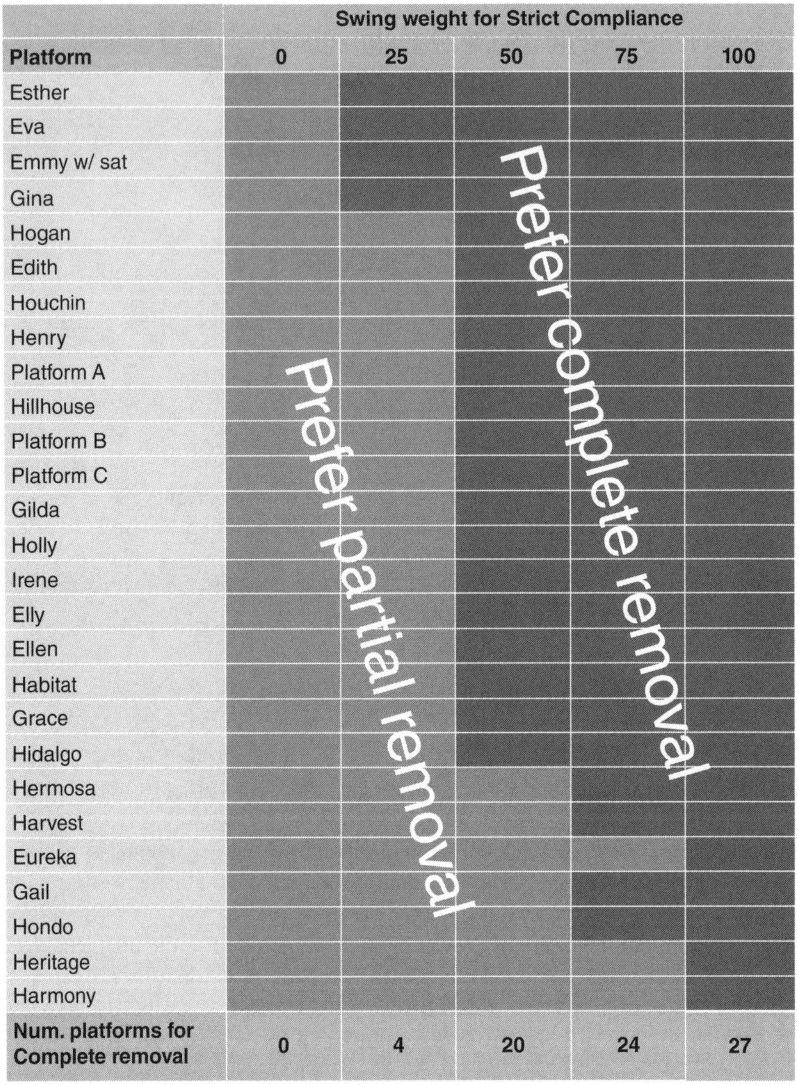

Platform	Swing weight for Strict Compliance				
	0	25	50	75	100
Esther					
Eva					
Emmy w/ sat					
Gina					
Hogan					
Edith					
Houchin					
Henry					
Platform A					
Hillhouse					
Platform B					
Platform C					
Gilda					
Holly					
Irene					
Elly					
Ellen					
Habitat					
Grace					
Hidalgo					
Hermosa					
Harvest					
Eureka					
Gail					
Hondo					
Heritage					
Harmony					
Num. platforms for Complete removal	0	4	20	24	27

Prefer partial removal / *Prefer complete removal*

Figure 14.35 Preferred decision, partial removal or complete removal for each platform according to the swing weight set for Strict Compliance. The bottom row shows the number of platforms recommended for complete removal. The platforms are ordered by depth (Source: http://www.lumina.com/case-studies/a-win-win-solution-for-californias-offshore-oil-rigs. Reproduced with permission of Max Henrion, Lumina Decision Systems, Inc)

PRODUCT RETIREMENT DECISION TRADE STUDY

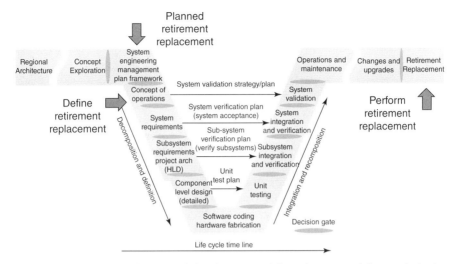

Figure 14.36 The Vee diagram and the placement of the retirement and decommissioning activity in the project life cycle (Source: U.S. Department of Transportation 2013)

Organizations must develop a plan to decommission legacy systems that are no longer relevant to the main business functions. The decommission strategy should be identified during the planning phase of the system's life cycle. An estimated cradle to grave product life cost should be determined upfront and analysis performed to allocate funds and resources for its graceful retirement at the end of the system service life cycle.

14.4.6 Conclusion

As demonstrated in this section, organizations must regularly review, evaluate, and analyze their systems portfolios to weigh the costs and benefits of each system to decide which should be retired, which should be upgraded, and which should be replaced. The portfolio review is a basis to evaluate and make informed decisions that may lead to improved systems efficiencies, a simplified technology infrastructure with increased software, hardware, and system maintenance cost savings. Retiring redundant systems and technology lowers infrastructure costs and removes the operational risks and threats these legacy systems pose. Retiring redundant systems allows the organization to reallocate those resources onto the support of systems that deliver greater ROI and business value. In making a retirement decision, the authors evaluated several methods identified in the research. Given the complexity of systems and the underlying decisions to continue maintaining or retire these same systems, stakeholders and managers need a variety of tools to assist in making rational decisions supported by data. NPV and Multiattribute Decision Analysis are techniques identified and leveraged in the decommissioning and retiring of aging systems from legacy software systems to aging oil and gas rigs to nuclear-powered submarines and air craft carriers.

14.5 KEY TERMS

Acceptance Decision: The decision to accept, reject, or delay the acceptance of a product.

Alternative Hypothesis: The alternative to the null hypothesis. It is often taken to be that which is to be shown. For the purposes of Section 14.2, it is the hypothesis that the system is good.

Commissioning: The process of assuring that all systems and components of a building or industrial plant are designed, installed, tested, operated, and maintained according to the operational requirements of the owner or final client.

Confidence: This term has two meanings that depend on context. **Statistical confidence** is the probability that the method used to calculate the one-sided confidence interval has probability C of including the actual mean of a measurement (Section 14.2.2). The **confidence of a test** is the probability that the test will correctly accept the null hypothesis (Section 14.2.3). It is equal to 1 minus the probability of a Type I error.

Confusion Matrix: Also known as a contingency table or an error matrix, it is a specific table layout that allows visualization of the performance of an algorithm. It is often used to describe the performance of a classification model on a set of test data for which the true values are known.

Consumer Risk: The probability of accepting a bad system.

Cost Benefit Analysis: A systematic process for calculating and comparing benefits and costs of a project, decision, or government policy. Cost–Benefit Analysis has two purposes: to determine if it is a sound investment or decision and to provide a basis for comparing projects.

Decision Tree: A graphical technique for modeling a decision in which nodes represent decisions, uncertain events, or outcomes, and arcs connecting nodes represent decision options or uncertain event outcomes.

Decommissioning: Planned shutdown or removal of a building, equipment, plant, system, ship, nuclear reactor, and so on, from operation or usage.

False Negative: A test result that (incorrectly) indicates that a condition was not present, when it was.

False Positive: A test result that (incorrectly) indicates that a condition was present, when it was not.

Hypothesis Test: A test designed to determine whether an effect is present (the Alternative Hypothesis) or not (the Null Hypothesis) and which permits one to calculate the probability of a Type I and Type II error, based on the results and the threshold established for accepting the alternative hypothesis.

Influence Diagram: Also called a relevance diagram, decision diagram, or a decision network, it is a compact graphical and mathematical representation of a decision situation. It depicts what is known or unknown at the time of making a choice and the degree of dependence or independence (influence) of each variable on other variables and choices.

Integrated Decision Model: A decision model that includes mathematical models for the system performance and system life cycle cost implications of each decision and uncertain event.

Legacy System: Outdated computer systems, programming languages, or application software that are used instead of available upgraded versions. A legacy system may be problematic, due to compatibility issues, obsolescence, or lack of security.

Multiattribute Decision Analysis: A method of analysis for determining the optimum decision strategy to take when there are multiple (competing or conflicting) criteria for what constitutes a desirable outcome.

Net Present Value: The value in the present of a sum of money, in contrast to some future value it will have, when it has been invested at compound interest. NPV is used in capital budgeting to analyze the profitability of a projected investment or project.

Null Hypothesis: The reference hypothesis. For the purposes of Section 14.2, it is the hypothesis that the system is bad.

Predictive Analytics: Is the practice of extracting information from existing data sets in order to determine patterns and predict future outcomes and trends.

Power (of a Test): The power of a test is the probability that the test will correctly reject the null hypothesis. It is equal to 1 minus the probability of a Type II error.

Project Life Cycle: A series of activities that are necessary to fulfill project goals or objectives.

Project Management: The process of managing several related projects, often with the intention of improving an organization's performance.

Producer Risk: It is the probability of rejecting a good system.

***p*-Value (stat):** The probability of observing a result that is more extreme than a given value, assuming that the null hypothesis is true.

Receiver Operator Characteristic (ROC) Curve: A graphical plot that shows the performance of a binary classifier system as its discrimination threshold is varied. Generally, it is a plot of the probability of correctly detecting the presence of a condition (i.e., present) versus the probability of incorrectly detecting its presences when it is not.

Regression Analysis: A statistical process for estimating the relationships among variables. It includes techniques for modeling and analyzing several variables. The focus is on the relationship between a dependent variable and one or more independent variables.

Retirement Strategy: Documents the way the system or subsystem will be retired or decommissioned.

Software: The programs, programming languages, and data that direct the operations of a computer system.

Software Engineering: The application of a systematic, disciplined, quantifiable approach to the development, operation, and maintenance of software, and the study of these approaches.

Systems Engineering: An interdisciplinary approach and means to enable the realization of successful systems that must satisfy the needs of their customers, users, and other stakeholders.

System Retirement: Also called application decommissioning and application sunsetting, it is the practice of shutting down redundant or obsolete business applications while retaining access to the historical data.

True Negative: A test result that (correctly) indicates that a condition was not present, when it was not.

True Positive: A test result that (correctly) indicates that a condition was present, when it was.

Type I Error: Rejecting the Null Hypothesis when it is true.

Type II Error: Accepting the Null Hypothesis as true when it is false.

14.6 EXERCISES

14.1. Consider the Acceptance Decision Model provided in Section 14.2.1 and the following changes to the inputs provided in Table 14.4.

Probabilities	
Prob Syst Good, Given Passed 1st Test:	0.50
Prob Syst Bad, Given Passed 1st Test:	0.50
Prob Syst Passes 2nd Test (assumes meets Mdg):	0.85
Prob Syst Fail 2nd Test:	0.15
Prob Syst Good, Given Passed 2nd Test:	0.80
Prob Syst Bad, Given Passed 2nd Test:	0.20

(a) Find Expected cost for each decision option.
(b) What decision option should adopt and why?

14.2. Consider the Acceptance Decision Model provided in Section 14.2.1 and the following changes to the inputs provided in Table 14.4.

Costs	
Cost to Produce System ($M):	100
Cost of Fix/Test Cycle ($M):	20
Annual Delay Cost ($M)	20

Probabilities	
Prob Syst Good, Given Passed 1st Test:	0.30
Prob Syst Bad, Given Passed 1st Test:	0.70
Prob Syst Passes 2nd Test (assumes meets Mdg):	0.70
Prob Syst Fail 2nd Test:	0.30
Prob Syst Good, Given Passed 2nd Test:	0.80
Prob Syst Bad, Given Passed 2nd Test:	0.20

(a) Find Expected cost for each decision option.

(b) What decision option should adopt and why?

14.3. Suppose one runs an acceptance test for 5000 operational hours and observes 45 failures.

(a) What is the observed MTBF (M_{ob})

(b) Given this, what confidence can one have that the actual MTBF is greater than $M_r = 120\,h$?

14.4. Find an expression for the confidence that the actual failure rate (Fa) is less than some required failure rate (Fr), given one observes N_{fo} failures out of Nt trials. Hint: this is a Poisson process.

14.5. Suppose one is performing a quality test on an assembly line. The quality requirement is that the defect rate shall not exceed 10%. One takes a sample of 20 items. One finds 1 defect. What confidence does one have that the assembly line meets or exceeds its quality requirement?

14.6. Suppose the required MTBF for a system is 150 h and that one sets the failure threshold for the acceptance test at 40 failures.

(a) Develop a table that shows how the C and P of a test will vary with the Test Duration (T_d).

(b) Suppose you believe your system design provides an MTBF of 200 h. What test duration should you select in order to have a Confidence of 0.8 and a Power of 0.8? Explain your answer.

14.7. Suppose one believes one's system design provides and MTBF of 200 h and that one only has resources to perform a reliability test that last for 5000 h and one requires a Confidence of 0.8 and a Power of 0.8. Provide a graph showing the trade between N_{fat} and the system requirement (M_r) that can be demonstrated with the required Confidence and Power.

14.8. Consider the cost model developed in Section 14.2.4. Suppose that the schedule delay for redesign (S_{dpdes}), realizing the redesign (S_{dpdm}), developmental testing (S_{dpdt}), and for setting up the OT (S_{ptf2}) is increased by 1 month and the number of sm required to fix the design (S_{mfd}) increases by 4 sm.

14.9. Suppose the uncertainty associated with each input value is 10%.

(a) Find the four variables that have the greatest effect on the total delta cost.

(b) Develop an MC simulation for the total fix and test cost that treats these four variables as random variables with a triangular distribution.

(c) Run the MC Simulation 10,000 times and plot the resulting distribution. Does it look more like a triangular or a normal distribution?

(d) Find the mean value for the total cost and the standard error associated with this mean. Does the mean differ significantly from the total cost calculated using the reference values. Explain.

14.10. Given the example inputs provided in Section 14.2.5,
- (a) How do the expectation values of the decision tree vary as one changes the duration of the second acceptance test (T_{d2}) from 2000 to 10,000 h?
- (b) For what values of T_{d2} does the recommended decision change and what is the change (from what to what)?
- (c) How does the delta cost change as a function of T_{d2}?
- (d) How does the expected value of the recommended decision change as a function of T_{d2}?
- (e) At what point does it cease to be cost-effective to increase T_{d2}? Explain your answer.

14.11. Collect the necessary data on a project in industry and determine its probability of failure using Predictive Model B.

14.12. What mitigation strategies can you implement to change the project's failure outcome?

14.13. Identify other areas in which the predictive model can be leveraged in industry.

14.14. Develop a Decision tree showing the decommissioning alternatives considered by the Environment Assessment of the USS Enterprise.

14.15. Develop an Influence diagram showing the qualitative and quantitative aspects of the decommissioning decision for the USS Enterprise.

14.16. Describe how the Navy might have used NPV to decide which vessel to retire in portfolio of aging nuclear vessels.

14.17. Perform multiattribute decision analysis on the Navy's fleet of aging vessels to decide which vessel to retire after USS Enterprise.

REFERENCES

AcqNotes. (n.d) *Contracts & Legal: Clinger-Cohen Act*. From AcqNotes: A simple Source of DoD Acquisition Knowledge for the Aerospace Industry: http://acqnotes.com/acqnote/careerfields/clinger-cohen-act (accessed 01 July 2016).

Arandjelovic, P., Bulin, L., and Khan, N. (2015) *Why CIOs Should Be Business-Strategy Partners*, From McKinsey and Company Insights and Publications: http://www.mckinsey.com/insights/business_technology/why_CIOs_should_be_business-strategy_partners?cid=other-eml-alt-mip-mck-oth-1502 (accessed 01 July 2016).

Baldridge (2012) *Baldridge Core Values-Management by Fact*. From Baldridge.

Brooks, F. (1995) *The Mythical Man-Month*, Addison-Wesley.

Department, U. S. (2015) *Treaties and Agreements*. From U.S. Department of State Deplomacy in Action: http://www.state.gov/t/isn/c18882.htm (accessed 01 July 2016).

DoD (2015) *DoD Directive 5000.02: Operation of the Defense Acquisition System*, USD(AT&L), Washington, DC.

DOT (2013) *System Retirement* XE "System Retirement" */Replacement*. From U.S. Department of Transportation Federal Highway Administration California Division: http://www.fhwa.dot.gov/cadiv/segb/views/document/sections/section3/3_8_1.cfm (accessed 01 July 2016).

Ebeling, C.E. (2010) *An Introduction to Reliability and Maintainability Engineering*, 2nd edn, Waveland Press.

Eisner, H. (2005) *Managing Complex Systems: Thinking Outside the Box*, John Wiley & Sons, Hoboken.

Everitt, B. and Hothorn, T. (2006) *A Handbook of Statistical Analyses Using R*, Taylor & Francis Group, Boca Raton.

Ewusi-Mensah, K. (2003) *Software Development Failures: Anatomy of Abandoned Projects*, MIT Press, Cambridge.

Fan, R.-E., Chang, K.-W., Hsieh, C.-J. et al. (2008) LIBLINEAR: a library for large linear classification. *Journal of Machine Learning Research*, **9**, 1871–1874.

FAR (2014) *Federal Acquisition Regulation*.

Fitzgerald, M. (2014) *The Four Traps of Predictive Analytics*. From MIT Sload Managment Review: http://sloanreview.mit.edu/article/the-four-traps-of-predictive-analytics/.

GAO (1992) *Patriot Missile Defense: Software Problem Led to System Failure at Dhahran, Saudi Arabia, GAO-IMTEC-92-26*, Government Accountability Office, Washington, DC.

GAO (2006) *Information Technology: Agencies and OMB Should Strengthen Process for Identifying and Overseeing High Risk Projects, GAO-06-647*, Government Accountability Office, Washington, DC.

GAO (2009) *Polar-Orbiting Environmental Satellites, GAO-09-564*, Government Accountability Office, Washington, DC.

GAO (2011) *Information Technology: Critical Factors Underlying Successful Major Acquisitions, GAO-12-7*, Government Accountability Office, Washington, DC.

GAO (2014) *Patient Protection and Affordable Care Act: Preliminary Results of Undercover Testing of Enrollment Controls for Health Care Coverage and Consumer Subsidies Provided Under the Act, GA0-14-705T*, Government Accountability Office, Washington, DC.

Hawkins, T. (2013) *Navy Prepares to Commision 10th Virginia-class Submarine*. From America's Navy: http://www.navy.mil/submit/display.asp?story_id=76317 (accessed 23 September 2016).

Hendrix, H. (2013) At What Costs a Carrier? *Disuptive Defense Papers*, From http://www.cnas.org/files/documents/publications/CNAS%20Carrier_Hendrix_FINAL.pdf 1–16.

Henrion, M., Bernstein, B., and Swamy, S. (2015) A Multi-attribute Decsion Analysis for Decommissioning Offshore Oil and Gas Platforms. *Integrated Environmental Assessment and Management*, 594–609. From Lumina Case Studies: http://www.lumina.com/case-studies/a-win-win-solution-for-californias-offshore-oil-rigs (accessed 23 September 2016).

Higgins, J. and Green, S. (2011) Cochrane Handbook for Systematic Reviews of Interventions. The Cochrane Collaboration. From http://handbook.cochrane.org/chapter_12/12_4_2_p_values_and_statistical_significance.htm (accessed 23 September 2016).

Hoover, C., Rosso-Llopart, M., and Taran, G. (2010) *Evaluating Project Decisions: Case Studies in Software Engineering*, Addison-Wesley, Upper Saddle River.

INCOSE (2010) *Systems Engineering Handbook*, Version 3 edn, INCOSE, Seattle, WA.

ISO/IEC/IEEE. (2008) 12207. *Systems and Software Engineering: Software Life Cycle Processes*. doi: 10.1109/IEEESTD.2008.4475826. ISBN: 978-0-7381-5663-7.

Istiaque, N., Guillaume, M., and Reid, J. (2014) Department of Treasury Human Resources Application and Feasibility Analysis. *Unpublished Class Paper for Masters Program at UMBC*. Baltimore County, Maryland: Unpublished Document.

Knot, J.A. and Allen, J.R. (2012) *Final Environment Assessment on the Disposal of Decommissioned, Defueled Naval Reactor Plants from USS Enterprise (CVN 65)*, US Deparment of the Navy.

Kopte, S. (1997) Paper 12: Nuclear Submarine Decommissioning and Related Problems. *Bonn International Center for Conversion (BICC)*, 1–61.

Kuhn, M. (2008) Building predictive models in R using the caret package. *Journal of Statistical Software*, **28**, 1–26.

Lattin, J., Carroll, J., and Green, P. (2003) *Analyzing Multivariate Data*, Brooks/Cole, Belmont.

Lehtinen, T., Mantyla, M., Vanhanen, J. et al. (2014) Perceived cause of software project failures - an analysis of their relationships. *Information and Software Technology*, **56**, 623–643.

Lunden, I. (2013) *Forrester: $2.1 Trillion Will Go Into IT Spending in 2013: Apps and the U.S. Lead The Charge*. Retrieved from TechCrunch: techcrunch.com.

Macreadie, P.I., Fowler, A.M., and Booth, D.J. (2011) Rigs-to-reefs: will the deep sea benefit from artificial habitat? *Frontiers in Ecology and the Environment*, **9**, 455–461.

Madni, A. and Sievers, M. (2014) Systems Integration: key perspectives, experiences, and challenges. *Systems Engineering: The Journal of The International Council on Systems Engineering*, **17** (1), 37–51.

MIL-HDBK-781A (1996) In *MIL-HDBK-781A: DoD Handbook for Reliability Test Methods, Plans, and Environments for Engineering, Development Qualification, and Production*, Section 5.10.8 (p. 49).

NASA (1995) *Systems Engineering Handbook*, National Aeronautics and Space Administration (NASA), Washington, DC.

NASA (1999) *Mars Climate Orbiter Mishap Investigation Board Phase I Report*, NASA.

Naur, P. and Randell, B. (1969) *Software Engineering: Report on Conference Sponsored by NATO Science Committee*, NATO Science Committee, Garmisch.

Navidi, W. (2011) *Statistics for Engineers and Scientists*, 3rd edn, McGraw Hill.

Navy (2000) *Submarine Chronology*. Retrieved from Chief to Naval Operations Submarine Warfare Division: http://www.navy.mil/navydata/cno/n87/history/chrono.html (accessed 23 September 2016).

Navy (2014) *United States Navy Fact File*. Retrieved from U.S. Navy Fact Sheet: http://www.navy.mil/navydata/fact_print.asp?cid=4100&tid=100&ct=4&page=1 (accessed 23 September 2016).

Navy, U. (2015) *The Legend of ENTERPRISE*. Retrieved from USS ENTERPRISE (CVN 65): http://www.enterprise.navy.mil/ (accessed 23 September 2016).

O'Connor, P. and Kleyner, A. (2012) *Practical Reliability Engineering*, 5th edn, Wiley.

Pettey, C. and van der Muelen, R. (2013). *Gartner Newsroom*. Retrieved from Gartner Executive Program Survey of More Than 2,000 CIOs Shows Digital Technologies Are Top Priorities in 2013: http://www.gartner.com/newsroom/id/2304615 (accessed 23 September 2016).

PMI. (2015) *What is Project Management?* Retrieved from Project Management Institute: http://www.pmi.org/About-Us/About-Us-What-is-Project-Management.aspx (accessed 23 September 2016).

R.-D. T. (2011) *Package 'stats': The R Stats Package.*

Roy, G. (2013) Prototype Development of Dynamic Predictive Models for Disease Prevention. *2013 International Conference on Advances in Computing, Communications and Informatics (ICACCI)*, 1681–1685.

Ryan, J., Sarkani, S., and Mazzuchi, T. (2014) Leveraging variability modeling techniques for architecture trade studies and analysis. *Systems Engineering: The Journal of INCOSE*, **17** (1), 10–25.

Singer, J. (2014) Systems Science is Fundamental:How Systems Science can Help Improve SE Practices. (D. J. Martin, Interviewer) *INCOSE International Symposium.*

Stevens, R. (2011) *Engineering Mega-Systems: The Challenge of Systems Engineering in the Information Age*, Taylor & Francis Group, Boca Raton.

Tan, S. (2011) *How to Increase Your IT Project Success Rate*, Gartner Research.

The Standish Group (2013) *The Chaos Manifesto*, The Standish Group.

Toppan, A. (2000) *sci.military.naval FAQ Part G Submarines*. From Science Military Naval FAQ: http://www.hazegray.org/faq/smn7.htm (accessed 23 September 2016).

Tsao, Y.-C., Hsu, K., and Tsai, T.-T. (2012) Using Content Analysis to Analyze the Trend of Information Technology Toward the Academic Researchers at The Design Departments of University of Taiwan. *2012 2nd International Conference on Consumer Electronics* (pp. 3691–3694). Yichang: IEEE.

Wade, D. J. (2014) Introduction to the SE Vision 2025. (P. Martin, Interviewer) Henderson, NV: *INCOSE International Symposium.*

Walls, C. (2006) *Embedded Softwarecp*, Elsevier/Newnes, Amsterdam.

15

SUMMARY AND FUTURE TRENDS

GREGORY S. PARNELL

Department of Industrial Engineering, University of Arkansas, Fayetteville, AR, USA

SIMON R. GOERGER

Institute for Systems Engineering Research, Information Technology Laboratory (ITL), U.S. Army Engineer Research and Development Center (ERDC), Vicksburg, MS, USA

> Anybody who has the reins of power has to look at practical limitations and trade-offs - the fact that you can focus at most on one or two things at a time, that resources are limited.
>
> Barton Gellman

> Strategy is about making choices, trade-offs; it's about deliberately choosing to be different.
>
> Michael Porter

15.1 INTRODUCTION

In this final chapter, we summarize the major themes of the book and identify some potential trends that may impact trade-off analyses in the future. The major themes were also the guiding principles that were used in the development of the INCOSE Decision Management Process in the Systems Engineering Handbook (INCOSE, 2015) and the Decision Management Process section in the SEBoK (SEBok, 2015). Next, we consider future trends in systems engineering that may provide opportunities for improved trade-off studies. This section discusses the trends concerning people, tools, and processes.

Trade-off Analytics: Creating and Exploring the System Tradespace, First Edition. Edited by Gregory S. Parnell.
© 2017 John Wiley & Sons, Inc. Published 2017 by John Wiley & Sons, Inc.
Companion website: www.wiley.com/go/Parnell/Trade-off_Analytics

15.2 MAJOR TRADE-OFF ANALYSIS THEMES

As we wrote this book, we had several themes or guiding principles. In this section, we describe each of these major themes and where they are emphasized in the book.

15.2.1 Use Standard Systems Engineering Terminology

In addition to systems engineering, many engineering disciplines are involved in trade-off analyses. Therefore, it should not be surprising that many different terms are used. We have attempted to use terminology from the ISO standard (ISO/IEC/IEEE 15288, 2015), the Systems Engineering Handbook (INCOSE, 2015), and the Systems Engineering Body of Knowledge (SEBok, Systems Engineering Body of Knowledge (SEBoK) wiki page, 2015). We believe these are useful references, and we encourage our readers to use these to expand their systems engineering knowledge. Chapter authors have introduced new terms when they believed an important distinction or insight could be obtained with the new term. Of course, the key terms at the end of each chapter provide the definitions of these terms.

15.2.2 Avoid the Mistakes of Omission and Commission

There are several papers that identify the problems with trade-off studies. Using the INCOSE Decision Management Process, the most common trade-off study mistakes of omission and commission are identified and explained in Chapter 1 (Parnell et al., 2014). The mistakes of omission are errors made by not doing the right things. The mistakes of commission are errors made by doing the right things the wrong way. For each step in the decision process, Table 1.4 provides a list of the trade-off mistakes, the type of mistake (omission or commission), and the potential impacts.

These mistakes can have significant consequences in system design and management and, ultimately, on the program and the system. Unfortunately, it is quite common to find multiple mistakes made in trade-off studies with some errors leading to or cascading with other errors. These cascading mistakes can lead to adverse impacts for the trade-off study team, decision-makers, stakeholders, and, ultimately, the system or program. In addition, repeating these trade-off mistakes can undermine the credibility of the SE organization/enterprise.

15.2.3 Use a Decision Management Framework

As we described in Chapter 1, trade-off decisions are required throughout the system life cycle. Many systems engineering decisions are difficult decisions that include multiple competing objectives, numerous stakeholders, substantial uncertainty, significant consequences, and high accountability. These decisions can benefit from a structured decision management process. The purpose of a decision management process is "to provide a structured, analytical framework for objectively identifying, characterizing, and evaluating a set of alternatives for a decision at any point in the life cycle and selected the most beneficial course of action." The decision management process uses the systems analysis process to perform the assessments (ISO/IEC/IEEE 15288, 2015). The process uses decision analysis to quantify the benefits and resource analysis to quantify the costs.

Chapter 5 introduces and illustrates the INCOSE Decision Management Process using a UAV example. The Decision Management Process provides the essential steps that should be considered for all major trade-off studies. Some of these steps may not be required in a particular study. We only illustrate one technique for each of the steps. However, we know there is no one way to perform trade-off analysis for all systems for all life cycle stages. Therefore, the rest of the book provides examples of trade-off analysis best practices in different life cycle stages using a variety of sound techniques. As the life cycle stages progress, the decision space decreases and there is an increase in the amount of data as design information, test, and operational data become available. Techniques used in later life cycle stages take advantage of this new information.

15.2.4 Use Decision Analysis as the Mathematical Foundation

We believe a credible trade-off analysis should be based on a sound mathematical foundation. Ad hoc methods and unsound mathematics provide an unsound foundation for providing insights for the decision-makers. Since trade-off studies involve complex alternatives, multiple objectives, and major uncertainties, we believe that decision analysis is the operations research technique that provides this sound mathematical foundation for trade-off analyses.

Decision analysis can quantify the value across the tradespace. Chapter 2 provides the foundations of decision analysis. The chapter reviews the mathematical assumptions of single and multiple values and utility. If all objectives can be converted to a single objective, usually dollars, then single-objective decision analysis is appropriate. If not, then multiple-objective decision analysis is the appropriate technique. However, we recommend the use of other analysis techniques to help explore the tradespace including design of experiments and optimization.

While decision analysis is the foundation for analyzing the tradespace, operational data, test data, and modeling and simulation data are critical for obtaining the scores on the value measures. This provides an important linkage between proposed systems assessed for trades and the analytical models and simulations that provide tradespace data for which real-world data does not yet exist, is not safe to collect, or is too expensive to obtain.

15.2.5 Explicitly Define the Decision Opportunity

Every trade-off study begins when an implicit understanding of the problem or opportunity. In our experience, the initial problem is never the final problem. Also, we believe that every problem can be viewed as a broader decision opportunity. Failure to identify the opportunity can result in rework of the study, the decision made without the benefit of the study, or even a lost opportunity for the organization or enterprise. Therefore, we recommend that the trade-off analysis team explicitly define the decision opportunity for each study and vet that definition with the decision-makers and key stakeholders.

Chapter 6 provides a summary of the best practices for opportunity definition. The chapter identifies the types of knowledge required for opportunity definition, the techniques for stakeholder analysis, and the tools to describe the opportunity from the systems engineering and decision analysis literature and practical applications.

15.2.6 Identify and Structure Decision Objectives and Measures

Once the opportunity is explicitly identified, the next step is to identify and structure the decision objectives of the decision-makers and stakeholders. The decision opportunity and our values determine the objectives. Identifying the objectives can be especially challenging if the opportunity is new to the organization or enterprise. Generally, the more complex and important the opportunity, the more decision-makers and stakeholders are faced with conflicting objectives. Structuring the objectives provides a visual framework that helps decision-makers and stakeholders understand and validate the completeness of the objectives.

Chapter 7 describes the best practices for defining and structuring objectives and then developing value measures that measure attainment of the objectives. Defining the value of the system's products and services is one of the most critical tasks of systems engineering. In systems engineering, the objectives include the business/mission objectives, the stakeholder objectives, and the system objectives. The business/mission objectives are derived from the system purpose for the organization and its customers (strategic level). The stakeholder objectives include the goals of the other important stakeholders in the system life cycle (operational level). Finally, the system objectives include the technical objectives that are necessary for a system to meet the business/mission and stakeholder objectives in the system life cycle (tactical level).

15.2.7 Identify Creative, Doable Alternatives

The key to trade-analysis is developing good alternatives that span the tradespace. A trade-off analysis should not be an advocacy for a predetermined alternative. If we analyze poor alternatives, we may only select the best of a bad set of alternatives. If we only consider the baseline or the baseline and a couple ad hoc alternatives, we may miss the opportunity to create value. The development of alternatives is a critical trade-off analysis task that requires participation of the entire trade-off analysis team and support from decision-makers, stakeholders, and subject matter experts.

Chapter 8 provides the best practice for alternative development. The best practice to develop alternatives is a two-phase process. The first phase is the creative or divergent phase. The purpose of this phase is to identify potential ideas that would create great value. The second phase is the analytical or convergent phase. The purpose of this phase is to develop doable alternatives. The chapter also identifies the techniques that provide explicit procedures to develop creative, doable alternatives.

15.2.8 Use the Most Appropriate Modeling and Simulation Technique for the Life Cycle Stage

In early life cycle stages, the alternatives are very different and the data available on new concepts is limited. As we proceed through the life cycle, the range of alternatives narrows and the data available on the alternatives increase as a concept is selected, an architecture is defined, a design is performed, development test data becomes available, systems are produced, systems are deployed, and operational data becomes available. As a result, different modeling and simulation techniques may be more appropriate in different life cycle stages. The trade-off analysis techniques used in the illustrative examples in Chapters 10–14 illustrated modeling and simulation techniques the authors have found to be the most useful in each stage.

15.2.9 Include Resource Analysis in the Trade-Off Analysis

Organizations do not have unlimited resources. Therefore, resource analysis is almost always a part of the trade-off analysis. Chapter 4 presents the important resource analysis techniques and also presents the key concepts of affordability analysis. We believe that all systems engineers should understand the cost analysis techniques in this chapter. However, for large trade-off analyses, we recommend that one or more cost analysts be assigned to the team.

This book provides numerous examples of comparing the cost versus the value of the alternatives. In most cases, the cost will be the life cycle cost. This plot provides essential information that senior decision-makers can use to determine the affordability of the alternatives.

15.2.10 Explicitly Consider Uncertainty

Systems development, deployment, operation, and retirement involve many uncertainties. The systems life cycle may be years to decades. For longer life cycles, we should expect more uncertainty. The major uncertainties include technology performance, integration with other systems, markets/missions, environments, and the actions of competitors/adversaries. Surprisingly, many trade-off analyses use only deterministic methods and do not explicitly consider uncertainty. According to our literature review and reinforced by our experience, cost analysis makes the most use of uncertainty analysis. Monte Carlo simulation is the most used technique.

Uncertainty analysis is a major topic of this book. Chapter 3 provides an introduction to uncertainty modeling for systems decisions. Several of the chapters address uncertainty models including Chapters 4, 5, and 8–14. Chapters 5, 13, and 14 provide uncertainty analysis with Monte Carlo simulation using an Excel add-in. Chapter 9 provides uncertainty analysis using Probability Management.

15.2.11 Identify the Cost, Value, Schedule, and Risk Drivers

The purpose of a trade-off analysis is to provide insights to aid in system decision-making. Decision-makers need to understand the cost, value, schedule, and risk drivers of the system. When properly focused on the decision objectives, trade-off analyses can identify these insights. These insights are critical to cost, value, schedule, and risk management.

Many of the chapters in this book identify, describe, and illustrate techniques to identify cost, value, and risk drivers. When combined with schedule models, these techniques can also be used to identify schedule drivers.

15.2.12 Provide an Integrated Framework for Cost, Value, and Risk Analyses

Unfortunately, most of the current systems engineering practice develops and performs separate cost, value, and risk analyses. The cost and value models are usually quantitative models. Many times, the risk analysis framework is only a likelihood consequence matrix. This practice may help summarize insights to decision-makers; however, it does not identify the critical dependences. Therefore, it does not provide the critical information that decision-makers need to understand the impacts of the cost, value, risk, and schedule drivers.

We recommend an integrated framework for cost, value, and risk analysis. In Chapter 9, we illustrate an integrated framework where the same decision factors and uncertainties are simultaneously propagated through value and cost models. We believe that this method provides the most promise to provide credible, traceable insights to decision-makers and stakeholders.

15.3 FUTURE OF TRADE-OFF ANALYSIS

There are several opportunities that offer the promise to significantly improve trade-off analysis in the era of big data and analytics. Analytics can be defined at three levels: descriptive, predictive, and prescriptive (INFORMS). As decision-makers are presented with the complexities associated with the increasing amounts of data collected from real-world observations (descriptive analytics) and generated by models and simulations (predictive analytics), the more they must rely on others to analyze the data and present recommendations (prescriptive analytics) that will ensure the best solution. This requires systems engineers to be familiar with proven methods and tools to collect and analyze data to produce actionable information. As new types and increasing amounts of data are produced, so must new methods, processes, and tools to analyze this data be developed. This necessitates that systems engineers dedicate themselves to a lifetime of education and training to stay current of the profession's evolving body of knowledge and, perhaps, to create some of the advancements. The remainder of this section discusses the education and training of systems engineers as well as summarizes some of the methodologies and processes discussed in this book or on the horizon for those working trade-off analyses.

15.3.1 Education and Training of Systems Engineers

This book is dedicated to providing a resource to improve the education and training of systems engineers and systems analysts who need to perform trade-off analyses. It provides the basic descriptions of methods, techniques, and tools as well as examples of practical applications using illustrative examples. The book can be used in numerous ways to further the education of the students and training of practitioners. Examples can be found in the "Trade-off Analysis Course Outlines" section of the preface. We have attempted to provide the necessary mathematical background for most of the techniques: for example, decision analysis (Chapter 2), probability and Monte Carlo analysis (Chapter 3), and resource analysis (Chapter 4). However, we have assumed an understanding of optimization (Chapters 10–13 and design of experiments (Chapter 12).

Education provides systems engineers with the understanding of basic and cutting-edge systems engineering methods, techniques, and tools. However, it takes exposure to real-world applications to develop and solidify one's skills as a systems engineer. Capstone projects in education programs, summer coops, as well as on-the-job training and mentoring are some of the best environments for honing one's systems engineering skills. Many organizations have professional continuing education and training programs to help increase the skill sets of their systems engineers. This book could be used to develop one or more modules for use in these programs. Figure 15.1 is an example layout for an Army organization's training program used to guide the development of systems engineers. It is an adaptation of a program used to train operations research analysts at the Center for Army Analysis.

15.3.2 Systems Engineering Methodologies and Tools

Modeling and simulation have been discussed as a method for analysts to test theories and explore concepts and tradespaces. Other methods used by systems engineers in various SE processes include functions-based systems engineering, object-oriented systems engineering, prototyping, interface management, integrated product and process development, lean systems engineering, agile systems engineering, and model-based systems engineering (MBSE) (INCOSE, INCOSE Systems Engineering Handbook, 2015, pp. 180–210). MBSE is a method that has been gaining in popularity since the publication of Wymore's book in 1993 entitled, *Model-Based Systems Engineering*.

MBSE is defined in the INCOSE Systems Engineering Vision 2025 as "the formalized application of modeling to support system requirements, design, analysis, verification, and validation activities beginning in the conceptual phase and continuing throughout development and later life cycle phases" (INCOSE, Systems Engineering Vision 2020, 2007, p. 15). Although the implementation of systems engineering practices varies by organization and system type, the use of model-based engineering has helped with cross-fertilization and standardization of methods, techniques, and

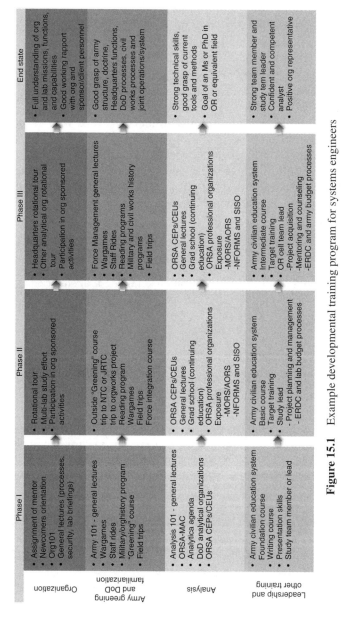

Figure 15.1 Example developmental training program for systems engineers

tools. This has occurred as collaboration between engineers increases as they leverage information and data generated from domain models to feed models of other subsystems or systems. To accomplish this, engineers link models and the interfaces between models and data sets to help simulate the complexities of larger systems.

This linkage requires integration across organizations, disciplines, and phases of system development. Although this helps to provide insights into how a system might integrate with other systems or perform in the real world, it creates additional challenges for systems engineers (INCOSE, SE Vision 2025, 2014, p. 22). Systems engineers are often used to perform the human and model integration. When done effectively, the integration of data, algorithms, and process results in a tradespace visualization that enables insights into viable alternatives for decision-makers.

As systems engineers must continue to leverage modeling and simulation to generate more complex and viable tradespaces, the term MBSE will likely become synonymous with systems engineering. Although it will be only one of the methods used by systems engineers, its broad application will make it part of the standard tool kit along with stakeholder and statistical analysis.

To be clear, MBSE is neither new nor is it the panacea of systems engineering. Long discusses four basic myths of MBSE: (i) models are new; (ii) documents are disappearing; (iii) there is only one kind of model; and (iv) Systems Modeling Language (SysML) equals MBSE (Long, 2015). This book has illustrated numerous types of models that have been used for decades (e.g., cost models and value models) and enhanced or new models (e.g., probability management and tradespace exploration) developed to meet the emergent needs of analysts to answer the questions of decision-makers. Although static documents capture a point in time and are difficult to maintain when changes are required, it is unlikely that they will disappear anytime soon. It is more likely that they will be replaced by dynamic documents generated as required from a digital thread of information as required. This will facilitate one's ability to understand the process and decisions made during the life cycle of a system while facilitating the development or enhancement of evolving systems. The use of SysML, or a similar system description language, to facilitate discussions between engineers and the interface between models is essential; however, it is merely a component of MBSE. MBSE requires systems descriptions to help to identify the linkage between models, network(s), data sets, analytical tools, and data/information visualization tools in order to generate the data thread used to create viable tradespaces and information. Furthermore, the MBSE requires a repeatable process to ensure its soundness and validity as a systems engineer methodology.

New methodologies often require new or enhanced analytical tools to help implement. Tradespace tools reside in this paradigm. Numerous tools are under development by academia, industry, and government to help generate, analysis, and visualize tradespace data. Some of the statistical or tradespace tools developed by industry and/or academia include the following: R, Analytics, CyDesign Studio, DecisionTools Suite 6.0, Excel®, SAS®, JMP®, Relational-Oriented Systems Engineering and Technology Trade-Off Analysis (ROSETTA), and Risk Solver Pro. Government sponsored tools include the Advanced Systems Engineering Capability (ASEC), ARL Trade Space Visualizer (ATSV), Capability Portfolio Analysis Tool

(CPAT), Engineered Resilient Systems (ERS) Tradespace Tool, Framework for Assessing Cost and Technology (FACT), Tradespace Analysis for Capabilities, Effectiveness, and Resources (TRACER), and Whole System Trades Analysis Tool (WSTAT).

The DoD-sponsored ERS Tradespace Tools package is an example of a cooperative effort between industry, academia, and DoD. It is dedicated to developing decision support methods as well as a collaborative MBSE Tradespace toolset based on an architecture open to organizations that work in partnership on DoD acquisition efforts. (Ender et al., 2015, p. 7)

15.3.3 Emergent Tradespace Factors

The increasing complexity, time to develop new systems, and cost of systems often require senior leaders to seek additional insights about systems under development. Previously, many of these insights were virtually impossible or too time-consuming to assess or not relevant to leader's decision-making process. However, as funding becomes more constrained and systems are used for extended periods of time or expected to operate in numerous environments, systems engineers are asked to expand the tradespace to include factors such as *resilience* and *flexibility*.

Depending on the community of interest, resilience can be defined in numerous ways. For example, fixed facilities or systems may define resilience as the ability to withstand or quickly recover from a natural disaster. For a nonstatic system, such as a transportation or DoD weapon(s) platform, "a resilience system is trusted and effective out of the box, can be used in a wide range of contexts, is easily adapted to many others through reconfiguration and/or replacement, and has a graceful and detectable degradation of function" (Goerger et al., 2014, p. 871).

Research is underway by industry, academia, and the federal government to help develop standard definitions within communities of interest for metrics such as resilience. This is important as analysts seek valid data to assess measures for system objectives. Industry is interested in these efforts as it seeks to understand the needs of its clients. This is true for its commercial as well as its government clients. Thus, industry is working with academia and government researchers on these efforts. An example of this is ERS.

Other system characteristics will be added to the tradespace in the future depending on the business model or purpose of the organization. Possible metrics include flexibility, manufacturability, deployability, sustainability, and easy of modification. Many of these metrics are still being defined and validated for their impact on system viability. Research also seeks to better understand the interrelationships between the metrics to help generate more valid tradespaces. For instance, the ease of building a system using additive manufacturing techniques may have a positive correlation to the maintainability and sustainability of the system. However, the initial cost of each copy of the system may be higher than systems built using more traditional methods. Conversely, the long-term cost to maintain the less expensive to build system could exponentially be more expensive than the system built using additive manufacturing. Of course, the exact opposite relationship may exist. Further research will help analysts better understand these new relationships and use them to help inform decision-makers of the most viable trades.

15.4 SUMMARY

The techniques and methods for trade-offs analysis discussed in this book expand on the Decision Management Process outlined in the INCOSE Systems Engineering Handbook (INCOSE, INCOSE Systems Engineering Handbook, 2015, pp. 110–114). However, they can also be used in any Decision Management Process, which "provide(s) a structured, analytical framework for objectively identifying, characterizing and evaluating a set of alternatives for a decision at any point in the life cycle and select the most beneficial course of action" (ISO/IEC/IEEE 15288, 2015). For complex systems, the selection of a course of action beneficial to the goals of the decision-makers and stakeholders will likely require trade-off analysis.

This trade-off analysis book has used the INCOSE Decision Management Process as the bedrock. Chapters 1 to 4 provide the foundational material for the process–decision analysis, uncertainty analysis, and resource analysis. An expanded discussion of the decision management process is provided in Chapter 5 with an illustrative UAV example. Chapters 6 and 7 provide important information on the defining the opportunity and identifying objectives and measures. Chapter 8 provides an overview and an assessment to techniques to generate and explore the tradespace. Chapters 9 to 14 provide key considerations and illustrative examples using a variety of sound techniques appropriate for the different life cycle stages.

Adoption of the Decision Management Process offers a foundation for continuous improvement of trade-off analyses within the organization. It provides a framework for which methods such as MBSE and new trade-off analysis techniques are leveraged to create better system solutions.

REFERENCES

Ender, T.R., Ake, B., Balestrini-Robinson, S. *et al.* (2015) *Engineered Resilient Systems: Tradespace Tools* Georgia Tech Research Institute, Systems Engineering Research Center, Atlanta, GA.

Goerger, S.R., Madni, A.M., and Eslinger, O.J. (2014) Engineered Resilient Systems: A DoD Perspective, in *Procedia Computer Science*, vol. **28** (eds A.M. Madni, B. Boehm, M. Sievers, and M. Wheaton), Elsevier B.V., pp. 865–872.

INCOSE. (2007). *Systems Engineering Vision 2020*, http://oldsite.incose.org/ProductsPubs/pdf/SEVision2020_20071003_v2_03.pdf (accessed 30 November 2015).

INCOSE. (2014). *A World in Motion - Systems Engineering Vision 2025*. INCOSE: http://www.incose.org/AboutSE/sevision (accessed 12 December 2015).

INCOSE (2015) *INCOSE Systems Engineering Handbook*, 4th edn, Wiley, Hoboken, NJ.

INFORMS (nd) Analytics Society: https://www.informs.org/Community/Analytics (accessed 20 December 2015).

ISO/IEC/IEEE 15288 (2015) *Systems and Software Engineering – System Life Cycle Processes*, International Organization for Standardization (ISO)/International Electrotechnical Commission (IEC)/Institute of Electrical and Electronics Engineers (IEEE), Geneva, Switzerland.

Long, D. (2015) *4 Myths and Misconceptions of MBSE*, Community.Vitechcorp.Com: http://community.vitechcorp.com/home/post/4-Myths-and-Misconceptions-of-MBSE.aspx (accessed 2 December 2015).

Parnell, G., Cilli, M. and Buede, D. (2014) *Tradeoff Study Cascading Mistakes of Omission and Commission. International Symposium.* June 30–July 3. Las Vegas, NV. INCOSE.

SEBok (2015) *Systems Engineering Body of Knowledge (SEBoK) wiki page.* from SEBok: http://www.sebokwiki.org (accessed 30 December 2015).

INDEX

acceptance decision, 484–486, 490, 499, 504, 507, 511, 512
acquisition costs, 129
activity based costing, 103
adjustment bias, 71
adversary, 303
affinity diagrams, 245, 246
affordability analysis, 135, 138, 139, 142
Affordability Analysis Framework, 142, 143, 147
affordability analysis outcomes, 139
alternative development, 263
alternative evaluation, 258, 271, 276–290
alternative generation, 197, 270–271
alternative hypothesis, 491, 493, 562
alternatives, 297, 298, 300, 301, 303, 306, 309, 310, 314–316, 318, 320, 322, 327, 330
analysis, *vs.* assessment, 387
Analytic Hierarchy Process (AHP), 287
anchoring bias, 71
annual cash flow, 126
architecture, 30, 35, 36, 47
    attributes, 399
    definition, 377
    vs. design, 383
architecture analysis
    criteria, 394
    method, 394, 396
architecture evaluation
    approach, 389
    definition, 379
    objectives, 390
architecture tradespace, 35
Assessment Flow Diagrams (AFDs), 180
assessment, *vs.* analysis, 387
attribute, 38, 43
availability, 437–454, 459, 464–471
Axiomatic Approach to Design (AAD), 277, 288

Bayes nets, 72, 73, 81
Bayes rule, 55, 65, 82, 89

Beta distribution, application, 65
Big A, 140–142, 147
binomial distribution, 59
Box–Behnken design, 405, 420, 431
break-even, 103
business knowledge, 205, 228

cash flow, 126, 127, 129
    after inflation, 127, 128
Central Composite Designs, 406, 431
coefficient of determination, 125
cognitive biases, 82
cold standby, 443–445, 451
combined standard, 246
commissioning, 562
communication, 162, 171, 174, 176, 177, 179, 183
competition, 303
complex structure, definition, 442
complex systems, 341, 350, 356, 371
composite learning curves, 123
concept, 29, 30, 35, 36, 44, 45, 47
concept maps, 219
concept of operations, 378, 391
concepts and associated concepts of operation (CONOPS), 338, 344–346, 357, 358, 370, 372
concept tradespace, 35
concern, 240, 248, 250
conditional probability, 78, 223, 280, 493, 494
confidence
    statistical, 493
    of a test, 491
confusion matrix, 526
consequence scorecard, 180, 183
constant node, 302, 310
constructed measure, 243
consumer risk, 493
convergent thinking, 263, 266
cost, 309, 310, 315, 320, 329
cost analysis, 98, 99, 138, 147
cost benefit analysis, 484, 539

cost estimate
 classification matrix, 104
 expert judgment, 103, 109
cost estimating relationships, 109, 111, 476
 exponential, 112
 linear, 112
 logarithm, 113
 power, 112
cost estimation, 102, 148
cost modeling, 454
cost objective, 148
creativity, 258, 264, 275, 290
criterion, 17, 182, 272
critical failure, 441, 442, 444–446, 450, 451, 458, 477
cumulative distribution function, 329
cumulative probability distribution, 69

decision, 2, 3, 8, 10–16, 18–20, 23–26
 frame, 11, 13, 235
decision hierarchy, 220, 228, 230
decision management process, 2, 8–10, 12, 13, 21, 24
decision node, 300–303
decision opportunity, 30–34, 41, 45, 47
decision space, 378, 379, 400
decision-theory based, 276
design tradespace, 35–36
decision traps, 207, 228
decision tree, 484, 487, 488, 511, 552, 553, 566
decommissioning, definition, 562
decomposition bias, 82
design, 20, 30, 31, 36
design of experiments, 79, 280, 288, 406, 411, 420, 432
design space, 408, 412, 421, 422, 425, 433
deterministic dominance, 82
deterministic model, 472
deterministic Pareto chart, 315
deterministic sensitivity analysis, 469–470
development cost, 477
direct costs, 98, 148
discounted cash flow (DCF), 127
discrete probability distributions, 325, 327
divergent thinking, 263, 266
domain knowledge, 205, 206, 229
down time, 438, 441–448, 450, 452, 459, 477, 479

Eagle's Beak, 3, 5
economic rate of return (ERR), 126
effect size, 58, 87
effects plot, 421
eliciting probabilities from experts, 82
expert judgement, 109
expert knowledge, 206, 225, 229
exploration, 337, 338, 341, 342, 346, 351–354, 356–359, 365, 368, 370

factor fractional factorial design, 433
false negative, 493
false positive, 494, 562
feasible region, 410, 412, 427, 428, 434
fixed costs, 98
focus groups, 213–215, 229, 241, 252
frameworks, 341, 348, 349, 371
function, 301, 305, 310, 316, 318, 324–327, 329–331, 335
functional allocation, 330
functional value hierarchy, definition, 244
fundamental objectives, 163, 164, 172, 173, 180, 200, 201
 definition, 238

gold standard, 246, 252
good decision, 341, 347, 370
good outcome, 341, 347, 370

hard skills, 92, 93, 149
historical data, 124
hurdle rate, 126
hurricane chart, 130
hypothesis testing, 493

ignoring base rates bias, 82
ilities, 341, 360
indirect costs, 98, 129
inflation rate, 127, 149, 150
influence diagram, 32, 33, 44, 45, 47, 48, 222, 229, 230, 442, 471, 553, 554, 566
initial vision statement, 204, 205, 219, 226, 227, 230
integrated approach, 300, 317
integrated decision model, 504, 562
integrated performance and cost models, 476
Integrated Systems Engineering Decision Management (ISEDM) Process, 158, 199
integrated trade-off analysis, 316, 330
interest rate, 126, 127, 137, 149, 150
internal rate of return (IRR), 126
interviews, 210, 229, 240, 252
issue list, 227

joint probability, 64, 71, 74

k-out-of-n cold standby, 444

laws of probability, 82
Lean Six Sigma Value Stream Mapping Activity, 147
learning curves, 103, 120
learning rate, 120–125, 150, 151
legacy system, 538, 539, 541–543, 563
life cycle cost, 100, 241, 242, 439, 454, 455, 459, 460, 462–463, 465–467, 478, 481
 analysis, 37, 44

INDEX

breakeven, 103
cost estimating relationships, 105, 110
expert judgment, 103, 109, 110
learning curves, 120
model, 316
replacement analysis, 103
uncertainty and risk analysis, 15, 20, 103
life cycle stage, 3, 10, 15, 21, 25, 26
linear regression, 113, 125, 151
linear relationship, 110, 112, 113, 116
linear transformation, 114
little a, 140, 141

maintainability, 438, 477
Markov process, 477
means objectives, 238, 247
mean time between critical failures, 439, 441, 444, 445
measures, types of, 165
  direct, 243, 247
  indirect, 166
  proxy measure, 37, 166, 243
mistakes of commission, 10, 15, 26
mistakes of omission, 9, 12, 15, 18
models, 298, 302, 303, 305, 309, 316, 322, 324–326, 329
Monte Carlo simulation, 75–78, 82, 130, 315, 322, 325, 329, 453, 454, 471, 473, 479–481
Morphological Box, 266, 275, 291, 292
motivational bias, 71
multi-attribute, 341, 351, 362, 371
  analysis, 553–556
  utility, 340
  value analysis, 41
  value, 22, 23, 39, 41, 166, 352
multi-objective decision analysis, 39, 166, 168, 250

net cash value, 126–128
net present value, 125, 127, 128–130, 148, 149, 241, 242, 253, 484, 539, 541–543, 561, 563, 566
nonrecurring costs, 98
normal distribution, 60, 76
normal probability plot, 116
null hypothesis, 491, 493, 562, 563

objectives, 30, 31, 33, 34, 37–39, 41–45, 47, 48, 300–303, 306, 310
  hierarchy, 244, 245, 254
  identification, 240, 252, 253
  structuring, 253
operating charachteristic curve, 23, 496, 497, 499
operational availability, 438, 439, 442, 447, 448
operations and support (O&S) costs, 455
opportunity space, purpose, 204
optimism bias, 82

optimization, 341, 356, 357, 371

Pareto chart, 316, 318, 330
Pareto Frontier, 315, 330, 342, 343, 351, 356, 365
performance models, 476
platinum standard, 245, 253
portfolio management, 144, 145
power, 486, 493–496, 498, 504, 511, 563, 565
predictive analytics, 519
present value, 127
priorities, 303, 305, 306, 314
probability assessment, 71
probability density function, 133
probability management, 324, 330
procurement costs, 455
producer risk, 493
Program (or Project) Evaluation and Review Technique (PERT), 148
project lifecycle, 515, 517, 529, 561
project management, 515, 517, 527, 530, 534, 538
Pugh method, 270–271, 275, 276, 288, 292
purpose, 244
p-value, 409, 433, 491, 563

quality attributes, 341–343, 347, 348, 357, 371
quality function deployment, 283, 289
qualitative value model, 245

real options, 348
recapitalization, 129
Receiver Operating Characteristic (ROC), 494
recurring costs, 98
recurring space, 149
redundancy, 83
regression analysis, 501, 520
regression plot, 114
relevance diagram see influence diagram
reliability, 437–439, 442, 443, 452–453, 461, 475, 476, 478, 479
reliability
  block Diagram, 429, 443
  testing, 565
replacement analysis, 103
representativeness bias, 83
requirements, 301, 303
residual plot, 116
resources, 303, 310, 322, 329
  facility, 95
  people, 91–93, 95, 99, 129, 139, 148, 149
resource space, 91, 92, 98, 99
retirement/disposal costs, 455
retirement strategy, 550
return on investment, 126
risk
  analysis, 6, 15, 20
  preference, 43–45, 48
  tolerance, 283, 355

roles for people resources
  budget, 96
  communication, 96
  executive, 96
  logistics, 96
  maintenance, 96
  management, 96
  operations, 96

sales, 129
scenarios, 302, 303, 305
screening criteria
S-curve, 318, 322
sensitivity analysis, 134
shareholder value, 236, 237, 243
silver standard, 246
soft skills, 92
software engineering, 512, 515, 519, 528, 533, 534, 537, 538
stakeholder concerns, 377, 382, 392, 402
  examples, 383
stakeholders, 2–4, 7, 8, 10, 12–13, 15, 16, 18, 20, 24, 25, 382
  knowledge, 206, 225, 229
stakeholder value, 236, 253
stakeholder value scatterplot, 189, 196, 197, 200
  with uncertainty, 194, 200
standby element, 443, 477
stochastic dominance, 58
  definition, 58
stochastic information packet (SIP), 324, 330
stochastic model, 478
stochastic Pareto chart, 316, 320
stochastic sensitivity analysis, 473
storytelling, 341, 371
structure, 442–445, 451, 453, 456, 476, 477
structured creativity methods, 264
subject matter experts, 210, 211, 229
surveys, 211, 215, 229, 241, 252
swing weight matrix, 168, 171
swing weights, 306, 314
system availability, definition, 438
system boundary, 220, 229, 230
system design, 30, 35, 36
system features, 306
system retirement, 559, 564
system safety, 70
systems engineering, 337, 338, 341, 343, 344, 347–349, 369, 370, 484, 512, 514, 515, 528, 534, 536, 538, 539, 548

Taguchi approach, 282, 289
technical knowledge, 205, 229
technology, 306, 315, 330
technology maturity, 7, 36, 37
Tornado diagrams, 319, 320, 470, 474, 475, 478

total life cycle cost, 314, 315, 330
total value, 314, 315, 325, 330, 331
trade-off, 339, 341, 343–345, 348–351, 369, 371
trade-off study, 9, 13–15, 18–20, 25, 40, 42, 169, 199, 234, 247, 298, 572, 573
tradespace, 32, 34–37, 39–41, 43, 45, 46, 48, 338–346, 351–359, 361, 364–368, 370–372, 378, 379, 388, 400
triangular distributions, 61, 84, 131, 132, 325, 327
TRIZ, 271, 274, 275, 280, 288
  for alternative development, 271
  for alternative evaluation, 280
true negative, 493, 494, 564
true positive, 493, 494, 564
Type I, 491, 493, 494, 562, 564
Type II, 491, 493, 494, 562–564

uncertainty, 6, 8, 10, 12, 14, 15, 18, 20, 21, 24, 26, 36, 37, 43, 45, 46, 338, 341, 344–347, 349, 351, 354–358, 370–372
  nodes, 303
  and risk analysis, 103
unit learning curve formula, 120
updating probabilities
use cases, 221
utility, 350
  functions, 33, 43, 44

value, 338–342, 344, 347, 351–353, 355, 356, 358, 359, 362, 364, 369, 371
value assessment criteria, 393
value assessment methods, 391
value component graph, 189
value-focused thinking, 234, 250, 253
value functions, 40, 166, 200
  continuous, 132
  discrete, 132
  inflection points, 172
value hierarchy, 16, 33, 34, 234, 244–248, 252, 303, 314, 320
value identification, 3, 6, 26
value measures, 30, 33, 34, 37–41, 309, 311
value node, 314–316
value realization, 3, 6, 26
value scorecard, 183, 187, 190
variable costs, 98
vision statement, 221, 230
visualization, 353, 356, 357, 365

Weibull distribution, 60, 61
  application, 64
weighted average cost of capital (WACC), 126
work breakdown, 105, 107, 111
work breakdown structure, 106, 107, 111